W9-AGV-190

ANNUAL REVIEW OF PHYTOPATHOLOGY

EDITORIAL COMMITTEE (1973)

ANNUAL REVIEW OF PHYTOPATHOLOGY

KENNETH F. BAKER, *Editor*
The University of California, Berkeley

GEORGE A. ZENTMYER, *Associate Editor*
The University of California, Riverside

ELLIS B. COWLING, *Associate Editor*
North Carolina State University

VOLUME 11

1973

ANNUAL REVIEWS INC.   4139 EL CAMINO WAY   PALO ALTO, CALIFORNIA 94306

581.205873
ANT
88671
May 1974

ANNUAL REVIEWS INC.
Palo Alto, California, USA

International Standard Book Number: 0–8243–1311–9
Library of Congress Catalog Card Number: 63–8847

Assistant Editor	Virginia Hoyle
Indexers	Mary Glass
	Leigh Dowling
Subject Indexer	Donald C. Hildebrand

PRINTED AND BOUND IN THE UNITED STATES OF AMERICA
BY GEORGE BANTA COMPANY, INC.

PREFACE

With this volume we begin the second decade of Annual Review of Phytopathology. An increasing number of inquiries makes it desirable to explain the procedures by which the content of each volume is determined.

The authors and topics for a given volume are decided upon at a meeting of the Editorial Committee held 24–26 months prior to its August 15 publication date. The five Editorial Committee members, three Editors, and two guest committeemen submit suggestions for topics and authors, based on a year-long scouting of papers, seminars and meetings, consultations with foreign and domestic colleagues, and suggestions from readers and previous authors. This continuing survey assures a broad subject input for consideration at each meeting. The list of potential authors tends to lengthen, and it may take 2 years or more for a name to reach the invitee list. Choices are influenced by the timeliness and importance of the topic, adequacy of available information, skill and knowledge of the author, subject balance of the volume, and worldwide distribution of authors.

This procedure, typical for the Annual Review series, leads to several points of interest to readers. We do not encourage volunteered papers because an offered manuscript would have to wait two years for publication after approval by the Editorial Committee. An author with a paper already written rarely will wait this long, and likely will resent the delay if he does. In any case, revision and up-dating inevitably result. Even so, some unsolicited papers are accepted by the Editorial Committee. If you wish to submit a review, the best procedure is to write one of the Editors or Committee Members before preparing the paper, telling him that you would like to write a chapter, and presenting justification for the topic and reasons why you should be the one to write on this subject. He will present this information for consideration at the annual committee meeting.

Because Annual Reviews Inc. publishes 20 Reviews, tight scheduling is enforced on the office in Palo Alto, the editors, and the printer. Late papers and defaults stress the system, and papers that exceed the allotted length may require excision, which authors rarely welcome. Book length, as well as scheduling, are determined by Annual Reviews Inc., and the editors must, therefore, put out a volume according to specifications. The editors deeply appreciate the efforts of authors to produce timely and interesting papers that determine, in large measure, the success of each volume.

We welcome R. G. Grogan as the new member of the Editorial Committee, replacing R. R. Nelson, who contributed much to the Review during his 5-year tenure, and we thank Gordon A. Brandes and John F. Fulkerson, guest committeemen who helped plan this volume.

The Editorial Committee

CONTENTS

GENETICS OF HOST-PATHOGEN INTERACTION

EPIDEMIOLOGY

INFLUENCE OF ENVIRONMENT

CHEMICAL CONTROL

BIOLOGICAL AND CULTURAL CONTROL

BREEDING FOR RESISTANCE

SPECIAL TOPICS

INDEXES

ERRATA

Volume 10 (1972)

Page 51, line 4: 24 m/ha *should read* 24 m³/ha

Page 64, line 36: acre *should read* hectare

Page 92, second heading: The *P. facilis-delafieldii* DNA Homology Group *should read* The *P. diminuta* DNA homology group

C. W. Bennett

A CONSIDERATION OF SOME OF THE FACTORS IMPORTANT IN THE GROWTH OF THE SCIENCE OF PLANT PATHOLOGY

❖ 3562

C. W. Bennett

Collaborator, Agricultural Research Service, U. S. Department of Agriculture, Salinas, California

When one has the temerity to accept an invitation to write a Prefatory Chapter for such a prestigious production as Annual Review of Phytopathology, it would be assumed that he might have something of interest or value to say. However, when I review more than fifty years of association with the science of plant pathology, and consider the many advances that have taken place in this period—sometimes by short steps, sometimes by giant strides, and even sometimes by false steps—I become cognizant of the magnitude of the challenge to anyone who may attempt to present facts, interpretations, or philosophies that will be of significant value in a world that seems at times to be well supplied—and at times perhaps overly supplied—with reviews, monographs, advice, and reminiscences.

Be that as it may, there is a strong temptation for one who has been privileged to witness the development of a branch of science from a near-embryonic stage to one in which thousands of people are employed, to pause and ponder some of the events, steps, and factors involved in a tremendous expansion of interests, demands, and needs that has taken place in this particular field in a relatively short period of time.

In the early years of the twentieth century, there were relatively few plant pathologists, and those who were so classified devoted much of their efforts to other branches of botanical science. Indeed, some of the most significant earlier contributions were made by people primarily interested in things other than plant diseases. Research was often a side interest combined with teaching or some other related activity. Early concerns of the field, aside from taxonomical and mycological interests, were related largely to disease control, with emphasis on seed treatments and use of fungicides, although other phases, such as sanitation and selection or production of resistant varieties, were not wholly neglected.

1

Interest in plant pathology, however, continued to grow and its potential importance in crop production became increasingly more evident with the passing years.

In recognition of the need for a more united approach to the problems presented by plant diseases, The American Phytopathological Society was organized on December 30, 1908, by a group of fifty persons who met at the Eastern High School in Baltimore, Maryland (10). The modesty of the financial requirements of this new organization is well illustrated by the following paragraph from a secretarial letter of invitation to membership, dated October 25, 1909: "I am instructed by the council of the American Phytopathological Society to invite you to become a charter member of the Society. Please signify the acceptance of the invitation by remitting to the secretary-treasurer 50 cents by post-office money order, to be used in defraying the expenses of the society this year." The growth of this society from a charter membership of 130 to its present membership of more than 2,700 is one measure of the tremendous increase in the importance of plant pathology in the United States over the past sixty years.

The first number of the new journal, *Phytopathology,* initiated by the society, appeared in February, 1911. The first volume consisted of only 204 pages, exclusive of index. It was not expected at that time that this journal would attract the most outstanding papers, but rather that it would serve a useful purpose in bringing research results of a more practical and utilitarian nature to fellow scientists and to the general public.

It is perhaps noteworthy that the earlier plant pathologists apparently had less inclination to publish results of experimental work than has become evident in more recent years. There was less economic and prestige pressure for publication, and research scientists felt a great sense of freedom in pursuing their investigations for extended periods before presenting results in published form. The extent of the tendency to procrastinate in publishing during this early period was emphasized in 1919 by Dr. L. R. Jones (9), among others, in the following statement: "Reddick (Phytopathology 8:182) has wisely warned against one of the most insidious temptations that deceives the investigator—to let his zeal in research excuse his delay in publication. Too often have we seen this pass from the stage of the natural and worthy modesty of the youth, unable quite to make his maiden manuscript meet his ideals, to that of the hopeless scientific recluse, as unfitted to bring himself to the exchange of results through the channel of publication as the miser is to barter his gold. The miser's coin were better unminted since all it represents is selfishly perverted human effort. How much more commendable or less selfish is the hoarding of ideas? And with our ideas as with our gold, gain normally follows legitimate exchange."

In the early days of the present century the image of the plant pathologist professionally was somewhat ill defined, both with his colleagues in other disciplines and with the general public. With the farmer, horticulturist, and gardener, the plant pathologist suffered from certain disadvantages inherent in the nature of his profession. The idea that plants could have diseases was not always easy to promote successfully. In some respects, the entomologist had a distinct advantage. He could usually locate an insect and point out that "this is the bug that did the damage," whereas the plant pathologist frequently had difficulty in pointing to anything readily visible that could be charged as a culprit.

On the other hand, plant pathologists have never been the subjects of humorous quips that have been so common with botanists and entomologists, probably because the organisms with which the plant pathologist dealt were not well enough known to the general public to serve as subjects of humor to the writers of popular articles. However, to a greater or lesser degree, people devoted to all branches of natural science have been expected to show a certain degree of eccentricity, particularly if they had a high degree of commitment to their chosen interest.

I recall visiting many years ago the birthplace of Liberty Hyde Bailey in South Haven, Michigan, in company with a representative of the Greening Nursery Company. While we were looking at the Bailey home—a modest but substantial frame building—an elderly gentleman came out of one of the neighboring homes to inquire about our presence. Upon being advised that we were interested in the birthplace of the great horticulturist, he told us that he had known Liberty Hyde Bailey in his youth and added, "You know, when he was a boy, people around here didn't think Liberty was right bright. He kept to himself a lot and he spent a great deal of his time examining plants, rocks, and insects, and I have even seen him in the wintertime scratching in the dead leaves along fence rows." Fortunately, there has been a great change for the better in the general understanding and appreciation of the value of agricultural research since these earlier days.

Well into the early part of this century, the plant diseases considered to be most destructive were those caused by fungi. In 1908, Duggar (5) published one of the early treatises on plant diseases, which he entitled, "Fungous Diseases of Plants." Had he chosen at that time to publish a general book on plant diseases, Duggar could have included certain bacterial diseases of great economic significance, such as "fire blight" and "crown gall," but much of the research on bacterial diseases still remained to be completed. It was some years later, in 1920, that Erwin F. Smith (12) published "An Introduction to Bacterial Diseases of Plants," in which he gave extensive description of fifteen bacterial diseases.

There would have been little to include on virus diseases, for at that time this group of diseases was largely unknown. I remember when, coming fresh from a one-semester course in plant diseases to graduate study at Michigan State University (then Michigan Agricultural College) under Dr. G. H. Coons in 1917, I was asked what I knew about virus diseases. Since my reply did not indicate any appreciable familiarity with this group of maladies, Dr. Coons reached into his file of reprints and handed me a bulletin case marked "Virus Diseases" and suggested that I immediately begin to remedy this glaring deficiency in my training. Accordingly, in one afternoon, I read most of the world's literature on virus diseases of plants.

It is perhaps worthy of note that publications on peach yellows were not included in this collection of bulletins and reprints, although Erwin F. Smith (11) had done extensive work on this disease before the turn of the century. It was so different from tobacco mosaic, the classical virus disease of the time, that it was hardly suspected that there could be any relationship between the causal agents of the two diseases. As late as 1922, in a discussion at Michigan State University, Dr. Smith himself stated that he still suspected that peach yellows could be caused by a bacterium. He might yet be proven right, for there is no conclusive evidence at this time that its long-time classification as a virus disease is correct.

As one looks back, it would seem that the potential of virus diseases for causing plant damage was slow in being recognized. Realization of their significance probably first began to dawn with the vegetatively-propagated crops in the "degeneration" diseases of potato varieties and the "running out" of raspberry varieties. No one had the remotest estimate of the multiplicity of causal agents involved in the production of this group of diseases, and little was known of their methods of spread or of their relationships to insect vectors.

It is significant that as late as the early 1920s a paper entitled, "The virus fad," was presented at a scientific meeting, in which it was predicted that plant pathologists would soon become disillusioned with the study of viruses and return to more profitable fields of research in fungus and bacterial diseases. As late as 1925, an extensive research paper was published suggesting that one virus—that of tobacco mosaic—might be the cause of most of the mottling known in a wide range of plants (7).

However, in time, it became abundantly evident that the study of viruses and virus diseases, instead of being a passing fad, was destined to develop into one of the most important branches of plant pathology. Many diseases of unknown etiology were shown to be caused by viruses or virus-like agents; previously unknown virus diseases were recognized, and many viruses spread from places of origin to new areas or to new crop plants. Some of the introduced viruses are among our most destructive plant-pathogenic agents.

During this period of expansion of virus research, one of the most intriguing challenges to investigators was the determination of the nature of the causal agents of this group of diseases. Research efforts continued to increase the fund of information on methods and agents of transmission, host ranges, and other characteristics of the causal agents and, even before the era of the electron microscope, much was already known or suspected regarding the general nature of viruses.

The available information led to endless discussion and speculation. At this time we knew the structure of chemical entities, extending from the molecule to organic acids, and the nature of living things, extending in size from small bacteria to the elephant, but in between the two there was a wide region of obscurity in which evidently viruses belonged. Perhaps Duggar & Armstrong (6) best presented the concept of the nature of viruses most generally held in this period in the following interesting assessment published in 1923, couched in the precise syntax so characteristic of the senior author: "Taking into consideration all of the facts, we cannot avoid the impression, tentatively, that the causal agency in mosaic disease may be, in any particular case, a sometime product of the plant; not a simple product such as an enzyme, but a particle of chromatin or of some structure with a definite heredity, a gene perhaps, that has, so to speak, revolted from the shackles of coordination, and being endowed with the capacity to reproduce itself, continues to produce disturbances and 'stimulation' in its path, but its path is only the living cell."

Perhaps the beginning of the first solid awareness of the true nature of viruses came with the "crystallization" of tobacco mosaic virus, first fully reported by Stanley (13) in 1936. Following this, virus particles were revealed by the electron microscope, and a whole new world of pathogenic agents was revealed, and is still unfolding in the discovery of disease-inducing mycoplasma and possibly other minute pathogenic agents.

As already indicated, the growth in the importance of plant pathology over the past seventy years has been enormous, not only in the expansion of interest in virus and virus-like diseases, but also in many other phases of the science. Probably at the beginning of this century, no one could have envisioned the great expansion that has occurred at a seemingly accelerated pace. This expansion has not resulted solely from a greater recognition and awareness of the importance of diseases, although this has been a factor. Of greater importance has been an actual increase in the economic importance of plant diseases. This increase has been associated with unparalleled industrial and social changes that have occurred over this period. These changes have resulted, not only in subjecting plants to a multitude of destructive chemical agents carried in air, water, and soil, but also in bringing about widespread increases in severity and diversity of attacks by pathogenic agents in many parts of the world.

Diseases undoubtedly have been factors in plant growth since long before the beginning of a recognizable agriculture. Pathogenic agents, no doubt, have originated on plants throughout the world over a long period of time. However, during many hundreds of years, as agriculture slowly increased in complexity, they had limited opportunity to spread to other areas from their places of origin, largely because of the isolation of agricultural communities and the limited transport of plants and plant products. There is reason to believe that, even now, plant diseases may be relatively less important in parts of the world where primitive agricultural methods are still in use than they are in areas agriculturally more advanced.

Wellman (14), in 1938, for example, recorded the interesting observation that in certain parts of Asia Minor, where agricultural practices had changed little over the centuries and where crops were grown in small more or less isolated areas, plant diseases were usually very scarce and caused little crop damage. This condition was attributed in part to practices of plant utilization in which diseased plants and plant parts were removed from the field for use as fuel or food, but it was pointed out, also, that farmers in these areas saved their own seed, which they planted year after year, and that disease resistance probably was a factor in reducing losses.

It may well be that, under conditions of primitive agriculture, disease resistance was highly important in reducing crop losses. Undoubtedly, as a general practice, selected strains of many crop plants were grown continuously for long periods under conditions of a relatively high degree of isolation in innumerable local agricultural areas. It seems probable, therefore, that a process of both conscious and unconscious selection for resistance to diseases may have been carried forward in many localities for centuries. Under such conditions, cultivated plants of all kinds would be expected to acquire high degrees of resistance to pathogenic agents native to any particular region. Thus, a state of equilibrium between parasites and hosts could have been established and maintained, in which the host plants were rarely seriously damaged by local diseases.

However, an equilibrium that probably was maintained with minor fluctuations over many centuries, began in more recent times to be upset by improvements in the efficiency of transport of plants and plant products; first slowly by sailing ships, and then at a markedly accelerated pace by steamships, railroads, and ultimately by the airplane, which has made it possible to transport pathogenic agents from one point on the earth's surface to any other point in a matter of hours.

It is well known that much of the recent expansion in the distribution of plant diseases has resulted from the introduction of pathogenic agents into new areas. In such areas of introduction the pathogen not infrequently attacked new species, often related to their native hosts, but which, in the absence of previous exposure to the pathogen, had not acquired an appreciable degree of resistance.

Increase in destructiveness of plant diseases has resulted, also, from extensive introduction of crop plants into new geographic areas. In many instances newly introduced plants have been attacked by previously unknown plant pathogens harbored by native plants, which by long association had reached a state of equilibrium with their pathogenic agents, in which little host damage was produced. Once they escaped from native host plants, however, and became established on introduced susceptible hosts, many pathogens have caused extensive damage to the introduced crop plant. Moreover, after such pathogens became established on an introduced plant, their likelihood of being transported to other areas was greatly increased.

Introduced pathogens have been especially destructive in North America. The American chestnut, for instance, survived its native enemies and thrived for thousands of years in eastern United States. However, the introduction of the fungus, *Endothia parasitica*, first noted in the New York Zoological Gardens in 1904, and which causes a minor canker disease on relatives of the chestnut in the Orient, has brought the American chestnut almost to the point of extinction.

Equally tragic have been the results of the introduction of the Dutch elm disease, caused by *Ceratocystis ulmi,* into eastern United States and Canada, where it has destroyed or is in the process of destroying the American elm throughout its zone of natural occurrence.

One of the most spectacular instances of introduction and dissemination of a highly destructive plant pathogen involves the spread of the tristeza virus of citrus. The disease produced kills sweet-orange trees on sour-orange rootstocks, but does not seriously damage sweet-orange trees on certain other rootstocks. The pathogen apparently had its origin in the Orient, where its presence was recognized only after it had killed thousands of trees in other parts of the world. It was apparently introduced into South Africa very early in the development of a citrus industry in that country, since it was reported by horticulturists many years ago that sweet orange and sour orange were "incompatible" in South Africa, although sour orange was considered one of the best rootstocks for sweet orange in other parts of the world. The pathogen was introduced into Brazil on nursery stock about 1937, and within twelve years it had killed upward of six million orange trees in the state of São Paulo alone (3). It has destroyed or threatens millions of trees in tropical and subtropical areas in many other parts of the world.

Numerous other instances of spread of introduced pathogens could be cited. Sugarcane mosaic virus, for instance, has spread around the world in little more than three-quarters of a century. It has caused damage in many varieties of sugarcane that were once of great value in their respective areas of culture and has resulted in the abandonment or elimination of a number of prized varieties that were very low in resistance.

Curly top in North America, thought for many years to be caused by a native pathogen, produced heavy losses in western United States in sugar beet and several other crop plants and still causes a disease of considerable economic importance. It now appears that this pathogen had its origin in the Mediterranean region, but its presence in its probable native home was not recognized until more than fifty years after it produced a destructive disease in the areas into which it was presumably introduced. Its discovery in the Mediterranean region was directly related to the introduction of the sugar beet into that area from northern Europe.

Many plant pathogens were discovered only after they had attacked introduced plants. The peach, for example, is relatively free of destructive diseases in its place of origin, but when it was introduced into North America, it encountered peach yellows, which has intermittently produced widespread destruction in orchards in northeastern United States over a period of many years. Peach yellows was known for more than one hundred years before it was discovered that the causal agent probably had its origin in native species of *Prunus* in northeastern United States, on which it produces little or no damage. Probably more than forty other pathogens native to North America have been found to attack peach and other species of stone fruits. Some of these have caused extensive damage on introduced species and varieties.

The sugar beet, which comes from northern Europe, is notorious for picking up pathogens from native plants in areas into which it has been introduced. When attempts were made to grow the sugar beet in the Rio Negro Valley of Argentina, it developed a destructive disease called yellow wilt, which caused the abandonment of a beet-sugar industry in that area (4). The disease is now known in Chile, where it greatly restricts the area in which sugar beets can be profitably grown. Except for the introduction of the sugar beet into the areas where it occurs, the existence of yellow wilt probably would still be unknown, and its causal agent probably would still be confined to native plants of southern South America.

An attempt to grow sugar beets at Tucumán in Argentina revealed another hitherto unknown disease, called Argentine curly top, first reported by Fawcett (8) in 1925. Two other curly top diseases later were found in Brazil on tomato (2). And, of course, when sugar beet was introduced into western United States, it developed a curly top disease that not only practically destroyed the beet-sugar industry of parts of the irrigated West, but also caused extensive damage to many other introduced plants such as tomato, cucurbits, and bean.

One may well wonder how many unknown pathogenic agents still exist in native plants in various parts of the world awaiting the introduction of a host plant on which they may be capable of producing destructive diseases, or awaiting the time when they may be transported to new geographic areas where they may encounter new hosts that have not acquired resistance to them.

We know that some such pathogenic agents do indeed exist. A virus found in native plants of southern California, which thus far has been transmitted experimentally only by species of *Cuscuta,* is capable of causing reductions in yield of a number of crop plants, including sugar beet, celery, potato, cantaloupe, and buckwheat (1). The introduction of an efficient vector of this agent could lead to the spread of a pathogen highly destructive to both cantaloupe and buckwheat. Other viruses, as yet not fully described, are known to exist in native plants of the same area.

As one reviews the diseases that have called for greatest effort in control through the years, it becomes evident that much of the expansion in the science of plant pathology over the past three-quarters of a century has resulted from the increased attention required in dealing with introduced and newly discovered pathogens.

Disease-inducing entities have been more rapidly and extensively disseminated over this period, and the process of dissemination is still in progress. Many diseases, of course, have achieved virtually worldwide distribution, or maximum distribution within the limits of their potential; many others, however, still have possibilities for invasion of extensive areas. Some of the diseases of this latter type hold great possibilities for more extensive damage in the future.

None of the curly-top diseases of South America, for instance, has reached its possible maximum geographic distribution. Each of these diseases has the potential for a high degree of destructiveness in many areas where it does not now occur. Yellow wilt, now confined to Argentina and Chile, has an especially high potential for damage. Introduction of this disease into western United States could seriously cripple the beet-sugar industry over a large area. Also, introduction of curly top from Argentina, or from North America, into the sugar beet-producing areas of Chile would force changes in the beet-sugar industry, and probably would adversely affect yields of several other crop plants in that country. Many other examples of diseases capable of further spread could be cited.

It would seem that, with the present efficient systems of transport, in which plants and plant products may be moved in quantity over any desired distance in short periods, we have facilities and conditions available for rapid spread of any particular pathogenic agent to all parts of the world where such an agent is capable of surviving.

In fact, the world has now shrunken, in terms of pathogen spread, to proportions comparable to those of the isolated valley of previous centuries where plants evolved in association with their natural enemies and where, over long periods of natural selection, they established and maintained balances that permitted them, in most instances, to thrive in association with their disease-inducing entities.

In the efforts that will be required in the future to provide adequate food supplies to an ever-increasing world population, one of the major challenges in agricultural production must inevitably continue to be reduction in losses from plant diseases. Great progress has been made in meeting the increasing problems of recent years posed by the increase in number, distribution, and severity of plant diseases. For example, resistant sugar beet varieties have made possible a thriving beet-sugar industry in western United States, varieties resistant to black stem rust have greatly reduced losses in cereals from this disease, new rootstocks for citrus varieties have largely eliminated the menace posed by tristeza disease, and varieties resistant to sugar cane mosaic have reduced losses from this disease in many sugar cane-producing areas of the world.

However, there have also been instances of little success or even total failure thus far, in attempts to control certain specific diseases. One may wonder, for example, whether the American chestnut is destined to become extinct because of the ravages of the chestnut-bark disease, and whether the American elm can ever be restored to its former glory and usefulness by development of adequate control measures for the Dutch elm disease.

Many other problems concerning plant diseases still are unsolved or have only partial solutions. These problems, along with those that will be posed by new or introduced pathogens, development of more virulent strains of pathogenic agents, and many other factors, promise a continued increase in the demands on plant pathologists. It may be expected, therefore, that the science of plant pathology will continue to expand, especially in the field of plant protection. In this expansion there will undoubtedly be increased profit from cooperation and collaboration with scientists in related disciplines, particularly with those in agronomy, entomology, and genetics.

Literature Cited

1. Bennett, C. W. 1944. Latent virus of dodder and its effect on sugar beet and other plants. *Phytopathology* 34:77–91
2. Bennett, C. W. 1971. The curly top disease of sugarbeet and other plants. *Am. Phytopathol. Soc.* Monograph 7:1–81
3. Bennett, C. W., Costa, A. S. 1949. Tristeza disease of citrus. *J. Agr. Res.* 78:207–37
4. Bennett, C. W., Hills, F. J., Ehrenfeld, K. R., Valenzuela, B. J., Klein, K. C. 1967. Yellow wilt disease of sugar beet. *J. Am. Soc. Sugar Beet Tech.* 14:480–510
5. Duggar, B. M. 1909. *Fungous Diseases of Plants.* Ginn & Company, New York. 508 pp.
6. Duggar, B. M., Armstrong, J. K. 1923. Indications respecting the nature of infective particles in the mosaic disease of tobacco. *Annals Mo. Bot. Gardens* 10:191–212
7. Elmer, O. H. 1925. Transmissibility and pathological effects of the mosaic disease. *Iowa Agr. Exp. Sta. Res. Bull.* 82:1–91
8. Fawcett, G. L. 1925. Encrespami-ento de las hojas de la remolacha azucarera. *Rev. Ind. Agr. Tucumán* 16:39–46
9. Jones, L. R. 1919. Our journal, Phytopathology. *Phytopathology* 9:159–64
10. Shear, C. L. 1919. First decade of the American Phytopathological Society. *Phytopathology* 9:165–70
11. Smith, E. F. 1891. Additional evidence on the communicability of peach yellows and peach rosette. *U.S. Dept. Agr. Div. Veg. Path. Bull.* 1:1–65
12. Smith, E. F. 1920. *An Introduction to Bacterial Diseases of Plants.* W. B. Saunders Company. Philadelphia, London. 688 pp.
13. Stanley, W. M. 1936. Chemical studies on the virus of tobacco mosaic. VI. The isolation from diseased Turkish tobacco plants of a crystalline protein possessing the properties of tobacco-mosaic virus. *Phytopathology* 26:305–20
14. Wellman, F. L. 1938. Poverty of human requisites in relation to inhibition of plant diseases. *Science* 87:64–65

THE GREAT BENGAL FAMINE

❖ 3563

S. Y. Padmanabhan

Central Rice Research Institute, Cuttack 6, India

Bengal, which prior to partition of India covered the state of West Bengal in India and Bangladesh, suffered from a calamitous famine in 1943, when it was estimated that two million people died of starvation. The author was appointed as Mycologist in Bengal in October 1943 when the famine was at its height. When he travelled to join his new assignment on 18th of October 1943, he could see dead bodies and starving and dying persons all along the way from Bahudurabad Ghat on the Brahmaputra to Dacca. This horrendous situation of several thousands of men, women, and children dying of starvation continued throughout October, November, and December in and around all the important cities in Bengal, especially Calcutta and Dacca.

There was a war raging in many theaters in the world. The British empire was visibly crumbling. The victorious Japanese army, in collaboration with Indian National Army, was knocking at the eastern gates of India. It is in this context that a serious shortage in rice production occurred in 1942. As there was very little marketable surplus from 1942 harvest, the price of rice started rising from the beginning of 1943 in all parts of Bengal. The civil administration could not and did not cope with the situation created by the shortage. Soon the cost of rice was beyond the reach of ordinary people. Most of the rural population migrated to the cities in the hope of finding employment and rice. Finding neither, they slowly died of starvation.

Though administrative failures were immediately responsible for this human suffering, the principal cause of the short crop production in 1942 was the epidemic of helminthosporium disease which attacked the rice crop in that year. This was caused by *Helminthosporium oryzae* Breda de Haan[=*Cochliobolus miyabeanus* (Ito & Kuribayashi) Drechsler *ex* Dastur. Nothing as devastating as the Bengal epiphytotic of 1942 has been recorded in plant pathological literature. The only other instance that bears comparison in loss sustained by a food crop and the human calamity that followed in its wake is the Irish potato famine of 1845.

The loss sustained by the rice crop in 1942 might be judged from Table 1, which gives the yield of the principal rice varieties widely grown in Bengal as recorded at the rice research stations of Chinsura and Bankura for the years 1941 and 1942. It may be seen that the loss sustained by the early maturing

11

Table 1 Yield of rice per hectare at the Rice Research Stations at Bankura and Chinsurah in the epiphytotic year (1942) compared with yield per hectare in the stations in a normal year (1941).

Variety of Paddy		BANKURA			CHINSURAH		
		Yield kg/ha 1941	1942	Percentage Loss in Yield	Yield kg/ha 1941	1942	Percentage Loss in Yield
Bhutmuri	Aus (Early)	1289	1242	6.81	372	1252	
Kataktara	Aus (Early)	1421	1205	15.1	1250	1215	
Tilakkachri	Aus (Early)	1867	1328	28.9	1713	965	43.7
Marichbati	Aus (Early)	1365	723	46.9	1365	674	50.3
Dharial	Aus (Early)	1323	669	49.5	1323	669	49.6
Charnok	Aus (Early)	1208	443	59.2	762	446	41.5
Dudsar	Aman (med late)	2105	559	73.5	2102	1274	39.5
Badkalamkatti	Aman (med late)	1504	909	39.5	1737	686	60.5
Indrasail	Aman (med late)	2962	755	74.5	3094	755	75.6
Nonaramsail	Aman (med late)	1693	426	74.7	1691	424	74.8
Chinsurah	Aman (med late)	3778	880	76.7	2501	713	79.1
Sundermukhi	Aman (med late)	2599	267	76.9	2362	272	88.5
Latisail	Aman (med late)	5427	1122	79.3	2906	1125	61.3
Ajan	Aman (med late)	3168	600	81.1	3173	561	82.3
Badshabhog	Aman (med late)	1938	316	83.2	1189	757	59.9
Juijasail	Aman (med late)	2499	331	85.5	2252	306	86.4
Boldar	Aman (med late)	2426	306	87.3	2426	309	87.3
Raghusail	Aman (med late)	2563	328	87.4	—	—	—
Rupsail	Aman (med late)	2156	284	87.7	2166	284	86.9
Patnai	Aman (med late)	2751	336	87.8	2256	336	85.1
Dandkhani	Aman (med late)	1722	152	91.2	1725	159	91.1

"Aus" varieties, though substantial, was in general less than that of medium late "Aman" varieties.

Certain abnormal trends in weather conditions have always been reported as being associated with severe outbreaks of helminthosporium disease. For instance, according to Sundararaman (25), unusually heavy rains at the time of ear formation followed by flooding and poorly drained conditions had favored the epiphytotic of 1918–19 in the delta of the Krishna and Godavari. Barat (2) reported that the rice crop suffered serious damage due to *H. oryzae* in Indochina, and mentioned that the outbreak followed untimely rainfall, late in the cold season. Similarly, a serious outbreak of the disease in Ceylon was reported to have been caused by wet humid weather, presence of sufficient inoculum, and poor growth of plants (1). Agricultural scientists in Bengal noted that weather conditions were unusual during 1942. Continuous cloudiness at the time of flowering and maturity of the crop was emphasized.

The nature and extent of the abnormal weather trends in 1942 were studied to see whether they could explain the outbreak. Relevant meteorological data, on (*a*) maximum and minimum temperature, (*b*) average range of daily temperature, (*c*) relative humidity at 8 hours and 17 hours, (*d*) cloud amount at 8 hours and 17 hours, (*e*) number of rainy days, and (*f*) rainfall for the months of October, November, and December 1941, 1942, 1943, and

1944 were obtained from the Director of Agricultural Meteorology, Poona, for three centers in Bengal where the disease was severe in 1942. Data of sunshine hours were available only from the observatory at Alipur in Calcutta. A comparative study of these data was made and the significant departures in weather conditions of 1942 deduced.

Secondly, the biology of the host and pathogen was examined to determine how the disease could have grown to epiphytotic proportions under the conditions that existed in 1942. During this exercise it was seen that on certain aspects of the phenomenon of infection sufficient data were not available; for example, the effect of weather conditions on spore release, leaching of nutrients, disease development, age of host on infectivity of *Helminthosporium oryzae*, etc. Some of these aspects were taken up for study at the Central Rice Research Institute, Cuttack, India by the writer and his colleagues, and have been integrated with the evidence in pathological literature in order to determine how the weather conditions of 1942 had favored the epiphytotic.

A Comparison of the Meteorological Factors of 1942 with Those of Other Years

MAXIMUM TEMPERATURE   The maximum temperature varied from 25.0–32.2°C in 1942 and showed more or less the same trend in the other three years, though it was slightly less in the months of September and October in 1942 at two of the centers, compared to 1943 and 1944.

MINIMUM TEMPERATURE   There was a difference between the monthly average minimum temperatures of 1942 for the period under study compared with those of 1941, 1943, and 1944. The average minimum for September and October were not different, but that of November was much higher in 1942. The minimum temperature in November 1942 on Sagour Islands was 21.1°C, whereas in other years it ranged from 18.9–20.0°C. The temperature in Calcutta was 20°C in November 1942, whereas in other years it ranged from 17–18°C. In Burdwan, the minimum was 19.4°C in November 1942; in other years it ranged from 16.7–17.2°C. This increase in average minimum temperature was also reflected in the average range of daily temperature which was much less in November 1942 (Table 2). The higher average minimum temperature of November 1942 and the lowering of the average daily range of temperature represent, therefore, a highly significant departure of weather conditions in 1942.

CLOUDINESS AT 8 HOURS AND 17 HOURS   The cloudiness in daytime was greater in September 1942 than in 1941, 1943, and 1944, but it was not very different in October. There was an increase in cloudiness in November 1942 as compared to 1941, 1943, and 1944 (Fig. 1), at all the three centers for 17 hours and also at 8 hours, except in Burdwan where there was an increase in November at 8 hours.

Table 2   Mean minimum temperature and range of daily temperature (centigrade) in 3 locations during 1941, 1942, 1943, & 1944 in Bengal.

Mean Minimum Temperature	1941				1942				1943				1944			
	Sept.	Oct.	Nov.	Dec.	Sept.	Oct.	Nov.	Dec.	Sept.	Oct.	Nov.	Dec.	Sept.	Oct.	Nov.	Dec.
Sagour Islands	26.7	24.4	19.4	17.2	25.6	25.0	21.1	16.7	26.7	25.6	20	17.2	26.7	25.0	20	17.2
Calcutta	26.7	23.3	18.3	15.6	25.6	23.9	20.0	13.9	26.1	24.4	17.8	15.0	26.1	24.4	17.2	14.4
Burdwan	26.1	23.3	17.8	14.4	25.6	23.3	19.4	13.3	26.1	24.4	17.2	14.4	26.1	23.3	16.7	13.9
Range of Daily Temperature																
Sagour Islands	13.9	12.2	10.0	9.4	13.3	12.2	11.1	8.9	14.4	12.8	8.9	8.9	13.9	12.8	8.9	8.3
Calcutta	11.1	9.4	6.7	5.0	12.2	9.4	8.3	4.4	11.7	9.4	5.0	5.0	11.1	10.0	4.4	3.9
Burdwan	11.1	10.0	6.7	4.4	11.7	9.4	7.8	4.4	11.7	9.4	5.0	4.4	11.1	10.0	3.9	3.3

Table 3   Total hours of bright sunshine in 1942 as compared to 1941, 1943, & 1944, in Calcutta (Bengal).

Month	1941		1942		1943		1944	
	No. of sunshine hours	Percentage of total sunshine hours	No. of sunshine hours	Percentage of total sunshine hours	No. of sunshine hours	Percentage of total sunshine hours	No. of sunshine hours	Percentage of total sunshine hours
September	189.0	51	121.0	33	130.8	37	174	71
October	239.8	67	250.9	70	242.0	67	213	59
November	247.2	75	208.8	63	295.4	89	278	84
December	271.8	81	271.8	81	274.6	82	278	84

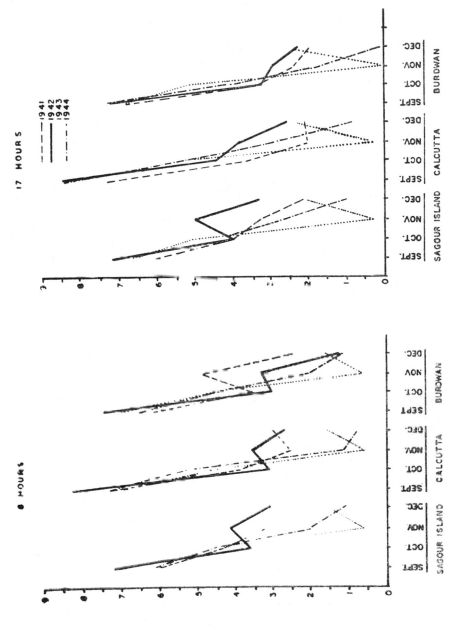

Figure 1 Degree of cloudiness at three localities in Bengal, September through December, 1941 through 1944, at 8 and 17 hours, expressed as hours of solar radiation.

NUMBER OF HOURS OF BRIGHT SUNSHINE   The data on the "hours of bright sunshine" gives a better insight into the cloudy conditions than the data on cloudiness which are recorded at fixed times of the day over different sectors of the sky. The number of "bright sunshine hours" was lower in September, 1942 than in other years, but the November record of 63% out of possible sunlight of 208.8 hours for the whole month was the lowest record for November during the years under review (Table 3). Thus the low number of bright sunshine hours of September and November 1942 constituted a striking abnormality in that year.

NUMBER OF RAINY DAYS   There were no rainy days in November during 1943 and 1944, but there were 2–6 rainy days in November 1942 in the 3 centers and 2–3 days in 1941. In this respect also, the weather factors of 1942 were somewhat different from those of the other two years (Fig. 2).

RAINFALL   September 1942 was characterized by an exceptionally heavy rainfall of 40.6–65 cm whereas the rainfall in 1941, 1943, and 1944 during September varied from 8.9–33.8 cm. Similarly in October 1942 the rainfall was 17.5–20.0 cm while in the other 3 years it was 5.1–12.7 cm during this month, except in Calcutta where there was 26.9 cm. The most significant difference was seen in the figures of November. November and December were rainless during 1943 and 1944 whereas there were 2–6 days of rain with 0.8–10.4 cm of rainfall in November 1942 (Fig. 2). In 1941 there was some rainfall, especially in Burdwan (2–10.7 cm).

RELATIVE HUMIDITY AT 8 HOURS   The average relative humidity was much higher during the whole of September 1942 in the morning, and similarly the relative humidity at 17 hours also was higher in the month of September 1942 when compared to other years.

To summarize, the epiphytotic year 1942 was different from the nonepiphytotic years of 1941, 1943, and 1944 in (a) unusually heavy rainfall in September, (b) unusual and prolonged cloudy weather in November, with very low sunshine hours and occasional rains, (c) higher minimum temperatures than normal (i.e., 19.4–21.4°C in November, whereas the normal average range of minimum temperature was 16.7–20.0°C in other years).

The Normal Crop Season and Disease Cycle in Bengal

There are three rice crop seasons per year in Bengal, *Aus, Aman,* and *Boro. Aus* is an upland direct-seeded crop maturing in 110–120 days. The seeds are sown broadcast towards the end of June with the commencement of monsoon, and the crop is harvested in September or early October. This crop occupies about 10–11% of rice area in Bengal. The *Aman* crop is transplanted in July-August and harvested in November-December. In waterlogged low-lying areas, a spring crop *Boro* (December-April) is grown when water recedes in November-December.

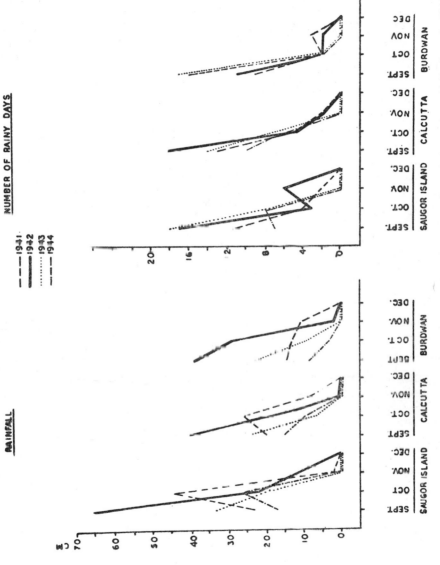

Figure 2 Rainfall (cm) at three localities in Bengal, September through December, 1941 through 1944.

In a normal season the monsoon starts in the third week of June, strengthening into heavy and continuous rainfall in July-August, and tails off towards the end of September, with a few occasional showers in October. November and December are rainless months. The total rainfall is about 152–200 cm except in North Bengal near the foothills of the Himalayas, where it is about 250 cm. The clear sky and favorable maximum of 28–30°C and the minimum of 17–20°C, November in Bengal represents an optimum condition for grain filling and maturity of the main *Aman* varieties.

Infected seeds (19, 22), alternate hosts (6, 21), and occasional air-borne conidia (4) are the probable sources of infection for the new crop sown in June-July.

The seed-bed infection is seen in a very mild form in June-July but is quite severe in the *Boro* seedlings raised during December-January. Very little infection is noticeable in *Aman* seedlings in transplanted fields during July-August and part of September. Toward the end of September, leaf-spots appear in the maturing *Aus* and in the *Aman* crop, and in some years the disease may be seen in a fairly severe form in the late *Aman* types towards November and December (Fig. 3).

Factors Favoring an Epiphytotic Outbreak of a Plant Disease

For intensification of cycles of infection of an air-borne pathogen like *H. oryzae* to produce an epiphytotic, there must be (*a*) multiple foci of infection from which an enormous quantity of propagules are released over a (*b*) prolonged continuous period of favorable conditions for establishment of infection on the host. (*c*) This process is greatly accelerated when the host is in the susceptible stage, preferably predisposed to infection by earlier conditions.

Critical Examination of the Meteorological Factors of 1942 in Relation to Development of the Epiphytotic

SOURCE OF INFECTION   The crop season may have started normally with only slight incidence of *Helminthosporium oryzae* disease till the end of August.

The *Aus* crop that matures in normal seasons towards the end of September and first week of October, shows heavy infection at maturity. This infected crop provides a good source for spread of infection to the adjacent *Aman* crop which flowers and matures in November and December. During 1942 the early maturing varieties must have been heavily infected (compare data on loss of yield, Table 1) and must have provided the necessary multiple foci for spread and infection on the later *Aman* varieties.

FACTORS FAVORABLE FOR SPORE RELEASE OF CONIDIA OF H. ORYZAE   The relation between production of conidia in nature and the weather conditions were studied at Cuttack (4). Once a week greased slides were exposed in 6 aeroscopes in 6 different locations at the Institute farm for a period of 24

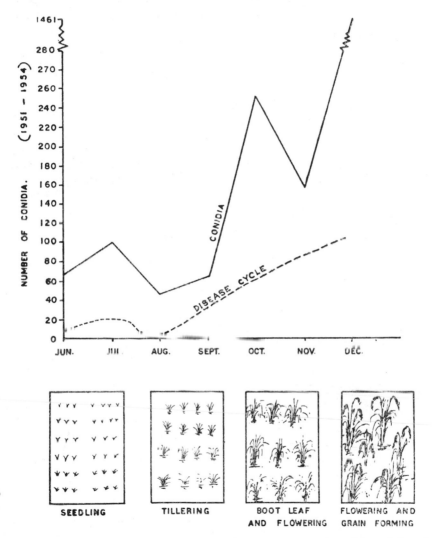

NORMAL DISEASE CYCLE OF HELMINTHOSPORIUM ORYZAE
ON RICE AND PATTERN OF CONIDIAL RELEASE

Figure 3   (Top) Average number of conidia of *H. oryzae* released from infected *Aus* and *Aman* rice, June through December, 1951–1954 in Bengal. Progress of disease is also shown. (Bottom) Corresponding stages in the development of a rice crop.

hours and the number of spores of *H. oryzae* trapped in the slides was determined. The study was continued for a period of 218 weeks from January-August, 1950 and then from April 1951 till December 1954.

The number of spores caught was generally lowest in May through August, while maximum spore-catch was in October through December. The spore-catch was thus very low during the summer and heavy rainy season. There was a very low population of spores in the air during the seedling stages of the crop; they were practically absent in the vigorous growth stage after transplanting, but appeared at time of flowering and maturity of the early varieties (September-October); the maximum population was attained at the time of flowering and maturity of the principal transplanted *Aman* crop (November-December) (Fig. 3).

To study the correlation between the weekly spore-catch and the meteorological factors, three apparently important factors, range of daily temperature, relative humidity, and solar radiation were taken into consideration. Three distinct periods that probably have an influence on spore production were studied with respect to the above factors: (*a*) the week ending with the days when spores were caught, (*b*) the last 4 days of that week and (*c*) a week before that in which the slides were exposed. Though no exact correlation could be established for all years between the spore-catch and factors for the different periods studied, some significant relationship could be seen in some years with respect to average relative humidity, range of temperature, or solar radiation. There was a correlation between these meteorological factors and spore-catch for one or the other of 3 periods studied (Table 4). The meteorological factors of the week ending with spore-catch, of the last 4 days of that week, and of the previous week, apparently had some relation to the number of spores trapped. The relevant data are presented in Table 5.

Though it might be difficult to establish statistical correlation between meteorological conditions and variations in spore-catch with data collected for only 218 weeks, yet a few associations between weather trends and a sudden rise in spore-catch were seen. In these weeks (principally in October, November, December, and March) there was a sudden enormous increase in spore-catch on the slides.

The periods prior to the days of slide exposure were characterized by (*a*) a sudden rise in relative humidity followed by a few dry days, (*b*) fall in the daily temperature range (difference between the maximum and the minimum for each day), (*c*) low solar radiation and cloudiness, and (*d*) intermittent light showers.

Conidia of *Helminthosporium* closely resembling those of *H. oryzae* were collected and counted. As the slides were exposed over rice fields close to the infection site, the assumption of identity was considered to be justified.

Spore release data from Cuttack indicate that the occurrence of even a few days with clouds and slight drizzle during the period, October to March, would result in an enormously increased spore release. Since these favorable conditions occurred continuously over the whole month of November in

1942 in the three centers studied it is probable that an unprecedented spore release occurred continuously during November 1942, thus providing one of the most essential conditions for the development of an epiphytotic.

INFECTION OF RICE BY H. ORYZAE   The optimum temperature for germination of conidia, growth of the fungus in the culture, and infection of the host lies between 25–30°C (9, 13, 14, 17, 18).

Contact with water at the leaf surface is essential for germination and infection of rice leaves by the pathogen, depending upon the prevailing temperature. The period of continuous contact with water necessary for establishment of infection is 8 hours at 20°C, 4 hours at 25–30°C, and 6 hours at 35°C (9). Sherf et al (24) however, have stated that 10 hours of continuous contact with water was necessary for establishment of infection by *H. oryzae* at 20°C. Naito (16) and Imura (11, 12) found that penetration of rice

Table 4 Relation between weather conditions and spores of *H. oryzae* caught in different weeks. Correlation coefficient between weekly catch of spores of *H. oryzae* and weather factors. Significant figures shown by*.

Year	Number observations	Daily average relative humidity for the week	Daily average relative humidity for the last 4 days of the week	Daily average relative humidity for the previous week
1951–52	32	0.3553*	−0.5122*	0.0335
1952–53	34	0.2860	0.2308	0.2532
1953–54	34	0.1298	0.1788	−0.0766
1954–55	14	0.002278	0.4086	0.1439

Year	Number observations	Range of temperature for the week	Range of temperature for the last 4 days of the week	Range of temperature for the previous week
1951–52	32	−0.1312	−0.2353	0.01497
1952–53	34	−0.4950	−0.3775*	−0.395*
1953–54	34	0.1632	0.0751	—
1954–55	14	0.2224	0.0918	—

Year	Number observations	Solar radiation (hours) for the week	Solar radiation (hours) for the last 4 days of the week	Solar radiation (hours) for the previous week
1952–53	34	−0.5369*	−0.1252	−0.2705
1953–54	34	0.1437	0.0615	0.190
1954–55	14	0.0965	−0.0088	−0.5991*

Table 5   Daily weather record of weeks in which there was a sudden increase in the spore-catch of *Helminthosporium oryzae*.

No. of spores trapped	Date	Relative humidity	Range of temperature	Hours of solar radiation	Remarks
17	21-11-1951	90	21.0	—	Foggy
	22-11-1951	97	22.5	—	
	23-11-1951	73	25.5	—	
	24-11-1951	64	22.3	—	
	25-11-1951	75	13.3	—	Cloudy
	26-11-1951	95	7.0	—	Cloudy and rain
75	27-11-1951	92	9.0	—	Cloudy
	21-11-1951	92	18.5	—	
46	5-12-1951	73	23.5	—	
	6-12-1951	81	23.4	—	
	7-12-1951	80	20.0	—	
	8-12-1951	79	23.3	—	
	9-12-1951	57	18.0	—	Cloudy
	10-12-1951	57	17.0	—	Cloudy
450	11-12-1951	63	11.5	—	Cloudy light rain
	12-12-1951	61	15.0	—	Cloudy breezy
4	7-3-1952	—	—	—	
	8-3-1952	86	21.3	9	
	9-3-1952	90	27.0	9	
	10-3-1952	84	22.5	9	
	11-3-1952	84	17.5	9	
	12-3-1952	81	30.8	3½	Cloudy
54	13-3-1952	80	19.0	8	Cloudy
	14-3-1952	77	19.2	6½	Cloudy and dust storm on previous night
	8-10-1952	98	4.0	Nil	Cloudy
7	9-10-1952	98	4.5	Nil	Cloudy and drizzle
	10-10-1952	91	8.4	1.0	Cloudy and drizzle
	11-10-1952	91	5.0	Nil	Cloudy
	12-10-1952	83	12.0	6½	Cloudy
	13-10-1952	85	4.0	6	Cloudy
162	14-10-1952	87	10.0	7	Cloudy
	15-10-1952	91	11.5	5½	Cloudy
5	8-12-1954	71	28.8	10.0	Cloudy
	9-12-1954	90	24.8	9.0	Cloudy (slight)
	10-12-1954	89	23.5	9.0	Cloudy and foggy
	11-12-1954	75	19.3	8.5	Cloudy
	12-12-1954	56	13.5	0.5	Cloudy
	13-12-1954	48	14.5	4.25	Cloudy
138	14-12-1954	65	21.1	7.75	Cloudy
	15-12-1954	66	28.0	10.0	Scattered cloudy

leaves and development of lesions were favored by darkness or partial shade, and retarded by sunlight. According to Hemmi & Suzuki (10), both excessive soil moisture and drought conditions favor development of the disease.

From the multiple infection foci provided by the *Aus* crop and the stubble left in the field, spores would be continuously released in 1942, as in a normal season. The temperature factor (maximum-minimum being 24.4–29.4° C) was quite favorable for infection in the *Aman* crop, the few cloudy days in October affording more favorable conditions for spore release and infection than the corresponding month in other years. In November 1942 the weather condition—cloudy days, slight drizzle, and high minimum temperature—provided exceptionally favorable conditions for abundant continuous spore release and infection.

HOST FACTOR   Padmanabhan & Ganguly (23) have shown that the susceptibility of rice to *Helminthosporium oryzae* increases with age, and that it is most susceptible to infection at the time of flowering and maturity. Therefore, the third important condition for the development of an epiphytotic— the presence of a susceptible host—was apparently satisfied since the crop was at flowering and maturing stages throughout Bengal in November-December.

PREDISPOSING FACTORS   Leaching of nutrients from the soil (5), low potassium status (20), upsetting of the iron-manganese ratio (3), low nitrogen status of the soil (7, 15), excessive soil moisture (10), and, waterlogged soil (8) favor the disease.

Evidence for the amount of nutrient loss sustained in normal rice soil in monsoon season under moderate rainfall is given in Table 6.

From the above data it might be inferred that leaching by heavy rainfall in 1942 must have been particularly large. The waterlogged condition that must

Table 6   Average loss of nutrients through leaching in flooded soil cropped to rice in 10 weeks during *kharif* (June–December 1972) at the Central Rice Research Institute, Cuttack (Shinde, Vamadevan, and Asthana, Unpubl.)

Nutrient	Leaching loss kg/ha.	Rainfall received during the crop growth in mm.
Nitrogen[1]	45.64	724.4
Iron	46.48	
Manganese	17.00	
Potassium[2]	10.08[3]	

[1] Applied at 150 kg/ha level in 3 splits (1) 37.5 kg at transplanting, 60 kg at tillering, and 52.5 kg at panicle initiation.

[2] Applied at 40 kg K_{20}/ha at final puddle.

[3] Data for 2 months only.

have followed is another predisposing host factor. Therefore, the crop was in a susceptible stage in November and also heavily predisposed for infection.

The scales were apparently heavily loaded against the host and in favor of the pathogen in the interaction of the host, pathogen, and environment in the 1942 rice crop season in Bengal, with the result that the most devastating recorded epiphytotic of helminthosporium disease of rice broke out in that year.

The Possibility of Forecasting Outbreaks of Helminthosporiose in India

Though exact correlations of disease outbreaks with meteorological conditions (macro- and micro-climatic factors) could only be established by studying a number of such epiphytotics, the indication obtained in this study might be useful for watching for such weather conditions as those of 1942. Cloudiness, light rainfall, heavy dew, and temperature of 20–29.4°C throughout the days continuously for a week with reduced hours of sunshine (say below 75% of normal during October or November) should constitute a warning that a severe epiphytotic of helminthosporiose is likely to occur.

A comparative examination of the meteorological data of 1942 with those of 1941, 1943, and 1944 revealed that the 1942 season had excessive rainfall in September, uniformly favorable temperatures of 20–30°C continuously for two months, unusually cloudy weather and rains in November, and a higher minima than usual in November (20°C). Experimental evidence on the temperature relations of germination, growth, infectivity, spread of lesion, and spore production in nature, and predisposition of the host show how the unusual weather trends of 1942 would have helped the rapid spread of the disease and led to an epiphytotic.

Acknowledgment

My thanks are due to Dr. L. A. Ramdas, and Mr. P. Mallik, Agricultural Meteorology, Poona, for providing meteorological records for the years 1941 through 1944 for the different centers in Bengal, and to Dr. M. Gangopadhyaya, for furnishing the "sunshine records" of Calcutta from the Alipore Observatory.

Literature Cited

1. Anonymous. 1943. Brown spot of paddy. *Trop. Agr.* 99:150–51
2. Barat, H. 1931. Études de la division de phytopathologie (Section Sud-Indochinoise de l'Institut des Recherches Agronomiques) au cours de l'année 1930 II. Laboratoire de cryptogamie. *Bull. Écon Indochine, n.s.* 34:779B–96B (*Rev. Appl. Mycol.* 11:432)
3. Baba, I., Takahashi, Y., Iwata, I. 1958. Nutritional studies on the occurrence of Helminthosporium leaf spot and akiochi of the rice plant. *Bull. Nat. Inst. Agr. Sci. Tokyo, Ser. D.* 7:1–157
4. Chandwani, G. H., Balakrishnan, M. S., Padmanabhan, S. Y. 1963. Helminthosporium disease of rice. V. A study of the spore population of *Helminthosporium oryzae* over paddy fields. *J. Indian Bot. Soc.* 62:1–14
5. Chakrabarti, N. K., Padmanabhan, S. Y. 1962. Studies on Helminthosporium disease of rice. VI. Nutritional factors and disease expression. Effect of leaching of soil on incidence. *Proc. Indian Sci. Congr. 49th.* Pt. III:479 (Abstr.)
6. Chattopadhyay, S. B., Chakrabarti, N. K. 1953. Occurrence in nature of an alternate host (*Leersia hexandra* SW.) of *Helminthosporium oryzae* Breda de Haan. *Nature, London* 172:550
7. Chattopadhyay, S. B., Dickson, J. G. 1960. Relation of nitrogen to disease development in rice seedlings infected with *Helminthosporium oryzae* Breda de Haan. *Phytopathology* 50:434–38
8. Das, C. R., Baruah, H. K. 1947. Experimental studies on the parasitism of rice by *Helminthosporium oryzae* Breda de Haan and its control in field and storage. *Trans. Bose Res. Inst.* 16:31–46
9. Hemmi, T., Nojima, T. 1931. On the relation of temperature and period of continuous wetting to the infection of the rice plant by *Ophiobolus miyabeanus. Forsch. Gebiet Pflanzenkr., Kyoto* 1:84–89 (Engl. summary)
10. Hemmi, T., Suzuki, H. 1931. On the relation of soil moisture to the development of the Helminthosporium disease of rice seedlings. *Forsch. Gebiet Pflanzenkr.,* Kyoto 1:90–98 (Engl. summary)
11. Imura, J. 1938. On the influence of sunlight upon the lesion enlargement of the Helminthosporium disease of rice seedlings. *Ann. Phytopathol. Soc. Japan* 8:203–11, (Engl. summary)
12. Imura, J. 1940. On the influence of sunlight upon the incubation period and the development of the blast disease and the Helminthosporium disease of rice. *Forsch. Gebiet Pflanzenkr.,* Kyoto 1:16–26 (Engl. summary)
13. Katsura, K. 1937. On the relation of the atmospheric humidity to the infection of the rice plant by *Ophiobolus miyabeanus* Ito and Kuribayashi and to the germination of its conidia. *Ann. Phytopathol. Soc. Japan,* 7:105–24 (Engl. summary)
14. Miura, M. 1931. A comparative study of species and strains of Helminthosporium on certain Indian cultivated crops. *Trans. Brit. Mycol. Soc.* 15:254–93
15. Misawa, T. 1955. Studies on the Helminthosporium leaf spot on rice plant. *Jubilee Publ. Commem. 60th Birthdays Profs. Y. Tochinai & T. Fukushi. Sapporo, Japan.* 65–73. (Engl. summary)
16. Naito, N. 1937. On the effect of sunlight upon the development of the Helminthosporium disease of rice. *Ann. Phytopathol. Soc. Japan,* 7:1–13 (Engl. summary)
17. Ocfemia, G. O. 1924. The Helminthosporium disease of rice occurring in the southern United States and in the Philippines. *Am. J. Bot.* 11:335–408
18. Ocfemia, G. O. 1924. The relation of soil temperature to germination of certain Philippine upland and lowland varieties of rice and infection by the Helminthosporium disease. *Am. J. Bot.* 11:437–60
19. Padmanabhan, S. Y., Chowdry, K. R. R., Ganguly, D. 1948. Helminthosporium disease of rice. I. Nature and extent of damage caused by the disease. *Indian Phytopathol.* 1:34–47
20. Padmanabhan, S. Y., Abhichan-

dani, C. T., Chakrabarti, N. K., Patnaik, S. 1962. Studies on Helminthosporium disease of rice. VI. Nutritional factors and disease expression—Effect of potassium. *Proc. Indian Sci. Congr.* Pt. III:480 (Abstr.)

21. Padmanabhan, S. Y. 1965. *Studies on Rice Diseases.* Doctoral Thesis, Utkal Univ., Bhubaneswar, Orissa

22. Padmanabhan, S. Y., Ganguly, D., Balakrishnan, M. S. 1953. Helminthosporium disease of rice. VII. Source and development of seedling infection. *Indian Phytopathol.* 6: 96–105

23. Padmanabhan, S. Y., Ganguly, D. 1954. Relation between age of rice plant and its susceptibility to Helminthosporium and blast diseases. *Proc. Indian Acad. Sci.* (B) 39: 44–50

24. Sherf, A. F., Page, R. M., Tullis, E. C., Morgan, R. L. 1947. Studies on factors affecting the infectivity of *Helminthosporium oryzae. Phytopathology,* 27:281–90

25. Sundararaman, S. 1922. Helminthosporium disease of rice. *Bull. Agr. Res. Inst. Pusa,* 128:1–7

THREATENING PLANT DISEASES ❖ 3564

H. David Thurston
Department of Plant Pathology, Cornell University, Ithaca, New York

INTRODUCTION

Plant pathogens have, during the past decade, amply exhibited their ability to spread rapidly to new areas and threaten crops. Coffee rust (*Hemileia vastatrix*), recently discovered in Brazil where the fungus has rapidly spread over almost the entire coffee growing area, and southern corn leaf blight (race T of *Helminthosporium maydis*) which recently caused a severe epiphytotic on maize in the U.S.A., are outstanding examples (in retrospect) of threatening plant diseases. A long catalog of examples might be compiled, but that is not possible in a paper of this length, and information on internationally dangerous plant diseases already exists (106, 119, 142, 154, 163, 181, 201, 215, 221).

Emphasis will be given to plant diseases that have potential international importance, but are at present limited to a few countries or a continent. Emphasis will also be given to tropical plant diseases, especially those that are important in developing countries. No attempt is made to cover all threatening plant diseases, but rather to give examples that illustrate the importance of obtaining additional knowledge of threatening plant diseases. There are probably twice the number of crop plants grown in the tropics as are grown in temperate zones, and an awareness of potentially important diseases on these crops becomes increasingly difficult as information accumulates. Plant pathologists in tropical countries usually know the problems of their own country or continent well, but because of language barriers, inadequate library facilities, and lack of travel opportunities, often are not cognizant of problems in other countries or continents. One tragic consequence of such lack of knowledge has frequently been the inadvertent introduction of a plant pathogen into a new area from which it easily could have been excluded. New pathogens have repeatedly spread into countries or continents that were completely unprepared for them as regards methods of eradication, resistant varieties, and other control methods. Knowledge of threatening plant diseases can give a country or region time to prepare for a new disease and thus minimize its effect on crop production. The threat of new diseases is real, and in this time of exploding populations, hunger, and shrinking agricultural land it is vital to expand our knowledge to prevent future catastrophes.

27

The components that make a plant disease "threatening" are numerous and diverse. To be highly threatening, a disease should be characterized by the ability to spread rapidly, cause serious losses, and be difficult to control. A thorough knowledge of the etiology and epidemiology of a pathogen is necessary for meaningful predictions on its potential for spread. Unfortunately, too often only bits and pieces of the information needed for evaluating a pathogen's potential for spread are available. Information on the ability of a pathogen to cause losses is even more difficult to find. Too often terms such as highly destructive, catastrophic, devastating, etc. are the only basis for making judgments. The potentials for disease control are usually better known, but often controversial and backed by little field experience.

HIGHLY THREATENING PLANT DISEASES

The diseases discussed under this heading are considered the most threatening because of high potential for rapid spread to other continents or countries, demonstrated ability to cause serious crop losses, and difficulty of control. This classification is necessarily made on judgments based on information that is often incomplete, but the diseases so classified are clearly potentially dangerous internationally.

The danger in making such classifications is well illustrated by coffee rust. Little controversy exists on whether *Hemileia vastatrix* can cause serious losses, and there is agreement that the lack of resistant varieties of *Coffea arabica* available to growers, plus the high cost of spraying, make control difficult, but there is still controversy on its ability to spread. Some workers believe the fungus is normally spread only very short distances by rain splash (151, 152); others believe it might spread long distances by air (178) and that it was perhaps carried from Africa to South America in a tropical storm (18). How *H. vastatrix* arrived in Brazil will probably remain unknown, but its spread since its arrival in Brazil can only be called rapid (188).

PHILIPPINE DOWNY MILDEW OF MAIZE   Waterhouse (220) lists eight species of *Sclerospora* attacking maize (*Zea mays*): *Sclerospora graminicola, S. macrospora, S. maydis, S. oryzae, S. philippinensis, S. sacchari, S. sorghi, and S. spontanea. S. macrospora* has been reclassified as *Sclerophthora macrospora* (211) and Payak & Renfro (164) discovered an additional downy mildew on maize in India; *Sclerophthora rayssiae* var. *zeae*. The downy mildews are probably the most serious pathological problem in maize production in tropical Asia. Among the more destructive species are *Sclerospora philippinensis* in the Philippines (64) and India (28), *S. maydis* in Indonesia (189), *S. sacchari* in Taiwan (210), *S. sorghi* in Thailand (179), and *Sclerophthora rayssiae* var. *zeae* in India (197).

Of these, only *S. sorghi* occurs outside of Asia. It is found in Africa (61) and North America (51, 74). *S. sorghi* was first found in Texas and northern

Mexico in 1961, and since then it has spread to eight other southern states of the U.S.A. (74). In 1971 *S. sorghi* caused a 2–3 million dollar loss of the Texas maize and sorghum crops (73).

S. philippinensis, Philippine downy mildew of maize, will be discussed as an example. According to Weston (226), C. F. Baker was the first to describe the disease. W. H. Weston, Jr. was sent in early 1918 to the Philippines to study Philippine downy mildew of maize because of the threat to the maize plantings of the U.S.A. if the disease were ever introduced. His publications are still one of the best sources of information on the disease.

The disease is extremely destructive in the Philippines. Often whole fields are destroyed, and in some areas growers have been forced to stop growing maize. Losses of 40–60% are frequent and often reach 80–100% (64, 226). During a trip through the island of Mindanao in 1968 I saw hundreds of maize fields; the average incidence was 15–25% although one field had over 90%.

The severity of the disease is related to high amounts of rainfall (71). Weston (227) found that sporangia are produced from infected maize only in the dark with free water present. Conidia germinate at temperatures as low as 6.5°C (226), and 19–24°C was found to be optimal for sporangial germination (66). More information on the environmental conditions necessary for infection is needed. On Mindanao I saw severe downy mildew from near sea level to about 4000 ft. This information, although not conclusive, indicates that the organism can cause disease over a wide range of environmental conditions. Weston (226) found *S. philippinensis* can infect sorghum, wild sugar cane (*Saccharum spontaneum*), and other grasses. Experimental inoculation of sugar cane has resulted in infection (65). These widely spread tropical grasses may facilitate the spread of *S. philippinensis*.

Napi-Acedo & Exconde (148) report finding oospores produced by *S. philippinensis*. This means that the danger of long distance dissemination is quite real, as the oospores and perhaps hyphae could be carried in seed, vegetative cuttings of plant material, plant debris, or packing material. There are several reports of other *Sclerospora* species infecting maize seeds (112).

Resistance to *S. philippinensis* has been found in several maize lines grown in the Philippines for centuries. The same lines give resistance to *S. sacchari* in Taiwan and *S. sorghi* in Thailand. With few exceptions maize collections from the Americas tested in Asia have been susceptible to the various species of *Sclerospora*. Maize lines resistant in Asia to *S. philippinensis* and *S. sacchari* are resistant in the U.S.

Most countries have a strict quarantine on plant material (maize, sorghum, sugar cane) from Asia, but how long this will be successful in excluding the pathogen is unknown. Every effort should be made to eradicate *S. philippinensis* if it is introduced into other tropical areas.

The maize downy mildew problem is so large, and the gaps in knowledge of the host-parasite-environment interactions are so great, that greatly expanded multi-national programs are needed if this worldwide threat to maize production is to be adequately countered.

BACTERIAL LEAF BLIGHT OF RICE   The two most serious bacterial diseases of rice (*Oryza sativa*), bacterial leaf blight (*Xanthomonas oryzae*) and bacterial leaf streak (*Xanthomonas translucens* f. sp. *orizicola*), are known only in Asia. Bacterial leaf blight is considered one of the most destructive diseases of rice in Asia, and perhaps the most serious problem of rice in India. Bacterial leaf streak is not found in temperate rice growing areas, but in the Asian tropics under favorable weather conditions it can cause losses comparable to those caused by *X. oryzae* (159).

Bacterial leaf blight was first noted by farmers in Japan in 1884, and Takaishi in 1908 and Bokura in 1911 isolated the organism and successfully inoculated rice, according to Mizukami & Wakimoto (146).

The disease is serious in both the temperate and tropical rice growing countries of Asia. Few figures are available for most tropical countries where the disease is usually more severe, but in temperate Japan up to 400,000 acres are affected annually, with losses ranging from 20–30% in severely infected fields (159). According to Srivastava (200) most of the rice growing areas in India are affected, with losses in yield ranging from 6–60%.

The epidemiology of *X. oryzae* in Japan has been extensively studied (146, 230), but less is known about its epidemiology in the tropics. High populations of the bacterium are found all year in the tropics in the water of rice fields, irrigation canals, reservoirs, and streams (158). Rainstorms and typhoons aid rapid spread of the pathogen. High nitrogen increases disease severity (146).

Many sources of resistance to *X. oryzae* have been found (77, 146, 159, 160, 161), but the occurrence of numerous strains of the organism (146, 147, 162) complicates breeding. Although chemical control is used in Japan, it has not been successfully used in tropical areas (159).

As neither *X. oryzae* or *X. translucens* f. sp. *orizicola* occur outside of Asia, and the organisms might be carried on seed, straw, or plant debris, every precaution should be taken to prevent their introduction into the Americas or Africa.

AFRICAN CASSAVA MOSAIC   A mosaic disease of cassava (*Manihot esculenta*) was first reported by Warburg in 1894 in East Africa under the name Krauselkrankheit. Transmission by white flies (*Bemisia* spp.) was first reported by Ghesquiere in 1932, according to Storey & Nichols (208). Kitajima & Costa (117) found elongated particles (rods) associated with American cassava common mosaic from Brazil and the cassava brown streak virus, but none were found associated with African cassava mosaic. The American cassava common mosaic was easily transmitted by mechanical inoculation, while the

African cassava mosaic was not. Furthermore American cassava common mosaic has been identified and characterized as a virus (48). The African cassava mosaic had been reported to be caused by a virus based upon symptomatology and transmission by white flies.

The causal agent of African cassava mosaic has not yet been positively identified and characterized. Galvez & Kitajima (80) found no virus-like particles or mycoplasma-like bodies *in situ* in ultrathin sections prepared from infected leaf tissues of cassava collected in Nigeria. They suggest that the so-called white fly-transmitted viruses might have a different causal agent. However, with electron microscope observations Plasvic-Banjac & Maramorosch (168) found inclusions in cells of leaves infected with African cassava mosaic, which they suggested might be virus particles. A similar disease transmitted by *Bemisia tabaci* is present in India and other Southeast Asia countries (139). In conclusion, African cassava mosaic, American cassava common mosaic, and cassava brown streak are caused by different viruses or causal agents.

African cassava mosaic is found in eastern, western, and central tropical Africa and the surrounding islands (45, 111). Padwick (163) brought together the available information on losses and estimated that the yearly loss in yield due to the disease is equivalent to almost 11% of the crop in Africa. Yield losses on individual varieties of susceptible cassava range from 20–95% (9). Cassava is a major food crop in tropical Africa (113), thus the disease seriously affects the food supply of millions of Africans.

It is doubtful that mechanical transmission of African cassava mosaic is common in nature. The causal agent is carried in cuttings (commonly used for planting) from diseased plants. In addition, white flies are important vectors (208). The environmental factors that influence the spread of the disease are not well understood although symptom expression is more severe at low temperatures and vector activity is highest during rainy periods (111). Several workers in Africa have worked on breeding for resistance to cassava mosaic (62, 110, 111). Immunity has not been found, but various levels of resistance have been. Jennings (111) reported three clones, derivatives of *Manihot glazovii*, with high resistance and good agronomic characters, but these clones are not in widespread use.

As the disease is now found only in Africa, every effort should be made to prevent its spread to Asia and the Americas by strict quarantines.

SOUTH AMERICAN LEAF BLIGHT OF RUBBER   The most serious disease of rubber (*Hevea brasiliensis*) is caused by *Microcyclus ulei* (*Dothidella ulei*). Hevea rubber is native to the upper valley of the Amazon in Brazil where it was destructively tapped in scattered areas throughout the jungle. Until 1900, practically all of the world's rubber came from Brazil and Peru. Increased rubber usage in the late 1870s stimulated H. A. Wickam, an English botanist, to improve production by establishing rubber plantations. He collected seeds in Brazil, shipped them to the Royal Botanical Gardens at Kew, England, and sent the germinated and growing seedlings to Ceylon and Singapore (105). These seedlings were the source of clones for most of the subsequent Asian rubber plantings. The discovery of a nondestructive tapping system in Singapore greatly accelerated rubber production in S.E. Asia, which now produces 90% of the world's rubber. By germinating the seeds in England, Wickam probably prevented the introduction of *M. ulei* into Asia.

Most of the rubber in Brazil was tapped from trees growing in their natural habitat in the jungle and *M. ulei* was hardly noticed as a problem until 1904 when Hennings (99) first described the fungus. The disease did not become a serious problem until attempts were made, early in the 20th century, to grow rubber in large plantations in South America. Plantations established in Trinidad and the British and Dutch Guianas were a total failure because of *M. ulei* (105, 177). The Ford Motor Co. planted 8000 acres in Fordlandia, Brazil, and by 1933 the disease had destroyed a quarter of the planting. Similar failures occurred in Panama, Costa Rica, and Colombia (105). Hilton (103) has reviewed the literature on the damage caused by *M. ulei* in tropical America. *M. ulei* is known only in tropical America (45) from 18° N in Mexico to 24° S in Brazil (105).

Although the fungus produces perithecia (*M. ulei*), conidia (*Fusicladium macrosporum*), and pycnidia (*Aposphaeria ulei*), the conidia seems to be most important in spread (105, 122). Young leaves are most susceptible, but leaves over 16 days of age are quite resistant (105). The appearance of new leaves on rubber trees in "flushes" helps to explain frequent disease escapes and apparent disease resistance which later breaks down. Breeding for disease resistance has been a difficult task because of the linkage of susceptibility to high yield and, conversely, resistance to low yield. Clones in S. E. Asia are highly susceptible to the disease, and it is for this reason that workers are prepared to eradicate the disease at any cost, as environmental conditions in Asia (especially Malaya) are considered favorable for the disease (105). Holliday (105) has reviewed the knowledge to date on the etiology and epidemiology of *M. ulei*, but much additional information is needed to evaluate fully its potential for spread.

Chemical control (except in nurseries) has not been economical in the Americas and, although resistance has been found (17, 86, 105, 123), it is utilized in only a few plantings here. Physiologic races of *M. ulei* (105, 120, 122, 144) complicate breeding. Rands (177) describes a system of grafting *M. ulei* resistant tops on high-yielding clones which, although expensive, is potentially a useful means of control.

South American leaf blight illustrates the dangers of growing a formerly wild plant in a monoculture. In the native habitat of *Hevea* in the Amazon jungle there are about six *Hevea* trees per acre. Many of the native rubber trees are very low yielding and somewhat tolerant to the disease, thus their mortality is not as high as those of selected high yielding trees grown as a monoculture. In addition, each tree is screened from the other rubber trees by the foliage of trees of other genera, which serve as barriers to windborne spores. The disease was obscure and did little damage during the time when rubber was collected from the jungle (121). Higher yielding *Hevea* trees grown in large plantations were more susceptible and had more exposure to attacks by *M. ulei.* The disease thus became a serious problem and destroyed thousands of acres of rubber in Latin America.

Rubber growing countries in Asia have set up strict quarantine measures to prevent the introduction of *M. ulei* (10). Malaya has distributed a well-illustrated brochure in color on the disease to alert rubber growers of the danger from *M. ulei,* and has worked out a comprehensive plan for a campaign of eradication in case the disease is ever introduced (105, 108). Malayan scientists are working in the Americas on resistance to *M. ulei.* This approach can serve as a model to other countries faced with similar problems. Although it is hindsight, it is interesting to speculate on what might have been the history of coffee rust in the Americas if the Americas had been as well prepared.

MOKO DISEASE OF BANANAS AND PLANTAINS   The banana is only one of many economic crop plants attacked by *Pseudomonas solanacearum,* which causes the Moko disease. Bacterial wilt is the name of the disease on most crops, and according to Kelman (114) was first reported in 1890 on potatoes by Burrill. Smith (198) first conclusively demonstrated that *P. solanacearum* caused the disease. The literature has been reviewed by Kelman (114), Buddenhagen & Kelman (24), and Stapp (202). The bacterium is able to attack more than 200 species of plants in 33 different families (114). The isolates of *P. solanacearum* have been classified into three races; 1, 2, and 3, on the basis of hosts attacked and other characters (24, 25). In general terms, race 1 attacks solanaceous and other plants, race 2 bananas and Heliconias, and race 3 potato. Numerous subraces or pathotypes occur within each race. Hayward (96) classified 185 isolates of *P. solanacearum* into four major biochemical types.

Race 2 of *P. solanacearum* causes the Moko disease of bananas and plantains. Possibly the earliest reference to the Moko disease was by Schomburgk in 1840 in Guyana, according to Martyn (137). Rorer (185) proved that *P. solanacearum* caused the disease in Trinidad. *Heliconia* species have been found naturally infected in virgin jungles, and this is probably the origin of race 2. Different strains of *P. solanacearum* have been found on *Heliconia* in banana plantations (20, 192). Race 2 is currently widely distributed in Central and South America and Trinidad (21, 22, 75), but apparently is not present in Africa or Asia. Race 1 does not normally attack commercial bananas; symptoms caused by race 1, if they do occur, are not similar to the Moko disease induced by race 2 (21, 22, 231). Infection of bananas by *P. solanacearum* has been observed occasionally in localized areas in Africa, Asia, and the Pacific Islands, but the validity of these reports of the Moko disease has been questioned by Buddenhagen (21, 22), either because the symptoms described were not typical of the Moko disease, or because the pathogenicity of the bacterium was not properly established. There is a recent report that Moko disease now may be present in the Philippines (142, 143).

The Moko disease caused almost total destruction of the "moko" plantain in Trinidad in the late 19th Century (185), and serious losses as early as 1840 in Guyana (22). A highly virulent insect-disseminated strain (23) caused an epidemic that devastated plantains in Central America in the 1960s with a loss of millions of plants. Another insect-disseminated strain has spread rapidly in Colombia (128) and more recently into the Amazon jungle of Peru (75), where the disease threatens a major food source for thousands of people. Buddenhagen (22) estimates that the disease has eliminated susceptible plantains from thousands of square miles in Latin America. The bacterium can be spread rapidly by man through mechanical means, but the insect-disseminated strain of *P. solanacearum* can spread much more rapidly (22, 23, 190).

Control can be achieved by sterilization of tools (190, 193) and by removing the male infloresence before insect transmission can take place (23, 128). A desirable plantain with resistance to *P. solanacearum* has been reported (209). Every effort should be made by pathologists in Africa, Asia, and the Pacific Islands where plantains and bananas are important, to maintain a strict quarantine against the introduction of the destructive race 2 of *P. solanacearum*.

DISEASES OF INTERMEDIATE THREAT POTENTIAL

These are diseases that either do not clearly possess the ability for rapid spread to other continents or countries, or for which efficient, economical control measures are not clearly available. Additional knowledge could change this classification for any of the diseases discussed. They should all be considered highly dangerous.

STREAK DISEASE OF MAIZE   Streak disease is only one of many diseases of maize caused by viruses or mycoplasma (155, 216). Storey (204) states that it was first recorded in S. Africa in 1901, and he first showed that the causal agent was transmitted by the planthopper *Cicadulina mbila*. Streak disease is widely distributed in Africa and Mauritius (88) on maize and causes serious crop losses (50, 63, 76, 87, 100), but precise loss figures were not found.

There are several "forms" or strains of maize streak (87, 138, 199, 207) and three insect vectors, all *Cicadulina* spp. (205). The pathogen is also found on sugar cane and many wild grasses (87, 138, 199, 207), and can attack wheat (101), barley, oats, and rye (87). Disease spread is increased by warm temperatures and high rainfall, conditions that favor the vector (88).

The literature states that the disease is caused by a virus, although the virus has never been isolated. Recent electron microscope studies (167) support the hypothesis that streak disease is caused by a virus.

It has been impossible to control the disease by chemical control of the vector, and cultural methods such as clean cultivation are only partially successful (88). Most promising has been the development of resistant varieties (76, 87, 88, 206). Other areas of the world where maize is important, such as the Americas and Asia, should maintain vigilance to keep this destructive pathogen confined to its present area of infestation.

HOJA BLANCA OF RICE   Twelve or more virus and mycoplasma-like diseases attack rice (126), but only hoja blanca is found in the Americas. Hoja blanca (Spanish for white leaf) was first observed in Colombia in 1935 (11, 84). Its appearance was sporadic in Colombia until 1957, probably because the unimproved native varieties had some resistance (79). From Colombia the disease spread to Panama (1952), Cuba (1954), Venezuela (1956), the United States (1957), Costa Rica (1958), and into other countries in Central and South America (79). Malaguti, Diaz & Angeles (135) found that the planthopper *Sogatodes oryzicola* was the vector. In 1960 *Sogatodes cubanus* was reported to transmit the virus from rice to *Echinochloa colonum,* but not from *E. colonum* to rice (81). Detailed histories of the disease are given by Galvez (79), Ou (159), and Ling (126). Herold, Trujillo & Munz (102) and Shikata & Galvez (195) have described virus particles for hoja blanca, but they disagree on size and shape.

The disease now occurs in most of the major rice growing areas of Central and South America, the Caribbean, and the U.S. (45, 79, 159). Losses from hoja blanca have varied widely from year to year. In 1958 many fields in Colombia were completely lost (84). A 25% loss of the entire rice crop in Cuba and a 50% loss for the rice crop of Venezuela was estimated for 1956 (214), and many individual fields in Cuba and Venezuela had a 75–100% loss (6). Losses have greatly diminished with the increased use of resistant varieties.

The rapid spread of hoja blanca throughout the Americas during the 1950s and 1960s illustrates the potential for long-distance spread of the virus when the insect vector is present. Transmission, host range, and factors affecting disease development have been reviewed (78, 79, 126, 159). Most *indica* types of rice are susceptible to hoja blanca, while *japonica* types are resistant. Commercially adapted and accepted lines with resistance have been developed, thus the importance of the disease in the Americas has been drastically reduced. Nevertheless, efforts should be made to assure that hoja blanca does not move from the Americas to tropical Asia or other rice growing areas of the world where vast acreages of *indica* types of rice susceptible to hoja blanca are grown, and the Americas should be equally vigilant to assure that the many rice viruses present in Asia are not introduced.

SOYBEAN RUST   Soybean (*Soja max*) rust, caused by *Phakopsora pachyrhizi,* was first described from Japan in 1903 by Hennings (97). The disease has been reported from most countries of S.E. Asia and also in Japan, Manchuria, India, and Latvia (36, 104, 127). Much of the literature describes the disease as very destructive (107, 118, 125, 219), but there is little precise information on losses. Knowledge of etiology and epidemiology is also limited (118); however, it appears that the disease has spread widely since its initial discovery. Control is possible by fungicides (107, 219), but resistant varieties are probably more economical means of control (39). Considering the worldwide importance of soybeans it is obvious that more information is needed on what appears to be a potentially destructive pathogen, which could easily be introduced or spread into new areas.

STUNTING VIRUS DISEASE OF PANGOLA GRASS   Pangola grass (*Digitaria decumbens*) is one of the more important and productive improved tropical grasses for livestock. Until recently the extensive plantings in the tropical Americas all originated from vegetatively propagated material originating from 3–5 plants imported to Florida (60). This is an extremely narrow germplasm base. The disease was first reported in Surinam and the causal agent (assumed to be a virus) was shown to be transmitted by the planthopper *Sogata furcifera* (60).

Since 1959 the disease has been reported from Peru, Brazil, Guyana, Surinam, Fiji, Malaya, and Borneo (13, 60, 67, 180). These reports are made solely on the basis of symptom expression, and whether the disease is the same is conjectural.

In 1962 the disease was estimated to reduce yield by 60% or more on the 2800 acres of Pangola grass in Guyana (13, 14). Yield losses of 50% or more were reported in Surinam (60). The disease appears to spread slowly in the field (14), but long distance dissemination probably takes place by shipments of vegetative planting material.

Control of the insect vector does not seem to be feasible (14). Plowing and fertilizing to encourage new growth will give temporary yield increases; however, new growth is soon attacked (13). A line of Pangola grass (A 24) introduced into Taiwan from the Philippines and imported to Guyana in 1962 has been resistant for several years (13, 14). Little is known of the etiology or epidemiology of this disease, thus, considering the present and future importance of Pangola grass in the tropics, efforts should be made to restrict movement of infected material and learn more about the disease.

GOMOSIS OF IMPERIAL GRASS   Imperial grass (*Axonopus scoparius*), and micay (*Axonopus micay*), are important for livestock in the tropics of the Americas. The bacterial disease gomosis (English-gummosis) once threatened the existence of these grasses in Colombia. This vascular disease was probably first noted in Colombia in 1936 (83), and the causal agent was named *Xanthomonas axonoperis* by Starr & Garces (203).

The bacterium causes a striking overgrowth and contortions of some stems of the plant before the plant wilts and eventually dies. The disease has caused the abandonment of many plantings of imperial grass and micay in large areas of Colombia. A 50% reduction in yield may occur in a planting two years after initial infection, and within four years the planting is usually abandoned (35). The disease is most severe in areas of Colombia with average mean temperatures of 20–24°C (34).

The bacterium is transmitted mechanically by harvesting tools such as machetes, by livestock, insects, rain-splash, and vegetative planting material (34, 35). Disease free plantings can be maintained by using sanitation control methods (35).

As Colombia is the only country known to have gomosis, care should be taken to prevent its spread to other areas, especially by vegetative cuttings.

LETHAL YELLOWING OF COCONUT PALMS   Lethal yellowing of coconut palms (*Cocos nucifera*) has been recognized since the first report by Fawcett in Jamaica in 1891 (136). The disease is now widespread in the islands of the Caribbean, some of the countries of Central and South America adjoining the Caribbean, Florida in the U.S.A., and West Africa (30, 31, 44, 90, 129, 136, 172). Disease losses have been serious although precise figures are difficult to find. In recent years 100,000 trees were being lost in Jamaica each year (166). Since the first record of the disease in Ghana, 100,000 palms have been lost (124). Lethal yellowing killed 75% of the palms on the island of Key West between 1955 and 1968 (173), and more than 2000 palms have succumbed to lethal yellowing during the first year of the present epidemic in the Miami, Florida area (182). Carter (31) considers lethal yellowing to be as potentially important as cadang-cadang.

Considerable information has been accumulated on the etiology and epidemiology of the disease (30, 31, 32, 89, 90), but definite proof of the causal agent is lacking. Some investigators have shown mechanical transmission (145, 173) which suggests a virus, and recently mycoplasma-like bodies have been found associated with the lethal yellowing disease (166). The best hope for control appears to be the use of resistant varieties of coconuts such as Malayan Dwarf (40, 228). The serious losses caused by lethal yellowing and its apparent ability to spread to other countries and continents, indicate that other coconut producing countries should take precautions to prevent the introduction of this disease.

RED-RING DISEASE OF COCONUT PALMS   The nematode of *Rhadinaphelenchus cocophilus* is the causal agent of the red ring disease of coconut palms. The disease was first described in Trinidad in 1905 (15). Nowell (149, 150) proved the pathogenicity of *R. cocophilus* and Cobb (42) described the nematode. The disease is found in Central America, South America, and several islands of the Caribbean (15, 47). Losses of 20–25% of the trees in Trinidad plantings have been reported (68), and 60% losses were described by Nowell (149). Because coconuts take 5–6 years before they bear, and can then produce for 30–50 years, such losses are serious.

The coconut weevil (*Rhychophorus palmarum*) has been frequently found associated with infected palms, and frequent recovery of the nematode from the insect gives strong evidence that it is a vector (43, 91, 92, 93). Blair (15, 16) and Fenwick (69) have reviewed the knowledge of the etiology and epidemiology of the disease. Control is difficult, and eradication and sanitation are the major controls used (16, 69, 187). The disease is not present in Africa and Asia, and great care should be taken to prevent its introduction.

COCOA SWOLLEN SHOOT DISEASE   This destructive disease of cocoa (*Theobroma cacao*) is caused by the cocoa swollen shoot virus (CSSV) (115, 170). The disease was first identified in West Africa in 1936 (223). Several cocoa viruses are reported from other continents (19, 213) but CSSV is now found only in five countries of West Africa (19). Losses in Ghana have been devastating. Wellman (223) states that half of the mature trees in a 250,000 acre area of cocoa were killed by CSSV after its discovery. Over 140 million trees have been killed in Ghana in eradication programs for CSSV (19). An annual loss in cocoa production in Eastern Ghana of over 20,000 tons was reported in 1961 (165).

The host range of CSSV includes many trees occurring in the forests and these serve as reservoirs of infection (19, 116). In an attempt to prevent further spread of the disease, authorities in Ghana and Western Nigeria have eliminated many infected trees. In some areas of Ghana this policy has given satisfactory control, but in the Eastern Region it has not (116).

Mild and severe strains of CSSV are found (171, 213). Several species of mealybugs transmit the virus (116, 171) and are of importance on cocoa only as vectors of CSSV. Biological and chemical control of mealybugs has not been successful or encouraging (116). Removal of diseased trees is extremely expensive and not entirely successful, perhaps due to the reservoirs of infection in forest trees, and as controlling the vector has not been successful, disease resistance seems to be the most promising control. All cocoa lines tested have been susceptible, but lines with high levels of tolerance to CSSV have been found. These are being utilized in a breeding program, and may eventually replace the highly susceptible lines now grown (116).

Every effort should be made by cocoa producers in Asia and the Americas to exclude CSSV from their countries. Thresh (212) discusses the measures now being taken to maintain quarantines against CSSV.

MONILIA POD ROT OF COCOA   Monilia pod rot, also known as watery pod rot, is caused by *Monilia roreri* and was first seen near Quevedo, Ecuador in 1914. Rorer in 1918 gave the first description of the disease (2). Since that time it has spread to Peru, Colombia, Venezuela, and Southern Panama (2). In Colombia the disease causes yield losses of 30–40% (7) while losses reach 90% in some areas (156). Ampuero (2) estimates that 15–80% of the pods in Ecuador and Colombia are lost due to *M. roreri*. Many cocoa plantings have been abandoned because of the disease, and exports in Ecuador went from 47,000 tons in 1914 to 10,580 tons in 1933 due to Monilia pod rot and witches' broom (*Marasmius perniciosus*) (2). Colombia was a major exporter of cocoa in the early 1900s but now imports over 10 million dollars worth of cocoa each year.

The conidia of *M. roreri* are disseminated by wind, rain, and insects (56, 72). Much more work needs to be done to understand properly the dissemination of the fungus (58). The sexual stage of *M. roreri* has not been found. Chemical control of the disease is possible (52, 57), but expensive and time consuming. There is evidence that resistance to *M. roreri* exists in cocoa (53, 109). Partial control of Monilia pod rot can be achieved by cultural and sanitation methods (7, 94).

Other cocoa producing countries in the Americas and other continents should take every precaution to prevent the spread or introduction of this highly destructive pathogen into their areas.

BUNCHY TOP OF BANANAS   Bunchy top of bananas (*Musa* spp.) is assumed to be caused by a virus because the causal agent is transmitted by the banana aphid *Pentalonia nigronervosa* (130). The disease was first noted in Fiji in 1890, where production declined sharply until 1895 when a possibly resistant variety, Gros Michel, and planting in new areas, helped the industry to recover (134). The introduction of bunchy top via infected bananas into Australia in 1913 caused devastation of the young industry there, with over 5000 acres abandoned in New South Wales (130). The disease has also caused serious losses in Egypt, Ceylon, parts of India and Pakistan, the Philippines and various Pacific Islands (133, 134, 141).

Bunchy top is found in Egypt and the Congo in Africa (45) and several countries of Asia, Australia, and various Pacific Islands (45, 141). It was recently reported from Vietnam (217). The major spread of the causal agent is by the banana aphid *P. nigronervosa*, although diseased asexual planting material also spreads the virus. The disease was almost completely eradicated from Australia by a vigorous government program of inspecting plantations, destroying infected plants, and planting only virus free bananas (29, 131). Similar control measures were not so successful in Fiji because of scattered plantings, lack of grower interest, and the presence of susceptible wild bananas in forests near banana plantations (134). Some resistance in bananas to bunchy top has been reported (132, 229), but resistant clones are not commonly used in control programs.

Since this disease has the potential to be highly destructive, and docs not exist in the Americas and most of Africa, great care should be taken to prevent its introduction.

DISEASES OF LIMITED THREAT POTENTIAL

The diseases discussed under this heading are considered to have a relatively limited threat potential to world-wide agriculture because they either have spread very slowly (such as cadang-cadang of coconuts), or efficient, economical means of control are available in case they do spread. All of the diseases are described in the literature as causing serious crop losses.

Diseases described in this section may not alway have a limited threat potential. For example, although cadang-cadang has spread very slowly and has not been found outside the Philippines, coconuts are known to drift from island to island in the ocean, and it has even been suggested that coconuts could have drifted across the Pacific to Panama (176). If coconuts can survive such trips perhaps the causal agent of cadang-cadang could also, thus the classification of cadang-cadang as slow-spreading could change.

ENANISMO OF BARLEY, OATS, AND WHEAT   Enanismo (dwarfing) is a disease of unknown etiology which attacks barley, oats, and wheat. It has been present in southern Colombia at least since 1952 (82) and is found only in southern Colombia and in northern Ecuador (85). It causes extreme stunting and rosetting of the plants, which produce few or no spikes.

Frequently severe losses or total crop failures due to enanismo occur in Colombia and Ecuador, especially on barley. The barley collection of the U.S. Dept. of Agriculture was planted in southern Colombia, and of 6000 lines tested only 19 were found with appreciable resistance (183). Although the highest levels of resistance have been found primarily in barley and wheat varieties native to Colombia, some introduced varieties were resistant. Attempts to transmit the causal agent through soil, seed, and by mechanical means were unsuccessful, but individual insects of the leafhopper *Cicadulina pastusae* were able to transmit the disease (82). Probably it is caused by a virus or a mycoplasma. Little more is known about the etiology or epidemiology of the disease. Although control can be achieved through breeding resistant varieties, there is some evidence for races of the causal organism (184). Because spread of the disease out of its present limited area on three of the world's most important food crops is possible, every effort should be made to obtain more information on this disease.

POTATO RUST   Potato (*Solanum tuberosum*) rust is caused by two species of fungi: *Puccinia pittieriana,* a short-cycle form that produces only teliospores and sporidia, and *Aecidium cantensis,* an aecial form. Considering the economic importance of rusts on many crops, it is unusual that these rusts of a crop of major economic importance such as the potato have received so little attention. Both rusts also attack tomato (4, 8).

P. pittieriana is found in Mexico, Costa Rica, Colombia, Venezuela, Ecuador, and Peru. Some literature (37, 59) cites Hennings (98) as stating that *P. pittieriana* occurs in Paraguay, but this is an error. *A. cantensis* is reported from Peru (1) and Honduras (49). Information on losses due to these rusts is sketchy. Some authors (27, 33, 41, 59, 70) state that *P. pittieriana* causes serious losses while others (8, 95) state that the fungus causes no important losses. No data were found in the literature on the effect of either rust on yield. All authors agree that *P. pittieriana* is found at high elevations (9–11,000 ft) (3, 33, 38, 59, 70, 95, 157), but *A. cantensis* seems to occur at lower elevations of 8000–9000 ft (1). The most complete study of *P. pittieriana* is that of Buritica & Orjuela (27). They found that teliospores would germinate at temperatures from 4° to 20°C, with optimum germination between 8–12°C. Differences in degree of susceptibility were found among potato clones, but none were immune.

In summary, little is known about the ability of *P. pittieriana* to cause disease and less about its potential for spread. Essentially nothing is known about *A. cantensis*.

Whether these organisms that thrive at the cool temperatures prevalent at these high elevations would be able to spread and cause serious disease in the temperate regions of Europe and N. America would be pure conjecture, but there is little doubt that *P. pittieriana* could become established in other tropical countries where potatoes are grown at high elevations and could cause serious economic losses.

CADANG-CADANG DISEASE OF COCONUTS   The coconut palm (*Cocos nucifera*) is a major crop in the tropics and is especially important in the Philippines, where 165 million trees planted on over a million hectares produce 42% of the Philippine's foreign exchange earnings (12). Cadang-cadang (dying-dying in English) was first reported by Ocfemia in 1937 (153) on San Miguel Island, but it had been observed as early as 1928. By 1968 almost all of the original 250,000 trees on the island had been destroyed by the disease (174). Millions of trees have been killed in the limited area of the Philippines where the disease occurs, and yields have been seriously reduced. The causal agent of cadang-cadang is still unknown, in spite of extensive research (174). A virus is suspected (54, 169, 175) but definite proof is lacking. The disease moves very slowly, and fortunately it appears possible to continue coconut production in the affected areas (12, 196).

Although the disease moves slowly, other coconut producing areas should make every effort to prevent the spread of cadang-cadang.

AMERICAN LEAF SPOT OF COFFEE   The American leaf spot of coffee (*Coffea arabica*), also called "ojo de gallo" (cock's eye), is caused by a most interesting tropical fungus. The disease was first described by Saenz in 1876 in Colombia, according to Wellman (222), and the fungus was first described by Cooke in 1881 (46). The pathogen is the basidiomycete *Mycena citricolor*. Ashby (5) first grew the perfect and imperfect stages in pure culture. The literature on *M. citricolor* was recently reviewed by Salas & Hancock (186).

The disease is found in all the major coffee growing countries of the Americas (45, 222, 225) and in some areas severe losses by premature defoliation are caused by the fungus (194). Wellman cites losses of 75% in some areas, and states that Costa Rica once had a 20% annual loss due to *M. citricolor* (222, 225). Improved coffee culture and the use of eradicant sprays have reduced losses in recent decades.

The disease is most severe in the cool, moist regions found in many of the coffee growing areas of the Americas. Sequeira (191) lists 150 plants in 45 families attacked by *M. citricolor*, but states that the fungus does not produce reproductive bodies on all of them. Wellman (225) indicates that perhaps 550 species may be infected. Many of the hosts occur naturally in virgin forests and from there the fungus has spread to coffee (191).

The spread of *M. citricolor* is unusual as the fungus is primarily disseminated by imperfect reproductive structures (gemmae) produced on leaves under moist conditions. The gemmae are produced on gemmifers, sometimes called stilboids (26). The perfect stage, a small agaric, is seldom found in nature and plays little role in spread. An unusual characteristic of the mycelium of the fungus is its luminous properties. When wet, the gemmae swell, detach easily, and are disseminated by splashing rain, man, or other agents (26, 225). Disease spread is relatively slow as the gemmae are not spread long distances. Van der Plank (218) uses it as an example of a disease that spreads very slowly, and thus might be controlled by sanitation. No real resistance to *M. citricolor* has been found in coffee (191, 224, 225). Cultural methods, such as drastic pruning of heavy shade and its subsequent regulations will reduce losses, but will not give complete control of the disease (224). The use of eradicant fungicides (224, 225), especially arsenicals and coppers, gives the best control.

Although *M. citricolor* does not spread rapidly, it can be highly destructive and, once established, because of its wide host range, might be impossible to eliminate. The coffee industries of Africa and Asia should be careful to prevent the introduction of the American coffee leaf spot.

DISCUSSION

In this age of jet transportation and increasing ease and frequency of travel from continent to continent, no one would dispute the obvious fact that plant pathogens may also move intercontinentally with greater ease. I hope this article will stimulate greater awareness of some tropical plant diseases that will constitute a threat to other countries or continents if they move from their present distribution. Scientists need to remain alert for pathogens or races of pathogens that may seriously threaten crops in other areas of the world.   .

Sound information on which to base judgments on the potential of pathogens to cause serious losses is generally lacking. Without such information it is extremely difficult for pathologists and governments to determine which diseases are truly of potential importance. Much more research is needed to determine the ability of pathogens to spread. Sound studies of the etiology and epidemiology of many tropical pathogens are lacking. Phytopathological societies and international and national institutions need to encourage the coordinated study of these and similar pathogens on a worldwide basis. Increased exchange of information, of gene pools for crop resistance, and especially of personnel between different countries and continents are essential for increased recognition of threatening plant diseases. Quarantines need to be strengthened on a worldwide basis. It is worth noting however, that quarantines can stifle useful exchange of plant introductions unless knowledgeable people enforce them. A worldwide cooperative effort to monitor pathogens, and perhaps other pests, should be established.

The efforts of scientists and governments in S.E. Asia to prevent the introduction of South American leaf blight of rubber, and to be prepared for it if it does arrive, might be used as a model for dealing with a threatening plant disease.

In 1963 I accompanied Dr. A. J. Riker of the University of Wisconsin who was visiting various countries in South America, on visits to various national and international agencies in Colombia. He wanted to obtain support for a project on methods of eradicating coffee rust. The project, a modest $190,000.00, was not funded. Had it been funded, it might have accomplished what S.E. Asian governments are now doing relative to South American leaf blight of rubber. Today, less than a decade later, many millions of dollars are being spent on coffee rust research and control by South American countries. This example illustrates the fact that plant pathologists need sound information that will convince administrators and governments of the need to invest in research and education on threatening plant diseases before the diseases are introduced.

The advances in food production in developing countries made possible by the green revolution can be lost if proper attention is not given to plant diseases and other pests. The increased yields of the green revolution are based on the widespread use of a small core of germplasm, larger plant populations per hectare, increased pressure to obtain several crops per year from the same land, increased use of irrigation, and improved cultural methods. It is paradoxical and unfortunate that often these practices increase the chances of serious attacks by plant diseases. The recent tungro epidemic on the improved rice varieties in the Philippines (55) is an example of this dilemma.

The high susceptibility of Texas cytoplasmic male sterile maize lines to race T of *H. maydis* was known in the Philippines at least nine years before race T caused an epidemic in the U.S.A. (140). This last example illustrates the need to maintain vigilance against threatening plant diseases on an international basis.

Literature Cited

1. Abbott, E. V. 1931. Further notes on plant diseases in Peru. *Phytopathology* 21:1061–71
2. Ampuero, C. E. 1967. Monilia pod rot of cocoa. *Cocoa Grower's Bull.* 9:15–18
3. Arthur, J. C. 1920. Two destructive rusts ready to invade the United States. *Science* 51:246–47
4. Arthur, J. C. 1921. Origin of potato rust. *Science* 53:228–29
5. Ashby, S. F. 1925. The perfect stage of *Stilbum flavidum* Cke. in pure culture. *Bull. Misc. Inf. Roy. Bot. Gard.* 1925:325–28
6. Atkins, J. G., Adair, C. R. 1957. Recent discovery of hoja blanca, a new rice disease in Florida, and varietal resistance tests in Cuba and Venezuela. *Plant Dis. Reptr.* 41:911–15
7. Barros N., O. 1966. Valor de las practicas culturales como metodo de reducir la incidencia de Monilia en plantaciones de cacao. *Agr. Trop.* 22:605–12
8. Bazan de Segura, C. 1965. *Enfermedades de Cultivos Tropicales y Sub-tropicales.* Lima: J. D. Segura Montoya. 439 pp.
9. Beck, B. D. A., Chant, S. R. 1958. A preliminary investigation on the effect of mosaic virus on *Manihot utilissima* Phol. in Nigeria. *Trop. Agr. (Trinidad)* 35:59 64
10. Berg, G. H. 1970. Plant quarantine measures against South American leaf blight. *FAO Plant Prot. Bull.* 18:1–7
11. Bernal-Correa, A. 1939. La situacion patologica del arroz en el Valle de Cauca. *Rev. Fac. Nac. Agron., Univ. Antioquia* 1:234–38
12. Bigornia, A. E. 1966. Coconut cadang-cadang incidence and replanting in the Philippines. *Phillipp. J. Plant Ind.* 31:71–78
13. Bisessar, S. 1965. Review of work in plant pathology in British Guiana 1950–1964. *PANS* (Sect. B) 11:392–96
14. Bissessar, S. 1966. The stunting virus disease of Pangola grass and the reaction of *Digitaria* species to the virus in British Guiana. *FAO Plant Prot. Bull.* 14:60–62
15. Blair, G. P. 1969. Studies of the red ring disease of the coconut palm. 89–106. In *Proc. Symp. Trop. Nematol.* Univ. P.R. 1967. 169 pp.
16. Blair, G. P. 1969. The problem of control of red ring disease. In *Nematodes of Tropical Crops.* ed. J. E. Peachey. Farnham Royal, England: Commonw. Agr. Bur. 355 pp.
17. Blazquez, C. H., Owen, J. H. 1963. Histological studies of *Dothidella ulei* on susceptible and resistant *Hevea* clones. *Phytopathology* 53:58–65
18. Bowden, J., Gregory, P. H., Johnson, C. G. 1971. Possible wind transport of coffee leaf rust across the Atlantic Ocean. *Nature* 229:500–1
19. Brunt, A. A., Kenten, R. H. 1971. Viruses infecting cacao. *Rev. Plant Pathol.* 50:591–602
20. Buddenhagen, I. W. 1960. Strains of *Pseudomonas solanacearum* in indigenous hosts in banana plantations of Costa Rica, and their relationship to bacterial wilt of bananas. *Phytopathology* 50. 660 64
21. Buddenhagen, I. W. 1961. Bacterial wilt of bananas: history and known distribution. *Trop. Agr. (Trinidad)* 38:107 21
22. Buddenhagen, I. W. 1968. Banana diseases in the Pacific area, *FAO Plant Prot Bull.* 16:17–31
23. Buddenhagen, I. W., Elsasser, T. A. 1962. An insect-spread bacterial wilt epiphytotic of Bluggoe banana. *Nature* 194:164–65
24. Buddenhagen, I. W., Kelman, A. 1964. Biological and physiological aspects of bacterial wilt caused by *Pseudomonas solanacearum. Ann. Rev. Phytopathol.* 2:203–30
25. Buddenhagen, I. W., Sequeira, T., Kelman, A. 1962. Designation of races in *Pseudomonas solanacearum. Phytopathology* 52:726 (Abstr.)
26. Buller, A. H. R. 1934. *Researches of Fungi.* 6:397–443. New York: Longmans. 7 vols.
27. Buritica, P., Orjuela, J. 1968. Estudios fisiologicos de *Puccinia pittieriana* Henn. Causante de la roya de la papa (*Solanum tuberosum* L.). *Fitotec. Latinoamer.* 5:81–88

28. Butler, E. J., Bisby, G. R. 1960. *The Fungi of India*. New Delhi: Indian Counc. Agr. Res. 552 pp.
29. Cann, H. J. 1966. The insect pests and diseases of bananas. *Agr. Gaz. N.S.W.* 77:669–74
30. Carter, W. 1964. Present status of research in lethal yellowing disease of coconut palm in Jamaica. *FAO Plant Prot. Bull.* 12:67–69
31. Carter, W. 1966. Lethal yellowing disease of coconuts. *World Crops* 18:64–69
32. Carter, W., Suah, J. R. R. 1964. Studies on the spread of lethal yellowing disease of the coconut palm. *FAO Plant Prot. Bull.* 12:73–78
33. Castano, J. J. 1952. Roya de la papa. *Agr. Trop.* 9:47–48
34. Castano, J. J. 1961. Nuevos aspectos en la investigacion de la gomosis de los pastos micay e Imperial. *Agr. Trop.* 17:531–41
35. Castano, J. J., Thurston, H. D.. Crowder, L. V. 1964. Transmission de la gomosis en los pastos micay e Imperial. *Agr. Trop.* 20: 379–87
36. Chamberlain, D. W., Lipscomb, B. R. 1967. Bibliography of soybean diseases. *U.S. Dep. Agr. CR* 50–67
37. Chardon, C. E., Toro, R. A. 1930. Mycological explorations of Colombia. *J. Dep. Agr. P.R.* 14: 195–369
38. Chardon, C. E., Toro, R. A. 1934. Mycological explorations of Venezuela. *Univ. P.R. Monogr.* 2
39. Cheng, Y. W., Chan, K. L. 1968. The breeding of rust resistant soybean Tainung 3. *J. Taiwan Agr. Res.* 17:30–34 (English Summary)
40. Chona, B. L. 1965. Lethal yellowing of coconuts (Cape St. Paul Wilt). *Commonw. Phytopathol. News* 1965 (4):4
41. Christiansen G., J. 1967. *El Cultivo de la Papa en el Peru*. Lima: Editorial Juridica. 351 pp.
42. Cobb, N. A. 1919. A newly discovered nematode, *Aphelenchus cocophilus* n. sp., connected with a serious disease of coconut palm. *West Indian Bull.* 17:203–10
43. Cobb, N. A. 1922. Notes on the coconut nematode of Panama. *J. Parasitol.* 9:44–45
44. Coconut Industry Board, Jamaica. 1969. Progress in lethal yellowing research. *Farmer* 44:45–49
45. Commonw. Mycol. Inst. 1971. *Distribution Maps of Plant Diseases*. Kew: Commonw. Mycol. Inst.
46. Cooke, M. C. 1881. The coffee-disease in South America. *J. Linn. Soc. London Bot.* 18:461–67
47. Corbett, M. K. 1939. Diseases of the coconut palm. III Red ring. *Principes* 3:83–86
48. Costa, A. S., Kitajima, E. W. 1972. Cassava common mosaic virus. *Commonw. Mycol. Inst. Descr. Plant Viruses* 90
49. Cummins, G. B., Stevenson, J. A. 1956. A check list of North American rust fungi (Uredinales). *Plant Dis. Reptr. Suppl.* 240:1–193
50. Dadant, R. 1968. Mais: III-Maladies. 249–50. In *IRAT Reunion Rapp. Ann.*
51. De Leon, C. 1970. Advances in the selection of downy mildew resistant materials in Mexico. *Indian Phytopathol.* 23:339–41
52. Delgado, J. C. 1963. Effecto de diversas dosis de oxido cuproso y zineb aplicados a bajo volumen en el control de *Monilia* en el cacao. *Turrialba* 13:129–31
53. Delgado, J. C., Ampuero E., Doak, K. D. 1960. Posible evidencia de resistencia a *Monilia roreri* Cif. and Par. en algunos clones de la Estacion Experimental Tropical de Pichilingue. 184–91. In *Inter-Am. Cacao Confr. Proc.* 8th
54. Del Rosario, M. S., Quiaoit, A. R. 1962. Studies on the virus aspect of cadang-cadang disease of coconut. Transmission and serology. *Philipp. Agr.* 45:477–89
55. Denney, E. W. 1972. Typhoons and tungro cause Philippines to import more rice. *Foreign Agr.* 10(5):8–12
56. Desrosiers, R., Buchwald, A. von, Blanco, C. 1955. Effect of rainfall on the incidence of Monilia pod rot of cacao in Ecuador. *FAO Plant Prot. Bull.* 3:161–64
57. Desrosiers, R., Diaz, J. 1956. Efecto de diversas fungicides en el control de la Monilia. *Turrialba* 6:19–22
58. Desrosiers, R., Suarez, C. 1972. Monilia pod rot of cacao. In press
59. Diaz, J. J., Echeverria, J. 1963. Frecuencia de la roya de la papa, *Puccinia pittieriana* y su control quimico. *Cienc. Natur.* 6:12–18

60. Dirven, J. G. P., Van Hoof, H. A. 1960. A destructive virus disease of Pangola grass. *Tijdschr. Plantenzickten* 66:344–49
61. Doggett, H. 1970. Downy mildew in East Africa. *Indian Phytopathol.* 23:350–55
62. Doughty, T. R. 1958. Cassava breeding for resistance to mosaic and brown streak viruses. 48–54. In *East Afr. Agr. Forest. Res. Organ. Ann. Rept.*
63. Esenam, E. U. 1966. Some observations on the maize streak disease in Western Nigeria. *Nigerian Agr. J.* 3:38–41
64. Exconde, O. R. 1970. Philippine corn downy mildew. *Indian Phytopathol.* 23:275–84
65. Exconde, O. R. Personal communication
66. Exconde, O. R., Edralin, E. A., Advincula, B. A. 1967. *Some Factors Affecting Formation and Germination of Spores of Sclerospora philippinensis.* Presented at Inter-Asian Corn Improvement Workshop, 4th, Lyallpur, West Pak.
67. FAO. 1961. Destructive disease of Pangola grass. *Inform. Lett. FAO. S.E. Asia Pacific* 28 [*Rev. Appl. Mycol.* 45:362 (Abstr.)]
68. Fenwick, D. W. 1958. Preliminary investigations into the red ring disease of coconut. *J. Agr. Soc. Trinidad Tobago.* 56:253–75
69. Fenwick, D. W. 1969. Red ring disease of the coconut palm. In *Nematodes of Tropical Crops.* ed. J. E. Peachy. Farnham Royal, England: Commonw. Agr. Bur. 355 pp.
70. Fernow, K. H., Garces, O. C. 1949. Producion de semilla certificada de papa. *Rev. Fac. Nac. Agron. Univ. Antioquia* 10:257–95
71. Francis, C. 1967. *Downy mildew of maize in the Philippines.* M. S. thesis. Cornell University, Ithaca, N.Y. 166 pp.
72. Franco, T. H. 1958. Transmission de la moniliasis del cacao por el *Meoistorhinus tripterus* F. 130–36. In *Inter-Am. Cacao Confr. Proc.* 7th
73. Frederiksen, R. A. Personal communication
74. Frederiksen, R. A., Bockholt, A. J., Rosenow, D. T., Reyes, L. 1970. Problems and progress of sorghum downy mildew in the United States. *Indian Phytopathol.* 23:321–38
75. French, E. R., Sequeira, L. 1970. Strains of *Pseudomonas solanacearum* from Central and South America: a comparative study. *Phytopathology* 60:506–12
76. Fritz, J. 1968. Mais: I-Amelioration du Mais Local par Hybridation. 225–43. In *IRAT Reunion Rapp. Ann.*
77. Fujii, K., Okada, M. 1967. Progress in breeding of rice varieties for resistance to bacterial leaf blight in Japan. In *Proc. Symp. Rice Diseases and Their Control by Growing Resistant Varieties and Other Measures.* 51–61. Tokyo: Japan Agr. Forest. Fish. Res. Counc. 250 pp.
78. Galvez, G. E. 1967. Frequencia de *Sogata orizicola* Muir y *S. cubana* Crowf. en campos de arroz y Echinochloa en Colombia. *Agr. Trop.* 23:384–89
79. Galvez, G. E. 1967. Hoja blanca disease of rice. In *The Virus Diseases of the Rice Plant.* 35–49. Baltimore: Johns Hopkins Univ. Press. 354 pp.
80. Galvez, G. E. Personal communication
81. Galvez, G. E., Jennings, P. R., Thurston, H. D. 1960. Transmission studies of hoja blanca of rice in Colombia. *Plant Dis. Reptr.* 44:80–81
82. Galvez, G. E., Thurston, H. D., Bravo, G. 1963. Leafhopper transmission of enanismo of small grains. *Phytopathology* 53:106–8
83. Garces-Orjuela, C. 1947. Informe preliminar sobre la gomosis de los pastos micay e Imperial o gramalote en Colombia. *Rev. Fac. Nac. Agron. Univ. Antioquia* 7:1–28
84. Garces Orjuela, C., Jennings, P. R., Skiles, R. L. 1958. Hoja blanca of rice and the history of the disease in Colombia. *Plant Dis. Reptr.* 42:750–51
85. Gibler, J. W. 1957. "Enanismo," a virus disease of cereals in southern Colombia. *Phytopathology* 47:13 (Abstr.)
86. Goncalves, J. R. C. 1968. The resistance of Fx and IAN rubber clones to leaf diseases in Brazil. *Trop. Agr. (Trinidad)* 45:331–36
87. Gorter, G. J. M. A. 1953. Studies on the spread and control of the

streak disease of maize. *S. Afr. Dep. Agr. Forest. Sci. Bull.* 341

88. Gorter, G. J. M. A. 1959. Breeding maize for resistance to streak. *Euphytica* 8:234–40

89. Grylls, N. E., Hunt, P. 1971. A review of the study of the aetiology of coconut lethal yellowing disease. *Oleagineux.* 26:311–15

90. Grylls, N. E., Hunt, P. 1971. Studies on the aetiology of coconut lethal yellowing in Jamaica by mechanical and bacterial inoculations and by insect vectors. *Oleagineux* 26:543–49

91. Hagley, E. A. C. 1962. The palm weevil, *Rhynchophorus palmarum* L., a probable vector of red ring disease of coconuts. *Nature* 193: 449

92. Hagley, E. A. C. 1963. The role of the palm weevil, *Rhynchophorus palmarum* L., as vector of red ring disease of coconut. I. Results of preliminary investigations. *J. Econ. Entomol.* 56:375–80

93. Hagley, E. A. C. 1965. Tests of attractants for the palm weevil. *J. Econ. Entomol.* 58:1002–3

94. Hardy, F. 1961. *Manual de Cacao.* Turrialba, Costa Rica: Inst. Interam. Cienc. Agr. 439 pp.

95. Hawkes, J. G. 1951. Organization y planamiento para el mejoramiento de la papa II. *Agr. Trop.* 7: 11–16

96. Hayward, A. C. 1964. Characteristics of *Pseudomonas solanacearum. J. Appl. Bacteriol.* 27:265–77

97. Hennings, P. 1903. Einige neue Japanische Uredineen IV. *Hedwigia* 42:107–8

98. Hennings, P. 1904. Einige neue pilze aus Costa Rica und Paraguay. *Hedwigia* 42:147–49

99. Hennings, P. 1904. Uber die auf Hevea-arten bisher beobachteten parasitischen pilze. *Notizbl. Bot. Gart. Mus. Berlin* 4:133–38

100. Herd, G. W. 1956. Maize diseases during the 1954–55 season. *Rhodesia Agr. J.* 53:525–37

101. Herd, G. W. 1963. Irrigated winter wheat, Part IV. Diseases of wheat in Southern Rhodesia. *Rhodesia Agr. J. Tech. Bull.* 1

102. Herold, F., Trujillo, G., Munz, K. 1968. Viruslike particles related to hoja blanca disease of rice. *Phytopathology* 58:546–47

103. Hilton, R. N. 1955. South American leaf blight. A review of the literature relating to its depredations in South America, its threat to the Far East and the methods available for its control. *J. Rubber Res. Inst. Malaya* 14:287–337

104. Hiratsuka, N. 1935. Phakopsora of Japan I. *Bot. Mag.* 49:781–88

105. Holliday, P. 1970. South American leaf blight (*Microcyclus ulei*) of *Hevea brasiliensis. Commonw. Mycol. Inst. Phytopathol. Pap.* 12

106. Holliday, P. 1971. Some tropical plant pathogenic fungi of limited distribution. *Rev. Plant Pathol.* 50:337–48

107. Hung, C. H., Liu, K. C. 1961. Soybean spraying experiment for rust disease control. *J. Taiwan Agr. Res.* 10(1):35–40 (English Summary)

108. Hutchinson, F. W. 1955. Defoliation of *Hevea brasiliensis* by aerial spraying. *J. Rubber Res. Inst. Malaya* 15:241–74

109. Imle, E. P. 1966. Plant material distribution and quarantine measures for cocoa. *FAO Plant Prot. Bull.* 14:134–40

110. Jameson, J. D. 1964. Cassava mosaic disease in Uganda. *East Afr. Agr. Forest. J.* 29:208–13

111. Jennings, D. L. 1960. Observations on virus diseases of cassava in resistant and succeptible varieties. I. Mosaic diseases. *Emp. J. Exp. Agr.* 28:23–34

112. Jones, B. L., Leeper, J. C., Frederiksen, R. A. 1972. *Sclerospora sorghi* in corn: its location in carpellate flowers and mature seeds. *Phytopathology* 62:817–19

113. Jones, W. O. 1959. *Manioc in Africa.* Palo Alto, Calif.: Stanford Univ. Press. 315 pp.

114. Kelman, A. 1953. The bacterial wilt caused by *Pseudomonas solanacearum. N.C. Agr. Exp. Sta. Tech. Bull.* 99

115. Kenten, R. H., Legg, J. T. 1965. Observations on the purification and properties of cocoa swollen shoot virus. *Ghana J. Sci.* 5:221–25

116. Kenten, R. H., Legg, J. T. 1971. Varietal resistance of cocoa to swollen shoot disease in West Africa. *FAO Plant Prot. Bull.* 19:1–11

117. Kitajima, E. W., Costa, A. S. 1964. Elongated particles found associated with cassava brown

streak. *East Afr. Agr. Forest. J.* 30:28–30

118. Kitani, K., Ingue, Y. 1960. Studies on the soybean rust and its control measure. (Part I) Studies on the soybean rust. *Shikoku Agr. Exp. Sta. Bull.* 5:319–42 (English Summary)

119. Klinkowski, M. 1970. Catastrophic plant diseases. *Ann. Rev. Phytopathol.* 8:37–60

120. Langdon, K. R. 1965. Relative resistance or susceptibility of several clones of *Hevea brasiliensis* and *H. brasiliensis X H. Benthamiana* to two races of *Dothidella ulei. Plant Dis. Reptr.* 49:12–14

121. Langford, M. H. 1944. Science's fight for healthy *Hevea. Agr. Am.* 4:151–58

122. Langford, M. H. 1945. South American leaf blight of *Hevea* rubber trees. *U. S. Dep. Agr. Tech. Bull.* 882

123. Langford, M. H., Townsend, C. H. T., Jr. 1954. Control of South American leaf blight of *Hevea brasiliensis. Plant Dis. Reptr. Suppl.* 225:42–48

124. Leather, R. I. 1959. Further investigations into the "Cape St. Paul Wilt" of coconuts at Keta, Ghana. *Emp. J. Exp. Agr.* 27:67–78

125. Leppik, E. E. 1970. Gene centers of plants as sources of disease resistance. *Ann. Rev. Phytopathol.* 8:323–44

126. Ling, K. C. 1972. *Rice Virus Diseases.* Los Banos, Phillipp.: Int. Rice Res. Inst. 134 pp.

127. Ling, L. 1951. Bibliography of soybean diseases. *Plant Dis. Reptr. Suppl.* 204:111–73

128. Lozano T., J. C., Thurston, H. D., Galvez, G. E. 1969. Control del "Moko" del platano y banano causado por la bacterium *Pseudomonas solanacearum. Agr. Trop.* 25:315–24

129. Maas, P. W. T. 1971. A coconut abnormality of unknown etiology in Surinam. *FAO Plant Prot. Bull.* 19:80–85

130. Magee, C. J. 1927. Investigation on the bunchy top disease of the banana. *Aust. Counc. Sci. Ind. Res. Bull.* 30:1–64

131. Magee, C. J. 1936. Bunchy top disease of bananas. Rehabilitation of the banana industry in New South Wales. *J. Aust. Inst. Agr. Sci.* 2:13–16

132. Magee, C. J. 1948. Transmission of bunchy top to banana varieties. *J. Aust. Inst. Agr. Sci.* 14:18–24

133. Magee, C. J. 1953. Some aspects of the bunchy top disease of banana and other *Musa* spp. *J. Proc. Roy. Soc. N.S.W.* 87:3–18

134. Magee, C. J. 1967. Report to the government of Western Samoa. The control of banana bunchy top. *S. Pac. Comm. Tech. Pap.* 150

135. Malaguti, G., Diaz, H., Angles, N. 1957. La virosis "hoja blanca" del arroz. *Agron. Trop. (Maracay, Venez.)* 6:157–63

136. Martinez, A. P., Roberts, D. A. 1968. Lethal yellowing of coconuts in Florida. *Proc. Fla. State Hort. Soc.* 80:432–36

137. Martyn, E. B. 1934. A note on plantain and banana diseases in British Guiana with special reference to wilt. *Agr. J. Brit. Guiana* 5:120–23

138. McClean, A. P. D. 1947. Some forms of streak virus occurring in maize, sugar cane and wild grasses. *S. Afr. Dep. Agr. Forest. Sci. Bull.* 265

139. Menon, M. R., Raychaudhuri, S. P. 1970. Cucumber: A herbaceous host of cassava mosaic virus. *Plant Dis. Reptr.* 54:343–35

140. Mercado, A. C., Lantican, R. M. 1961. The susceptibility of cytoplasmic male-sterile lines of corn to *Helminthosporium maydis* Nisikado and Miy. *Philipp. Agr.* 45:235–43

141. Meredith, D. S. 1970. Major banana diseases: past and present status. *Rev. Plant Pathol.* 59:539–54

142. Meredith, D. S. 1972. Epidemiological considerations of plant diseases in the tropical environment. *Phytopathology In press*

143. Meredith, D. S. Personal communication

144. Miller, J. W. 1966. Differential clones of *Hevea* for identifying races of *Dothidella ulei. Plant Dis. Reptr.* 50:187–90

145. Miller, M. E., Roberts, D. A. 1970. Mechanical transmission of the coconut palm lethal-yellowing pathogen from frozen or fresh inocula prepared in two buffers. *Phytopathology* 60:1304 (Abstr.)

146. Mizukami, T., Wakimoto, S. 1969. Epidemiology and control of bac-

terial blight of rice. *Ann. Rev. Phytopathol.* 7:51–72

147. Murata, N. 1967. Genetic aspects on resistance to bacterial leaf blight in rice and variation of its causal organism. See Ref. 77, 39–49

148. Napi-Acedo, G., Exconde, O. R. 1967. Oospores of *Sclerospora philippinensis* Weston from corn leaf tissue. *Philipp. Agr.* 51:279–82

149. Nowell, W. 1919. The red ring disease of coconut palms. *West Indian Bull.* 17:189–92

150. Nowell, W. 1920. The red ring disease of coconuts—infection experiments. *West Indian Bull.* 18:74–76

151. Nutman, F. J. 1959. Evidence that spores of coffee leaf rust are not dispersed by wind. *Kenya Coffee* 24:451–53

152. Nutman, F. J., Roberts, F. M., Bock, K. R. 1960. Methods of uredospore dispersal of the coffee leaf rust fungus, *Hemileia vastatrix. Trans. Brit. Mycol. Soc.* 43:509–15

153. Ocfemia, G. O. 1937. The probable nature of cadang-cadang disease of coconut. *Philipp. Agr.* 26:338–40

154. O'Conner, B. A. 1969. *Exotic plant pests and diseases.* Noumea, New Caldonia: S. Pac. Comm. ca. 400 pp.

155. Ohio Agr. Res. Develop. Center. 1971. *Maize Virus Diseases and Corn Stunt.* Wooster, Ohio: Ohio Agr. Res. Develop. Center. 34 pp.

156. Orellana, R. G. 1954. Cacao diseases in Venezuela, Colombia, Ecuador, and Trinidad. *FAO Plant Prot. Bull.* 2:49–52

157. Orjuela, N. J. 1965. Indice de enfermedades de plantas cultivadas en Colombia. *Inst. Colomb. Agropec. Bol. Tec.* 11

158. Ou, S. H. 1968. *Rice diseases in the tropics.* Presented at Int. Congr. Plant Pathol. 1st, London

159. Ou, S. H. 1972. *Rice Diseases.* Kew: Commonw. Mycol. Inst. 368 pp.

160. Ou, S. H., Jennings, P. R. 1969. Progress in the development of disease resistant rice. *Ann. Rev. Phytopathol.* 7:383–410

161. Ou, S. H., Nuque, F. L., Silva, J. P. 1971. Varietal resistance to bacterial blight of rice. *Plant Dis.*

Reptr. 55:17–21

162. Ou, S. H., Nuque, F. L., Silva, J. P. 1971. Pathogenic variation among isolates of *Xanthomonas oryzae* of the Philippines. *Plant Dis. Reptr.* 55:22–26

163. Padwick, G. W. 1956. Losses caused by plant diseases in the colonies. *Commonw. Mycol. Inst. Phytopathol. Pap.* 1

164. Payak, M. M., Renfro, B. L. 1967. A new downy mildew disease of maize. *Phytopathology* 57:394–97

165. Phillips, M. P. 1961. Production in the Eastern Region of Ghana. *Rep. Cocoa Conf. London.* p. 178

166. Plavsic-Banjac, B., Hunt, P., Maramorosch, K. 1972. Mycoplasmalike bodies associated with lethal yellowing disease of coconut palms. *Phytopathology* 62:298–99

167. Plavsic-Banjac, B., Maramorosch, K. 1972. Electron microscopy of African maize streak. *Phytopathology* 62:671 (Abstr.)

168. Plavsic-Banjac, B., Maramorosch, K. 1973. Cassava mosaic virus. *Phytopathology* (Abstr.) In press

169. Plavsic-Banjac, B., Maramorosch, K., von Uexkull, H. P. 1972. Preliminary observation of cadang-cadang diseased coconut palm leaves by electron microscopy. *Plant Dis. Reptr.* 56:643–45

170. Posnette, A. F. 1940. Transmission of swollen shoot. *Trop. Agr. (Trinidad)* 17:98

171. Posnette, A. F. 1950. Virus diseases of cocoa in West Africa. VII. Virus transmission by different vector species. *Ann. Appl. Biol.* 37:378–84

172. Price, W. C., Martinez, A. P., Roberts, D. A. 1967. Mechanical transmission of coconut lethal yellowing. *FAO Plant Prot. Bull.* 15:106–8

173. Price, W. C., Martinez, A. P., Roberts, D. A. 1968. Reproduction of the coconut lethal yellowing syndrome by mechanical inoculation of seedlings. *Phytopathology* 58:593–96

174. Price, W. C., Nitzany, F. E. 1969. Research on cadang-cadang sponsored by FAO. *Oleagineux* 24:615–20

175. Protacio, D. B. 1965. Aphid transmission of a cadang-cadang-like disease to coconut seedlings. *Advan. Front. Plant Sci.* 10:141–56

176. Pursglove, J. W. 1968. *Tropical*

Crops. Dicotyledons 1. New York: Wiley 332 pp.

177. Rands, R. D. 1945. *Hevea* rubber culture in Latin America, problems and procedures, 183–99. In *Plants and Plant Science in Latin America*. ed. F. Verdoorn. Waltham, Mass: Chronica Botanica. 381 pp.

178. Rayner, R. W. 1960. Rust disease of coffee. 2. Spread of the Disease. *World Crops* 12:222–24

179. Renfro, B. Personal communication

180. Revilla M., V. A. 1966. Una virosis destructiva del pasto pangola en el Peru. *Estac. Exp. Agr. La Molina Bol.* 12

181. Riker, A. J. 1964. Internationally dangerous tree diseases and Latin America. *J. Forest.* 62:229–32

182. Roberts, D. A. Personal communication

183. Rockefeller Foundation. 1959. Barley improvement. 19–25. In *Colombian Agricultural Program. Director's Ann. Rep.* 1958–1959. New York: Rockefeller Found. 128 pp.

184. Rockefeller Foundation. 1963. Barley—Colombia. 85–88. In *The Rockefeller Foundation Program in the Agricultural Sciences. Ann. Rep.* 1962–1963. New York: Rockefeller Found. 310 pp.

185. Rorer, J. B. 1911. A bacterial disease of bananas and plantains. *Phytopathology* 1:45–49

186. Salas, J. A., Hancock, J. G. 1972. Production of the perfect stage of *Mycena citricolor* (Berk. and Curt.) Sacc. *Hilgardia* 41:213–34

187. Sanchez, P. A., Victorio, K. J. 1970. Control del anillo rojo. *Inst. Colomb. Agropec. Plegable de Divulgacion* 44

188. Schieber, E. 1972. Economic impact of coffee rust in Latin America. *Ann. Rev. Phytopathol.* 10:491–510

189. Semangoen, H. 1970. Studies of downy mildew of maize in Indonesia, with special reference to the perennation of the fungus. *Indian Phytopathol.* 23:307–20

190. Sequeira, L. 1958. Bacterial wilt of bananas: Dissemination of pathogen and control of the disease. *Phytopathology* 48:64–69

191. Sequeira, L. 1959. The host range of *Mycena citricolor* (Berk. & Cart.) Sacc. *Turrialba* 8:136–47

192. Sequeira, L., Averre, C. W. 1961. Distribution and pathogenicity of strains of *Pseudomonas solanacearum* from virgin soils in Costa Rica. *Plant. Dis. Reptr.* 45:435–40

193. Sequeira, L., Buddenhagen, I. W. 1958. Disinfectants and tool disinfection for prevention of spread of bacterial wilt of bananas. *Plant Dis. Reptr.* 12:1399–1404

194. Sequeira, L., Steeves, T. A. 1954. Auxin inactivation and its relation to leaf drop caused by the fungus *Omphalia flavida*. *Plant Physiol.* 29:11–16

195. Shikata, E., Galvez, G. E. 1969. Fine flexous threadlike particles in cells of plants and insect hosts infected with hoja blanca virus. *Virology* 39:635–41

196. Sill, W. H., Bigornia, A. E., Pacumbaba, R. P. 1963. Incidence of cadang-cadang in coconut trees of different ages. *FAO Plant Prot. Bull.* 11:49–58

197. Singh, J. P., Renfro, B. L., Payak, M. M. 1970. Studies on the epidemiology and control of brown stripe downy mildew of maize (*Sclerophthora rayssiae* var. *zeae*). *Indian Phytopathol.* 23:194–208

198. Smith, E. F. 1896. A bacterial disease of the tomato, eggplant and Irish potato (*Bacillus solanacearum* nov. sp.). *U.S. Dep. Agr. Div. Veg. Physiol. Pathol. Bull.* 12:1–28

199. Smith, K. M. 1957. *A Textbook of Plant Virus Diseases*. Boston: Little, Brown. 625 pp.

200. Srivastava, D. N. 1967. Epidemiology and control of bacterial blight of rice in India. See Ref. 77, 11–18

201. Stakman, E. C., Harrar, J. G. 1957. *Principles of Plant Pathology*. New York: Ronald Press. 581 pp.

202. Stapp, C. 1965. Die bakterielle schleimfaule und ihr erreger *Pseudomonas solanacearum*. *Zentralbl. Bakt. Parasit. Infektionskr. Hygiene* 119:166–90

203. Starr, M. P., Garces-Orjuela, C. 1950. El agente causal de la gomosis bacterial del pasto Imperial en Colombia. *Rev. Fac. Nac. Agron. Univ. Antioquia* 12:73–83

204. Storey, H. H. 1925. The transmission of streak disease of maize by the leafhopper, *Balclutha mbila*

Naude. *Ann. Appl. Biol.* 12:422–39

205. Storey, H. H. 1936. Streak disease of maize. *East Afr. Agr. J.* 1:471–75

206. Storey, H. H., Howland, A. K. 1967. Transfer of resistance to the streak virus into East African maize. *East Afr. Agr. Forest. J.* 33:131–35

207. Storey, H. H., McClean, A. P. D. 1930. The transmission of streak disease between maize, sugar cane and wild grasses. *Ann. Appl. Biol.* 18:691–719

208. Storey, H. H., Nichols, R. F. W. 1938. Studies of the mosaic diseases of cassava. *Ann. Appl. Biol.* 25:790–806

209. Stover, R. H., Richardson, D. L. 1968. "Pelipita," an ABB Bluggoe-type plantain resistant to bacterial and fusarial wilts. *Plant Dis. Reptr.* 52:901–3

210. Sun, M. H. 1970. Sugar cane downy mildew of maize. *Indian Phytopathol.* 23:262–69

211. Thirumalachar, M. J., Shaw, C. G., Narasimhan, M. J. 1953. The sporangial phase of the downy mildew *Elensine coracana* with a discussion of the identity of *Sclerospora macrospora* Sacc. *Bull. Torrey Bot. Club* 80:299–307

212. Thresh, J. M. 1960. Quarantine arrangements for intercepting cocoa material infected with West African viruses. *FAO Plant Prot. Bull.* 8:89–92

213. Thresh, J. M., Tinsley, T. W. 1959. The viruses of cocoa. *West Afr. Cocoa Res. Inst. Tech. Bull.* 7

214. U. S. Dep. Agr. 1960. Hoja blanca, serious threat to rice crops. *Agr. Res. Serv.* ARS 22–57

215. U. S. Dep. Agr. 1963. Internationally dangerous forest tree diseases. *U.S. Dep. Agr. Misc. Publ.* 939

216. U. S. Dep. Agr. 1968. Corn (maize) viruses in the continental United States and Canada. *Agr. Res. Serv.* ARS 33–118

217. Vakili, N. G. 1969. Bunchy top disease of bananas in the central highlands of South Vietnam. *Plant Dis. Reptr.* 53:634–38

218. Van der Plank, J. K. 1960. Analysis of epidemics. In *Plant Pathology: An Advanced Treatise.* eds. J. Horsfall, A. Dimond, 3:230–89. New York: Academic 3 vols.

219. Wang, C. S. 1961. Chemical control of soybean rust. *J. Agr. Assoc. China* 35:51–54 (English Summary)

220. Waterhouse, G. M. 1964. The genus *Sclerospora.* Diagnoses (or descriptions) from the original papers and a key. *Commonw. Mycol. Inst. Misc. Publ.* 17

221. Watson, A. J. 1971. Foreign bacterial and fungus diseases of food, forage and fiber crops: An anotated list. *U.S. Dep. Agr. Agr. Handb.* 418

222. Wellman, F. L. 1953. Some important diseases of coffee. 891–96. In *Yearbook of Agriculture* Washington: U.S. Dep. Agr. 940 pp.

223. Wellman, F. L. 1954. Some important diseases of cacao. *FAO Plant Prot. Bull.* 2:129–33

224. Wellman, F. L. 1961. *Coffee.* New York: Interscience. 488 pp.

225. Wellman, F. L. 1972. *Tropical American Plant Disease.* Metuchen, N. J.: Scarecrow. 989 pp.

226. Weston, W. H. 1920. Philippine downy mildew of maize. *J. Agr. Res.* 19:97–122

227. Weston, W. H. 1923. Production and dispersal of conidia in the Philippines. *Sclerospora* of maize. *J. Agr. Res.* 23:239–77

228. Whitehead, R. A. 1968. Selecting and breeding coconut palms resistant to lethal yellowing disease. *Euphytica* 17:81–101

229. Yang, I. L. 1970. Studies on the varietal resistance to bunchy top disease of banana. *J. Taiwan Agr. Res.* 19:57–63 (English Summary)

230. Yoshimura, S., Tagami, Y. 1967. Forecasting and control of bacterial leaf blight of rice in Japan. See Ref. 77, 25–38

231. Zehr, E. I. 1970. Isolation of *Pseudomonas solanacearum* from abaca and banana in the Philippines. *Plant Dis. Reptr.* 54:516–20

THE BDELLOVIBRIOS: BACTERIAL PARASITES OF BACTERIA

❖ 3565

Heinz Stolp

Institut für Allgemeine Botanik, Abteilung für Mikrobiologie,
Universität Hamburg, D 2000 Hamburg 36, Germany

The first representative of a new kind of microorganisms, now known as the bdellovibrios, was isolated in Berlin in 1962 (72). These organisms are tiny vibrioid bacteria, which have the unique property of being parasites of other bacteria. The bdellovibrios and their symbiotic relationships to the host bacteria have since been the subjects of intensive investigation. The burgeoning literature has recently been reviewed in two admirable treatments (57, 58), the first focussing on structural features, parasite-host interactions, taxonomy, and ecology, the second emphasizing the physiology and biochemistry. A series of earlier review articles (31, 36, 49, 50, 54, 59, 64, 67–71, 74), partly comprehensive treatments, partly brief summaries, has contributed to the general information on bdellovibrios and has possibly incited microbiologists to become interested in the phenomenon of bacterial parasitism of bacteria.

THE DISCOVERY OF BDELLOVIBRIO

The present essay aims to evaluate the scientific progress made during the past decade concerning knowledge about *Bdellovibrio* (particularly parasite-host relations) and to point to the many problems so far left unsolved. To this purpose, it seems opportune to recall the first report on this organism (72). I came across the bacterial parasite quite accidentally in an experiment designed to isolate bacteriophages active against phytopathogenic pseudomonads. I had exhausted my supply of membrane filters of small pore size diameter (0.2 μm or less, allowing passage of bacteriophage but retaining bacteria) commonly used for preparing "sterile" (but phage-containing) solutions, so I filtered a soil suspension—that previously had been inoculated with a *Pseudomonas* sp. for phage enrichment—using a sintered glass filter of relative large pore size diameter (G5 on G3, type M, Schott & Gen., Mainz; maximum pore size diameter 1.35 μm). The soil solution of the enrichment culture happened to be rather clear, and filtration was done without a prior centrifugation. The filtrate was then tested for the presence of bacteriophage, using bacterial lawns in Petri dishes and the common double-layer technique. The host bacterium in this particular experiment was the phytopathogen, *Pseudomonas phaseolicola* (ATCC 11355). Under the experimental conditions used, development of phage plaques was usually apparent in the lawns within 24 hours. These particular test plates did not show development of such phage plaques after overnight incubation. As a rule such negative test plates were discarded, because plaque development cannot be expected to begin in a lawn of bacteria that have already passed their logarithmic growth phase. In this experiment, however, the Petri dishes were not discarded and two days later were checked once more for the presence of phage plaques. Surprisingly, a few plaques had appeared in the aged lawns. Because the retarded plaque development was not in accordance with general experience in working with phage, the lytic action attracted my interest. When the material from a single plaque was diluted in PY solution (1% peptone, 0.3% yeast extract; pH 6.8) and checked again for plaque-forming ability, the result was conclusive: plaques developed again, and again plaque formation did not start before the second or third day. As with phage, concentrated preparations of the dilution series caused confluent lysis of the bacterial lawn; the more diluted preparations elicited the appearance of individual plaques, which suggested that lysis was connected with the action of an agent having a particulate nature. In contrast to phage plaques, the lytic zones under study increased in size for about a week. Both the delayed appearance of plaques and their continuous increase in size were regarded as being incompatible with bacteriophage action. The cause of the lytic reaction finally found an explanation when I examined, under the phase-contrast microscope, some material taken from a developing plaque. In addition to a few cells of the *Pseudomonas*, a great number of rapidly moving, tiny microbes were to be seen. The tiny organisms collided vigorously with (physically "attacked") the much larger cell of the pseudomonad, attached to its surface, and somehow seemed to induce lysis of the attacked bacterial cell.

The first report on the hitherto unknown organism (72) was mainly concerned with the description of the tiny microbe, its physical attack upon and attachment to the other larger bacterial cell, the lytic reaction caused in lawns of susceptible bacteria, and the spectrum of its activity. As the organism could not be propagated on the usual complex nutrient laboratory media, but rather required living bacteria as the substrate, it was characterized, with reservations, as an "obligate" parasite and the larger bacterium that was lysed as its host. It was demonstrated that saprophytically growing derivative clones, which produce colonies on nutrient agar, could be selected from a population of the parasitic wild-type. Because the accidental isolation of the first strain from soil could be successfully repeated, it was speculated that the organism represents a common component of the soil microflora.

A comprehensive study on the tiny parasite and its interactions with the bacterial host was made in 1963 (73), using 11 additional strains isolated from California soils. The results clearly showed that the parasite of bacteria is itself a bacterium; it was given the name *Bdellovibrio bacteriovorus*.[1] The individual strains of the parasite were shown to be restricted in their lytic activity to Gram-negative bacteria; some were able to attack a broad spectrum of host bacteria, others had a limited host-range. As with the original strain, saprophytic mutants could be selected from a population of the parasite. The saprophytic bdellovibrios had rather unusual morphological and physiological properties, and did not show a close relationship to any of the already known kinds of bacteria.

After several years of descriptive work on the structural features of the bdellovibrios and on structural changes occurring in the host, research is increasingly directed towards biochemical studies on the metabolism of the bdellovibrios and on the biochemical events in the parasite-host interaction.

In the early studies (72, 73), the bdellovibrios were divided into two groups: "parasitic" strains and "saprophytic" strains. These terms have the following meanings: (*a*) A *Bdellovibrio* cell that possesses the capability of attacking a living bacterium, attaching to its surface, penetrating the cell wall, multiplying inside the host, and causing lysis of the infested cell is called "parasitic." (*b*) A *Bdellovibrio* cell that is capable of multiplying on a laboratory nutrient medium is called "saprophytic." The division into parasitic and saprophytic strains is based on the general experience that the parasitic wild-type isolates do not grow on laboratory nutrient media, and that the saprophytic derivatives do not possess parasitic abilities. This classification does not imply that one condition necessarily excludes the other. Starr & Seidler (58) characterized the bdellovibrios as "HD" (host-dependent), "HI" (host-independent), or "FP" (facultatively parasitic). "HD" refers to dependency on other viable or nonviable bacteria; "HI" refers to the ability to live without other viable or nonviable bacteria. Whereas "parasitic" refers to the behavior of a *Bdellovibrio*, the term "HD" is defined nutritionally. "Saprophytic" and "HI," on the other hand, both refer to the ability of growing nonparasitically on a laboratory nutrient medium. "FP" characterizes a *Bdellovibrio* cell as being capable of multiplying either parasitically or saprophytically; a population which is truly "FP" would be expected to form plaques on a host lawn and produce colonies on a nutrient agar medium in a 1:1 ratio.

In this review the terms "parasitic" and "saprophytic" will be used preferentially. When referring to particular *Bdellovibrio* strains characterized by the properties of being HD, HI, or FP these will be marked accordingly.

[1] The generic name *"Bdellovibrio"* expresses the vibrioid shape and refers to the attachment of the parasite to its host bacterium in the initial host interaction; *"Bdello-"* means "leech". The specific name *"bacteriovorus"* describes the feeding on bacteria.

ISOLATION, PROPAGATION, AND MAINTENANCE OF PARASITIC BDELLOVIBRIOS

The isolation from soil, sewage, water, or other materials of bdellovibrios parasitic to other bacteria is necessarily connected with the use of a prospective host bacterium and appearance of lytic zones (plaques) in a lawn of that bacterium. In principle, the appearance of plaques requires a solution derived from natural material that does not contain an excess of "contaminants" as compared to the number of bdellovibrios. The presence of many bacteria of different kinds causes overgrowth by organisms unsusceptible to bdellovibrios, and thus interferes with the development of individual plaques. For this reason, the isolation procedure must be directed towards separation of the bdellovibrios from other microorganisms. Methods used for this purpose include centrifugation and filtration (63, 73). As with many other bacteria, the development of a specific enrichment technique has greatly facilitated the isolation of bdellovibrios (68). This enrichment method is based on the selective growth conditions provided for the bdellovibrios in a concentrated suspension of prospective host bacteria (10^{10} cells/ml) in highly diluted yeast extract (0.03%) or mineral salt solution. Under such conditions, the introduced host bacteria represent the only source of nutrients, thus avoiding overgrowth by the many microorganisms present in the inoculum. Since growth of host bacteria in the nutrient-free environment is not possible, bacteriophages that require growing bacteria for their development have no chance to develop and interfere with the formation of plaques. This procedure is, of course, different from a technique that allows multiplication of the prospective host bacteria. For this reason, enrichment culture comparable to bacteriophage enrichment was originally not recommended (73).

The techniques successfully applied for isolation of bdellovibrios do not allow quantitative estimations in a natural environment, mainly for the following reasons: (a) When a suspension of natural material is cleared by centrifugation, step-wise filtrations, or both, only part of the bdellovibrios will be recovered. (b) Use of the particular prospective host bacterium does not allow the development of bdellovibrios that do not attack the given host but might attack others. (c) Enrichment of bdellovibrios does not permit an estimation of the original concentration. The isolation and enumeration procedures have been treated in some detail (58).

Isolation procedures and cultivation methods are intimately connected because the isolation of any bacterium requires conditions suitable for propagation of the organism. Although the accidental appearance of the first strain of *Bdellovibrio* was observed in a complex nutrient medium, it soon became evident that development of the bacterial parasite is retarded, or even inhibited, under conditions allowing intensive multiplication of host bacteria. Stolp & Starr (73) introduced a medium of relatively low nutrient content (YP:

0.3% yeast extract, 0.06% peptone; pH 7.2). "Pure cultures" (individual strains; clones derived from a single cell) of *Bdellovibrio* are usually developed from individual plaques by a series (at least 3) of successive single-plaque isolations. The method is comparable to the procedure used for obtaining a "pure line" of bacteriophage.

In experiments that require quantitative estimation of bdellovibrios, the plaque technique is presently the method of choice. Use of YP medium for cultivating the host bacteria in most instances is appropriate. Plaque development may be favored by using ten-times diluted YP medium. *Bdellovibrio* on lawns in water agar with mineral salts, but without organic nutrients exhibits a pronounced plaque growth if host bacteria are supplied in a concentration that confers turbidity to the surface agar layer at time of inoculation.

As in lawns of bacteria, growth of *Bdellovibrio* with host bacteria in liquid cultures is influenced by the presence of organic nutrients. Low concentration of nutrients favors propagation of the bdellovibrios. It is highly probable that this results from limitation of the accumulation of metabolic products by host bacteria that might exert an inhibitory effect on development of *Bdellovibrio* (57). Study of biochemical reactions in the bdellovibrio-host system often requires exclusion of exogenous organic materials. Host cells consequently have been used as the sole source of organic nutrients (26, 30). A simple mineral salt solution consisting of $10^{-3}M$ Tris buffer (pH 7.5), $10^{-3}M$ $Ca(NO_3)_2$, $3.6 \times 10^{-5}M$ $FeSO_4$, $3 \times 10^{-3}M$ $MgSO_4$, $6.6 \times 10^{-5}M$ $MnSO_4$ (30), or of similar composition, is appropriate.

The maintenance of cultures of *Bdellovibrio,* as with other bacteria, is possible by (*a*) regular transfer of the organism under suitable culture conditions, and keeping the culture after incubation at room temperature or at low temperature (2 to 4°C), or (*b*) preserving the organism in a physiologically inactive state, i.e. by storage under deep frozen conditions (−30°C) or lyophilization. Because of rather rapid loss in viability, transfer of the parasite-host system is recommended at intervals of 2 weeks or a month (72, 73). Several authors reported a drastic decrease in number of viable cells after complete lysis of the host bacteria (1, 6, 72). Safe maintenance for a long period of time of a *Bdellovibrio*-host system is possible in deep-frozen condition. The following procedure has proved very effective: A lysate from a liquid culture containing actively motile bdellovibrios in great numbers (10^9 cells/ml or more) is supplied with fresh host bacteria (24-hr old; final concentration in the lysate approx. 10^9 bacteria/ml) and incubated on the shaker for 30 min. At the end of the 30 min incubation period, many host cells are "infected" and contain intracellular (intramural) parasites. After addition of glycerol (final concentration 10%), the culture is stored at −30°C. Cultures kept under such conditions for more than two years were easily revived by melting a part of the frozen material and transferring a bit to a host-cell suspension. Compared to glycerol, DMSO (dimethylsulfoxide), as used in bacteriophage preservation, did not give consistently satisfactory results.

PARASITE-HOST INTERACTIONS

The interaction (symbiotic association; 58) between *Bdellovibrio* and its host bacterium may be characterized by the development cycle of the parasite, its effects on the host cell, the mechanism of lysis with respect to both biochemical activities of the parasite and chemical and structural alterations of the host cell, and by the activity spectrum (host range) of the various *Bdellovibrio* isolates. It is now possible to give a detailed description of the parasite-host interaction in terms of morphological and ultrastructural events, whereas the biochemical steps involved in the intimate association are less well understood. *Bdellovibrio bacteriovorus* had been regarded in early studies as a predatory and ectoparasitic bacterium capable of lysing other bacteria (51, 72, 73). Entrance of the externally-attached parasite into the host cell, and the intracellular life stages of *Bdellovibrio,* were disclosed independently and almost simultaneously by Scherff, De Vay & Carroll (41) and by Starr & Baigent (56).

The various steps in the parasite-host interactions have been studied by phase-contrast microscopy (including cinematography) and by electron microscopy. These studies have demonstrated the following sequence of events (6, 7, 29, 41, 56, 65, 66, 68):

PRIMARY CONTACT   When actively motile parasites and susceptible host bacteria are brought together under suitable conditions, individual *Bdellovibrio* cells physically collide with individual host bacterial cells within seconds. The result of the collision is attachment of the parasite to the surface of the host cell. Motility of the *Bdellovibrio* is an essential prerequisite for attachment; the high velocity brings about a violent collision which may be recognized under the microscope. Primary interaction is reversible during the early period of attachment; the parasite can detach and attack another host cell. Although temporary attachment to nonsusceptible cells has been reported (50, 58), there is much evidence that attachment is specific for susceptible cells (65, 66, 68). Whether "recognition" of the suitable (congenial) host is mediated by chemotaxis is still unknown.

ATTACHMENT   Attack and attachment occur with the nonflagellated end of the parasite (7, 41, 56, 72, 73). Multiple attachments on the surface of a single host cell are possible; with a high ratio of numbers of *Bdellovibrio* cells to numbers of host cells, a sort of "lysis-from-without," devoid of intracellular multiplication of the parasite has been reported (50). The nature of attachment is still unknown. The existence of specific receptor spots and receptor substances such as are known in bacteriophage-host reactions is possible, but not very likely. Mutants of *Escherichia coli* and *Salmonella* spp. which differed in chemical composition of their cell walls were susceptible to the same strain of *Bdellovibrio,* although the kinetics of attachment were somewhat different among the mutants (82). There is also no difference in susceptibility to *Bdellovibrio* of S- and R-form cultures of *E. coli* (28). The presence or absence of the outmost structured layer of the cell wall of *Spirillum serpens*, however, has been reported to be decisive for susceptibility to *Bdellovibrio* (5). In contrast to bacteriophage-host systems in which bacteria regularly give rise to phage-resistant mutants by mutational change of the receptors, resistance to *Bdellovibrio* has never been observed in a susceptible population.

PENETRATION   Penetration of *Bdellovibrio* into the host bacterium necessarily requires a hole in the cell wall. The process of hole (pore) formation is not yet satisfactorily understood. Mechanical damage, enzymatic action, or a combination of both can be envisaged. Immediately after attachment, the parasite rotates around its long axis, as can be observed under the phase-contrast microscope (62, 73). High-speed cinematography has revealed that the parasite rotates with extreme rapidity, up to 100 revolutions per sec (65, 66). The ballistic "impact" connected with the primary contact, and the subsequent high-speed rotation have been regarded as an essential element in formation of the pore (68, 73). This hypothesis is supported by the appearance of localized wall damage (round spots) at the site of primary and abortive attachments as shown in negative stainings and by the existence of special spikelike filaments located at the pole of attachment of the *Bdellovibrio* cell (49, 50). The rotation around the bdellovibrio's long axis has also been interpreted as an arm-in-socket type of motion (56) or as a swiveling of the posterior end of the parasite without real rotation at the attachment tip (7), assuming that there is a tight bond between parasite and host cell. In electron-micrographs, a bulge on the cell wall is often present, the *Bdellovibrio* cell entering through the center of the bulge. Its formation may arise from primary damage to the host cell wall and turgor pressure (7). The penetration pore is smaller than the diameter of the entering parasite, as may be recognized from the constrictions observed during the process of penetration (7, 56, 68). Although there is some morphological and biochemical evidence that invasion of the parasite is mediated, in part at least, by enzymatic action (25, 57, 58), the relatively small diameter of the entrance pore may be interpreted in favor of the mechanical theory. An enzymatic action by the *Bdello vibrio* cell would be expected, because of diffusion, to cause a weakening beyond the area of contact given by the diameter of the cell tip attached to the host cell wall. The time required for complete entrance measured from primary attachment has been determined as 5–20 min (65, 66), 5–60 min (56), or 10–60 min (6). For the actual step of invasion after completion of the pore only seconds are required. It is conceivable that the *Bdellovibrio* cell is forced through the hole into the interior by its flagellar motion. From this point of view, it is unlikely that the flagellum is lost during attachment or invasion, although electron micrographs of thin-sections show no flagella. The process of penetration of the parasite is accompanied by a conversion of the host cell into a globular body which resembles a spheroplast, but which Starr & Seidler (58), with commendable caution, term a "spherical body." This morphological change occurs most quickly in physiologically young and morphologically short rods (41, 46, 56). Varon & Shilo (81) observed that penetration was blocked by inhibitors of protein synthesis (streptomycin, chloramphenicol), although attachment occurred. They postulated that formation of inducible enzymes is required for penetration, which is inhibited under such conditions. It cannot be ruled out, however, that antibiotics exert a direct effect on the metabolism and motility of *Bdellovibrio,* thus preventing formation of the penetration pore in the host's cell wall. Since the antibiotic effect does not express itself immediately, attachments can occur but subsequent processes are blocked.

MULTIPLICATION INSIDE THE HOST CELL AND RELEASE OF PROGENY   During the process of invasion of the parasite, the cytoplasmic membrane (CM) of the host bacterium is separated from the cell wall and is pushed aside by the entering *Bdellovibrio* cell, thus locating the parasite between the cell wall and the CM (6, 7, 41, 46, 56, 68). For this type of localization, the apt term "intramural" has been proposed (58). The parasite never crosses the CM and never enters the cytoplasm. After complete penetration, the "cell envelope" is closed again, leaving a scar at the site of the penetration pore (50). Closing of the pore becomes apparent in later stages of intracellular development. After disintegration of cell contents, the spherical body becomes sensitive to osmotic shock (68). The spheroplast-like bodies enlarge in size and explode in distilled water (personal observations). This behavior may vary with different host strains, as stability of the "spheroplasts" in hypotonic medium likewise has been reported (78). The intramural growth phase of *Bdellovibrio* generally requires one hour or more, depending on the system (41, 46, 56, 66, 68, 83). The parasite then develops into a spiral-shaped cell which is several times (5–10×) the length of the infecting cell (8, 29, 41, 46, 56), and by segmental fragmentation and genesis of flagella develops into motile daughter cells morphologically identical with the parental parasite. The *Bdellovibrio* progeny that occasionally have been observed to move actively within the ghosted cell (6, 56, 68) are finally released, leaving an empty envelope. The mechanism by which the ghosted host cells burst is still unknown. The number of daughter cells released depends primarily on the size of the host bacterium; with *E. coli* 5.7 (46), with *Pseudomonas fluorescens* 8–12 (56), and with *Spirillum serpens* 20–30 (65, 66) daughter cells have been reported to form.

EFFECTS ON THE HOST CELL

After attachment of the *Bdellovibrio,* motile host bacteria rapidly lose motility. With *Bdellovibrio* strain *Bd.* 100 and *Rhodospirillum rubrum,* motility ceased 5 sec after attachment (73); with the much larger *Spirillum serpens,* paralysis of motility required several minutes (68). The basis of the effect on motility is not known. Another effect recognizable under the phase-contrast microscope is the change of the host cell into a spherical body. This alteration necessarily requires weakening the rigid structures (murein sacculus) of the cell wall. It is not yet clear whether mureinolytic enzymes of the parasite, the host, or of both are involved in the process of "spheroplast" formation. Host activity may be involved, because young and metabolically active cells react very rapidly; under optimal conditions, the spheroplast-like body may be formed before the parasite has entered the cell (65, 66). Since nutrition of the parasite requires passage of metabolites from the cytoplasm into the intramural area an alteration of permeability of the host's CM would be expected. Indeed, drastic changes in permeability at very early stages of infection have been observed (9, 40). Rittenberg & Shilo (40) studied respiration of a lactose-cryptic mutant of *E. coli* (ML 35) producing β-galactosidase but lacking galactoside permease, and found that lactose respiration started 5 min after *Bdellovibrio* attack, as a consequence of increased permeability to small molecules. The respiratory potential of the host was destroyed within the first hour of the 3–4 hr development cycle. Varon et al (84) demonstrated interference of *Bdellovibrio* with synthesis of β-galactosidase-specific m-RNA of the host. RNA synthesis was inhibited 3 min after infection by *Bdellovibrio;* protein synthesis was inhibited after 8–9 min. These results easily explain the fact that host cells lose viability shortly after attachment, and they are in accordance with the extensive deterioration of the cytoplasm visible in electron micrographs (7, 56, 68) at early stages of intracellular infection. The drastic changes in host metabolism occur before growth and multiplication of the bdellovibrios start. Damage to the CM, allowing leakage of nutrients from the protoplast into the intramural space and passage of digestive enzymes (primarily proteases) from the parasite into the protoplast, seems to have a major function in the parasite-host interaction. The progressive degradation of the protoplast (under the control of the parasite) probably secures a continuous supply of nutrients, and the surrounding cell envelope probably prevents considerable leakage from the intramural space into the environment outside the infested cell, thus making the system of exploitation ingenious, highly effective, and economic. For better understanding of the process, additional data are needed on leakage of substances from the parasite-host complex into the medium, before release of the progeny.

Lysis of bacteria has been the subject of intensive research. The literature pertinent to this field has been reviewed (14–16, 22, 52, 74, 75). The lytic process induced by *Bdellovibrio* attack is not yet fully understood in mechanistic terms. In principal, lysis of bacteria by *Bdellovibrio* involves the same structural and biochemical alterations that occur in other lytic processes, including autolysis, and they all result in the dissolution of cellular constituents. The inducing reactions by both biological and nonbiological agents, however, may be different (74).

One important feature of lysis by *Bdellovibrio* is the fact that a *Bdellovi-*

brio lysate of host bacteria, after separation of the parasites by filtration, does not possess lytic activity against living host bacteria. However, it exhibits enzymatic activities that dissolve the cytoplasm of heat-killed host cells (51, 62, 73). Since the early studies on *Bdellovibrio*, it has been known that saprophytic bdellovibrios produce extracellular proteases that lyse heat-killed bacteria, dissolve casein, and liquefy gelatin (62, 73). There is now good evidence that both parasitic and saprophytic bdellovibrios produce extracellular proteolytic enzymes (17, 47), although certain strains have been reported to lack proteolytic activity (51). In a study on the possible enzymatic basis of bacteriolysis by *Bdellovibrio*, Huang & Starr (25) demonstrated that some mutants of *Bdellovibrio*, obtained by treatment with N-methyl-N'-nitro-N-nitrosoguanidine and selected by inability to clear casein agar, nevertheless possessed carboxypeptidase. Earlier reports on nonproteolytic bdellovibrios (51, 62) probably refer to strains that do not hydrolyze casein (exopeptidase-negative), but do have carboxypeptidase. In their study on the enzyme production by *Bdellovibrio*, Huang & Starr (25) used a system that allowed growth of the parasite in heat-killed cell suspensions of *Spirillum serpens* VHL, thus excluding interference from enzymatic activities of host bacteria. They isolated and characterized a lysozyme-like enzyme (mol wt 12,500; mol wt of egg-white lysozyme 12,400), three proteolytic enzymes, and a lipase. Azocollase, an enzyme that is able to hydrolyze Azocoll (a brand of collagen), was found to have a molecular weight of about 11,000; it acts also on hemoglobin, bovine serum albumin, and gelatin. Azocollase requires Ca^{2+} and Mg^{2+}; inhibition of protease activity by EDTA supports the notion of metallo-proteases in *Bdellovibrio* (17). It has been postulated that the lysozyme-like enzyme, the proteases, and the lipase may be relevant to digestion of the host cell by the parasite (25, 57, 58). It must be kept in mind, however, that the enzymes per se do not attack viable cells and that, in the process of penetration, mechanical forces may be more important than enzymatic activities (see sections on Attachment and Penetration). There are many bacteria producing mureinolytic, proteolytic, and lipolytic enzymes which do not have the parasitic capacity of a bdellovibrio! Although dissolution of heat-killed cells must necessarily be attributed to activities of the bdellovibrios, it cannot be excluded from experiments with dead cells that *Bdellovibrio* may also interfere in *living* bacteria with the mechanisms involved in autolysis of the host (74). Participation of the host's enzymes in disintegration of the infected cell seems to be possible, but has not yet been proved.

HOST RANGE OF BDELLOVIBRIO

The first strain of *Bdellovibrio* (*Bd.* 321) that appeared in a culture of *Pseudomonas phaseolicola* exhibited lytic activity against a variety of pseudomonads and xanthomonads (72). For evaluation of the activity spectrum, 120 cultures covering several categories of bacteria were tested. Apart from one exception (a strain of *Pectobacterium parthenii*), the activity of *Bd*. 321 was restricted to fluorescent saprophytic and phytopathogenic pseudomonads and xanthomonads. It was demonstrated that representatives of *Pseudomonas solanacearum* and *P. caryophylli* that were not closely related to the fluorescent pseudomonads, were not susceptible to this *Bdellovibrio* strain. Stolp & Starr (73) studied the host ranges of 11 additional *Bdellovibrio* strains, using 34 test cultures. Activity of these strains was likewise limited to Gram-negative bacteria, but not all Gram-negative bacteria tested were susceptible. The host ranges of individual *Bdellovibrio* strains were different; two had a comparatively restricted activity spectrum. The majority of known *Bdellovibrio* strains is active against pseudomonads and enterobacteria (57, 58). Susceptibility of *Azotobacter chroococcum* (76), *Rhizobium* spp., and *Agrobacterium tumefaciens* (37) has been reported. *Bdellovibrio* strain *Bd*. W has been claimed (6) to possess activity against Gram-positive bacteria (*Streptococcus faecalis* and *Lactobacillus plantarum*), but this could not be confirmed (58).

For determining susceptibility of a given bacterium, several procedures have been applied (57). Lysis in liquid culture, plaque formation in agar medium, or both, are commonly used. Reliable results, naturally, require experimental conditions that allow attachment, penetration, and lysis by *Bdellovibrio*. To increase optical contrast between plaques and growth of the host, addition of 0.1% 2,3,5-triphenyltetrazolium chloride (27, 47) or 2% skim milk (57) has been recommended; with very small plaques, use of such methods may be of advantage.

CHARACTERISTICS OF THE BDELLOVIBRIOS

Present information on *Bdellovibrio* is derived from study of parasitic strains isolated from natural habitats and from study of saprophytic strains isolated from populations of the parasitic type. While parasitic bdellovibrios have the capacity to grow in living host bacteria (they may also utilize but do not grow within heat-killed cells), saprophytic derivatives can be propagated on complex nutrient laboratory media. The following section refers primarily to general characteristics of the bdellovibrios including morphology, structure, and chemical composition; the interactions between *Bdellovibrio* and its host cell, the physiology, nutrition, and metabolic activities will be discussed in subsequent sections.

THE PARASITIC BDELLOVIBRIOS   The bdellovibrios are vibrioid to rod-shaped Gram-negative bacteria of relatively small size. Outside the host, they usually measure 1–2 μm in length and 0.25–0.40 μm in width. Slight differences in size and shape have been reported (1, 8, 49, 73). During growth inside the host, *Bdellovibrio* elongates into a rather large spiral-shaped cell (41, 46, 56). Segmentation of the spiral entity produces daughter cells that are identical in morphology with free-living bdellovibrios outside the host cell. All parasitic strains so far studied are extremely motile. High-frequency cinematography (200 pictures/sec) has revealed a relative speed of locomotion of up to 100 cell-lengths per sec (65, 66). The unusually thick flagellum has a diameter of 28 nm (45). It consists of an inner core (13 nm) and a surrounding sheath (7.5 nm) originating from the cell wall and being continuous with it (1, 7, 45). In the presence of 6M urea the sheath separates from the core of the flagellum; the cell wall, however, is not affected (45).

The aflagellated cell end that attaches to the host cell bears unique structures (spike-like filaments, ring structures, "holdfast" structures) possibly involved in attachment and penetration (41, 49). The "holdfast" also has been suggested to be an artifact (1, 8). The cell wall of *Bdellovibrio* is extremely susceptible to disruption by physical agents (1) and contains chemical components characteristic of the murein layer (77).

The fine structure of the cytoplasm of *Bdellovibrio* is comparable to that of other bacteria. Ribosomes of HI strains have co-sedimentation properties comparable to those found in *Escherichia coli* (44). The nucleoplasm occupies about two-thirds of the cytoplasm (56). Mesosomes located at the anterior end of the cell (7, 56) possibly are involved in attachment or penetration (7) and cell division (8). Thin sections often show electron-dense inclusion bodies (1, 7, 56), the chemical nature and function of which still are obscure. In one strain of *Bdellovibrio* (*Bd.* W), an encysted "resting stage" has been reported (6, 24).

THE SAPROPHYTIC BDELLOVIBRIOS   Saprophytic bdellovibrios (HI strains) can be isolated from populations of parasitic bdellovibrios. The wild-type is sensitive to streptomycin (72, 73), but streptomycin-resistant (sm^r) mutants can easily be selected from streptomycin-sensitive (sm^s) populations (62). Based on the relationship to streptomycin, Seidler & Starr (47) introduced an effective technique for isolation of HI (i.e., independent of both living and heat-killed bacteria) bdellovibrios. The procedure includes isolation of HD sm^r mutants, propagation on a sm^s host bacterium, and inoculation of the lysate into a selective medium consisting of PY + streptomycin. Under such conditions, residual host cells are killed, sm^r HD bdellovibrios do not grow, and sm^r HI mutants develop into colonies. The ratio of HI : HD cells in several strains was 1 : 10^6. The relatively high frequency indicates that only a few genetic loci are involved in this probably mutational event. In the first study on selection of saprophytic derivatives, the ratio was found to be 1 : 10^8 (72). As HI strains have been isolated from HD populations whenever tried, it has been suggested that such organisms also exist in natural habitats. So far, however, there are no reports on the isolation of saprophytic bdellovibrios directly from nature.

For propagation of the HI bdellovibrios, PY medium has been recommended (72). As the nutritional requirements are not fully known, propagation still requires a complex substrate and is not yet possible in a chemically-defined synthetic medium.

Compared to HD strains, HI derivatives generally live somewhat longer under common cultural conditions. Long-term preservation is effective with storage of young cultures in PY solution + 10% glycerol at −30°C, or with lyophilization.

Morphologically, HI bdellovibrios are similar to HD parental bacteria. A HI population is rather heterogeneous as to cell length. Spiral forms with up to 30 times (72) or even 80 times (38) the length of an individual parasitic cell occur. Although most HI strains possess 1 polar flagellum (57, 58), multiple flagellation (1–3 flagella) has been reported (72). In HI *Bdellovibrio stolpii* (strain *Bd*. UKi2), morphological sequences were observed during the development cycle which resembled those known from the development cycle of HD cells, that is, production of spirals that give rise to individual cells by segmental fragmentation (8).

The protein content of HI bdellovibrios is 60–65% of the dry weight, the DNA content 5% (48). In HI *Bdellovibrio starrii* (strain *Bd*. A3.12) and HI *Bdellovibrio bacteriovorus* (strain *Bd*. 109), d-TMP, UMP, d-CMP, ribose, adenine, and guanine (48) have been found. DNA base composition has been intensively studied (44, 48). Phosphatidylethanolamine and phosphatidylglycerol are the major phospholipids in *Bdellovibrio stolpii* (strain *Bd*. UK12); a sphingolipid, rare in bacteria, is also present (60, 61).

As a rule, HI strains no longer possess parasitic properties; reversion to the parental parasitic type has been reported (58, 72). Attempts have been made to select "parasites" from saprophytic bacteria of known systematic position (47, 73). These attempts were unsuccessful with several cultures of *Caulobacter*, *Vibrio*, and *Spirillum*, indicating that these known vibrioid organisms do not represent saprophytic derivatives of parasitic bdellovibrios.

DIVERSITY AMONG BDELLOVIBRIO ISOLATES   Stolp & Starr (73) proposed the establishment of a new genus, *Bdellovibrio*, for vibrioid bacteria endowed with the unique property of parasitizing other bacteria. The relatively few isolates then known were tentatively regarded as representatives of one species, *Bdellovibrio bacteriovorus*.[2]

During the past decade, many new isolates of *Bdellovibrio* have been studied. Because of the advantages of growth in axenic culture, HI strains were preferentially used in physiological and biochemical investigations (44, 47, 48, 51, 73). Comparative studies of DNA base composition, nucleic acid homogeneity or heterogeneity, and enzymes disclosed differences among bdellovibrios that necessitated their division into three species: *Bdellovibrio bacteriovorus*,[2] *Bdellovibrio starrii*, and *Bdellovibrio stolpii* (44).

[2] Strain *Bd*. 100, originally isolated with *Erwinia amylovora* (strain ICPB EA137), was designated the nomenclatural type culture (ATCC 15356). As the working type (cotype) HI *Bd*. 100 (ATCC 25622) has been proposed (44).

The majority of strains studied have 50–51% guanine + cytosine (GC); *Bd.* A3.12 and *Bd.* 321 have 43.5% GC (48). *Bd.* UKi2 (a HI isolate from the obligately parasitic strain *Bd.* UK) has 41.8% GC (44).

Seidler, Mandel & Baptist (44) studied the genetic relatedness among bdellovibrios on the basis of reassociation of their nucleic acids. When comparing the DNA of HI *Bd.* 109 (the reference DNA) with the ribosomal RNA of various HI *Bdellovibrio* isolates and *Escherichia coli,* the results were as follows: (*a*) HI *Bd.* B, HI *Bd.* 110, and HI *Bd.* 2484 Se-2 behaved as the homologous HI *Bd.* 109. Since reassociation of DNA and ribosomal RNA is known to have a high degree of conservation (13, 35), the results indicate identical nucleotide sequences among representatives of the GC group with 50%–51% GC. (*b*) HI *Bd.* A3.12 competed at a level of only 20–30%, indicating that *Bd.* A3.12 is but little related to *Bd.* 109 and the rest of the high GC group. (*c*) *E. coli* RNA did not compete with DNA of HI *Bd.* 109, indicating that the organisms are unrelated.

In further experiments on reassociation between DNA of *Bd.* 100 (the nomenclatural type of *B. bacteriovorus*) and DNAs of various bdellovibrios, the following results were obtained: (*a*) Perfect homology was observed with HI *Bd.* 109, HI *Bd.* 114, and HI *Bd.* 118. All of them belong to the group with 50–51% GC, so the result supports the view of close relationship among the organisms of this group. (*b*) DNA/DNA reassociation was not established in combinations HI *Bd.* 100/HI *Bd.* A3.12 and HI *Bd.* 100/HI *Bd.* UKi2. (*c*) The DNAs of HI *Bd.* A3.12 and HI *Bd.* UKi2 reassociated only at the 16% level. This result indicates that, in spite of the close GC percent values (43% GC in HI *Bd.* A3.12; 42% GC in HI *Bd.* UKi2), these two strains differ considerably in their genetic make-up. The level of molecular-genetic relationship is comparable to that observed between *Escherichia coli* and *Salmonella typhimurium* (3, 43). From considerations on genome size, studies on nucleic acids, and on zone electrophoresis of soluble enzymes (44), molecular-genetic heterogeneity of bdellovibrios is evident, although the various isolates represent a unique "type" of bacterium characterized by the property, common to wild-type isolates, of being parasitic to other bacteria. In consequence of the diversity, Seidler, Mandel & Baptist (44) proposed two new species of the genus *Bdellovibrio: Bdellovibrio starrii* [nomenclatural type specimen *Bd.* A3.12; cotype is HI *Bd.* A3.12 (ATCC 27110)] and *Bdellovibrio stolpii* [nomenclatural type specimen *Bd.* UKi2; cotype is HI *Bd.* UKi2 (ATCC 27111)]. *Bdellovibrio bacteriovorus* (nomenclatural type specimen *Bd.* 100) comprises the majority of strains so far studied (50–51% GC). On the basis of molecular characteristics and additional facts such as differences in cytochrome spectra (47) and enzyme migration rates (44), establishment of these three species within the genus *Bdellovibrio* seems fully justified. In my view, the study of many more strains will probably demonstrate the existence of additional molecular-genetic-cultural types requiring the establishment of further species.

NUTRITION AND METABOLISM OF BDELLOVIBRIO

The parasitic bdellovibrios meet their nutritional requirements by degrading and consuming viable host bacteria. Early attempts to propagate parasitic wild-type isolates on nonviable bacteria, on extracts of bacteria (sonic extracts, phage lysates), or on complex nutrient media, were unsuccessful (62, 72, 73). There is now evidence that many, if not all, "parasitic" bdellovibrios are capable of multiplying in suspensions of heat-killed cells of the congenial host or other bacteria, provided that Ca^{2+} and Mg^{2+} are present in sufficient concentrations (10, 25, 26, 57, 58). With growth of Bd. 6-5-S, Bd. 100, Bd. 109D, or Bd. A3.12 on heat-killed bacteria of Spirillum serpens VHL, the ions requirement for optimal multiplication was $2 \times 10^{-3}M$ for Ca^{2+} and $2 \times 10^{-5}M$ for Mg^{2+} (26). No intracellular life cycle occurred under such nutritional conditions; the population developing on heat-killed cells, however, retained its parasitic character. Obviously, the suspension of heated cells satisfics all nutritional requirements of parasitic bdellovibrios that make available the necessary nutrients by their own metabolic activities. Enzymatic activities displayed by bdellovibrios in the presence of heat-killed cells (25), certainly arc involved in the processes of degrading bacterial components. In contrast to saprophytic bdellovibrios (as defined above), parasitic bdellovibrios that develop on heat-killed bacteria do not grow on complex nutrient media. Even inoculations of high concentrations (10^9 to 10^{10} cells/ml) of such bdellovibrios into agar-medium (yeast extract plus peptone) did not give rise to colonies and, in the absence of Ca^{2+} and Mg^{2+} or by the addition of EDTA, multiplication on dead host cells was prevented (26). The discrepancies existing among results of experiments with heat-killed cells were at least partly caused by the absence or presence of calcium and magnesium ions at suitable levels. Since prolonged autoclaving renders host cells unsuitable as growth substrate for Bdellovibrio (26), this factor likewise might have caused inconsistent results.

Saprophytic bdellovibrios grow on a complex nutrient medium such as PY. They also develop in the presence of heat-killed cells (73). Selection of a saprophytic clone is possible by massive inoculation of parasitic bdellovibrios into a semi-solid (0.6% agar) PY medium (72, 73). A most effective isolation technique for saprophytic bdellovibrios (HI strains) has been developed (47).

When studying the potential for virulence of saprophytic strains, it has usually been observed that only a few individuals of the population have parasitic ability (72, 73). From a number of HI strains, Seidler & Starr (47) regularly succeeded in isolating such revertants when introducing large numbers of log-phase HI bdellovibrios into suspensions of host bacteria; direct inoculation into lawns of host bacteria, for some unknown reason, did not result in plaque formation, except with HI Bd. A3.12. Bdellovibrio starrii strain

Bd. A3.12 was reported as early as 1964 (4) to give rise to HI mutants that retain their parasitic ability. These mutants produced plaques on host lawns and colonies on solid media at a ratio of 1 : 1 (51). With successive transfers through host-free media, virulence was gradually lost, i.e. the proportion of virulent individuals continuously decreased (51). From obligately parasitic *Bd.* UK (host: *E. coli* B/r), a saprophytic derivative (*Bd.* UKi2) was selected that is facultatively parasitic (8, 12). However, not every derivative of *Bd.* UK was FP; another HI strain from HD *Bd.* UK was devoid of parasitic activity (12). Saprophytic development of *Bdellovibrio stolpii* strain UKi2 shows morphological parallels to the intracellular development (comma-shaped cells grow into spirals that divide into 7–10 vibrioid cells). On lawns of viable host bacteria, *Bd.* UKi2 colonies are produced which are surrounded by a circular clear zone (12). This type of plaque formation by a *Bdellovibrio* strain had also been observed in a derivative of *Bd.* 321 (72). In HI *Bd.* UKi2, endoparasitic ability was retained after 250 transfers through host-free media (12). The final population was not examined quantitatively with respect to the proportion of parasitic individuals, and the data allow judgment only on the parasitic potential of the population. As in HI *Bd.* A3.12 (47), reselection of parasitic bdellovibrios has been reported to be difficult in *Bd.* UKi2 (58).

The available information on saprophytic bdellovibrios indicates that certain strains (such as HI *Bd.* A3.12 and HI *Bd.* UKi2) keep the parasitic potential when grown in the absence of living or dead host bacteria; that is, during growth on complex nutrient media. A population change in the absence of living host bacteria showing increase of nonparasitic individuals, however, seems to be a general phenomenon (58). Why a given population of bdellovibrios is statistically parasitic, saprophytic, or both cannot yet be answered. Differences in parasitic abilities probably do not reflect only different nutritional requirements, but are of a more complex nature.

Studies on physiology and metabolic activities of bdellovibrios have been facilitated by the existence of host-independent mutants allowing axenic culture.

Stolp & Petzold (72) isolated several saprophytic mutants from the first isolate of parasitic *Bdellovibrio*. The most striking properties of one of these derivatives (Sp. 19) may be summarized as follows: Gram-negative; aerobic; growth with PY-medium; pH range for growth 6.0–8.5; strongly proteolytic (liquefaction of gelatin, clearing of heat-killed bacteria); no catalase; neither oxidation nor fermentation of carbohydrates, organic acids, alcohols; production of a yellow pigment (73). Seidler & Starr (47) thoroughly examined 12 HI strains: they neither fermented nor oxidized 14 different carbon sources, but were positive in proteolytic activity and oxidase; none could reduce nitrate, although nitrate reduction has been reported for other strains (12, 73). Upon initial isolation, most strains were catalase-positive; this activity, however, was lost in serial transfer. All strains had a cytochrome a and c component; all except one were sensitive to the vibriostatic pteridine O/129.

Simpson & Robinson (53) demonstrated the presence of cytochromes, $NADH_2$- and $NADPH_2$-oxidase, ATPase, six enzymes of the tricarboxylic acid (TCA) cycle, and glutamate and alanine dehydrogenases in the host-dependent *Bd*. 6-5-S. Extracts did not contain glucose kinase, glucose 6-phosphate dehydrogenase, or 6-phosphogluconate dehydrogenase, but gave weak reactions in aldolase and glyceraldehyde 3-phosphate dehydrogenase, indicative for the glycolytic pathway. These latter two enzymes were not found in later studies with seven other HI *Bdellovibrio* strains (44). A HI derivative from *Bd*. A3.12 requires thiamine for maximum growth (4). It is not known whether thiamine requirement in *Bdellovibrio* is a general character.

Matin & Rittenberg (30) followed the fate of host DNA during the development cycle of *Bd*. 109D (39) in *Pseudomonas putida* strain N-15 and *Escherichia coli* strain ML35. Viable host cells were supplied as the sole source of nutrients (in Tris buffer, pH 7.5, and minimal salts). They observed an early and rapid destruction of host DNA, which was complete within 45–60 min after infection by *Bdellovibrio*. In *P. putida,* 70% of the host DNA was degraded after 30 min. The breakdown products (oligonucleotides) were utilized by *Bdellovibrio*. With ^3H-thymidine-labeled *E. coli,* 73% of the radioactivity initially present in the methyl group of host thymine was incorporated into *Bdellovibrio* DNA. Synthesis of DNA in *Bdellovibrio* started after complete degradation of host DNA. The mechanism of host DNA destruction is unknown, but possible causes are (*a*) action of host deoxyribonucleases in consequence of the early damage to the host membrane (9, 40, 84), and (*b*) release of *Bdellovibrio* DNase into the host cell. The fact that about 80% of host DNA is incorporated into *Bdellovibrio* DNA (if the efficiency of uptake of thymidine is representative for all DNA components) does not necessarily mean that the organism is unable to synthesize nucleotides. The take-over of nucleotides, however, is an effective and economic step. Since the data indicate that the rapid breakdown of host DNA yields oligonucleotides, and that monomeric units entering into biosynthesis of *Bdellovibrio* DNA are generated as they become needed, a precise control mechanism over the degradative process has been postulated (30).

Huang & Starr (25) recently studied the enzymatic activities of *Bd*. 6-5-S, using as the substrate autoclaved cells of *Spirillum serpens* VHL, thus excluding parasitism as well as metabolic activities of the host. HD bdellovibrios *Bd*. 100, *Bd*. 109, and *Bd*. A3.12 likewise multiplied nonparasitically in suspensions of heat-killed (10 min 70°C; 15 min 121°C) cells of *S. serpens* VHL. They demonstrated the production of a muramidase, of proteolytic enzymes, and of a lipase.

The inability of bdellovibrios to utilize carbohydrates and similar C-compounds as energy sources points to their metabolic specialization: digestion of bacteria consisting mainly of proteins, nucleic acids, and membrane and cell-wall components.

Although research on metabolic activities has clarified a number of biochemical reactions of the bdellovibrios, it is not yet possible to cultivate any

Bdellovibrio strains in a synthetic medium of exactly known composition. Growth under defined conditions is an essential prerequisite for certain metabolic or genetic studies, and a defined medium that meets the nutritional requirements is urgently needed.

BACTERIOPHAGES OF THE BDELLOVIBRIOS

Because isolation of bacteriophages by the common enrichment technique requires propagation of the prospective phage host, the existence of host-independent bdellovibrios has much facilitated phage isolation. Hashimoto, Diedrich & Conti (23) succeeded in isolating the first known *Bdellovibrio* bacteriophage (HDC-1) using HI *Bd.* UKi2 as the bacterial host. Phage HDC-1 is hexagonal in shape, in size between 60 and 70 nm, tailless, and possesses single-stranded DNA. With other host-independent bdellovibrios, 10 additional phage strains, 4 of which were different from one another, were isolated by Althauser et al (2). By comparative phage susceptibility tests with the 5 different phages, 19 HI *Bdellovibrio* strains could be divided into 6 groups. Two additional phages were isolated by Varon & Levisohn (79). Some of the known phages have been demonstrated also to be active against the HD parental strains of the HI *Bdellovibrio* phage host. The three-membered system (host bacterium, *Bdellovibrio*, bacteriophage) opens interesting perspectives for studies on host reactions caused by phage-infected bdellovibrios.

ECOLOGY OF THE BDELLOVIBRIOS

The bdellovibrios are ubiquitous in many natural habitats. Isolations have been successful from soil and sewage (11, 28, 32, 37, 72, 73, 87), rivers and lakes (18–21), and from seawater (34, 49). The techniques originally used for isolation purposes, based on centrifugation and differential filtrations (72, 73), are not suited for quantitative studies because of the low recovery. For this purpose, the method of dilution-to-extension (28) and other techniques have been applied (50, 80, 81). Quantitative enumeration of the bdellovibrios is influenced both by isolation procedure and by the prospective host bacterium used. *Bdellovibrio* enumerations obtained with one particular host bacterium must, as a rule, be regarded as smaller than the real numbers in the habitat because a single kind of host bacterium cannot be expected to be susceptible to every *Bdellovibrio* cell. Titers reported: 10^3–10^5 bdellovibrios per gram of soil (28), 10^5 per ml of sewage (19), 50 per ml of pond and seawater (49). The ecological significance of the bdellovibrios as an integral component of the microflora is not yet well understood. Their possible role under natural conditions in population changes (72, 73), particularly with respect to elimination of *Escherichia coli* in seawater (33, 34) and removal of *Salmonella* from polluted waters and raw sewage (11, 18, 20, 21, 29), has been discussed. For additional information the reviews by Starr & Seidler (58) and by Starr & Huang (57) may be consulted.

POSITION OF THE BDELLOVIBRIOS IN THE BACTERIAL WORLD

The bdellovibrios possess the unique ability of physically attacking a viable bacterial cell, attaching to its surface, penetrating the cell wall, and multiplying within the intramural space, accompanied by a progressive disorganization and lysis of the invaded cell. After completion of cellular dissolution, the progeny bdellovibrios are released and start new infection-cycles, provided suitable host bacteria are available. The parasitic action of *Bdellovibrio,* its intracellular development, and lysis of the host bacterium are the most striking features of the *Bdellovibrio*-host interaction.

The taxonomic position of the bdellovibrios has been analyzed in detail (58). Stolp & Starr (73) in 1963 established the genus *Bdellovibrio* primarily on the basis of the predatory, ectoparasitic, and bacteriolytic action of the organism. The original concept had to be modified with respect to the nature of parasitism, which is not ectoparasitic (72, 73), but rather endoparasitic (41, 56). Because the *Bdellovibrio* cell locates in an intramural position between cell wall and cytoplasmic membrane, Stanier (55) pointed out that *Bdellovibrio* is not an "endosymbiont" sensu strictu (i.c., comparable to cellular endosymbionts as known in eukaryotic cells). There is, indeed, no such endosymbiosis known in prokaryotes. Nevertheless, use of the term "endoparasite," in the more general and descriptive sense, seems admissible.

The genus *Bdellovibrio* as presently conceived comprises three species (44). With increasing knowledge of the bdellovibrios, particularly of representatives that live under specific conditions (e.g. obligately halophilic forms), it seems probable that establishment of additional species will follow. Based on available facts, Starr & Seidler (58) have given a definition of the genus *Bdellovibrio* in terms of a set of properties, possession of most of which is considered essential for inclusion in the taxon. These are the properties:

(*a*) generally vibrioid, sometimes spirillar; (*b*) unusually narrow (about 0.3–0.45 μm in width; (*c*) polarly flagellated (very rarely more than one flagellum but the flagellum is always sheathed and the insertion is always polar); (*d*) Gram-negative; (*e*) strictly aerobic; (*f*) chemo-organotrophic; (*g*) endowed with DNA which is highly homogeneous and contains 42–51 percent GC; (*h*) markedly and invariably proteolytic; (*i*) able to metabolize through the TCA cycle; (*j*) generally not capable of using carbohydrates; (*k*) fitted with a dimorphic developmental cycle: alternation of flagellated, predatory vibrioid swarmers (capable of attaching by the aflagellated tip to a host cell) and non-flagellated spirillar (usually intracellular) vegetative stages; and (*l*) actually capable of (or having the genetic potential for) entering the cells of certain other bacteria and developing and multiplying therein.

The taxonomic relationship of bdellovibrios to other bacteria is not yet well established. Starr & Seidler (58) expressed the opinion that the genus *Bdellovibrio* might be placed in the family Spirillaceae, as defined by Véron (85, 86). Although similar in cell shape, and having in common the inertness towards carbohydrates, a close relationship between *Campylobacter* and *Bdellovibrio* does not exist; the former have 35% GC and are nonproteolytic (42, 85). From a comparison of the properties of vibrios and bdellovibrios, Starr & Seidler (58) concluded that some degree of relationship between them, despite the fundamental differences in metabolism, must be considered.

CONCLUDING REMARKS

At the close of the first decade since bdellovibrios were discovered, there are still many problems that await solution—problems related to the taxonomy, cytology, physiology, biochemistry, and ecology of bdellovibrios, as well as to their interactions with host bacteria. Some of these unresolved problems are listed.

Certain stages in the parasitic life cycle of the bdellovibrios are not yet fully understood, and have been the subject of speculation and controversial hypotheses. (*a*) The primary contact between parasite and host: Does the contact occur by random collision or by a directed attack that might be mediated by chemotaxis? (*b*) The nature of attachment: What is the role of the special structures reported at the end of the parasite cell that attaches to the host bacterium? Is there a tight bond between the two organisms immediately after attachment, or is there a loose connection that allows rotation of the parasite, which thereby causes mechanical damage to the cell wall of the host? (*c*) The formation of the entrance pore: Does the hole in the host's cell wall result from mechanical forces (impact, rotation), from enzymatic action, or from both? Does the *Bdellovibrio* enter the host bacterium under the propelling force of its flagellum? How is the pore closed after penetration of the parasite? (*d*) The nature of structural and chemical changes in the host bacterium following attachment, penetration, and intramural growth of the parasite: Are enzymes of the host cell involved autolytically in the dissolution of its own cellular components, or are enzymes of the parasite exclusively responsible for breakdown of the host's polymers? (*e*) The release of the progeny: How is the remnant of the host's cell envelope opened so that the *Bdellovibrio* progeny can get out of the ghosted host cell?

The present knowledge concerning nutrition of the bdellovibrios is still fragmentary. There is now evidence that many, if not all, parasitic wild-type bdellovibrios are able to multiply and keep their parasitic ability when grown nonparasitically with heat-killed bacteria, provided that Ca^{2+} and Mg^{2+} are available at sufficient levels. Under such conditions, however, bdellovibrios do not enter nonviable cells. Since these parasitic bdellovibrios do not grow upon complex laboratory nutrient media, they obviously need one or more compounds present in heat-killed and in living cells. The nature of the "growth factor(s)" is still unknown. It is probably not one of the more common factors (vitamins, amino acids) which are present in yeast extract, but it may be an oligopeptide, an oligonucleotide, or some other factor. The situation with saprophytic bdellovibrios is similar; although they are able to grow on nutrient media, we do not know their exact nutritional requirements. Consequently, it is not yet possible to propagate parasitic or saprophytic bdellovibrios in a chemically defined synthetic medium. A problem connected with nutrition is the purported loss of parasitic ability in strains grown for extended periods in the absence of living host bacteria; reports on this point are contradictory.

Distribution of bdellovibrios in natural habitats, and their ecological significance there, are largely unknown. The same is true with respect to taxonomic relationships between bdellovibrios and other bacteria; knowledge might be advanced by comparative studies on serology and phage susceptibility.

The unique symbiotic interaction of these bacterial parasites with other bacteria has attracted increasing interest by several competent research groups. For this reason, I have high hopes that many of the problems presently posed by the bdellovibrios will be solved in the near future.

Acknowledgments

I am greatly obliged to Professor Mortimer P. Starr for many profitable and informative discussions about bdellovibrios, for critical readings of this essay in manuscript, and for substantial editorial and linguistic aid. However, this good friend and pioneer collaborator on *Bdellovibrio* should not in any way be held responsible for my errors of omission or commission!

Literature Cited

1. Abram, D., Davis, B. K. 1970. Structural properties and features of parasitic *Bdellovibrio bacteriovorus*. *J. Bacteriol.* 104:948–65
2. Althauser, M., Samsonoff, W. A., Anderson, C., Conti, S. F. 1972. Isolation and preliminary characterization of bacteriophage for *Bdellovibrio bacteriovorus. J. Virol.* 10:516–24
3. Brenner, D. J., Fanning, G. R., Johnson, K. E., Citarella, R. V., Falkow, S. 1969. Polynucleotide sequence relationships among members of *Enterobacteriaceae. J. Bacteriol.* 98:637–50
4. Bruff, B. S. 1964. *Studies on a predacious vibrio.* M.A. Thesis, University California, Berkeley
5. Buckmire, F. L. A., Murray, R. G. E. 1970. Studies on the cell wall of *Spirillum serpens.* 1. Isolation and partial purification of the outermost cell wall layer. *Can. J. Microbiol.* 16:1011–22
6. Burger, A., Drews, G., Ladwig, R. 1968. Wirtskreis und Infektionscyclus eines neu isolierten *Bdellovibrio bacteriovorus*-Stammes. *Archiv*

Mikrobiol. 61:261–79
7. Burnham, J. C., Hashimoto, T., Conti, S. F. 1968. Electron microscopic observations on the penetration of *Bdellovibrio bacteriovorus* into Gram-negative bacterial hosts. *J. Bacteriol.* 96:1366–81
8. Burnham, J. C., Hashimoto, T., Conti, S. F. 1970. Ultrastructure and cell division of a facultatively parasitic strain of *Bdellovibrio bacteriovorus. J. Bacteriol.* 101:997–1004
9. Crothers, S. F., Robinson, J. 1971. Changes in the permeability of *Escherichia coli* during parasitization by *Bdellovibrio bacteriovorus. Can. J. Microbiol.* 17:689–97
10. Crothers, S. F., Fackrell, H. B., Huang, J. C. C., Robinson, J. 1972. Relationship between *Bdellovibrio bacteriovorus* 6-5-S and autoclaved host bacteria. *Can. J. Microbiol.* 18:1941–48
11. Dias, F. F., Bhat, J. V. 1965. Microbial ecology of activated sludge. II. Bacteriophages, *Bdellovibrio,* coliforms, and other organisms. *Appl. Microbiol.* 13:257–61

12. Diedrich, D. L., Denny, C. F., Hashimoto, T., Conti, S. F. 1970. Facultatively parasitic strains of *Bdellovibrio bacteriovorus. J. Bacteriol.* 101:989–96
13. Doi, R. H., Igarashi, R. T. 1965. Conservation of ribosomal and messenger ribonucleic acid cistrons in *Bacillus* species, *J. Bacteriol.* 90: 384–90
14. Ghuysen, J.-M. 1968. Use of bacteriolytic enzymes in determination of wall structure and their role in cell metabolism. *Bacteriol. Rev.* 32: 425–64
15. Ghuysen, J.-M., Strominger, J. L., Tipper, D. J. 1968. Bacterial Cell Walls. In *Comprehensive Biochemistry*, M. Florkin, E. H. Stotz, eds. 26A:53–104, Elsevier: New York
16. Ghuysen, J.-M., Tipper, D. J., Strominger, J. L. 1966. Enzymes that degrade bacterial cell walls. In *Methods in Enzymology*, S. P. Colowick, N. O. Kaplan, eds. 8: 685, Academic: New York
17. Gloor, L., Seidler, R. J. 1972. Properties of some *Bdellovibrio* extracellular enzymes. *Ann. Meet. Am. Soc. Microbiol.* 1972, 1:38
18. Guélin, A., Lamblin, D. 1966. Sur le pouvoir bactéricide des eaux. *Bull. Acad. Nat. Méd. (Paris)*, 150: 526–32
19. Guélin, A., Lépine, P., Lamblin, D. 1967. Pouvoir bactéricide des eaux polluées et role de *Bdellovibrio bacteriovorus. Ann. Inst. Pasteur,* 113: 660–65
20. Guélin, A., Lépine, P., Lamblin, D. 1968. Sur l'autoépuration des eaux. *Rev. Int. Oceanogr. Med.* 10:221–27
21. Guélin, A., Lépine, P., Lamblin, D. 1969. Microorganismes responsables du pouvoir bactéricide des eaux. *Verhandl. Int. Verein. Limnol.* 17: 744–46
22. Guze, L. B. 1968. *Microbial Protoplasts, Spheroplasts and L-Forms.* Williams & Wilkins: Baltimore
23. Hashimoto, T., Diedrich, D. L., Conti, S. F. 1970. Isolation of a bacteriophage for *Bdellovibrio bacteriovorus. J. Virol.* 5:97–98
24. Hoeniger, J. F. M., Ladwig, R., Moor, H. 1972. The fine structure of "resting bodies" of *Bdellovibrio* sp. strain W developed in *Rhodospirillum rubrum. Can. J. Microbiol.* 18:87–92
25. Huang, J. C.-C., Starr, M. P. 1973. Possible enzymatic bases of bacteriolysis by bdellovibrios. *Archiv Mikrobiol.* 89:147–67
26. Huang, J. C.-C., Starr, M. P. 1973. Effects of calcium and magnesium ions and host viability on growth of bdellovibrios. *Antonie van Leeuwenhoek, J. Microbiol. Serol.* 39: 151–67
27. Jackson, H. R. 1967. Differential staining of bacteriophage plaques for photography. *Can. J. Microbiol.* 13:117–19
28. Klein, D. A., Casida, L. E., Jr. 1967. Occurrence and enumeration of *Bdellovibrio bacteriovorus* in soil capable of parasitizing *E. coli* and indigenous soil bacteria. *Can. J. Microbiol.* 13:1235–41
29. Lépine, P., Guélin, A., Sisman, J., Lamblin, D. 1967. Étude au microscope électronique de la lyse de *Salmonella* par *Bdellovibrio bacteriovorus. C. R. Acad. Sci. (Paris)*, 264:2957–60
30. Matin, A., Rittenberg, S. C. 1972. Kinetics of deoxyribonucleic acid destruction and synthesis during growth of *Bdellovibrio bacteriovorus* strain 109 D on *Pseudomonas putida* and *Escherichia coli. J. Bacteriol.* 111:664–73
31. Maugeri, T. L. 1969. *Bdellovibrio bacteriovorus. Boll. ist. sieroterap. Milan,* 48:84–95
32. Mishustin, E. N., Nikitina, E. S. 1970. Infection and lysis of Gram-negative bacteria by parasitic bacteria *Bdellovibrio bacteriovorus. Bull. Acad. Sci. USSR, Biol. Ser.* 3:423–26
33. Mitchell, R., Morris, J. C. 1969. The fate of intestinal bacteria in the sea. 811–17. In *Advances in Water Pollution Research.* S. H. Jenkins, ed. Pergamon Press, Oxford & New York
34. Mitchell, R., Yankofsky, S., Jannasch, H. W. 1967. Lysis of *Escherichia coli* by marine microorganisms. *Nature (London)*, 215:891–93
35. Moore, R. L., McCarthy, B. J. 1967. Comparative study of ribosomal ribonucleic acid cistrons in enterobacteria and myxobacteria. *J. Bacteriol.* 94:1066–74
36. Ottow, J. C. G. 1969. *Bdellovibrio bacteriovorus*, een unieke predator onder de bodem-bacteriën. *Landbouwk. Tidschr.* 81:275–80
37. Parker, C. A., Grove, P. L. 1970.

Bdellovibrio bacteriovorus parasitizing *Rhizobium* in Western Australia. *J. Appl. Bacteriol.* 33:253–55

38. Reiner, A. M., Shilo, M. 1969. Host-independent growth of *Bdellovibrio bacteriovorus* in microbial extracts. *J. Gen. Microbiol.* 59:401–10

39. Rittenberg, S. C. 1972. Nonidentity of *Bdellovibrio bacteriovorus* strains 109 D and 109 J. *J. Bacteriol.* 109:432–33

40. Rittenberg, S. C., Shilo, M. 1970. Early host damage in the infection cycle of *Bdellovibrio bacteriovorus*. *J. Bacteriol.* 102:149–60

41. Scherff, R. H., De Vay, J. E., Carroll, T. W. 1966. Ultrastructure of host-parasite relationships involving reproduction of *Bdellovibrio bacteriovorus* in host bacteria. *Phytopathology* 56:627–32

42. Sébald, M., Véron, M. 1963. Teneur en base de l'ADN et classification des vibrions. *Ann. Inst. Pasteur,* 105:897–910

43. Seidler, R. J., Mandel, M. 1971. Quantitative aspects of DNA renaturation: Base composition, state of chromosome replication, and polynucleotide homologies. *J. Bacteriol.* 106:608–14

44. Seidler, R. J., Mandel, M., Baptist, J. N. 1972. Molecular heterogeneity of the bdellovibrios: Evidence of two new species. *J. Bacteriol.* 109:209–17

45. Seidler, R. J., Starr, M. P. 1968. Structure of the flagellum of *Bdellovibrio bacteriovorus*. *J. Bacteriol.* 95:1952–55

46. Seidler, R. J., Starr, M. P. 1969. Factors affecting the intracellular parasitic growth of *Bdellovibrio bacteriovorus* developing within *Escherichia coli*. *J. Bacteriol.* 97:912–23

47. Seidler, R. J., Starr, M. P. 1969. Isolation and characterization of host-independent bdellovibrios. *J. Bacteriol.* 100:769–85

48. Seidler, R. J., Starr, M. P., Mandel, M. 1969. Deoxyribonucleic acid characterization of bdellovibrios. *J. Bacteriol.* 100:786–90

49. Shilo, M. 1966. Predatory bacteria. *Sci. J.* 2:33–37

50. Shilo, M. 1969. Morphological and physiological aspects of the interaction of *Bdellovibrio* with host bacteria. *Curr. Top. Microbiol. Immu-nol.* 50:174–204

51. Shilo, M., Bruff, B. 1965. Lysis of Gram-negative bacteria by host-independent ectoparasitic *Bdellovibrio bacteriovorus* isolates. *J. Gen. Microbiol.* 40:317–28

52. Shockman, G. D. 1965. Symposium on the fine structure and replication of bacteria and their parts. IV. Unbalanced cell-wall synthesis: Autolysis and cell-wall thickening. *Bacteriol. Rev.* 29:345–58

53. Simpson, F. J., Robinson, J. 1968. Some energy-producing systems in *Bdellovibrio bacteriovorus* strain 6-5-S. *Can. J. Biochem.* 46:865–73

54. Sourek, J. 1968. *Bdellovibrio bacteriovorus*—the first case of parasitism of bacteria on other bacteria. *Cesk. Epidemiol. Mikrobiol. Immunol.* 17:364–71

55. Stanier, R. Y. 1970. Some aspects of the biology of cells and their possible evolutionary significance. In: Organization and control in prokaryotic and eukaryotic cells. Ed. H. P. Charles, B. C. Knight. *Symp. Soc. Gen. Microbiol.* 20:1–38

56. Starr, M. P., Baigent, N. L. 1966. Parasitic interaction of *Bdellovibrio bacteriovorus* with other bacteria. *J. Bacteriol.* 91:2006–17

57. Starr, M. P., Huang, J. C.-C. 1972. Physiology of the bdellovibrios. *Advan. Microbial Physiol.* 8:215–61

58. Starr, M. P., Seidler, R. J. 1971. The bdellovibrios. *Ann. Rev. Microbiol.* 25:649–78

59. Starr, M. P., Skerman, V. B. D. 1965. Bacterial diversity: The natural history of selected morphologically unusual bacteria. *Ann. Rev. Microbiol.* 19:407–54

60. Steiner, S., Conti, S. F. 1970. Phospholipids of *Bdellovibrio bacteriovorus*. *Bacteriol. Proc.* 1970:74

61. Steiner, S., Lester, R. L., Conti, S. F. 1971. Phosphonolipids, unique lipids in a facultatively parasitic strain of *Bdellovibrio bacteriovorus*. *Bacteriol. Proc.* 1971:143

62. Stolp, H. 1964. *Bdellovibrio bacteriovorus*, a new kind of parasitic microorganism that attacks and lyses bacteria. In: *Lectures on Theoretical and Applied Aspects of Modern Microbiology*, Univ. Maryland, College Park, Maryland

63. Stolp, H. 1965. Isolierung von *Bdellovibrio bacteriovorus*. In: An-

reicherungskultur und Mutantenauslese. *Zentralbl. Bakteriol. Parasitenk., Abt. I, Suppl. 1*, 52–56

64. Stolp, H. 1965. Lysis von Bakterien durch räuberische Ektoparasiten (*Bdellovibrio bacteriovorus*). *Zentralbl. Bakteriol. Abt. 1, Referate* 204:174–76

65. Stolp, H. 1967. *Bdellovibrio bacteriovorus* (Pseudomonadaceae). Parasitischer Befall und Lysis von *Spirillum serpens*. Film E-1314. Göttingen, Institut für den wissenschaftlichen Film

66. Stolp, H. 1967. Lysis von Bakterien durch den Parasiten *Bdellovibrio bacteriovorus*. Film C 972. Göttingen, Institut für den wissenschaftlichen Film. Begleittext in *Publ. Wiss. Film, Bd. AII*, 695–706

67. Stolp, H. 1967. Neues über den Bakterienparasiten *Bdellovibrio bacteriovorus. Umschau*, 2:58–9

68. Stolp, H. 1968. *Bdellovibrio bacteriovorus*, ein räuberischer Bakterienparasit. *Naturwissenschaften*, 55:57–63

69. Stolp, H. 1969. *Bdellovibrio bacteriovorus*—ein Raubbakterium. *Anz. Schädlingskunde Pflanzenschutz*, 42:13–14

70. Stolp, H. 1969. *Bdellovibrio bacteriovorus*—a predatory and bacteriolytic parasite. *Image, Medical Photo Reports Roche*, 30:2–5

71. Stolp, H. 1969. *Bdellovibrio bacteriovorus:* Ein intrazellulärer Bakterienparasit. *Ärztl. Praxis*, 21:2273, 2293–97

72. Stolp, H., Petzold, H. 1962. Untersuchungen über einen obligat parasitischen Mikroorganismus mit lytischer Aktivität für *Pseudomonas*-Bakterien. *Phytopathol. Z.* 45:364–90

73. Stolp, H., Starr, M. P. 1963. *Bdellovibrio bacteriovorus* gen. et sp. n., a predatory, ectoparasitic, and bacteriolytic microorganism. *Antonie van Leeuwenhoek J. Microbiol. Serol.* 29:217–48

74. Stolp, H., Starr, M. P. 1965. Bacteriolysis. *Ann. Rev. Microbiol.* 19:79–104

75. Strominger, J. L., Ghuysen, J.-M. 1967. Mechanisms of enzymatic bacteriolysis. *Science* 156:213–21

76. Sullivan, C. W., Casida, L. E., Jr.

1968. Parasitism of *Azotobacter* and *Rhizobium* species by *Bdellovibrio bacteriovorus. Antonie van Leeuwenhoek J. Microbiol. Serol.* 34:188–96

77. Tinelli, R., Shilo, M., Laurent, M., Ghuysen, J.-M. 1970. De la présence d'un glycopeptide dans la paroi de *Bdellovibrio bacteriovorus. C. R. Soc. Biol. (Paris)*, 270:2600–02

78. Varon, M. 1968. *Interaction of Bdellovibrio with its host*. Ph.D. Thesis: Hebrew University, Jerusalem, Israel

79. Varon, M., Levisohn, R. 1972. Three-membered parasitic system: a bacteriophage, *Bdellovibrio bacteriovorus*, and *Escherichia coli. J. Virol.* 9:519–25

80. Varon, M., Shilo, M. 1966. Methods for separation of *Bdellovibrio. Israel J. Med. Sci.* 2:654

81. Varon, M., Shilo, M. 1968. Interaction of *Bdellovibrio bacteriovorus* and host bacteria. I. Kinetic studies of attachment and invasion of *Escherichia coli* B by *Bdellovibrio bacteriovorus. J. Bacteriol.* 95:744–53

82. Varon, M., Shilo, M. 1969. Attachment of *Bdellovibrio bacteriovorus* to cell wall mutants of *Salmonella* spp. and *Escherichia coli. J. Bacteriol.* 97:977–79

83. Varon, M., Shilo, M. 1969. Interaction of *Bdellovibrio bacteriovorus* and host bacteria. II. Intracellular growth and development of *Bdellovibrio bacteriovorus* in liquid cultures. *J. Bacteriol.* 99:136–41

84. Varon, M., Drucker, I., Shilo, M. 1969. Early effects of *Bdellovibrio* infection on the synthesis of protein and RNA of host bacteria. *Biochem. Biophys. Res. Commun.* 37:518–25

85. Véron, M. 1965. La position taxonomique des vibrios et de certaines bactéries comparables. *C. R. Acad. Sci. Paris*, 261:5243–46

86. Véron, M. 1966. Taxonomie numérique des vibrions et de certaines bactéries comparables. *Ann. Inst. Pasteur*, 111:671–709

87. Wood, O. L., Hirsch, P. 1966. Lysis and colony formation by *Bdellovibrio* isolates parasitical on budding bacteria and *Pseudomonas fluorescens. Bacteriol. Proc.* 1966:25

PYTHIUMS AS PLANT PATHOGENS ❖ 3566

F. F. Hendrix, Jr. and W. A. Campbell
Department of Plant Pathology and Plant Genetics and School of Forest Resources, University of Georgia, Athens, Georgia

INTRODUCTION

The genus *Pythium* includes a number of readily recognized species with wide distributions and host ranges. The taxonomic position of the genus and its relationship to other Phycomycetes were well established during the latter part of the 19th century. In the early 1900s pathologists found *Pythium* spp. consistently associated with root diseases and it soon became apparent that these fungi were important plant pathogens. Certainly, while not all isolates of *Pythium* species are capable of causing diseases of plants, many are soil-borne pathogens that cause serious economic loss on a wide variety of hosts, while others are more limited in host and geographic range or affect plants only under special environmental conditions. New species are being discovered as pathologists investigate soil organisms associated with plant growth problems.

Rands & Dopp (123) in their classic investigation of sugarcane root diseases established the symptoms and determined the conditions needed for these fungi to become destructive. This work, and others of a similar nature, are part of the extensive literature on diseases caused by species of *Pythium*. This review will concentrate on examples of the readily available literature of the past 50 years and will emphasize the pathology of the genus, with only brief treatment of other aspects such as taxonomy and control. It will be confined to *Pythium* spp. as they affect economic plants and will not deal with those affecting algae, other marine plants, fungi, and unusual hosts.

MYCOLOGICAL HISTORY AND TAXONOMY   The genus *Pythium* was established by Pringsheim in 1858 and placed in the Family *Saprolegniaceae*. *P. monospermum* Pringsh. is the type species. By the end of the century taxonomic details of the genus had been clarified and many new species had been described. The relationship with other Oomycetes was established and the genus was included in a new family, *Pythiaceae*, by Schroter in 1897. The genus, its botanical position, and relationship to other fungi have remained unchanged since that time.

Species of *Pythium*, and the closely related genus *Phythophthora*, were

77

originally of interest solely to taxonomic mycologists. Eventually these fungi were found to cause root disease in many plants and their identification became important to pathologists. Sideris (131, 132) published extensively on his taxonomic studies of *Pythium* spp. isolated from pineapple and other plants in Hawaii. His proposed classification system generally was not accepted by other mycologists because his species separations were vague and indefinite, and later workers found that he had described the same species under different names.

Matthews (98) in 1931 did much to clarify the taxonomic position of known *Pythium* spp., named new ones, and devised a workable key. During this same period, Drechsler (32, 33) described a number of new species, illustrating many with superb examples of botanical art.

The proliferation of new species soon made Matthews' treatment obsolete and Middleton (104) compiled an extensive monograph of the genus *Pythium*, with complete mycological treatment, host records, and illustrations of the more important species, which was published in 1945. His key to the species reflects the important morphological characters and offers a good picture of relationships based on broad groupings. It is not good for distinguishing species whose sole claim to distinction rests on some minor variation or obscure cultural quirk. He was restricted to the study of single isolates of some species, or was forced to depend upon his interpretation of published descriptions. Hence, it was impossible for him to check the range of variation within or between species in many instances. With the exception of these defects, and excluding species described since 1940, this monograph remains the chief taxonomic reference on the genus.

Frezzi (41) in 1956 working in Argentina produced a good taxonomic treatment for 17 species. This work is well-illustrated and in general follows Middleton's concept of the genus. Waterhouse (160, 161) compiled the original descriptions and illustrations of all described species and published a key to the genus in 1967. Her key, like all keys, works well with species with clear-cut morphological characters. It is difficult to use when dealing with abnormal forms, or those without a full complement of taxonomic characters. Her review of published species shows that many are inadequately described and totally useless in mycological treatments. They only clutter the literature with names that cannot be applied to any recognizable fungi.

When lists of fungi isolated from nurseries, agronomic crops and forests are surveyed, it is exceptional to find *Pythium* spp. included. Yet, more recent studies have shown that they are common in these habitats. *Pythium* spp. are generally difficult to isolate from soil or plant material by the usual nutrient agar techniques because these media favor the more competitive, ubiquitous saprophytes. While these fungi can be readily isolated from plant material using water agar or corn meal agar, they were usually isolated only infrequently prior to the development of selective media.

Practically all host references to *Pythium* spp. compiled before 1960 are based on isolations from diseased host tissue and hence do not indicate actual distribution as to soil type or locality. With the development of specialized techniques for isolating Phycomycetes from soil, new data were obtained on species distribution and relative abundance. The isolation of numerous representatives of different species permitted studies on cultural variations and relationships between species. After examining over 10,000 *Pythium* isolates from 2,100 soil samples from various parts of the United States and Canada, Hendrix & Campbell (61) concluded that *P. debaryanum* Hesse *sensu* Middleton and *P. irregulare* Buis., the most abundant species in the literature, really form a species complex with characters distinct only at the extremes but which merge at a median point. Other species likewise proved better subjects for species complexes than futile exercises in trying to determine which of several names to use. Biesbrock & Hendrix (16) in a taxonomic study of *P. irregulare* and related species demonstrated how morphological characters of the oogonia, antheridia, and sporangia varied with differences in temperature, light, and nutrient media. Since pathologists encounter these fungi only in culture, when attempting identification, laboratory environment requires careful standardization for comparative studies.

Further research should suggest a limited number of species complexes, or species groupings probably including *P. dissotocum* Drechs.—*P. perniciosum* Serbinow, *P. oligandrum* Drechs.—*P. acanthicum* Drechs., *P. helicoides* Drechs.—*P. oedochilum* Drechs., *P. graminicola* Subra.—*P. arrhenomanes* Drechs. Each complex grouping should suffice for the particular use of taxonomic names. Most differences between species, as well as individuals, are physiologic in origin as they relate to parasitism. Hence, any conclusions drawn from pathogenicity studies are restricted to the isolates used in the studies and may or may not indicate the parasitic ability of the species or the genus as a whole.

HETEROTHALLISM IN PYTHIUM   Middleton considered the species of *Pythium* as universally homothallic since no evidence of heterothallism had appeared in the literature up to the early 1940s, and since his own cultural studies also suggested that these fungi were entirely homothallic. Taxonomic separation of species is based largely on characters of the sporangial and oogonial forms. A number of described species lacked one or the other of these structures, or they developed too infrequently in culture to be useful in taxonomic treatments. In 1967, Campbell & Hendrix (24) described a new species, *P. sylvaticum* from southern United States which was heterothallic. Further studies revealed that this fungus was widely distributed, sometimes very abundantly in forest soils (25). The nature of the heterothallic reaction of the two mating forms isolated from soil has been demonstrated (118, 122). Extensive search for other heterothallic forms uncovered numerous isolates that produced only sporangia or chlamydospores in culture. While many of these retained their original condition and would not mate with other similarly appearing isolates, two other mating populations were recognized. One was described as *P. heterothallicum* by Hendrix & Campbell (59). The other

proved to be *P. intermedium* d By., a species known for years as a nonoo-
spore producing example of the genus (156). Later studies with species that
rarely or reluctantly formed oospores in culture demonstrated the heterothallic condition of *P. catenulatum* Matthews (60) and *P. splendens* Braun
(157).

Failure to recognize heterothallism in *Pythium* spp. by earlier workers
probably was related to the cultural media in general use. The mating reaction rarely occurs on cornmeal, potato-dextrose, or similar agars. Hemp seed
extract agar or similar agars containing sterols are needed to stimulate the
mating reaction.

Classification of Pythium-Caused Diseases

Most *Pythium* spp. infect mainly juvenile or succulent tissues. This restricts
their parasitism to seedlings or the feeder roots or root tips of older plants,
and to watery fruits or stem tissues. They do not spread widely throughout
host cells and are quickly followed by more aggressive or faster growing
fungi. With certain hosts, such as grass, tomato transplants, peanuts, and
chrysanthemums, these fungi attack stems and foliage of nonseedling plants.
They also cause fruit rots of crops such as beans, squash, and watermelon.

Juvenile or succulent host tissue is very susceptible to infection and *Pythium* spp. commonly infect seed and the radicals causing seed rot and pre-
emergence damping-off. The fungi also infect newly emerged seedlings at
ground level, causing them to collapse or topple over, a common symptom of
post-emergence damping-off. This disease is most noticeable in nursery beds,
greenhouse flats, and row crops because symptoms develop suddenly, killing
large numbers of seedlings in local areas (15, 30, 37, 45, 54, 58, 91, 92,
124). At a later stage, when cells of stems and main roots have developed
secondary wall thickenings, infection is restricted to feeder roots, causing
seedlings to become stunted and chlorotic. This early root rot is an important
cause of poor growth and yield in many agricultural crops, since plants frequently fail to recover even if conditions become unfavorable for further disease development. *Pythium* spp. are widely distributed in undisturbed soils
and sites (62). In forest and nonagricultural soils, the death of seedlings by
damping-off or early root rot normally goes unnoticed. These pathogens,
through the selective elimination of susceptible plant species, may act as potent determinants of forest and plant vegetational types. They may be as important as other soil factors in the ultimate natural selection of plants that
occupy a given ecological situation.

In recent years a great deal of attention has been given to decline diseases
of perennial plants. A common symptom in peaches, turf, citrus, apples,
pines, and other perennials has been the gradual to sudden deterioration of
established plants and poor survival of replants. Declining plants lack a
healthy root system (67, 127). The classic example of this is found with
shortleaf and loblolly pines affected by littleleaf disease (26, 27, 69, 94).
This disease is caused mainly by *Phytophthora cinnamomi* Rands which in-

fects and destroys only nonmycorrhizal feeder roots and root tips. *Pythium* spp. are also involved in the destruction of the fine roots and root tips of trees (94). This gradually curtails the extent of the root system, leading to inability to absorb sufficient nitrogen from the soil (23). Campbell & Copeland clarified the sequence of events causing decline symptoms and defined a type of root disease now known as fine or feeder root necrosis. In most cases, only the root tips or newly formed roots are killed, mainly by species of *Pythium* and *Phytophthora* (23). Peach decline, which kills thousands of trees each year, has been shown to be a feeder root problem caused by a combination of factors, including *Pythium* spp. (67). Citrus decline in many instances is also associated with high soil populations of pathogenic Phycomycetes (134). The same associations have been observed in apple (111) and other perennial crops in various parts of the world (8, 18, 25, 28, 36, 57, 61–63, 65, 81, 86, 110, 114, 121, 127, 149, 153). Susceptible feeder roots are almost constantly present under perennial plants, and are infected when environmental conditions are favorable. This leads to a build-up of propagules of pathogenic *Pythium* spp. (69). These remain in the soil after the death or removal of diseased plants. Replants in the same area often fail to grow satisfactorily, or are quickly killed by concentrated infections of the root tips and feeder roots (67, 69). The only solution, other than soil fumigation, is to plant on new land or to rotate with plants not susceptible to the root pathogens (127).

While *Pythium* spp. are generally considered to be root pathogens, they can infect stems and foliage of some plants. In greenhouses and nurseries, where plants are crowded, well-fertilized, and watered, the closely spaced, succulent leaves and stems are subject to infections by *Pythium* (1, 19, 31, 46, 95, 109). Disease development depends upon environmental conditions, host susceptibility, and the presence of virulent *Pythium* spp. Succulent fruits such as squash, watermelons, peppers, oranges, and cucumbers are also attacked by *Pythium* spp. *Pythium* also affects fruits and vegetables in storage and transit, and bulbs and tubers of horticultural plants before and after harvesting (119, 149, 159).

Etiology

SURVIVAL *Pythium* spp. survive in soil by saprophytic growth and by resistant resting structures. They are not vigorous competitors, and their saprophytic activities are greatly restricted (14). Generally, they grow saprophytically only under circumstances where other organisms either are not present, or have greatly reduced activity due to environmental conditions. Barton's definitive results with saprophytic growth of *P. mamillatum* Meurs probably apply equally well to many other *Pythium* spp. *P. mamillatum* is restricted to a pioneer position in an ecological succession, and does not compete well for food bases already colonized by a well-established flora (13, 14). Soil moisture is important in saprophytic growth of *Pythium* spp. *P. ultimum* Trow saprophytically colonized 8–12% of dead, excised peach roots in soil at 30–90% moisture holding capacity (MHC). *P. irregulare,* on the other hand, colonized only 5% of the root pieces at 30% MHC, but 45 % at 90% MHC. *P. vexans* d By. colonized from 55–72% of the roots at the same MHC (107). The conclusion was that high soil moisture favored saprophytic activity of these fungi. Griffin (50) found that *P. ultimum* grew well under conditions where the pore space was filled with either water or air. He concluded that toleration of high soil moisture and conditions of poor gas exchange were an ecological advantage of this fungus. Gardner & Hendrix, studying the effect of soil atmosphere on saprophytic growth, found that elevated levels of CO_2 in soil favored saprophytic activity of *P. irregulare* and *P. vexans* provided that the O_2 level was not reduced. Reduction of the O_2 level could be offset by increased CO_2 levels to the extent of about 50% of the saprophytic activity (42). When field soil was stored at various temperatures, populations of species with high optimum temperatures increased at high temperatures, while species with lower optimum temperatures decreased. This may occur without a change in total population of *Pythium* spp. in the soil (48). The role of saprophytic growth as a survival mechanism for *Pythium* spp. is important only because the fungus is able to germinate and colonize virgin substrata rapidly and because it grows under conditions unfavorable for growth of other organisms. One must conclude from the available literature that *Pythium* spp. probably are unable to compete on any other basis.

Survival by resistant resting structures is considerably more important than saprophytic persistence. As is the case with many other soil fungi, mycelium of *Pythium* spp. lyses as food sources are depleted, or become colonized by competing organisms (14, 137). The chief mechanism of survival is by means of zoospores and sporangia for short and intermediate periods, and oospores for longer periods.

P. ultimum survived at −18°C for 24 months, and in air-dried soil for 12 years (72, 112). Sterols and calcium stimulate oospore formation, presumably leading to greater survival potential (56, 68, 165). Zoospores of *P. aphanidermatum* (Edson) Fitz. survived in field soil for 7 days, even though they failed to grow saprophytically (96). Sporangia of *P. ultimum* were found to survive in soil for 11 months (137). Garren found that *P. myriotylum* Drechs. survived longer in soils in which it was not indigenous than in soil in which it was indigenous, and postulated that this was due to lack of antagonists (44).

GERMINATION   In order to attack plants, the survival structures must germinate, which they do by germ tubes or zoospores. Some species have the capability of germinating both ways, while others characteristically produce only germ tubes (34). Dormant spores frequently must be activated before they will germinate. Hoppe found that spores in air-dried soil for 6 years would germinate only after the soil had been kept moist for 15 days. After 12 years, a 90-day activation period was required (72). Stimulation of germination of resting structures by seed or root exudates has been shown for *P. aphanidermatum* (29, 88), *P. irregulare* (6), *P. afertile* Kanouse & Humph. (5), *P.*

vexans (17), and *P. mamillatum* (12). Barton found that root exudates stimulated germination of oospores of *P. mamillatum* (12). Stimulation of germination occurred as far as 10 mm from the surface of seed at 28% soil moisture; this distance decreased as the moisture level decreased (13). Stanghellini & Hancock found that sporangia of *P. ultimum* stored in moist or dry soil for 11 months germinated in 1.5 hr and reached a maximum 3 hr after nutrients were added to soil (138). Mycostasis prevents germination of sporangia of *P. ultimum,* but this can be overcome by dilution of soil with distilled water, sterilization of soil, or by the addition of nutrients to the soil (4). Once germination has occurred, germ tubes grow vigorously as long as there is available food.

Since *Pythium* spp. are unable to compete with other organisms for food bases, and since their resting spores germinate and grow rapidly, they are generally in a mycelial state when they are no longer able to compete. At this time, there are 3 possible fates for the germlings: (*a*) lysis, (*b*) infection of a susceptible host, or (*c*) formation of new resting structures. Lysis of mycelium following formation of resting structures is common in the absence of susceptible hosts or other virgin food bases. *P. ultimum* grew more abundantly in the rhizosphere of susceptible plants than in that of nonsusceptible plants (7). A great deal of research has elucidated specific substances present in root exudates which serve as nutrient sources for *Pythium* germlings (5, 6, 12, 29, 87, 88, 126). These generally are sugars or amino acids.

INFECTION   Factors influencing infection include inoculum density, soil moisture, soil temperature, pH, cation composition, light intensity, and presence and numbers of other microorganisms. Soil temperature and moisture are the most important factors. Which factor is more important in given instances is sometimes dependent on the *Pythium* species involved. Generally, high soil moisture is reputedly necessary for disease development. Dying of plants in low areas of fields, which are frequently water-logged, is often attributed to *Pythium* spp. Since high soil moisture is not a definitive term, we can hardly take issue with it. However, in experiments to evaluate the role of soil moisture in the development of *Pythium* diseases, maintaining soil for extended periods at saturation capacity or above has not favored disease development. Biesbock & Hendrix found that *P. vexans,* a species readily producing zoospores, responded more to soil moisture than to soil temperature, particularly to a wet-dry cycle involving saturated soil. *P. irregulare,* a species not readily producing zoospores, responded more to soil temperature than to soil moisture. Using constant soil moistures, they found that neither species caused disease at 90% saturation capacity (SC). *P. vexans* caused disease at 70% SC and *P. irregulare* at 50% SC (17, 18).

Temperature has as much effect on disease development as soil moisture. In some cases, the optimum temperature for growth in culture and disease development is similar, while there is no relationship with other species. For instance, a *Pythium* spp. with an optimum temperature of 30° in culture caused about 5 times more disease on rice at 20° than at 30° (162). With *P. myriotylum* soil moisture had less effect than soil temperature (45). Klisiewicz found that certain *Pythium* spp. caused more damping-off and root rot at low temperatures, while others were more damaging at high temperatures (82). Generally, *P. irregulare, P. spinosum* Sawada, *P. ultimum,* and related species are more damaging at lower temperatures, while *P. myriotylum, P. aphanidermatum, P. arrhenomanes, P. polytylum* Drechs., *P. carolinianum* Matthews, *P. volutum* Vanterpool & Truscott, and related species are more damaging at higher temperatures. As is usually the case when one makes generalizations, there are exceptions to this grouping, usually associated with particular hosts (92).

Pathogenicity of *Pythium* spp. on nonseedling plants is sometimes difficult to demonstrate. As is the case with all pathogens, *Pythium* spp. are not pathogenic under all environmental conditions, and possibly are more demanding than many other pathogens, particularly so far as temperature and moisture are concerned. In cases where temperature and moisture have been ignored, researchers find themselves unable to reproduce results in existing literature (108). Another factor making evaluation of greenhouse inoculations difficult is the drastic reduction of root growth in checks when they become pot bound. Inoculated plant roots continue to grow slowly, since some roots are not infected and killed. Eventually, root weight of check and inoculated plants will be equal.

Other factors have much less effect, usually indirect. Hydrogen ion concentration has been shown to affect diseases caused by *Pythium* spp. (49, 125). Generally, the main effect is reduction of vigor of the host at unfavorable hydrogen ion concentrations, keeping it in a susceptible state for a longer period of time. Low light intensity also affects disease, again probably an effect mainly on the host (82).

Nitrate nitrogen and potassium reduced damping-off in moist soil, whereas ammonium nitrogen and phosphate did not (164). Since this effect did not occur in dried soil, it was associated with microbial activity. Nitrogen deficiency in cucumber plants resulted in more disease, as compared with plants having adequate nitrogen (102). The amount of damping-off caused by *P. ultimum* decreased on both tomato and cucumber as the strength of Knop's solution increased above normal (15). Cottony blight of bent grass responds to calcium fertilization. Deficient or excessive levels of calcium increased susceptibility of bent grass to *P. ultimum* (110).

Many histological studies of penetration have been made. Spencer & Cooper found that infection of cotton roots by germ tubes from zoospores occurred in about 2 hr, while infection by mycelial fragments took about 12 hr (135). Penetration of peach roots by *P. ultimum* occurs 5–8 hours after inoculation, mainly at junctions of epidermal cells, by means of an infection peg formed from an appresorium. Colonization of the cortex occurs within 24 hr, and stelar invasion by 36 hr. Cell collapse and separation follow. Colonization of root cells is limited to those cells lacking secondary thickening of the cell walls (106). Lignification of cell walls limits advance of *P. splendens* in cucumber seedlings (102).

Since most of the research on diseases caused by *Pythium* spp. has been on seedling diseases, little information is available concerning interaction of these fungi with other parasitic organisms. Sugarcane was damaged more by *P. graminicola* and sugarcane mosaic virus than by either pathogen alone. The effect was either additive or synergistic, depending on the cultivar of sugarcane tested (85). Inoculation of cucumber with *P. ultimum* and cucumber mosaic virus (CMV) resulted in sudden wilt, a symptom much less severe with the fungus alone, and not caused by CMV alone (117).

There have been many studies of *Pythium*-nematode interactions. Concurrent inoculation of corn roots with *Tylenchus agricola* or *T. claytoni* and *P. ultimum* resulted in less damage to the roots than when *P. ultimum* alone was used. The population of both nematodes decreased during the study (79). Neither *Criconemoides quadricornus* nor *P. irregulare* caused disease to pecan roots at 27°C. When combined, they reduce root weight by about 50% (73). *P. aphanidermatum* causes root rot of chrysanthemum, and the amount of disease increases when *Belonolaimus longicaudatus* or *Meloidogyne incognita* is present. The fungus suppresses egg production by *M. incognita* but not reproduction of *B. longicaudatus* (77).

Pythium spp. frequently are components of complexes with other fungi. A complex consisting of *P. myriotylum* and other soil fungi, particularly *Fusarium spp.*, causes pod rot of peanuts, while individual organisms alone cause little or no rot (39). *P. myriotylum,* in other circumstances, has been shown to cause extensive pod rot (43). *P. tardecresens* Vanterpool and *P. graminicola* form a complex on sugarcane which results in less or the same amount of disease as that caused by *P. graminicola* alone (84) Kerr (78) found that *P. ultimum* and *F. oxysporum* caused more severe damage to peas than either fungus alone.

Isolation of Pythium Spp.

Abundance and significance of *Pythium* in soil were recently reviewed by Schmitthenner (129).

When collecting soil samples, researchers need to consider that *Pythium* spp. generally exist at higher populations in the zone where feeder root development is greatest, and decline with depth of sampling below this zone. Knaphus & Buckholtz found little *P. graminicola* in the top 7.5 cm or below 67.5 cm, as determined by isolation from roots, and by disease development in seedlings planted in soil collected from various depths (83). Hendrix & Powell found similar distribution. Using a quantitative technique, they found that population of *Pythium* spp. decreased with depth below the root zone, but that the fungus was still present at 90 cm (unpublished data).

The number of propagules reported in soil depends as much or more on the techniques used for estimation as on the numbers actually there. These figures are valuable for prediction of degree of disease outbreaks only after a researcher has had enough experience with his technique to correlate numbers of propagules with disease development.

Qualitative isolation of *Pythium* spp. from soil has been accomplished with various baiting techniques for many years (11, 22, 51, 80, 83). Basically, these involve placing living plant material in soil, and vice versa. The characteristic resistance of the plant to saprophytic organisms, and susceptibility of living plants to pathogenic organisms are used to separate *Pythium* spp. from the more abundant fungi, bacteria and actinomycetes. The infected plant material is plated on agar media to complete the isolation process. Use of these techniques is the simplest way to isolate *Pythium* spp. from soil if only qualitative data are needed. No one baiting technique is suitable for isolation of all *Pythium* spp. (61). For instance, the Campbell apple trapping technique is excellent for isolating *P. ultimum, P. vexans, P. splendens,* and the *P. helicoides–P. oedochilum* complex, but not for the *P. irregulare–P. debaryanum* complex, the heterothallic *Pythium* spp., *P. spinosum,* etc. The citrus leaf (51) and hemp seed baiting techniques work well for those *Pythium* spp. which readily form zoospores. Other fruits, including cucumbers, oranges, and lemons, as well as roots of avocado and lupine seedlings, have worked for particular species. A bait should be selected for the *Pythium* spp. of interest to the individual researcher.

With the advent of widespread use of antibiotics in the late 1950s selective media for the isolation of *Pythium* spp., first from plant material, then from soil were reported (10, 35, 78, 128, 133, 136, 142, 151). These media involved the use of polyene antibiotics, combined with various chemicals, usually pentachloronitrobenzene, streptomycin, and rose bengal. Other factors that influenced recovery of *Pythium* spp. included agar concentration, pH, incubation temperature and time, and temperature of the medium prior to seeding (38, 64). Gallic acid was found to be a suitable inhibitor for fungi other than *Pythiaceace* (38). The more commonly used media result in counts of less than 250 propagules per gram. Newer techniques have resulted in counts of 500 and above (150). Numbers of propagules per gram cannot be compared unless the same techniques are used. A review of selective media for isolation of *Pythium* spp. is available (150).

Our experience is that use of media for the selective isolation of any organism is a difficult procedure for anyone except the inventor of a particular technique. Researchers experienced in the use of selective media seldom are able to use a new medium successfully for the first few trials. None of these media will isolate all species of *Pythium* equally. For instance, the Flowers & Hendrix gallic acid medium isolates the *P. acanthicum–P. oligandrum* group, *P. ultimum,* and *P. vexans,* while modified Kerr's medium is best for *P. afertile, P. aphanidermatum,* and the *P. dissotocum–P. perniciosum* group (61).

None of the media described are suitable for all *Pythium* spp., and none give a total count. These techniques are valuable in specific situations for making comparisons between treatments. To approach a total qualitative estimate of the *Pythium* flora, we use the following techniques on each sample:

modified Kerr's medium, gallic acid medium, Campbell's apple trapping technique, hemp seed baits, and the citrus leaf trapping technique. There is currently no way to approach a total quantitative estimate. The relationship between propagules per unit of soil and disease is largely undefined. The greatest value of these counts is to establish that a minimum inoculum level for disease development is present. If environmental conditions suitable for disease development occur, then outbreaks can be predicted. With perennial crops such as turf, orchard crops, etc., the presence of minimal inoculum levels assures that there will eventually be disease outbreaks. With annual crops, the necessary environmental conditions may not occur prior to maturity of the crop.

Pythium *Diseases of Specific Hosts*

During the past several decades, research on plant diseases causing economic loss in specific crops has stressed the role of *Pythium* spp. as root pathogens. Examples of some of these diseases are presented to emphasize the importance of individual *Pythium* spp. in relation to certain hosts.

Coniferous forest trees are widely grown throughout the world in nurseries for reforestation or afforestation purposes, and root diseases and damping-off are the main disease problems (55). *Pythium* spp. are among the more important pathogens and considerable research has been done on their identification, mode of operation, and control in the United States (25, 26, 37, 58, 70, 125, 141), Canada (4–6, 152–154), and Australia (155). *P. ultimum* and the *P. debaryanum–P. irregulare* complex, are world-wide in forest tree nurseries while other species occur less frequently. In general, *Pythium* spp. operate as part of a disease complex involving other root pathogens that attack tree species of all ages under nursery culture. *Pythium* spp. commonly found in nurseries infect and damage seedlings. Millions of seedlings, grown in contaminated soil, have been planted and have grown to economic size, suggesting that these diseases are only nursery problems. Recently, nursery-men have begun to doubt this, but definite data are not available.

During World War II, an attempt was made to grow guayule on a large scale as a source of rubber. The plants were first grown in forest tree nurseries and then transplanted to the field. Seedlings on former agricultural land in the Salinas Valley of California were affected by damping-off and a collar rot caused by *P. ultimum* (28). In older plants, infections were confined to the younger and more succulent feeder roots causing stunting and chlorosis. Severely diseased plants rarely attained a size considered minimal for field planting.

Pythium spp. are also abundant in horticultural nurseries and much unthrifty planting stock is sold to homeowners each year. When sanitation practices are poor or nonexistent in nurseries and greenhouses, pathogenic Phycomycetes become established in plant beds, and potted plants (9, 18, 19, 31, 76, 77, 95, 148, 149). A survey of soil fungi present in horticultural nurseries in Georgia disclosed the presence of over 15 species of *Pythium* and

Phytophthora. It was interesting to note the nearly constant association of some pathogenic Phycomycetes with specific horticultural plants. Apparently certain species become concentrated in the soil around the roots of susceptible hosts (57, 61).

With the widespread development of golf greens planted to various hybrid bermuda and bent grasses, serious disease problems have become a limiting factor in the maintenance of satisfactory turf. *P. aphanidermatum* has been recognized as the cause of a cottony blight of turf grasses such as winter rye, blue grass, bent grasses, and bermuda grass during periods of excessive moisture during winter and spring months throughout the south (63, 99, 110). Soil isolation techniques disclosed a varied *Pythium* flora (63). *P. catenulatum* and *P. torulosum* Coker & Patterson, both infrequently isolated species in other situations, proved to be common and destructive, along with the more generally distributed *P. ultimum* and the *P. debaryanum–P. irregulare* complex. Oddly enough, *P. aphanidermatum* was encountered infrequently in the soil assays. Until recently *P. catenulatum* had been associated mainly with root rot and poor growth of sugarcane, especially in Taiwan, and was not considered a widely distributed species (74). Research on turf grasses proved its general distribution in the southern United States and related its abundance to the distribution of suitable hosts (63).

Grain crops of various kinds have a distinctive *Pythium* flora. Root rots caused by *Pythium* spp. resulting in stunting and reduced yields are fairly common. Species of the *P. graminicola–P. arrhenomanes* complex, *P. ultimum*, and *P. aphanidermatum* are common pathogens. The latter is an important cause of poor stands of small grains in Florida and south Georgia. These same fungi are also generally associated with root diseases of corn and sugarcane (8, 20, 21, 40, 54, 75, 84, 85, 109, 130, 140, 146, 158).

P. ultimum is a very widely distributed species and is a major root pathogen affecting a large number of plant species, especially in the seedling stage. If it does not kill outright it leads to poor root development, stunting, and reduced yields. In Washington and Wisconsin, *P. ultimum* is the common cause of a root rot of beans and peas (1, 2, 90). In the South, it is one of the main causes of poor emergence and stunting in cotton (101, 124). *P. ultimum* also affects soybeans, lettuce, sweet potatoes, and other crop plants (82, 91, 119, 147). *P. ultimum* and *P. debaryanum* have been associated with root diseases of safflower, red clover, and alfalfa (30, 52, 145). *P. ultimum* was the only pathogen that could be recovered from the diseased roots of declining citrus trees in Winterhaven, Texas (134). When associated with fruit tree decline of such trees as citrus, peaches, pears, and the like, *P. ultimum* is always confined to infections of the fine feeder roots. An interesting disease of muckgrown carrots, characterized by brown rot and forking, proved to be caused by *P. ultimum* and *P. debaryanum* (103, 105).

P. sylvaticum and *P. intermedium*, two heterothallic species, along with *P. ultimum* have been reported as pathogenic to the primary roots of apple seedlings in the Netherlands (111). *P. sylvaticum* is also pathogenic to sweetgum seedlings under nursery culture in Mississippi (37). This species is widely and abundantly distributed in forest soils, but its effect on other forest plant species has not been investigated.

Strawberries are subject to a variety of root pathogens. Recent investigations in Illinois have demonstrated the association of *P. sylvaticum* with certain types of root disease and the widespread distribution of the fungus in strawberry fields in the southern part of the state. Representatives of the *P. dissoticum–P. perniciosum* complex were also encountered (114–116).

Pythium spp. were investigated in the 1920s as the cause of diseases of sugarcane, pineapples, and other plants in Hawaii. These same species continue to plague pineapples and a 1964 report lists 8 *Pythium* spp. common to pineapple soils (81). However, of these 8, only *P. arrhenomanes, P. graminicola, P. splendens, P. torulosum,* and *P. irregulare* were pathogenic in greenhouse tests.

Despite the large number of reported *Pythium* spp. only a relatively few are world-wide in distribution and are generally associated with root diseases of crop plants.

Some *Pythium* spp. show temperature adaptations. *P. myriotylum* has a high temperature optimum and is restricted in distribution and economic damage to warm climates where it causes wilt, root and pod rot of peanuts, and stem rot of snap beans (43, 45, 46, 120). It has also been reported infrequently on other hosts. *P. aphanidermatum* also has a relatively high optimum temperature and affects a variety of hosts in southern United States and other areas of the world with a warm climate (77, 99, 100). It is common in Iran on certain legumes and has been associated with damping-off on tomato seedlings in hydroponic culture in Arizona (139).

Many *Pythium* spp. have been reported as mycological curiosities from restricted or unusual habitats and hosts and are unimportant to economic agriculture. Certain ones such as *P. ultimum* and the *P. debaryanum–P. irregulare* complex are nonspecific as to host while others such as *P. catenulatum* affect mainly grasses or grass-related genera. Others are temperature-restricted such as *P. myriotylum.* In a sense, *P. splendens* falls in this category since it is most common under greenhouse and nursery conditions in the south or associated with fine root necrosis in the same general area (57). It has been reported on safflower in the southwest (82, 144). Undisturbed forest and agricultural soils harbor large *Pythium* populations, many of which proved to be pathogenic to agricultural crops planted on cleared land (62). Certain of these pathogens build up under agriculture while others decline in numbers. In general, monocultures tend to increase the propagules of certain species, hence the hazard of economic loss from continued cropping by a single plant species.

Control

Most *Pythium* spp., once established in soil by virtue of parasitism on plant roots or saprophytic existence on organic material, produce oospores and chlamydospores that persist for many years. Once established in the soil, these resting structures are virtually impossible to eliminate except with wide spectrum soil fumigants. Hence, direct control of *Pythium* diseases on field scale is difficult and expensive (66). However, plant response to control is frequently great enough to justify this type of fumigation (163).

On a small scale, *Pythium* spp. can be eliminated from soil for greenhouse use by heating. The common method is by steam treatment or pasteurization. The latter method is preferred because it does not kill all organisms or drastically alter soil structure. Excessive heat treatment may so change the biotic flora as to render the soil unfit for plant growth.

Fumigation with chloropicrin or methyl bromide, or combinations of the two, is now standard practice in many nursery and horticultural operations (66, 113). It is effective in eliminating pathogens before planting turf grasses and has also been used for spot treatment of the planting site for peaches and other fruit trees (66). An important advantage of fumigation is the control of weeds, many of which also harbor pathogens.

Fumigation with methyl bromide and chloropicrin does not kill all soil organisms. During the first few weeks following fumigation bacterial populations increase rapidly followed by other soil organisms (71). Ideally, a biotic equilibrium is reached in which saprophytic fungi predominate, often providing effective competition to plant pathogens. After fumigation, however, the soil flora is altered by temporary removal of competitors to *Pythium* spp. and other pathogens; these multiply rapidly if re-introduced, and offer greater threats to plants than before fumigation (47, 66, 89, 152). This is especially true where spot fumigation leaves an island in the soil with reduced competitor populations surrounded by soil with abundant pathogens. Pathogens move quickly into the fumigated area often building up populations greater than before fumigation (47, 66, 89, 152).

Dexon and other soil fungicides, applied at low rates, have been used successfully to retard the rate of reinfestation of fumigated soils by *Pythium* (66). Soil type, moisture content, and other factors influence the effectiveness of Dexon and other fungicides (113).

Soil amendments such as sawdust, bark, and other crop residues as well as green manuring have been suggested and tried as a control for *Pythium* root rots in nurseries and field crops (155). These are sometimes effective, mainly because they encourage soil flora antagonistic to *Pythium* spp. and other plant pathogens. Improvement in plant growth and survival may be more related to promoting good host response, thus stimulating root formation that may counterbalance roots killed by pathogens. In nurseries, soil amendments may pose nutrient problems that often prove more difficult to correct than problems associated with *Pythium* spp. and other pathogens. The standard recommendation to control diseases caused by pathogenic Phycomycetes is the proper regulation of the environmental factors such as site selection, moisture control, proper fertilization, and temperature, along with fumigation or use of soil fungicides (53, 89, 93, 148). While the environmental factors sometimes can be controlled to a reasonable extent in greenhouses and nurseries, they generally cannot be regulated in field planting and chemicals must be used more extensively.

Crop rotation is another method of reducing the populations of soil pathogens, but because the more common and destructive pathogenic *Pythium* spp. have wide host ranges, this procedure is not usually successful. However, as different crops have different moisture requirements and provide differing degrees of soil shading, they affect the soil environment and hence influence the conditions that *Pythium* spp. require for optimum growth and parasitism.

Recently, chemicals have been formulated that when sprayed on the foliage are translocated to different parts of the plant and affect pathogens attacking specific areas. For example, foliar applications of chloroneb (1, 4-dichloro-2, 5, dimethoxy-benzene) have controlled *P. aphanidermatum* in tomato, pepper, bean, and rye grass in greenhouse tests. Soil drenching also proved effective, but whether such applications would be effective under field conditions still needs investigation (93). Captan, ferbam, thiram, and zineb have also been tried to control *Pythium* root diseases with limited success in greenhouse tests (143). Thiram–captan mixtures have given excellent control of *P. aphanidermatum* wilt of chrysanthemum in greenhouses in Florida when used as a soil drench (31).

There is considerable evidence of varietal or cultivar differences in susceptibility to *Pythium*-caused root diseases. Rands & Dopp (123) demonstrated varietal resistance in sugarcane, one of the early reports of this nature. Some degree of control of *Pythium* diseases through resistant cultivars has been reported for beans (3), safflower (145), and a number of other plants (100). In other crops such as cotton, varietal resistance to *Pythium* root diseases could not be demonstrated (97). In the long run, control of *Pythium* diseases by resistant cultivars may be a promising and enduring approach to effective control. Due to the wide host range of many *Pythium* spp., this will be a difficult breeding job.

Literature Cited

1. Adegbola, M. O. K., Hagedorn, D. J. 1969. Symptomatology and epidemiology of *Pythium* bean blight. *Phytopathology* 59:1113–18
2. Adegbola, M. O. K., Hagedorn, D. J. 1969. Host-parasite relations in *Pythium* bean blight. *Phytopathology* 59:1484–87
3. Adegbola, M. O. K., Hagedorn, D. J. 1970. Host resistance and pathogen virulence in *Pythium* blight of bean. *Phytopathology* 60:1477–79
4. Agnihotri, V. P., Vaartaja, O. 1967. Effects of amendments, soil moisture contents, and temperatures on germination of *Pythium* sporangia under influence of soil mycostasis. *Phytopathology* 57:1116–20
5. Agnihotri, V. P., Vaartaja, O. 1968. Seed exudates from *Pinus resinosa* and their effects on growth and zoospore germination of *Pythium afertile*. *Can. J. Bot.* 46:1135–41
6. Agnihotri, V. P., Vaartaja, O. 1970. Effect of seed exudates of *Pinus resinosa* on the germination of sporangia and on the population of *Pythium irregulare* in the soil. *Plant Soil* 32:246–49
7. Alicbusan, R. V., Ichitani, T., Takahashi, M. 1965. Ecologic and taxonomic studies on *Pythium* as pathogenic soil fungi. III Population of *Pythium ultimum* and other microorganisms in rhizosphere. *Bull. Univ. Osaka Pref.* (Series B) 16:59–64
8. Apt, W. J., Koike, H. 1962. Pathogenicity of *Helicotylenchus nannus* and its relation with *Pythium graminicola* on sugarcane in Hawaii. *Phytopathology* 52:798–802
9. Ark, P. 1950. Destructive diseases of orchids in California. *Orchid Digest* Sept-Oct. 174–77
10. Averre, C. W. III. 1966. Isolating *Pythium* and *Fusarium* from a limestone soil in subtropical Florida. *Soil Crop Sci. Soc. Fla.* 26:279–85
11. Banihashemi, Z. 1970. A new technique for isolation of *Phytophthora* and *Pythium* species from soil. *Plant Dis. Reptr.* 54:261–62
12. Barton, R. 1957. Germination of oospores of *Pythium mamillatum* in response to exudates from living seedlings. *Nature* 180:613–14
13. Barton, R. 1958. Occurrence and establishment of *Pythium* in soils. *Trans. Brit. Mycol. Soc.* 41:207–22
14. Barton, R. 1961. Saprophytic activity of *Pythium mamillatum* in soils. II. Factors restricting *P. mamillatum* to pioneer colonization of substrates. *Trans. Brit. Mycol. Soc.* 44:105–18
15. Beach, W. S. 1949. The effects of excess solutes, temperature and moisture upon damping-off. *Pa. State Agr. Exp. Sta. Bull.* 509
16. Biesbrock, J. A., Hendrix, F. F. 1967. A taxonomic study of *Pythium irregulare* and related species. *Mycologia* 59:943–52
17. Biesbrock, J. A., Hendrix, F. F. 1970. Influence of soil water and temperature on root necrosis of peach caused by *Pythium* spp. *Phytopathology* 60:880–82
18. Biesbrock, J. A., Hendrix, F. F. 1970. Influence of continuous and periodic soil water conditions on root necrosis of holly caused by *Pythium* spp. *Can. J. Bot.* 48:1641–45
19. Braun, H. 1924. Geranium stem-rot caused by *Pythium complectens*, n. sp. *J. Agr. Res.* 29:399–419
20. Bruehl, G. W. 1953. *Pythium* root rot of barley and wheat. *U.S. Dept. Agr. Tech. Bull.* 1084
21. Bruehl, G. W. 1955. Barley adaptation in relation to *Pythium* root rot. *Phytopathology* 45:97–103
22. Campbell, W. A. 1949. A method of isolating *Phytophthora cinnamomi* from soils. *Plant Dis. Reptr.* 33:134–35
23. Campbell, W. A., Copeland, Otis L. 1954. Littleleaf disease of shortleaf and loblolly pines. *USDA Circ.* 940
24. Campbell, W. A., Hendrix, F. F., Jr. 1967. A new heterothallic *Pythium* from southern United States. *Mycologia* 59:274–78
25. Campbell, W. A., Hendrix, F. F. 1967. *Pythium* and *Phytophthora* species in forest soils in the southeastern United States. *Plant Dis. Reptr.* 51:929–32

26. Campbell, W. A., Hendrix, F. F., Powell, W. M. 1972. *Pythium* and nematode species implicated in root rot. *Tree Planters' Notes* 23:5–7

27. Campbell, W. A., Hendrix, F. F. 1973. Feeder root system diseases. *The Plant Root and its Environment.* Ed. E. W. Carson

28. Campbell, W. A., Sleeth, B. 1945. A root rot of guayule caused by *Pythium ultimum. Phytopathology* 35:636–39

29. Chang-Ho, Y. 1970. The effect of pea root exudate on the germination of *Pythium aphanidermatum* zoospore cysts. *Can. J. Bot.* 48: 1501–14

30. Chi, C. C., Hanson, E. W. 1962. Interrelated effects of environment and age of alfalfa and red clover seedlings on susceptibility to *Pythium debaryanum. Phytopathology* 52:985–89

31. Cox, R. S. 1969. Control of *Pythium* wilt of Chrysanthemum in South Florida. *Plant Dis. Reptr.* 53:912–13

32. Drechsler, C. 1934. *Pythium scleroteichum* n. sp. causing mottle necrosis of sweetpotatoes. *J. Agr. Res.* 49:881–90

33. Drechsler, C. 1940. Three species of *Pythium* associated with root rots. *Phytopathology* 30:189–213

34. Drechsler, C. 1946. Zoospore development from oospores of *Pythium ultimum* and *Pythium debaryanum* and its relation to rootlet-tip discoloration. *Plant Dis. Reptr.* 30:226–27

35. Eckert, J. W., Tsao, P. H. 1962. A selective antibiotic medium for isolation of *Phytophthora* and *Pythium* from plant roots. *Phytopathology* 52:771–77

36. Fatemi, J. 1971. *Phytophthora* and *Pythium* root rot of sugar beet in Iran. *Phytopathol. Z.* 71: 25–28

37. Filer, T. H., Jr. 1967. Damping-off of sweetgum by *Pythium sylvaticum. Phytopathology* 57:1284

38. Flowers, R. A., Hendrix, J. W. 1969. Gallic acid in a procedure for isolation of *Phytophthora parasitiva* var. *nicotianae* and *Pythium* spp. from soil. *Phytopathology* 59:725–31

39. Frank, Z. R. 1968. *Pythium* root rot of peanut. *Phytopathology* 58: 542–43

40. Freeman, T. E., Luke, H. H., Sechler, D. T. 1966. Pathogenicity of *Pythium aphanidermatum* on grain crops in Florida. *Plant Dis. Reptr.* 50:292–94

41. Frezzi, M. J. 1956. Especies de *Pythium* fitopathogenas identificados en la Republica Argentina. *Rev. Invest. Agr., T.X. No. 2.* Buenos Aires. 241 pp.

42. Gardner, D. E., Hendrix, F. F., Jr. 1973. Carbon dioxide and oxygen concentrations in relation to survival and saprophytic growth of *Pythium irregulare* and *Pythium vexans* in soil. *Can. J. Bot.* In press

43. Garren, K. H. 1970. *Rhizoctonia solani* versus *Pythium myriotylum* as pathogens of peanut pod breakdown. *Plant Dis. Reptr.* 54:840–43

44. Garren, K. H. 1971. Persistence of *Pythium myriotylum* in soils. *Phytopathology* 61:596–97

45. Gay, J. D. 1969. Effects of temperature and moisture on snap bean damping-off caused by three isolates of *Pythium myriotylum. Plant Dis. Reptr.* 53:707–09

46. Gay, J. D., McCarter, S. M. 1968. Stem rot of snap bean in southern Georgia caused by *Pythium myriotylum. Plant Dis. Reptr.* 52:416

47. Gill, D. L. 1970. Pathogenic *Pythium* from irrigation ponds. *Plant Dis. Reptr.* 54.1077–79

48. Golden, J. K., Powell, W. M., Hendrix, F. F., Jr. 1972. The influence of storage temperature on recovery of *Pythium* spp. and *Meloidogyne incognita* from field soils. *Phytopathology* 62:819–23

49. Griffin, D. M. 1958. Influence of pH on the incidence of damping-off. *Trans. Brit. Mycol. Soc.* 41: 183–90

50. Griffin, D. M. 1963. Soil physical factors and the ecology of fungi. II. Behaviour of *Pythium ultimum* at small soil water suctions. *Trans. Brit. Mycol. Soc.* 46:368–72

51. Grimm, G. R., Alexander, A. F. 1970. Citrus leaf pieces as trap for soil-borne *Phytophthora* spp. *Phytopathology* 60:1294

52. Halpin, J. E., Hanson, E. W., Dickson, J. G. 1952. Studies on the pathogenicity of seven species of *Pythium* on red clover seed-

lings. *Phytopathology* 42:245–49
53. Hanan, J. J., Langhans, R. W., Dimock, A. W. 1963. *Pythium* and soil aeration. *Proc. Am. Soc. Hort. Sci.* 82:574–82
54. Harper, A. M., Smith, A. D., Harper, F. R. 1969. Low emergence of "prolific" spring rye caused by a *Pythium* species in soil. *Can. J. Plant Sci.* 49:531–33
55. Hartley, Carl. 1921. Damping-off in forest nurseries. *USDA Bull.* 934
56. Haskins, R. H. 1965. Sterols and temperature tolerance in the fungus *Pythium*. *Science* 150:1615–16
57. Hendrix, F. F., Jr., Campbell, W. A. 1966. Root rot organisms isolated from ornamental plants in Georgia. *Plant Dis. Reptr.* 50:393–95
58. Hendrix, F. F., Jr., Campbell, W. A. 1968. Pythiaceous fungi isolated from southern forest nursery soils and their pathogenicity to pine seedlings. *Forest Sci.* 14:292–97
59. Hendrix, F. F., Jr., Campbell, W. A. 1968. A new heterothallic *Pythium* from the United States and Canada. *Mycologia* 60:802–05
60. Hendrix, F. F., Jr., Campbell, W. A. 1969. Heterothallism in *Pythium catenulatum*. *Mycologia* 61:639–41
61. Hendrix, F. F., Jr., Campbell, W. A. 1970. Distribution of *Phytophthora* and *Pythium* species in soils in the continental United States. *Can. J. Bot.* 48:377–84
62. Hendrix, F. F., Jr., Campbell, W. A., Chein, C. Y. 1971. Some *Phycomycetes* indigenous to soils of old growth forests. *Mycologia.* 63:283–89
63. Hendrix, F. F., Jr., Campbell, W. A., Moncrief, J. B. 1970. *Pythium* species associated with golf turf grasses in the south and southeast. *Plant Dis. Reptr.* 54:419–21
64. Hendrix, F. F., Jr., Kuhlman, E. G. 1965. Factors affecting direct recovery of *Phytophthora cinnamomi* from soil. *Phytopathology* 55:1183–87
65. Hendrix, F. F., Jr., Powell, W. M. 1968. Nematode and *Pythium* species associated with feeder root necrosis of pecan trees in Georgia. *Plant Dis. Reptr.* 5:334–35

66. Hendrix, F. F., Jr., Powell, W. M. 1970. Control of root pathogens in peach decline sites. *Phytopathology* 60:16–19
67. Hendrix, F. F., Jr., Powell, W. M., Owen, J. H. 1966. Relation of root necrosis caused by *Pythium* species to peach tree decline. *Phytopathology* 56:1229–32
68. Hendrix, J. W. 1964. Sterol induction of reproduction and stimulation of growth of *Pythium* and *Phytophthora*. *Science* 144:1028–29
69. Hine, R. B. 1961. The role of fungi in the peach replant problem. *Plant Dis. Reptr.* 45:426–65
70. Hock, W. K., Klarman, W. L. 1967. The function of the endodermis in the resistance of Virginia pine seedlings to damping-off. *Forest Sci.* 13:108–12
71. Hodges, C. S. 1959. *Studies on black root rot of pine seedlings.* Ph.D. Thesis, University of Georgia, Athens, Ga. 120 pp.
72. Hoppe, P. E. 1966. *Pythium* species still viable after 12 years in airdried muck soil. *Phytopathology* 56:1411
73. Hsu, Dien-Sie, Hendrix, F. F., Jr. 1973. Influence of *Criconemoides Quadricornis* on Pecan feeder root necrosis caused by *Pythium irregulare* and *Fusarium solani* at different temperatures. *Can. J. Bot.* In press
74. Hsu, S. C. 1965. The morphology and physiology of *Pythium catenulatum* on sugarcane root rot. *Rept. Taiwan Sugar Exp. Sta.* 37:89–104
75. Ito, S., Tokunaga, Y. 1933. Studies on the rot-disease of rice seedlings caused by *Pythium* species. *J. Faculty Agr.* Hokkaido Imp. Univ. 32:201–27
76. Jackson, C. R., McFadden, L. A. 1961. Chrysanthemum diseases in Florida. *Univ. Fla., Agr. Exp. Sta. Bull.* 637
77. Johnson, A. W., Littrell, R. H. 1970. Pathogenicity of *Pythium apanidermatum* to Chrysanthemum in combined inoculations with *Belonolaimus longicaudatus* or *Meloidogyne incognita*. *J. Nematol.* 2:255–59
78. Kerr, A. 1963. The root rot-Fusarium wilt complex of peas. *Aust. J. Biol. Sci.* 16:55–69
79. Kisiel, M., Deubert, K., Zucker-

man, B. M. 1969. The effect of *Tylenchus agricola* and *Tylenchorhynchus claytoni* on root rot of corn caused by *Fusarium roseum* and *Pythium ultimum*. *Phytopathology* 59:1387–90

80. Klemmer, H. W., Nakano, R. Y. 1962. Techniques in isolation of Pythiaceous fungi from soil and diseased pineapple tissue. *Phytopathology* 52:955–56

81. Klemmer, H. W., Nakano, R. Y. 1964. Distribution and pathogenicity of *Phytophthora* and *Pythium* in pineapple soils of Hawaii. *Plant Dis. Reptr.* 48:848–52

82. Klisiewicz, J. M. 1968. Relation of *Pythium* spp. to root rot and damping-off of safflower. *Phytopathology* 58:1384–86

83. Knaphus, G., Buchholtz, W. F. 1958. Vertical distribution of *Pythium graminicolum* in soil. *Iowa State College J. Sci.* 33:201–07

84. Koike, H. 1971. Individual and combined effects of *Pythium tardicrescens* and *P. graminicola* on sugarcane: a first report. *Plant Dis. Reptr.* 55:766–70

85. Koike, H., Yang, S. 1971. Influence of sugarcane mosaic virus strain H and *Pythium graminicola* on growth of sugarcane. *Phytopathology* 61:1090–92

86. Kraft, J. M., Endo, R. M., Erwin, D. C. 1967. Infection of the primary roots of bentgrass by zoospores of *Pythium aphanidermatum*. *Phytopathology* 57:86–90

87. Kraft, J. M., Erwin, D. C. 1967. Stimulation of *Pythium aphanidermatum* by exudates from mung bean seeds. *Phytopathology* 57:866–68

88. Kraft, J. M., Erwin, D. C. 1968. Effects of inoculum substrate and density on the virulence of *Pythium aphanidermatum* to Mung bean seedlings. *Phytopathology* 58:1427-28

89. Kraft, J. M., Haglund, W. A., Reiling, T. P. 1969. Effect of soil fumigants on control of pea root rot pathogens. *Plant Dis. Reptr.* 53:776–80

90. Kraft, J. M., Burke, D. W. 1971. *Pythium ultimum* as a root pathogen of beans and peas in Washington. *Plant Dis. Reptr.* 55:1056–60

91. Laviolette, F. A., Athow, K. L. 1971. Relationship of age of soybean seedlings and inoculum to infection by *Pythium ultimum*. *Phytopathology* 61:439–40

92. Leach, L. D. 1947. Growth rates of host and pathogen as factors determining the severity of pre-emergence damping-off. *J. Agr. Res.* 75:161–79

93. Littrell, R. H., Gay, J. D., Wells, H. D. 1969. Chloroneb fungicide for control of *Pythium aphanidermatum* on several crop plants. *Plant Dis. Reptr.* 53:913–15

94. Lorio, P. L. 1966. *Phytophthora cinnamomi* and *Pythium* species associated with loblolly pine decline in Louisiana. *Plant Dis. Reptr.* 50:596–97

95. Lumsden, R. D., Haasis, F. A. 1964. *Pythium* root and stem diseases of *Chrysanthemum*. *North Carolina Agr. Exp. Sta. Tech Bull.* 158

96. Luna, L. V., Hine, R. B. 1964. Factors influencing saprophytic growth of *Pythium aphanidermatum* in soil. *Phytopathology* 54:955–59

97. Mathre, D. E., Otta, J. D. 1967. Sources of resistance in the genus *Gossypium* to several soil-borne pathogens. *Plant Dis. Reptr.* 51:864–66

98. Matthews, V. D. 1931. Studies on the genus *Pythium*. University of N. C. Press. Chapel Hill. 136 pp.

99. McCarter, S. M., Littrell, R. H. 1968. Pathogenicity of *Pythium myriotylum* to several grass and vegetable crops. *Plant Dis. Reptr.* 52:179–83

100. McCarter, S. M., Littrell, R. H. 1970. Comparative pathogenicity of *Pythium aphanidermatum* and *Pythium myriotylum* to twelve plant species and intraspecific variation in virulence. *Phytopathology* 60:264–68

101. McCarter, S. M., Roncadori, R. W. 1971. Influence of low temperature during cotton seed germination on growth and disease susceptibility. *Phytopathology* 61:1426–29

102. McClure, T. T., Robbins, W. R. 1942. Resistance of cucumber seedlings to damping-off as related to age, season of year, and level of nitrogen nutrition. *Bot. Gaz.* 103:684–97

103. McElroy, F. D., Pepin, H. S., Ormond, D. J. 1971. Dieback of car-

rot roots caused by *Pythium debaryanum*. *Phytopathology* 61: 586–87

104. Middleton, J. T. 1943. The taxonomy, host range and geographic distribution of the genus *Pythium*. *Mem. Torrey Bot. Club.* 20:1–171

105. Mildenhall, J. P., Pratt, R. G., Williams, P. H., Mitchell, J. E. 1971. *Pythium* brown rot and forking of muck-grown carrots. *Plant Dis. Reptr.* 55:536–40

106. Miller, C. R., Dowler, W. M., Peterson, D. H., Ashworth, R. P. 1966. Observations on the mode of infection of *Pythium ultimum* and *Phytophthora cactorum* on young roots of peach. *Phytopathology* 56:46–49

107. Mircetich, S. M. 1971. The role of *Pythium* in feeder roots of diseased and symptomless peach trees and in orchard soils in peach tree decline. *Phytopathology* 61: 357–60

108. Mircetich, S. M., Fogle, H. W. 1969. Role of *Pythium* in damping-off of peach. *Phytopathology* 59:356–58

109. Moore, L. D., Couch, H. B. 1961. *Pythium ultimum* and *Helminthosporium vagans* as foliar pathogens of *Gramineae*. *Plant Dis. Reptr.* 45:616–19

110. Moore, L. D., Couch, H. B., Bloom, J. R. 1963. Influence of environment on diseases of turfgrasses. III. Effects of nutrition, pH, soil temperature, air temperature and soil moisture on *Pythium* blight of highland bentgrass. *Phytopathology* 53:53–57

111. Mulder, J. 1969. The pathogenicity of several *Pythium* species to rootlets of apple seedlings. *Neth. J. Plant. Pathol.* 75:178–81

112. Munnecke, D. E., Moore, B. J. 1969. Effect of storage at −18°C of soil infested with *Pythium* or *Fusarium* on damping-off of seedlings. *Phytopathology* 59:1517–20

113. Munnecke, D. E., Moore, B. J., Abu-El-Haj, F. 1971. Soil moisture effects on control of *Pythium ultimum* or *Rhizoctonia solani* with methyl bromide. *Phytopathology* 61:194–97

114. Nemec, S. 1970. Fungi associated with strawberry root rot in Illinois. *Mycopathol. Mycolog. Appl.* 41:331–46

115. Nemec, S. 1970. *Pythium sylvaticum*—pathogenic on strawberry roots. *Plant Dis. Reptr.* 54:416–18

116. Nemec, S. 1971. Mode of entry by *Pythium perniciosum* into strawberry roots. *Phytopathology* 61: 711–14

117. Nitzany, F. E. 1966. Synergism between *Pythium ultimum* and cucumber mosaic virus. *Phytopathology* 56:1386–89

118. Papa, K. E., Campbell, W. A., Hendrix, F. F. 1967. Sexuality in *Pythium sylvaticum*: heterothallism. *Mycologia* 59:589–95

119. Poole, R. F. 1934. Sweet-potato ring rot caused by *Pythium ultimum*. *Phytopathology* 24:807–14

120. Porter, D. M. 1970. Peanut wilt caused by *Pythium myriotylum*. *Phytopathology* 60:393–94

121. Powell, W. M., Owen, J. H., Campbell, W. A. 1965. Association of Phycomycetous fungi with peach tree decline in Georgia. *Plant Dis. Reptr.* 49:279

122. Pratt, R. G., Green, R. J., Jr. 1971. The taxonomy and heterothallism of *Pythium sylvaticum*. *Can. J. Bot.* 49:273–79

123. Rands, R. D., Dopp, E. 1938. *Pythium* root rot of sugarcane. *USDA Tech. Bull.* 666

124. Roncadori, R. W., McCarter, S. M. 1972. Effects of soil treatment, soil temperature, and plant age on *Pythium* root rot of cotton. *Phytopathology* 62:373–76

125. Roth, L. F., Riker, A. J. 1943. Influence of temperature, moisture, and soil reaction on the damping-off of red pine seedlings by *Pythium* and *Rhizoctonia*. *J. Agr. Res.* 67:273–93

126. Royle, D. J., Hickman, C. J. 1964. Analysis of factors governing in vitro accumulation of zoospores of *Pythium aphanidermatum* on roots. II. Substances causing response. *Can. J. Microbiol.* 10: 201–19

127. Savory, B. M. Specific replant diseases. *Commonwealth Agr. Bur. Res. Rev.* No. 1

128. Schmitthenner, A. F. 1962. Isolation of *Pythium* from soil particles. *Phytopathology* 52:1133–38

129. Schmitthenner, A. F. 1970. Significance of populations of *Pythium* and *Phytophthora* in soil. *Root diseases and soil-borne Pathogens,*

ed. T. A. Toussoun, R. V. Bega, P. V. Nelson, Univ. Calif. Press, Berkeley 25–27

130. Sechler, D., Luke, H. H. 1967. Stand loss of small grains in Florida. *Plant Dis. Reptr.* 51:919–22

131. Sideris, C. P. 1931. Taxonomic studies in the family *Pythiaceae* I. *Nematosporangium.* *Mycologia* 23:252

132. Sideris, C. P. 1932. Taxonomic studies in the family *Pythiaceae* II. *Pythium.* *Mycologia* 24:14–61

133. Singh, R. S., Mitchell, J. E. 1961. A selective method for isolation and measuring the population of *Pythium* in soil. *Phytopathology* 51:440–44

134. Sleeth, B. 1953. Winter Haven decline of citrus. *Plant Dis. Reptr.* 37:425–26

135. Spencer, J. A., Cooper, W. E. 1967. Pathogenesis of cotton (*Gossypium hirsutum*) by *Pythium* species: Zoospores and mycelium attraction and infectivity. *Phytopathology* 57:1332–38

136. Stanghellini, M. E., Hancock, J. G. 1970. A quantitative method for the isolation of *Pythium ultimum* from soil. *Phytopathology* 60:551–52

137. Stanghellini, M. E., Hancock, J. G. 1971. The sporangium of *Pythium ultimum* as a survival structure in soil. *Phytopathology* 61:157–64

138. Stanghellini, M. E., Hancock, J. G. 1971. Radial extent of the bean spermosphere and its relation to the behavior of *Pythium ultimum*. *Phytopathology* 61:165–68

139. Stanghellini, M. E., Russell, J. D. 1971. Damping-off of tomato seedlings in commercial hydroponic culture. *Prog. Agr. Arizona* 23:15–16

140. Summers, T. E., Buchholtz, W. R. 1958. Time and frequency of recovery of *Pythium debaryanum* and *Pythium graminicolum* from roots of growing barley seedlings. *Iowa State College J. Sci.* 33:209–17

141. Sutherland, J. R., Adams, R. E., True, R. P. 1966. *Pythium vexans* and other conifer seedbed fungi isolated by the apple technique following treatment to control nematodes. *Plant Dis. Reptr.* 50:545–47

142. Takahashi, M., Ozaki, T. 1965. Ecologic and taxonomic studies on *Pythium* as pathogenic soil fungi. The isolation methods of *Pythium*. *Bull. Univ. Osaka Prefecture* 17:1–10

143. Tammen, J., Muse, D. P., Hass, J. H. 1961. Control of *Pythium* root diseases with soil fungicides. *Plant Dis. Reptr.* 45:858–63

144. Thomas, C. A. 1970. Effect of temperature on *Pythium* root rot of safflower. *Plant Dis. Reptr.* 54:300

145. Thomas, C. A. 1970. Effect of seedling age on *Pythium* root rot of safflower. *Plant Dis. Reptr.* 54:1010–11

146. Thomason, I. J., Dickson, J. G. 1960. Influence of soil temperature on seedling blight of smooth bromegrass. *Phytopathology* 50:1–7

147. Thomson, T. B., Athow, K. L., Laviolette, F. A. 1971. The effect of temperature on the pathogenicity of *Pythium aphanidermatum, P. debaryanum* and *P. ultimum,* on soybean. *Phytopathology* 61:933–35

148. Tompkins, C. M., Middleton, J. T. 1950. Etiology and control of poinsettia root and stem rot caused by *Pythium* spp. and *Rhizoctonia solani. Hilgardia* 20:171–82

149. Tompkins, C. M. 1950. *Pythium* rot of pink and yellow calla corms and its control. *Hilgardia* 20:183–90

150. Tsao, P. H. 1970. Selective media for isolation of pathogenic fungi. *Ann. Rev. Phytopathol.* 8:157–86

151. Vaartaja, O. 1960. Selectivity of fungicidal materials in agar cultures. *Phytopathology* 50:870–73

152. Vaartaja, O. 1967. Reinfestation of sterilized nursery seedbeds by fungi. *Can. J. Microbiol.* 13:771–76

153. Vaartaja, O. 1968. *Pythium* and *Mortierella* in soils of Ontario forest nurseries. *Can. J. Microbiol.* 14:265–69

154. Vaartaja, O., Salisbury, P. J. 1961. Potential pathogenicity of *Pythium* isolates from forest nurseries. *Phytopathology* 51:505–07

155. Vaartaja, O., Bumbieris, M. 1964. Abundance of *Pythium* species in nursery soils in South Australia. *Aust. J. Biol. Sci.* 17:436–45

156. Van der Plaats-Niterink, A. J. 1968. The occurrence of *Pythium*

in the Netherlands. I. Heterothallic species. *Acta Bot. Neerl.* 17: 320–29

157. Van der Plaats-Niterink, A. J. 1969. The occurrence of *Pythium* in the Netherlands. II. another heterothallic species: *Pythium splendens* Braun. *Acta. Bot. Neerl.* 18:489–95

158. Vanterpool, T. C., Truscott, J. H. L. 1932. Studies on browning root rot of cereals II. Some parasitic species of *Pythium* and their relation to the disease. *Can. J. Res.* 6:68–93

159. Wager, V. A. 1931. Diseases of plants in South Africa due to members of the *Pythiaceae*. *Union S. Africa, Dept. Agr. Bull.* 105

160. Waterhouse, G. M. 1967. Key to *Pythium* Pringsheim. *Commonwealth Mycol. Inst.* Mycol. Papers 109

161. Waterhouse, G. M. 1968. The genus *Pythium* Pringsheim. *Commonwealth Mycol. Inst.* Mycol. Papers 110

162. Webster, R. K., Hall, D. H., Heeres, J., Wick, C. M., Brandon, D. M. 1970. *Achlya klebsiana* and *Pythium* species as primary causes of seed rot and seedling disease of rice in California. *Phytopathology* 60:964–68

163. Wilhelm, S. 1965. *Pythium ultimum* and the soil fumigation growth response *Phytopathology* 55:1016–20

164. Yale, J. W., Vaughan, E. K. 1962. Effects of mineral fertilizers on damping-off of table beets. *Phytopathology* 52:1285–87

165. Yang, C. Y., Mitchell, J. E. 1965. Cation effect on reproduction of *Pythium* spp. *Phytopathology* 55: 1127–31

NEMATODES AND FOREST TREES— ❖ 3567
TYPES OF DAMAGE TO TREE ROOTS

John L. Ruehle

Southeastern Forest Experiment Station, U.S.D.A. Forest Service,
Forestry Sciences Laboratory, Athens, Georgia

INTRODUCTION

The presence of plant-parasitic nematodes and their importance as pests of agricultural and ornamental crops have long been recognized in the developed nations of the world. The injury and crop losses caused by the root-knot nematode, the sugarbeet nematode, and the golden nematode have been obvious for many years because these parasites are limiting factors in economic crop production in certain areas. It is now becoming obvious, particularly in perennial ornamentals and fruit and nut crops, that still greater injury may be caused by nematodes whose presence and damage is less easily measured.

Since nematodes cause serious damage to citrus, peaches, walnuts, cherries, almonds, and many other horticulturally important trees, it is not surprising that they also damage forest trees, yet nematode diseases of forest trees remain virtually unknown. Disregard of rhizosphere ecology by forest scientists probably accounts for past failures to recognize nematodes as important soil and site factors in forestry. Root losses due to nematodes on forest trees have generally gone unnoticed. In fact, forest soils are probably the least known portion of the forest environment.

Plant-parasitic nematodes are found in practically all regions of the world in which forest trees grow (48). And yet to the present day, nursery seedlings not withstanding, few records implicate these parasites as causes of growth loss. This is due in large measure to difficulties encountered in accurately measuring growth loss in trees resulting from damage to feeder roots. Approaches to soil-borne diseases of any perennial crop are limited by the complexity of the problems. The fact that the research is laborious and time consuming has probably caused many investigators to avoid this area. Nevertheless, information concerning host-parasite relations of nematodes and tree roots is essential to provide a basis for recognition and diagnosis of diseases. Until such information is available, accurate estimates of yield losses of trees due to nematodes will be lacking.

Recently, plant-parasitic nematodes have been recognized as causal agents

99

capable of producing economic losses in tree nurseries. Nursery seedlings are particularly vulnerable to nematode damage because of the continuous cultivation of the same or closely related plant species within the same area, the use of irrigation to maintain soil moisture levels for optimum plant growth, and the maintenance of high soil fertility levels. All of these create an environment favorable for nematode development. We are now aware that nematodes cause losses in a tree nursery crop as readily as they do in horticultural crops.

Evidence of pathogenicity is lacking for the majority of plant-parasitic nematodes associated with forest trees in plantations and natural stands. Most of these nematodes have been identified during routine surveys of stands of trees displaying only symptoms of general decline, a condition easily attributed to causes other than nematodes. The lack of rapid decline symptoms ascribed to nematode damage, such as those identified in many cultivated crops, presumably causes most investigators to dismiss nematodes as of no consequence to forest trees. A better understanding of the effect of nematodes on tree roots, however, would lead to the realization that continuing and insidious deterioration of succulent feeder roots of forest trees by plant-parasitic nematodes causes considerable yield loss, and would establish a greater appreciation for the need to expand our knowledge in forest nematology. Such information is available for fruit and nut trees from studies comparing yields of naturally infected plants with those of noninfected plants in the same field, and from studies comparing yields of infected plants with those artificially protected from infection by chemical treatment. Since these crops are of great value, and yield is more easily defined, loss estimates under field conditions have been fairly accurate and continued work on nematode control has been economically justified.

Foresters have traditionally emphasized only the primary need for wood and wood products. Recent public awareness of the need for preserving forest ecosystems has caused a shift in thinking. Today, foresters are becoming ecologically and socially oriented. Forest resources are now being evaluated by combining timber and environmental values. When we consider only dollar values per acre per year, based on yield of timber alone, forest crops have much lower value than fruit and nut crops. However, by computing values from integrated uses of forest lands, combining both timber and nontimber values, we find that forest lands are valued much higher. This reappraisal, resulting in a significant increase in the social and economic value of forests, will probably justify intensive studies of soil-borne problems, including those caused by nematodes. It is also highly probable, particularly in forest monoculture as practiced in the southern United States, that nematode damage in some locations will eventually justify direct control methods.

Information needed first in forest nematology is determination of the types of damage produced on roots of forest trees. A proper understanding of this host-parasite relationship requires analysis of the nature of nematode interference with normal growth and development of roots. Nematodes can attack feeder roots directly and cause mechanical or physiological damage. Some act as vectors for soil pathogens and others serve as predisposing agents for invasion by other soil organisms in root-disease complexes. Surveys of timber stands indicate that not all of the plant-parasitic nematodes recovered from forest soils feed on the roots of trees. Some are subsisting on understory shrubs and weeds, and others are mycophagous. Since fungal symbionts convert a portion of the roots of all trees growing under natural conditions into mycorrhizae, nematodes parasitic on fungi must be considered in any study of root ecology.

PRIMARY PATHOGENS

Root Symptoms

Pathogenic symptoms caused by nematode parasites of tree roots may be separated into necrotic, hyperplastic, and hypoplastic tissue reactions. In many cases a single symptom may be observed, but in others two or more reactions may be seen. For example, galling (hyperplastic reaction) may be the dominant symptom, but devitalized root tips (hypoplastic reaction) may occur with galling and, in older infection, browning and shriveling of feeder roots (necrotic reaction) may develop.

Symptoms on tree roots parasitized by nematodes are usually nonspecific since galling, restricted root development, and root necrosis may be produced by other organisms or soil conditions. An understanding of the symptoms resulting from nematode infection is important in assessing possible damage by these organisms.

NECROTIC SYMPTOMS   Damage caused by nematodes ranges from surface necrosis, exhibited as superficial browning, to splitting, cracking, and blackening of the entire feeder root. Only slight damage to feeder roots of trees is caused by certain ectoparasites that feed over large areas of the root system and kill the epidermal cells, causing superficial discoloration. Christie (6) reported that *Xiphinema americanum* was associated with laurel oak (*Quercus laurifolia*) roots with superficial necrosis. An intensive browning is caused by *Tylenchorhynchus claytoni* on *Acacia* sp. (21). Distinct necrotic lesions result when nematodes penetrate the root tissue and congregate in limited areas. These lesions usually originate internally and develop as distinctly circumscribed necrotic areas (Fig. 1A). *Pratylenchus brachyurus* causes such lesions on the roots of yellow-poplar (*Liriodendron tulipifera*).[1] This nematode also causes lesions and fissures in the cortical tissue of feeder roots of *Eucalyptus* spp. (30). It usually is confined to the cortex. *Pratylenchus vulnus,* pathogenic on walnuts (*Juglans hindsii*), produces black lesions that extend through the cortex into the phloem and xylem (31). *Pratylenchus*

---

[1] Ruehle, J. L., unpublished data.

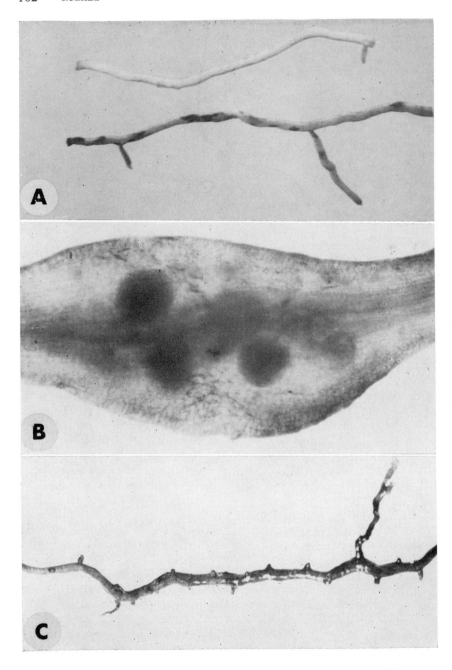

penetrans also causes damage to the cortex of *Cryptomeria japonica* which extends into the endodermis and stele (32).

General cortical necrosis caused by *Hoplolaimus galeatus* has been reported on sycamore (*Platanus occidentalis*) (7), loblolly pine (*Pinus taeda*), and slash pine (*P. elliottii*) (52). *Rotylenchus robustus* induces similar cortical necrosis in feeder roots of seedlings of *Pinus* spp. (54), *Picea* spp., *Acer* spp. (18), *Thuja* sp., *Taxus* sp., *Cedrus* sp., and *Pseudotsuga* spp. (9).

HYPERPLASTIC SYMPTOMS   The location of root galls induced by nematodes on tree roots is determined by the feeding site of the parasite involved. The galls vary in size and shape, depending on the particular nematode and host, density of nematode populations, and soil environment. Although galling of roots is a common reaction to nematode feeding, it is not specific to nematodes—bacteria, fungi, and insects also produce root galls.

The simplest types of root galls on forest trees are those caused by *Xiphinema, Longidorus,* and *Belonolaimus*. These nematodes feed on the root apex as ectoparasites and stimulate hyperplasia and hypertrophy that result in a gall composed of a compact mass of parenchyma cells. *Longidorus maximus* causes thickenings and contortions of root tips and poor overall development of root systems of nursery seedlings (56). *Xiphinema chambersi* stimulates similar symptoms on root tips of sweetgum (*Liquidambar styraciflua*). Parasitism by this pathogen produced symptoms of "coarse root" and "curly tip," characteristic of damage caused by other *Xiphinema* spp. on several other perennial hosts (51). *Belonolaimus longicaudatus* parasitizes roots of sycamore (*Platanus occidentalis*) and causes swollen, blunted root tips (49).

The most common, best-known gall-inducing nematode genus is *Meloidogyne* (Fig. 1B). The host response is complex in susceptible host roots following infection. Second stage larvae of *Meloidogyne* spp. penetrate feeder roots near the root tip, move to the pro-vascular region, become sedentary, feed, and mature. Multinucleate giant cells form at the feeding site about the larval head. This is followed by hypertrophy and hyperplasia in the pericycle, cortex, and epidermis adjacent to the enlarging nematode. *Meloidogyne incognita* stimulates gall formation on roots of *Cornus florida* (29), *Anthocephalus chinensis* (17), and *Acacia* sp. (21). *Meloidogyne ovalis* produces

Figure 1   Typical damage to roots of forest trees resulting from parasitism by plant-parasitic nematodes. (A) Comparison of healthy (above) and nematode-damaged (below) roots of yellow poplar. Note dark lesions caused by *Pratylenchus brachyurus;* (B) Typical root-knot gall on catalpa root containing several females of *Meloidogyne incognita;* (C) Lateral root of slash pine seedling with feeder roots stunted as a result of parasitism by *Belonolaimus longicaudatus.*

spherical galls on root tips of *Acer saccharum, Ulmus americana,* and *Fraxinus americana* (46).

Meloidogyne spp. rarely cause large, distinct galls on conifers, but a few reports list infections on feeder roots of conifers caused by species of this nematode (10, 21, 36, 37, 45). Symptoms on conifers appear as thickened roots slightly larger than their normal diameter.

The pine cystoid nematode, *Meloidodera floridensis,* closely related to the genus *Meloidogyne,* commonly infects pines. Its feeding results in formation of giant cells and limited hypertrophy and hyperplasia with almost no root-gall formation (47).

HYPOPLASTIC SYMPTOMS   Devitalization of root tips or inhibition of apical growth may accompany galling or necrosis. In some cases, however, inhibition of growth is the only visible response to nematode feeding. Certain ectoparasites feed on the tips of tree roots causing them to stop growing. This form of root injury on trees is seen merely as an underdeveloped root system, with no signs of necrosis or swellings (Fig. 1C). *Trichodorus christiei* has been associated with this type of injury on pine seedlings in nurseries (27). This nematode feeds on the root tips of seedlings of *Pinus elliottii, P. palustris,* and *P. taeda,* devitalizes the meristematic tissue in the root tip, and produces a severely underdeveloped root system consisting of numerous short, stubby roots (50). *Tylenchorhynchus claytoni* and *T. ewingii* are also capable of causing stubby-root symptoms on seedlings of *P. elliottii* in forest nurseries (26).

Host Responses

Most nematodes cause disease in trees by decreasing the water- and nutrient-absorbing area of feeder roots. These harmful effects are enhanced by adverse environmental conditions, such as mineral deficiency, drought, and non-optimal pH. Thus, most symptoms of nematode injury to forest trees are nonspecific, in that above-ground symptoms resemble those of any plant having malfunctioning roots, regardless of the cause.

In nursery situations the environment is greatly modified by man. Intensive monoculture is very conducive to the build-up of nematode populations and consequent damage to tree seedlings. Nematode injury to root systems of seedlings is commonly indicated by decreased development of foliage (25, 26, 27). In nurseries, if suitable soil-moisture conditions and high soil fertility are maintained, foliage on infected seedlings often retains its normal green color, and the seedlings are simply smaller than those grown in noninfested soil. Frequently, large populations of nematodes are necessary to cause marked top symptoms when all other factors are favorable for growth of nursery seedlings.

In the more natural habitat of forest plantations, populations of nematodes generally are in equilibrium with the environment (35) and rarely reach the densities found in nurseries. Since water and fertilizer are sometimes below optimum, however, damage may result even from small populations of nematodes. Also, population densities in plantations interrelate with soil characteristics, host composition, weather patterns, and timber management practices to cause noticeable loss of timber yield and quality.

In the southern United States, many forest plantations are established on abandoned farm lands. Population densities of plant-parasitic nematodes are generally high in these areas. The polyphagous nature of many parasitic species retards the development of plantations of trees that are susceptible hosts for the nematodes involved. Although population densities become stabilized in plantations after a number of years, the high densities at planting time undoubtedly result in significant initial root damage, stunting of newly planted seedlings, and mortality in certain adverse environments. In such cases, a direct relationship between population density prior to planting and seedling damage can normally be demonstrated (39).

Nematodes may adversely affect young tree seedlings in a short time, but mature trees may take years to exhibit a response to these pathogens. In situations of balanced parasitism the initial damage may be slight—almost undetectable—on young, actively growing seedlings. In later years as trees mature they lose their juvenile vigor and lack the ability to regenerate feeder roots rapidly. Also, in later years when the maturing trees have reduced the originally high concentration of soil nutrients, they require more nutrients to support the increased total plant volume and hence need a more efficient root system. Insidious root damage to feeder roots caused by plant-parasitic nematodes may account for slow decline and deterioration of mature trees in plantations.

The phenomenon of parasitism in natural woodlands is controlled by age and species of parasite and tree host, soil temperature, soil moisture, and soil pH. Natural woodlands usually contain a mixture of tree species of uneven age classes. Such stands rarely contain damaging populations of parasitic nematodes. This basically stable biotic community, nevertheless, is modified from time to time by nature or man. Feeding by nematodes on tree roots over long periods of time, especially on trees adversely affected by changes in climate, may weaken the trees and make them more vulnerable to environmental changes, and to attack by other pathogens and insects. Many aspects of the relationships among host, parasite, and environment affect plant parasitic nematodes (63) in natural habitats; as a consequence, evaluation of nematode effects on growth is difficult to separate from those of other related factors.

MULTIPLE ASSOCIATIONS

Root-disease complexes are probably more common on forest trees than diseases caused by single organisms. In some instances, nematodes act as vectors for viruses and possibly bacteria. They may also act synergistically with—or simply provide a means for entry of—fungus pathogens into the roots. Since a portion of the roots of all trees growing in natural areas are mycorrhizal, nematodes definitely are associated with, and in some cases influenced by, symbiotic fungi in feeder roots.

VECTORS   Many soil-borne viruses are transmitted by nematodes (61). The nematode-transmitted polyhedral-shaped viruses (NEPO), i.e., tobacco ringspot virus, are transmitted by species of *Xiphinema* and *Longidorus,* whereas the nematode-transmitted tubular-shaped viruses (NETU), i.e., tobacco rattle virus, are transmitted only by species of *Trichodorus.* Species of all three of these nematode genera are commonly found in forest soils. Although it is now well established that viruses infect forest trees, little information is available regarding their characteristics, vectors, and pathogenicity (55). Information regarding diseases of forest trees caused by soil-borne viruses is almost nonexistent. In the first account of the natural occurrence of tobacco ringspot virus in forest trees, Hibben & Bozarth (24) reported isolation of the virus from leaves of declining *Fraxinus americana. Xiphinema americanum,* a common associate of the roots of declining *F. americana* in New York, was previously found to be a vector for this virus (23).

Although information is lacking regarding soil-borne virus diseases of forest trees, there are several accounts of controlled experiments in which trees became infested with viruses following feeding by viruliferous nematodes. Harrison (20) found roots of *Chamaecyparis lawsoniana* infected with arabis mosaic virus after feeding by *Xiphinema diversicaudatum,* and roots of *Picea sitchensis* infected with tomato blackring virus after feeding by *Longidorus attenuatus.* Fulton (14) found that *Xiphinema americanum* transferred tomato ringspot virus to roots of *Cupressus arizonica.* The virus was not recovered from seedling tops and no symptoms were noted. This plant may act as a reservoir host for tomato ringspot virus.

Robinia and *Ulmus* can be infected by NEPO viruses transmitted by *Xiphinema* spp. (5). *Acer pseudoplatanus, Chamaecyparis lawsoniana,* and *Fraxinus excelsior* are infected by NEPO viruses transmitted by *Xiphinema diversicaudatum* (62). Reinking & Radewald (41) suspected that Cadang-Cadang disease of coconuts in Guam was caused by a soil-borne virus spread by a *Xiphinema* sp.

There are presently no accounts of virus infections of forest trees by NETU viruses. This is surprising, since *Trichodorus* spp., vectors that commonly transmit this group of viruses, are frequently found in forest rhizosphere soil.

DISEASE COMPLEXES   Injury to trees by other root pathogens is often increased by plant-parasitic nematodes (16, 38). Recent studies also indicate that certain rhizosphere organisms are destructive only when they occur in combination with nematodes parasitizing feeder roots (40). Nematodes interact with fungi, and possibly with bacteria, within the feeder roots of trees and in locations where root rots of forest trees are found. When parasitic nematodes are recovered from the soil or roots, consideration should be given to nematode-fungus complexes.

The earliest reports of nematode-fungus complexes in forestry came from nursery research. Henry (22) suspected that a root rot of pine seedlings in Mississippi was caused by a nematode-fungus complex. He found that soil fumigation controlled the disease, but he failed to identify the causal agents. He found several endoparasitic nematodes in small numbers in the feeder roots of *Pinus elliottii, P. palustris, P. echinata,* and *P. taeda* with symptoms of a fungus root rot, and controlled the disease by applying a nematicidal fumigant, ethylene dibromide (EDB), to the nursery bed soil prior to planting. Unfortunately, no mention was made of assaying the soil in the root zone of the diseased seedlings for fungi and nematodes. It is possible that a disease complex involving ectoparasites and pathogenic fungi may have been overlooked. Foster (13), working on black root-rot in forest nurseries, isolated *Fusarium* sp. and *Sclerotium* sp. from the characteristic black swellings on the taproot and larger laterals. He concluded that the parasitic nematodes recovered from the root zones of diseased seedlings could not cause the severe symptoms found in later stages of the disease. He failed to mention the possibility that nematodes and fungi damage feeder roots in the early stages of disease development. Unlike Henry (22), he failed to control this disease in a Georgia nursery by soil fumigation with EDB, but conceded that this may have been due to low initial nematode densities.

When *Tylenchorhynchus claytoni* (58), *Helicotylenchus dihystera* (52), and *Rotylenchus pumilis* (43) were tested in soil fumigated or steamed to kill fungi, they fed and reproduced on tree seedlings, but this feeding did not significantly decrease root growth of seedlings. These nematodes should not be considered unimportant parasites until further study shows that they are not involved with fungus pathogens in nature. Hsu (28) investigated the combined effects of two fungi, *Fusarium solani* and *Pythium irregulare,* and the nematode *Criconemoides quadricornis* on the development of feeder-root necrosis on pecan seedlings at different temperatures. At 21°C she found that *P. irregulare* alone caused significant loss of root weight compared to the controls, whereas at 27°C, a temperature less favorable for this pathogen, it caused no significant loss. At both temperatures, *C. quadricornis* alone had no effect. Adding the nematode to the fungus at the lower temperature failed to produce any loss of root weight beyond that caused by the fungus alone. At the higher temperature, however, the nematode and fungus together caused a 44% decrease in root weight, whereas neither the fungus nor the nematode alone decreased root weight below that of checks; the nematode apparently extended the environmental range over which the fungus pathogen caused disease.

In a survey of nurseries in British Columbia, Sutherland & Dunn (60), found *Xiphinema bakeri* consistently associated with a "corky root"—a roughened, thickened, and brown root condition of *Pseudotsuga menziesii*

(4). After isolating *Cylindrocarpon radicicola* from the diseased roots, they suspected that a nematode-fungus complex was involved in this problem. De-Maeseneer (9) isolated *C. radicicola* and *Fusarium oxysporum* from nematode-infected conifer seedlings in Belgium, but concluded that parasitic nematodes were the primary pathogens and expressed doubt that a nematode-fungus complex was involved. It would be interesting to reinvestigate the above problems, using the hypothesis that the nematodes modified the host feeder roots to permit invasion by the associated fungi. Such investigations, conducted under controlled conditions, may prove that root damage ascribed to nematodes or fungi alone is, in fact, damage resulting from an interaction between nematodes and soil fungi.

The fungus component of a complex occasionally has a marked effect on the nematode population. Some fungal pathogens have a depressing effect on nematode activity while others seem to enhance nematode reproduction (40). In roots of American elm (*Ulmus americana*) and sugar maple (*Acer saccharum*), reproduction by *Pratylenchus penetrans* increased in the presence of *Verticillium dahliae* (11).

Several reports suggest that nematodes predispose tree roots to invasion by fungi incapable of colonizing nematode-free roots. Blake (3) found *Fusarium oxysporum* readily established in feeder roots of banana invaded by *Radopholus similis,* but seldom recovered this fungus from nematode-free roots. Lesions formed after inoculation by both *R. similis* and *F. oxysporum* were more necrotic and extensive than when *R. similis* alone was the inoculum. In such plants, *F. oxysporum* grew through the endodermis into the stele, killing the vascular cells. Pathogenic fungi, such as *Fusarium, Sclerotium,* and *Thielaviopsis,* are commonly found in lesions formed by *R. similis* in citrus roots soon after nematode invasion (12). Subsequently, bacteria and tissue-decomposing fungi, such as *Penicillium* and *Aspergillus,* cause additional destruction in root lesions. As root damage becomes extensive, the nematodes generally migrate to new healthy roots. Investigators studying damaged feeder roots of trees frequently isolate organisms involved only in the later sequence of events following initial invasion by plant-parasitic nematodes. This has led to misleading reports of the cause of feeder-root necrosis. In many investigations, isolations were made only from the necrotic roots, ignoring the rhizosphere. Failure to consider rhizosphere ectoparasites as primary inciting agents also adds to the controversy and misunderstanding of the etiology of feeder-root necrosis in forest trees.

MYCORRHIZAE   The feeder roots of most forest trees growing in a natural environment become mycorrhizal. Ectomycorrhizae and endomycorrhizae are the two general types of mycorrhizae on tree roots (33). The fungal symbionts of ectomycorrhizae, generally *Basidiomycetes,* penetrate juvenile feeder roots intercellularly; partially replace the middle lamellae between cortical cells, developing a hyphal arrangement termed the Hartig net; and eventually produce a continuous hyphal network over the exterior of the root.

Most coniferous tree species and hardwood species in the *Salicaceae, Betula-
ceae,* and *Fagaceae* form ectomycorrhizae in nature. Endomycorrhizal symbionts are *Phycomycetes* that penetrate cortical cells intracellularly, form
large vesicles and arbuscules in cortical cells, and unlike ectomycorrhizae, do
not form a dense mantle on the surface of feeder roots. Endomycorrhizae are
normally found on roots of tree species that do not form ectomycorrhizae. It
is generally accepted that most forms of mycorrhizae are beneficial to plants,
and, in the case of pine trees, are indispensable for their growth under natural conditions because they increase the absorbing surface of roots and provide a more rapid absorption of essential elements (33).

Marx & Davey (34) found ectomycorrhizal symbionts capable of protecting feeder roots of pines from attack by soil pathogens, and concluded that
ectomycorrhizae are physically or chemically resistant. Hence, they function
as a biological deterrent to infection by such pathogens as *Phytophthora* and
Pythium species.

Occasionally, soil samples from the root zone of forest trees contain large
numbers of mycophagous nematodes, such as *Tylenchus* and *Aphelenchus*
spp. These "fungus feeders" may parasitize mycorrhizal symbionts and indirectly affect the health of trees. Riffle (42) found an undescribed species of
Aphelenchoides associated with mycorrhizae in natural stands of *Pinus
ponderosa, P. edulis,* and *Juniperus monosperma* in the southwestern United
States. He isolated the common symbiont *Suillus granulatus* from sporophores growing in these natural stands, and in laboratory culture found that
the nematodes decreased growth of mycelium of this fungus. Sutherland
(57) found *Aphelenchus avenae* commonly in forest nursery soils in Canada.
In studies utilizing both aseptic and septic techniques, he showed that this
nematode failed to parasitize the root tissue of *Pinus* and *Picea* seedlings.
Nematode populations, however, were maintained in nonsterile nursery soil
planted with seedlings, suggesting that the nematodes fed and reproduced on
mycorrhizal symbionts. A later study by Sutherland & Fortin (59) showed
that *A. avenae* fed and reproduced on seven species of mycorrhizal fungi
grown in vitro. High inoculum densities of nematodes decreased the growth
of the fungal symbionts, and in pure culture, *A. avenae* prevented the formation of the mycorrhizal relationship between *Suillus granulatus* and *Pinus resinosa* seedlings.

Tylenchorhynchus claytoni damages cortical tissue of newly emerging
short roots of *P. taeda* seedlings.[2] Such damage is slight because nematodes
cause no evident root necrosis (Fig. 2) nor significant decrease in weight of
roots or foliage of seedlings growing in pots in a greenhouse. Such damage,
consequently, prevents normal ectomycorrhizal development because fungal
symbionts require healthy cortical tissue to complete the symbiotic relationship. This type of root damage, albeit a subtle one, also harms trees by preventing development of physiologically and biologically beneficial mycorrhizae.

[2] Ruehle, J. L., unpublished data.

Figure 2   Short roots of loblolly seedling lacking mycorrhizal development because of cortical damage caused by *Tylenchorhynchus claytoni* (below) compared with healthy mycorrhizae formed on noninfected short roots (above).

Nematodes also directly attack mycorrhizae. Ruehle (47) studied histologically the mycorrhizae on *Pinus elliottii* and *P. taeda* seedlings infected by *Hoplolaimus galeatus* and *Meloidodera floridensis*. Larvae and adults of *H. galeatus* attacked the roots of both pine species, penetrated the mycorrhizal mantle, and caused extensive internal damage to the cortex. In several instances, nematodes penetrated the endodermis and fed on the xylem cells. Larvae of *M. floridensis* also entered mycorrhizae of both pine species, migrated through the cortex causing little disruption, and became sedentary after reaching the stele. Giant cells in contact with the head of young females formed in the protophloem and protoxylem, and also in differentiating tissues behind the root tips of mycorrhizae. Limited hyperplasia occurred in the cortical and vascular parenchyma tissues surrounding the giant cells. Enlarging females disrupted the vascular elements, compressed and collapsed the cortical cells, and often erupted through the surface of the roots, leaving only the head and neck embedded in the root. Infected mycorrhizae lacked hypertrophy and no galls formed.

Riffle (43) found that *H. galeatus,* as well as *Rotylenchus pumilis, Tylenchus exiguus,* and *Xiphinema americanum* directly attacked ectomycorrhizae on container-grown seedlings of *Pinus edulis* and *Juniperus monosperma.*

Hoplolaimus galeatus and *R. pumilis* fed endoparasitically, while *X. americanum* and *T. exiguus* failed to penetrate the roots. Parasitism by these nematodes caused ectomycorrhizae to become dark brown to black, and feeding by *X. americanum* almost completely destroyed the ectomycorrhizae on these seedlings after 9 months.

Riffle & Lucht (45) recovered an undescribed *Meloidogyne* species that parasitized ectomycorrhizae of mature *Pinus ponderosa* in southwestern New Mexico. Recently, Riffle (44) described the effects of this nematode on the anatomy of ectomycorrhizae of *P. ponderosa*. He found that larvae entered near the mycorrhizal root tips, migrated through undifferentiated tissues in the apical meristems or through cortical tissue, and became established in the vascular tissues. Host reaction to the developing female nematodes produced giant cells in clusters in the stele, and hypertrophy and hyperplasia of both cortical and vascular tissues forming spherical galls. As females enlarged and matured, they compressed and collapsed the adjacent cortical cells with the associated Hartig net, and commonly ruptured the outer mantle.

Ruehle & Marx (53) evaluated mycorrhizal and nonmycorrhizal roots of *Pinus echinata* and *P. taeda* as feeding sites for *Hoplolaimus galeatus*. They employed aseptic techniques to synthesize *Pisolithus tinctorius* ectomycorrhizae on roots of *P. taeda*. They also used open pot culture to form *Thelephora terrestris* ectomycorrhizae on *P. echinata* seedlings. Following inoculation of mycorrhizal and nonmycorrhizal roots, they found by histological techniques that, irrespective of pine host or mycorrhizal symbiont, both male and female nematodes penetrated the mantle and migrated through the cortical tissue of mycorrhizae. Ectomycorrhizae were more favorable feeding sites for this parasite than nonmycorrhizal roots.

Apparently the mechanisms operating in ectomycorrhizae that prevent infection by fungus pathogens fail to inhibit infection by plant-parasitic nematodes. In fact, the role of ectomycorrhizae as biological deterrents to pathogenic root infection by fungi may be modified by nematode parasitism. Barham et al (1) studied the association of plant-parasitic nematodes, mycorrhizae of pines, and *Phytophthora cinnamomi*, a pathogen commonly associated with shortleaf pine showing symptoms of littleleaf disease. Using a root-cell technique previously described by Marx & Davey (34), they inoculated intact ectomycorrhizae of shortleaf pine seedlings with *Helicotylenchus dihystera* and zoospores of *P. cinnamomi*. Intracellular hyphae and vesicles of the fungus pathogen were observed in cortex cells surrounded by the Hartig net in several of these ectomycorrhizae (Fig. 3). It was concluded that the inhibitors contributing to the resistance of nonparasitized ectomycorrhizae were probably inactivated following parasitism by *H. dihystera*.

The effect of plant-parasitic nematodes on endomycorrhizal roots is not fully understood. Endomycorrhizae formed by phycomycetous fungi occur on more plant species than any other type, a fact not generally recognized, as this kind of infection produces little, if any, change in external root morphology. Attempts to isolate fungi from roots generally obtain soil- and root-in-

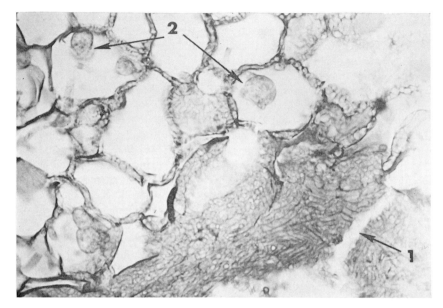

Figure 3   Phytophthora cinnamomi infection of shortleaf pine mycorrhizae parasitized by *Helicotylenchus dihystera*. Note break in fungus mantle caused by nematodes (1) and hyphae and vesicles in cortical cells surrounded by Hartig net (2). (R. O. Barham, M. S. thesis, Dept. Plant Pathology & Plant Genetics, Univ. Georgia, Athens.)

habiting fungi rather than endophytes. The vesicular-arbuscular mycorrhizae formed by *Endogone* spp., the most common endophyte formed in tree roots, require special techniques only recently devised for isolation of the symbiont (15). Past investigators of nematode-fungus disease complexes of feeder roots have generally overlooked the role of endophytes.

There is now evidence that endomycorrhizae increase plant growth (19). Clark (8) found that endomycorrhizae stimulated seedlings of yellow-poplar (*Liriodendron tulipifera*) to grow 6 times larger than nonmycorrhizal seedlings in the 12 weeks following inoculation. Baylis (2), investigating trees growing in New Zealand, concluded that in forest soils low in available phosphorus, endomycorrhizae are essential for uptake of adequate phosphorus.

Since endomycorrhizae are the common nutrient-gathering organs of trees without ectomycorrhizae, and are essential for normal growth of trees in natural woodlands, we must, therefore, evaluate the role of these endophytes in any study dealing with the etiology of feeder root damage caused by nematodes. In a recent study[3] of the effect of *Pratylenchus brachyurus* on feeder

[3] Ruehle, J. L., unpublished data.

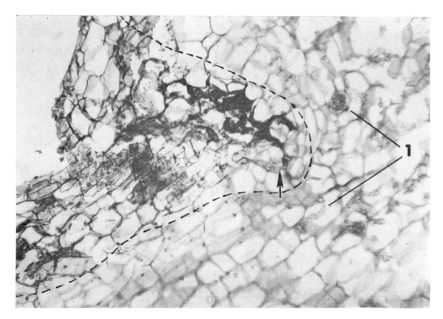

Figure 4 Lesion (bordered by dash line) in yellow-poplar root caused by *Praty-lenchus brachyurus*. Note hyphae and spore (1) of endomycorrhizal fungus in cells adjacent to damaged cortical cells; arrow denotes collapsed hypha of fungus.

roots of *Liriodendron tulipifera* seedlings, nematodes invaded cortical tissue extensively colonized by endophytic fungi of the vesicular-arbuscular type (Fig. 4). The presence of endomycorrhizal fungi failed to prevent these nematodes from causing large lesions in the cortical tissue of feeder roots. As invading nematodes collapsed the cortical cells at the periphery of lesions, vesicles of the endophyte within these cells also deteriorated. In a study of peach roots from orchards in Georgia, Marx[4] found feeder roots infected with *Endogone* spp. and galled by *Meloidogyne* sp. Endomycorrhizal fungi were readily observed in the cortex of roots free of nematodes, but were absent in gall tissue. Since there were no remnants of vesicles or arbuscules in gall tissue, he speculated that the host reaction in feeder roots infected by nematodes altered host tissue sufficiently to prevent colonization by endomycorrhizal fungi.

CONCLUSIONS

Foresters tend to judge growth and development of trees on the basis of the appearance of the crowns, and generally underestimate the importance of healthy roots. They frequently blame decreased growth and yield of forest

[4] Marx, D. H., unpublished data.

trees on poor soil fertility, drought and other seasonal deficiencies, and unfavorable soil conditions due to erosion or compaction. In many instances plant-parasitic nematodes may be the direct or indirect cause of these pathological symptoms. Unfortunately, many aspects of the parasite-host relations between nematodes and forest trees never have been explored adequately. We still lack adequate knowledge about the effects of plant-parasitic nematodes on roots of forest trees. We need to analyze the nature of nematode damage to tree roots to learn if they directly impede normal root function mechanically, chemically, or physiologically, or if nematodes act as vectors or predisposing agents for invasion by other soil organisms.

The few existing reports concerning host-parasite relations between forest tree roots and plant-parasitic nematodes deal, for the most part, with single infections and pure populations. Under field conditions, such infections are the exception and multiple infections and mixed nematode populations of both endo- and ectoparasites occur generally. Analyses of population dynamics and the distribution of certain species of nematodes in forest stands generally show that some species are more prevalent than others; however, the most numerous ones do not necessarily produce the greatest damage. In certain forest stands, particularly in southern pine plantations of trees with inherently sparse root systems, population densities for any one nematode species are usually low compared with populations of these same species associated with cultivated crops. Even though individual species are few in number in many forest stands, the aggregate in any one locality usually constitutes a high population density. There is a need for definitive studies on multiple infections interrelated with adverse environmental factors as they affect root development.

It is not surprising that soil-borne viruses transmitted by nematode vectors to forest trees are still relatively unknown. It was not until 1958 that experimental evidence proved that nematodes act as vectors of soil-borne viruses (61). Since then, nematodes have been demonstrated as vectors of a number of soil-borne virus diseases of cultivated crops. But nematode-virus relationships still have not been proved for diseases of forest trees.

Virus diseases of forest trees are potentially serious because once perennial hosts are infected, they remain infected throughout their long life span. As we shift from natural forest culture to tree plantations, often using vegetative propagation, which perpetuates viruses and encourages their widespread distribution, we increase the threat and potential importance of virus diseases in forestry. When species of any of the three dorylamoid genera known to be vectors for viruses are found associated with trees showing any of the typical virus-like leaf or general decline symptoms, a nematode-virus vector relationship should be suspected.

Soil-borne diseases of feeder roots of forest trees are more often etiological complexes involving nematodes and certain fungi rather than diseases involving single nematode pathogens. In some instances, their combined pathologic potential is greater than the sum of their individual effects. Frequently, fungi incapable of directly parasitizing feeder roots become important pathogens following nematode infection. In future work on feeder-root necrosis of tree roots, research must recognize that root-rot complexes are widespread and more important than single pathogen infections.

Any approach to soil-borne diseases of forest tree roots cannot logically overlook mycorrhizal symbionts. The relationships of ectomycorrhizae on conifers and certain hardwood tree species in the complex interactions leading to feeder-root necrosis have recently been investigated. Symbiotic ectomycorrhizal fungi form structures on the root system of trees which are readily visible and difficult to overlook in any histological study of the etiology of root pathogenesis. On the other hand, symbiotic fungi forming endomycorrhizae are often overlooked or regarded as unimportant in root studies because they are difficult to isolate and fail to produce significant changes in external root morphology.

In almost all forest locations, the mycorrhizal condition in tree roots is the rule rather than the exception (15). Nematodes parasitic on fungi suppress the fungus symbionts prior to formation of mycorrhizae, and certain plant-parasitic nematodes attack and destroy mycorrhizae directly. Hence, nematodes not only can harm the tree by preventing development of physiologically beneficial mycorrhizae, but can destroy already established mycorrhizae that serve as biological barriers to root infection by soil pathogens. Since the main feeding structures of forest trees in natural environments are mycorrhizal, any study of the complex etiology of feeder-root necrosis must consider the interrelationship between the host, nematodes, associated rhizosphere organisms, and mycorrhizal symbionts.

Literature Cited

1. Barham, R. O., Marx, D. H., Ruehle, J. L. 1973. Infection of ectomycorrhizal and nonmycorrhizal roots of shortleaf pine by nematodes and *Phytophthora cinnamomi*. *Phytopathology*. In press
2. Baylis, G. T. S. 1967. Experiments on the ecological significance of phycomycetous mycorrhizas. *New Phytol.* 66:231–43
3. Blake, C. P. 1966. The histological changes in banana roots caused by *Radopholus similis* and *Helicotylenchus multicinctus*. *Nematologica* 12:129–37
4. Bloomberg, W. J. 1968. Corky-root of Douglas-fir seedlings. *Can. Dep. For. Rural Dev., Bi-Mo. Res. Notes* 24:8
5. Cadman, C. H. 1963. Biology of soil-borne viruses. *Ann. Rev. Phytopathol.* 1:143–72
6. Christie, J. R. 1952. Some new nematode species of critical importance to Florida growers. *Soil Sci. Soc. Fla. Proc.* 12:30–39
7. Churchill, R. C., Ruehle, J. L. 1971. Occurrence, parasitism, and pathogenicity of nematodes associated with sycamore (*Platanus occidentalis* L.). *J. Nematol.* 3:189–96
8. Clark, F. B. 1963. Endotrophic mycorrhizae influence yellow-poplar seedling growth. *Science* 140:1220–21
9. De Maeseneer, J. 1964. De betekenis van vrijlevende wartelaaltjes bij het wortelrot van coniferen. *Meded. Landb Hogesch. Opzoek Stns Gent* 29:797–809
10. Donaldson, F. S., Jr. 1967. *Meloidogyne javanica* infesting *Pinus elliottii* seedlings in Florida. *Plant Dis. Reptr.* 51:455–56
11. Dwinell, L. D., Sinclair, W. A. 1967. Effects of N, P, K and inoculum density of *Verticillium dahliae* on populations of *Pratylenchus penetrans* in roots of American elm and sugar maple. *Phytopathology* 57:810 (Abstr.)
12. Feder, W. A., Feldmesser, J., Walkinshaw, C. H., Jr. 1956. Microorganisms isolated from feeder roots of citrus seedlings affected by spreading decline. *Proc. Soil Crop Sci. Soc. Fla.* 16:127–29
13. Foster, A. A. 1961. Control of black root rot of pine seedlings by soil fumigation in the nursery. *Ga. Forest Res. Comm. Rep.* 8, 5 pp.
14. Fulton, J. P. 1969. Transmission of tobacco ringspot virus to the roots of a conifer by a nematode. *Phytopathology* 59:236
15. Gerdemann, J. W. 1971. Fungi that form the vesicular-arbuscular type of endomycorrhizae. In *Mycorrhizae,* ed. E. Hacskaylo, 9–18. U.S. Dep. Agr. Misc. Publ. 1189, 255 pp.
16. Gill, D. L. 1958. Effect of root-knot nematodes on *Fusarium* wilt of mimosa. *Plant Dis. Reptr.* 42:587–90
17. Glori, A. V., Postrado, B. T. 1969. Control of root-knot nematode on Kaatoan Bangkal by soil fumigation. *Reforestation Admin. Philippines Res. Note 3,* 6 pp.
18. Goodey, J. B. 1965. The relationships between the nematode *Hoplolaimus uniformis* and sitka spruce. *Great Brit. Forest Comm. Bull.* 37:210–11
19. Gray, L. E. 1971. Physiology of vesicular-arbuscular mycorrhizae. In *Mycorrhizae,* ed. E. Hacskaylo, 145–50. U.S. Dep. Agr. Misc. Publ. 1189, 255 pp.
20. Harrison, B. D. 1964. Infection of gymnosperms with nematode transmitted viruses of flowering plants. *Virology* 24:228–29
21. Hashimoto, H. 1962. Some observations on damage of nematodes in forestry nurseries. (Jap.) *Jap. Forest Soc. J.* 44:248–52
22. Henry, B. W. 1953. A root rot of southern pine nursery seedlings and its control by soil fumigation. *Phytopathology* 43:81–88
23. Hibben, C. R., Walker, J. T. 1971. Nematode transmission of the ash strain of tobacco ringspot virus. *Plant Dis. Reptr.* 55:475–78
24. Hibben, C. R., Bozarth, R. F. 1972. Identification of an ash strain of tobacco ringspot virus. *Phytopathology* 62:1023–29
25. Hijink, M. J. 1969. Groeivemindering van fijnspar veroorzaakt door *Rotylenchus robustus. Mededel. Rijksfakulteit LandbouwWettensch. Gent.* 34:539–49
26. Hopper, B. E. 1959. Three new spe-

cies of the genus *Tylenchorhynchus* (Nematode: Tylenchida). *Nematologica* 4:23–30

27. Hopper, B. E., Padgett, W. H. 1960. Relationship of nemas (nematodes) with the root rot of pine seedlings at the E. A. Hauss State Farm Nursery, Atmore, Alabama. *Plant Dis. Reptr.* 44:258–59

28. Hsu, D. 1972. *Influence of Criconemoides quadricornis on pecan feeder root necrosis caused by Pythium irregulare and Fusarium solani at different temperatures.* Ph.D. thesis, Univ. of Georgia, Athens, 14 pp.

29. Johnson, A. W., Ratcliffe, T. F., Freeman, G. C. 1970. Control of *Meloidogyne incognita* on dogwood seedlings by chemical dips. *Plant Dis. Reptr.* 54:952–55

30. Lordello, L. G. E. 1967. A root-lesion nematode found infesting Eucalyptus trees in Brazil. *Plant Dis. Reptr.* 51:791

31. Lownbery, B. F. 1956. *Pratylenchus vulnus*, primary cause of the root-lesion disease of walnuts. *Phytopathology* 46:376–79

32. Mamiya, Y. 1970. Parasitism and damage of *Pratylenchus penetrans* to Cryptomeria seedlings. (Jap.) *J. Jap. Forest Soc.* 52:41–50

33. Marx, D. H. 1972. Ectomycorrhizae as biological deterrents to pathogenic root infections. *Ann. Rev. Phytopathol.* 10:429–54

34. Marx, D. H., Davey, C. B. 1969. The influence of ectotrophic mycorrhizal fungi on the resistance of pine roots to pathogenic infections. III. Resistance of aseptically formed mycorrhizae to infection by *Phytophthora cinnamomi. Phytopathology* 59:549–58

35. Minderman, A. 1956. Aims and methods in population researches on soil-inhabiting nematodes. *Nematologica* 1:47–50

36. Nemec, S., Struble, F. S. 1968. Response of certain woody ornamental plants to *Meloidogyne incognita. Phytopathology* 58:1700–03

37. Nemec, S., Morrison, L. S. 1972. Histopathology of *Thuja orientalis* and *Juniperis horizontalis plumora* infected with *Meloidogyne incognita. J. Nematol.* 4:72–74

38. Newhall, A. G. 1958. The incidence of Panama disease of banana in the presence of the root-knot and the burrowing nematodes (*Meloidogyne* and *Radopholus*). *Plant Dis. Reptr.* 42:853–56

39. Oostenbrink, M. 1966. Major characteristics of the relation between nematodes and plants. *Meded. LandbHoogesch. Wageningen* 66 (4), 46 pp.

40. Powell, N. T. 1971. Interaction of plant parasitic nematodes with other disease-causing agents. In *Plant Parasitic Nematodes*, Vol. 2, eds. B. M. Zuckerman, W. F. Mai, R. A. Rohde, 119–36. Academic: N.Y., 347 pp.

41. Reinking, O. A., Radewald, J. D. 1961. Cadang-Cadang disease of coconuts in Guam may be caused by a soil-borne plant virus spread by dagger nematodes (*Xiphinema* sp.) *Plant Dis. Reptr.* 45:411–13

42. Riffle, J. W. 1967. Effect of an *Aphelenchoides* species on the growth of a mycorrhizal and pseudomycorrhizal fungus. *Phytopathology* 57:541–44

43. Riffle, J. W. 1972. Effect of certain nematodes on the growth of *Pinus edulis* and *Juniperus monosperma* seedlings. *J. Nematol.* 4:91–94

44. Riffle, J. W. 1973. Histopathology of *Pinus ponderosa* ectomycorrhizae infected with a *Meloidogyne* species. *Phytopathology.* In press

45. Riffle, J. W., Lucht, O. D. 1966. Root knot nematodes on ponderosa pine in New Mexico. *Plant Dis. Reptr.* 50:126

46. Riffle, J. W., Kuntz, J. E. 1967. Pathogenicity and host range of *Meloidogyne ovalis. Phytopathology* 57:104–07

47. Ruehle, J. L. 1962. Histopathological studies of pine roots infected with lance and pine cystoid nematodes. *Phytopathology* 52:68–71

48. Ruehle, J. L. 1967. Distribution of plant-parasitic nematodes associated with forest trees of the world. *U.S. Dep. Agr., S.E. Forest Exp. Sta.*, Asheville, N.C., 156 pp.

49. Ruehle, J. L. 1968. Pathogenicity of sting nematode on sycamore. *Plant Dis. Reptr.* 52:524–25

50. Ruehle, J. L. 1969. Influence of stubby-root nematode on growth of southern pine seedlings. *Forest Sci.* 15:130–34

51. Ruehle, J. L. 1972. Pathogenicity

of *Xiphinema chambersi* on sweet-gum. *Phytopathology* 62:333-36

52. Ruehle, J. L., Sasser, J. N. 1962. The role of plant-parasitic nematodes in stunting of pines in southern plantations. *Phytopathology* 52:56-58

53. Ruehle, J. L., Marx, D. H. 1971. Parasitism of ectomycorrhizae of pine by lance nematode. *For. Sci.* 17:31-34

54. Rühm, W. 1959. Nematoden und Forstflanzen. *Merck. Blatter 9, Ser. 3*, 1-16

55. Seliskar, C. E. 1966. Virus and viruslike disorders of forest trees. *Food & Agr. Org./IUFRO Symp. Int. Dangerous Forest Dis. Insects,* Oxford, 1964, Vol. 1, Meeting V, FAO, Rome, 44 pp.

56. Sturhan, D. 1963. Der pflanzenparasitische Nematode *Longidorus maximus,* seine Biologie und Ökologie, mit Untersuchungen an *L. elongatus und Xiphinema diversicaudatum. Z. Angewandte Zool.* 50:129-93

57. Sutherland, J. R. 1967. Failure of the nematode *Aphelenchus avenae* to parasitize conifer seedling roots. *Plant Dis. Reptr.* 51:367-69

58. Sutherland, J. R., Adams, R. E. 1964. The parasitism of red pine and other forest nursery crops by *Tylenchorhynchus claytoni* Steiner. *Nematologica* 10:637-43

59. Sutherland, J. R., Fortin, J. A. 1968. Effect of the nematode *Aphelenchus avenae* on some ectotrophic, mycorrhizal fungi and on a red pine mycorrhizal relationship. *Phytopathology* 58:519-23

60. Sutherland, J. R., Dunn, T. G. 1970. Nematodes in coastal British Columbia forest nurseries and associations of *Xiphinema bakeri* with a root disease of Douglas-fir seedlings. *Plant Dis. Reptr.* 54:165-68

61. Taylor, C. E. 1971. Nematodes as vectors of plant viruses. In *Plant Parasitic Nematodes,* Vol. 2, eds. M. Zuckerman, W. F. Mai, R. A. Rohde, 185-211. Academic: N.Y., 347 pp.

62. Thomas, P. R. 1970. Host status of some plants for *Xiphinema diversicaudatum* (Micol.) and their susceptibility to viruses transmitted by this species. *Ann. Appl. Biol.* 65:169-78

63. Wallace, H. R. 1964. *The biology of plant parasitic nematodes.* St. Martin's Press, N.Y., 280 pp.

CYTOLOGICAL AND HISTOLOGICAL ABERRATIONS IN WOODY PLANTS FOLLOWING INFECTION WITH VIRUSES, MYCOPLASMAS, RICKETTSIAS, AND FLAGELLATES

❖ 3568

Henry Schneider

Department of Plant Pathology, University of California, Riverside, California

INTRODUCTION

To understand how a disease is produced by a systemic, intracellular submicroscopic (below ca 0.3μ) agent, it is necessary to study the complete sequence of pathological aberrations and host responses that occur following infection. In such studies of pathogenesis or pathogeny (113) one gathers information about: (*a*) how and where the causal agent enters the host; (*h*) which cells are infected and how they are affected by the initial infection; (*c*) the effect of infected cells on adjacent noninfected "cells" in producing diseased tissues; (*d*) the host response to the diseased tissues such as formation of wound periderm or tyloses; and (*e*) the maleffects of the tissues on other parts and functions of the plant. A standard set of steps in pathogenesis cannot be formulated for all diseases because the steps vary with the disease, the host, and the vector.

Most plant pathologists do not consider all of these steps but rather specialize in certain well defined areas of experimentation such as: (*a*) infection processes (93, 100); (*b*) synthesis of virus particles (5, 72, 82); (*c*) translocation of the agent (93); etc. This paper will consider cytological and anatomical aspects of pathogeny beginning with the first appearances of the agent and/or with the first pathologic changes in the host that are detectable with the light or electron microscope.

Until Doi et al (17) discovered mycoplasma associated with several plant diseases, and Goheen et al (40) discovered Rickettsia-like organisms associated with Pierce's disease, all graft-transmissible diseases with submicro-

119

scopic, intracellular causal agents were considered to be virus induced (19, 22, 25, 28).

Diseases induced by the following graft-transmissible agents will be considered: (a) Viruses; (b) Mycoplasmas; (c) Rickettsias; (d) Flagellates; (e) unknown agents. Flagellates are included because they produce symptoms similar to mycoplasmas and some viruses. Most reports of discoveries of "virus-like particles" and "mycoplasmalike bodies" in tissues will not be included. Diseases with associated mycoplasmalike bodies have been discussed recently (15, 44, 55, 66, 119).

In this paper, studies of cells and of tissues will be referred to as cytopathology and histopathology, respectively, and together as pathological anatomy. Outward effects of the disease on an organ or on the plant as a whole will be referred to as symptoms. The infectious agents with which we are here concerned multiply only in "cells." The term "cell" in quotation marks will include sieve-tube elements and xylem-vessel members. Pathosis or pathology will be used to denote diseased cells or tissues. For instance a pathological condition at the bud union will simply be called the bud-union pathology. It is proper to think of diseases as developing—especially in the cases of tumors and enations—hence the term "developmental study." In the case of diseases leading to necrosis and other degenerative aberrations the word development is not very applicable. The term developmental anatomy will be used for studies of the normal, growing plant and with care when used in connection with diseases.

Pathological anatomists early described some standard pathological phenomena but without reference to sequential aspects of pathogenesis (e.g. 61, see also 19, 22). They described necrosis, hypertrophy, hyperplasia, hypoplasia, lysis, wound-periderm formation, tyloses, suberization, as well as wound-gum and callose deposition. When intracellular agents intermingle with host cell protoplasts, it is difficult to know what stimuli caused what response. Necrosis and hypoplasia apparently result from the interference by the agent with the cell's normal physiological and biological processes. Phenomena such as callose deposition and wound-periderm formation may be host responses to the causal agent or to the pathology it has induced. Some phenomena such as degeneration of cells, callose deposition on sieve plates, and formation of tyloses occur as part of the normal seasonal degeneration of tissues or "cells." These phenomena are pathological when they are untimely or occur in tissues that should remain functional.

Certain characteristic pathologic phenomena may occur in cells when multiplication of the causal agent occurs in them. Among these are: increase in thickness of the cytoplasm; deterioration of organelles and membranes (lysis); and the abnormal appearance of structures such as inclusion bodies, virions aggregated into paracrystaline bodies, etc.; and ultimately even necrosis. The character of the cellular pathology that is induced by graft- and insect-transmitted causal agents usually is more specific for disease identification than histopathology and external symptoms. This is especially true if the causal agent's morphology may also be used for identification.

Unfortunately, pathologists frequently do not determine degenerative sequences in diseased plants. They tend to section plants in an advanced state of disease. By this time so many interactions have been superimposed upon each other that the description is of little value in helping to understand how the condition arose. The problem is vaguely comparable to studying the wreckage of two airplanes that collided, in an attempt to reconstruct the events leading to the crash. Although some evidence of the initial damage may remain in the rubble, knowledge of initial events is of great value in diagnosing the cause.

The degree of tissue maturation has an effect on disease development. Although most agents thrive in growing tissues, a few are found only in mature tissue; others incite disease in tissues of any age. In the case of tobacco mosaic virus, leaves that were mature at the time of inoculation showed little structural aberration, but leaves that were in the primordial stage of development did not differentiate into normal mature tissues (41). On the contrary, in pear-decline-diseased trees that resume growth after chilling to break dormancy, sieve-tube necrosis did not appear in the major lateral veins of leaves on the new growth that ensued until formation of secondary phloem began (102).

Based on symptoms of transmissible diseases, two categories were established: mosaics typified by tobacco mosaic, and yellows typified by peach yellows (113, p. 470). Some yellows affected plants also show witches brooms, small leaves, poorly formed fruit, and—especially in herbaceous hosts—floral deformities such as virescence and phyllody. Thomas (108) suggested a third category—rough bark diseases—to include such diseases as psorosis of citrus, stony pit of pear, and diamond canker of French prune. These diseases affect bark primarily, wood occasionally, and leaves and fruit infrequently. Many diseases fall nicely into these three categories, but many others do not, in part because symptom syndromes vary with the host. Some symptoms encountered singly or in combinations that do not fall into these categories are: pitting or grooving of the wood, vein clearing, chlorotic and necrotic ring spotting, wilting, decline, epinasty, enations, tumors, etc.

The direct approach for determining pathogenesis is to make anatomical studies just before onset of infection, during onset, and then periodically during development of the disease. With some diseases such as wound tumor of clover and Tristeza of citrus, inoculation attempts are predominantly successful, the incubation period is short and predictable, and developmental studies are relatively easy (62, 87). On the other hand, when successful inoculations occur erratically and the incubation period is long and unpredictable, as with the Rickettsia induced Pierce's disease of the vine, early stages of disease development are found only after sectioning many samples (24). Also, anatomical change may not always occur; for example, mycoplasmas may do part of their damage by producing toxins that merely suppress growth (16).

If the site of initial pathology is not known, exploratory sectioning should be carried out on the organ where symptoms first appear. In some diseases, however, visible symptoms are secondary, and one must search elsewhere for the initial events.

Various schemes have been proposed for classifying diseases caused by graft-transmissible causal agents. Terms such as phloem-limited viruses (22, 25), aphid-transmitted viruses (38), etc. have been used rather than "diseases with the agent limited to the phloem" or "diseases with the agent transmitted by a certain insect species."

Broad classifications of causal agents by systems heavily weighted on a single characteristic have not been successful. Such classifications have been based on: the type of symptoms they cause (50, 69); the species of insect

Table 1   Esau's 1948 and 1956 Proposed Classification

Kind of Virus	Example of Disease
I. Ubiquitous viruses (mosaics)	Tobacco mosaic
II. Phloem-limited viruses with primary pathology in the phloem. (Yellows)	
A. Death and collapse of sieve tubes	Peach Western X Potato leaf roll
B. Hyperplasia in phloem parenchyma. Daughter cells differentiate into sieve tubes. Then necrosis of the tissue	Curly top of sugar beet
III. Xylem-limited viruses, primary pathology in the xylem	Pierce's disease of the vine Phoney peach disease

that transmits the disease (38); and the tissue in which the infectious agent occurs. Before the discovery of Mycoplasmas and Rickettsias, Esau (22, 25) suggested the "broad" classification for viruses outlined in Table 1. The tissue relationships outlined were based on studies of pathological anatomy, experiments involving ringing and leaf shading, the feeding habit of the insect vector, inoculation experiments, etc.

We now recognize that ubiquitous viruses are true viruses, phloem limited viruses are a mixture of viruses and mycoplasmas, and the xylem-limited viruses appear to be Rickettsias. Very similar histopathology and symptoms can be produced by diverse causal agents. For instance, Esau's category II A contains both X-disease (buckskin) of peach, associated with mycoplasma-like bodies (42, 64, 73), and potato leaf roll, associated with hexagonal virus particles (e.g. 59). Obviously, it is not always sound to classify causal agents only on the basis of the tissue to which they are limited. With regard to structural aberrations very different cyto- and histopathology can be produced simply by changing the strain of a given virus (4), or by varying both the host and virus (87).

There is greater utility in grouping diseases than grouping causal agents. Diseases of woody plants may be classified descriptively as: phloem diseases, xylem diseases, rough bark diseases, stem pitting diseases, and tumorous diseases just as fungal diseases are categorized as: mildews, stem cankers, root rots, soft rots, and vascular wilts. Although the causal agents within categories may be diverse, the nature of the disease should be apparent from its category.

Woody plants live for many years and undergo far more secondary growth than do herbaceous plants. Secondary xylem accumulates in annual increments; in some trees, this is also true of the secondary phloem. In time, the rings of xylem and phloem lose their primary functions and some of their "cells" degenerate. Normally degenerated "cells" should not be confused with diseased necrotic "cells." As the xylem cylinder enlarges, the bark is forced to dilate. In certain trees dilation of bark is accommodated by radially-oriented, transitory, dilation meristems (27, 85) that form radially through the rays, the dilation tissue of previous years, and the phelloderm. Tangential sheets of phellogen simultaneously form in conjunction with them in the phelloderm just inside the phellem. When new periderms form, the periderms of previous years may either slough off or accumulate as rhytidome. In some woody plants such as the grape, the outer bark is exfoliated each year by a periderm that forms in the previous year's phloem (23). Diseases such as psorosis and exocortis affect citrus bark, which retains its original cortex and phloem (85). The bark pathology of these diseases appears with advancing years and hence with the aging of cells (90, 117).

Inclusions and virions aggregated into paracrystalline bodies are not often found in woody plants perhaps because these plants often accumulate tannins and similar substances (36) that make cellular studies difficult. Woody plants are conspicuous by their absence in the review of viral inclusions by McWhorter (70). In some species of trees and shrubs, tyloses occur in the xylem vessels of the older wood, a phenomenon that is pathological when it occurs in the current annual ring of xylem.

Desirable varieties of woody plants often are propagated from cuttings or by grafting them onto rootstocks of the same or a different species. Infectious agents, many of them latent, are carried in the propagative parts. Since scions and rootstocks usually differ genetically, they may differ in susceptibility to diseases, the causal agent being latent in one of the graft partners and virulent in the other.

In this review of cytological and histological aberrations, certain nonanatomical aspects essential to understanding each disease will be mentioned first. Then information will be given on the initial and subsequent location of the causal agent, initial injury to the host, secondary or remote sites of injury, and on how the structural aberrations ultimately produce symptoms. At least one disease will be reviewed to demonstrate each type of degenerative sequence.

Table 2 Key to Virus Diseases Having Various Virion Distributions and Degenerative Sequences

	Examples of Diseases
I. Virus Distribution Ubiquitous	
A. The mosaics	Tobacco mosaic
B. Trunk and stem diseases	
1. Rough bark diseases	Exocortis of *Poncirus*
2. Distortion of trunks and limbs[a]	
II. Phloem Affinity Viruses[b]	
A. Phloem-limited pathosis	
1. Necrosis is the primary tissue aberration	
a. Sieve-tube necrosis is primary. Adjacent parenchyma may hypertrophy. Cambial hyperactivity and copious replacement phloem may follow	Potato leaf roll; and Seedling-yellows tristeza of lemon
b. Phloem necrosis is primary	
2. Hyperplasia is the primary tissue aberration	
a. General hyperplasia in phloem is followed by necrosis	Sugar beet curly top[c]
b. Tumors arise from one or a few tumor cells	Wound tumor virus
B. Primary infection is in the phloem, but locally infections spread to other tissues	
1. Meristems adjacent to phloem invaded and disorganized	Tristeza of Mexican lime
2. Many nearby tissues invaded	Beet yellows
C. Diseases with more than one degenerative sequence	Various strains of tristeza in two host types

  [a] Examples of these are in a later section on "Transmissible diseases with associated agent undetermined."
  [b] Degenerative changes used for this classification are at the tissue level. Cytopathological changes precede them but their early appearance has only been reported for some of the diseases.
  [c] Evidence for the virus nature of the causal agent has not yet been published and placement with the viruses is tentative.

VIRAL DISEASES

Table 2 contains a tentative scheme for classifying plant virus diseases. It is based partly on virion distribution in the host and partly on types of pathogenesis. It has nothing to do with classification of viruses. Since varying the viral strain or the host variety or species may throw the disease into another degenerative sequence, these factors must be specified. If we regard a disease as being a plant interacting with a causal agent, then a disease that occurs in one host may differ from that in another infected with the same causal agent if certain tissues in the hosts show differences in susceptibility or if certain tissues or "cells" react differently or with varying intensity to the infection.

Most of the transmissible diseases of woody plants are classified as virus diseases because virus-like particles have been associated with them. In many cases, Koch's postulates have not been carried out for these diseases.

Diseases with Ubiquitous Virus Distribution

THE MOSAICS   Although the tobacco plant (*Nicotiana tabacum*) is only semiwoody, *tobacco mosaic* will be used as representative of the mosaic group because it is the most widely studied virus disease of plants. Leaves of susceptible varieties become chlorotic with buckled, elongated islands of normal green tissue usually along the major lateral and sublateral veins. The plants show some stunting with slightly shortened internodes. Leaves that are mature when infection occurs are only slightly blotched and undistorted while leaves infected in the primordial stage become severely mottled. In extreme cases the formation of laminae on enlarging leaf primordia may be so impeded that the resulting leaves are lobed, narrow bladed, or shoe strings (41, 107).

Nicotiana glutinosa is resistant to tobacco mosaic and attempts to inoculate it result in local lesions of very different tissue pathology than that of mosaic-susceptible plants (28).

Several strains or mutants of TMV have been described as "defective strains"; they produce variations in cyto- and histopathology as well as in symptoms.

Infections of epidermal cells apparently occur through wounds which can be created experimentally by rubbing the leaves and during natural spread of the disease by other abrasions (72, 100). Although early stages of infection have been studied intensively (5, 72, 82), most workers have emphasized the sites of virus formation. Upon infection of tobacco epidermal or hair cells, there is a build-up of virus RNA in the nucleus or nucleolus. After about 2 hours the RNA begins to move into the cytoplasm and by hour 7 it has spread throughout the cytoplasm. The site of virus-protein synthesis is not known; assemblage of TMV particles may occur in the cytoplasm (82). Several structures foreign to the cell appear sequentially in the protoplast. These include virus particles, monolayers of virions, crystallinelike virus bodies, x bodies, and tubules (broad proteinaceous filaments) occurring either in masses or in aggregates of three (5, 30, 31, 60, 96).

Anatomical aberrations occur in newly forming leaves of *Nicotiana tabacum* in all stages of their development. When marginal meristems of primordia become pathologic, deformed leaves designated "shoestrings," "narrow bladed" and "Frenched" result (41, 107). The mosaic pattern in the laminae results from maldifferentiation of the palisade and spongy parenchyma and of the chloroplasts. Differentiation of cells in these localized areas is impeded at whatever developmental stage the leaf happens to be when infection occurs. The affected areas remain yellow or light green. Dark green areas develop normally but they buckle due to confinement within the nonexpanding yellow areas. The vaguely blotched leaves that are histologically mature when infection occurs show only deterioration of chloroplasts (41). The anatomy of the phloem is not affected (21).

Infected cells may contain amoeboid-shaped x bodies and/or crystalline-like striate material. Chloroplasts are scarce, small, and either degenerate or poorly formed while the nuclei may be enlarged. X-bodies and striate bodies may be found in every kind of "cell" throughout the plant including guard cells and apical cells (20, 21, 31, 41). Electron microscope studies reveal that the x-bodies are an accumulation of endoplasmic reticulum, ribosomes, some dictyosomes, and enclaves of virus particles and broad, tubular, protein-aceous filaments (31, 96). The striate bodies observed by earlier workers are aggregates of virus rods with parallel arrangement. Virus particles occur most abundantly in the cytoplasm but have been found in nuclei, chloroplasts, vacuoles, and between plasma membranes and cell walls (29, 31).

Tobacco mosaic is a disease in which cells throughout the plant may contain the causal virus and become pathological. They do not die and are still capable of division (33). This is in contrast to diseases in which a tissue critical to the development or function of the plant, such as the cambium or phloem, becomes pathological and affects the plant through its disfunction. An exception is when the marginal meristem of developing leaves either fails to form, or forms and then malfunctions. As a result the laminae do not form in their entirety.

Pear vein yellows is a mosaic-type disease of *Pyrus communis* L. that has virus-like flexous rods (17–22 × 700–800 nm) associated with it. In tissues there exist both virus-like aggregates and inclusion bodies—unusual for woody plants (47). The yellow areas of the leaf show underdevelopment of the mesophyll cells and of intercellular spaces—a condition typical for mosaics. Chloroplasts are fewer and smaller than normal.

TRUNK AND STEM DISEASES   The *rough bark diseases* are characterized by symptoms such as flaking off of bark tissues, bark cracking, necrosis, and gum exudation. Leaf symptoms of a subtle and sometimes transitory nature may or may not be part of the syndrome. Most of the rough bark diseases such as psorosis of citrus, corky bark of vitis, and diamond canker of French prune are placed in the category of diseases for which a suspect causal agent has not been found, even though they are likely virus diseases.

Bark flaking is accomplished by wound periderms that form in variously oriented planes across living but pathological bark tissue. Some patches of bark may die to the cambium and not flake off. Water soluble gum may exude through splits in the bark. The source of gum in several diseases of viral or suspected viral origin is dissolution of immature xylem mother cells; gum pockets result. Gum pocket formation is a process common in species of both Citrus and Prunus that are suffering from several diseases and nutritional imbalances.

Exocortis of the trifoliate orange, *Poncirus trifoliata*. Bark scaling occurs on the trunk, limbs, and all but the smallest branches where chlorosis and mottling may occur. The disease is troublesome when the trifoliate orange is used as a rootstock for citrus. Large patches of outer bark scale off the rootstock portion and the trunk is more completely affected than the localized, flaky scaling of psorosis. The virus has been partially purified (95); it is mechanically transmitted (37).

The pathogeny of excortis disease has not been studied, but sections through scaling bark reveal that scaling is brought about by successive layers of wound periderm that form first in the outer bark and then increasingly closer to the cambium (90). The periderms apparently form immediately under necrotic tissues; or under ray and phloem parenchyma cells affected by necrosis, hypertrophy, and hyperplasia. Sieve-tube necrosis also occurs. Psorosis is a similar disease that has been more thoroughly investigated but for which virus particles have not been found.

In citron plants, *Citrus medica,* there are symptoms of browning and cracking on the leaf midribs when infected with the exocortis virus. Necrosis occurs in "cortical" cells of the midrib, accompanied by hypertrophy and hyperplasia (35).

Trunk and limb distortion diseases such as concave gum of citrus and flat limb of apple may eventually be included here, but at present a viral cause is not evident and anatomical studies are deficient. Flat limb is briefly considered in the section for diseases with undetermined causal agents.

Phloem Affinity Viruses

A number of graft transmissible diseases with phloem pathology have virus-like particles associated with them. Where known, their insect vectors are phloem feeders. Degenerative pathways for the diseases are outlined in Group II of the key to virus diseases (Table 2). Insect vectors possibly introduce these viruses into the sieve tubes and then primary infections occur in adjacent parenchyma cells. Infected cells of some diseases have denser, thicker, and more stainable cytoplasm. Pathosis may be phloem limited; or it may spread out from the phloem.

PHLOEM LIMITED PATHOSIS   *Sieve-tube necrosis is primary.* Both potato leaf roll and the seedling-yellows tristeza disease of citrus belong in this class and have virus-like particles associated with them. For the sake of clarity and organization tristeza will be discussed under the heading: "Diseases with more than one degenerative sequence."

Potato leaf roll occurs in two stages. Primary infection results when the causal agent is introduced by the peach aphid, *Myzus persicae* Sulz. (113); secondary infections occur when infected tubers are used as seed pieces. In the former, the margins of younger leaflets roll upward toward the midvein;

or if infections occur late in the season symptoms may not develop. When young shoots emerge from seed pieces, rolling occurs in the older leaves first and later in the younger ones. Strands of necrotic phloem are visible macroscopically as a network in tubers in the primary but not the secondary stages of the disease. Virus particles are 25 nm spheres (59).

An attempt has been made to diagnose several potato diseases anatomically. Three of these diseases involved necrosis of phloem; two diseases, acronecrosis and acropetal necrosis, involved other tissues in addition to phloem. "Leaf roll" was called "phloem necrosis" by these pathologists, and there is now general agreement that the necrosis occurring in the primary phloem is limited to the primary phloem strands and that affected strands become red when treated with phloroglucinol and hydrochloric acid (1, 2, 8, 77, 97).

The above papers mention that the phloem strands were necrotic, without going into detail as to which cells were involved, although Quanjer (77) mentioned sieve tubes and companion cells. His work has been taken as evidence (19, 34) that degeneration at least begins with necrosis of the sieve tubes and companion cells with parenchyma cells later becoming affected. Photomicrographs by Sheffield (97) indicate that secondary phloem, and xylem formation may also be involved.

Hyperplasia is the primary anatomical aberration. Hyperplasia is general. The causal agent of curly top of sugar beets has yet to be reported. The vector, *Circulifer tenelus,* is a phloem-feeding leaf hopper. Physiological studies indicate that the causal agent is phloem limited (12). Veins on the stunted, rolled leaves become prominent, sometimes with enations. Initially, in the primary phloem, parenchyma cells adjacent to sieve tubes stain prominently. Then disorderly divisions of parenchyma cells occur with derivatives differentiating into sieve tubes. All of the abnormal phloem thus formed eventually becomes necrotic (18).

Tumors arise from one or a few cells. These diseases are characterized by galls or tumors that form on leaves, stems, or roots. Tumors induced in clover shoots by the wound tumor virus (62), and in citrus leaves by a probable citrus-vein-enation virus (52), originate with the formation of darkly staining "tumor initial cells" from one or a packet of cells next to a primary-phloem sieve-tube element in a procambial strand where proto-phloem is differentiating. The cells divide repeatedly, giving rise to a mass of abnormal vascular tissue.

Wound tumor virus has a broad host range and is transmitted by leaf hoppers (several species of Agallia and Agalliopsis) or by grafting. An exceptionally well illustrated developmental study of the disease was made by Lee (62) in sweet clover plants. The tumor initial cells are derived from phloic procambial cells or more often from their derivatives destined to become parenchyma cells or phloem fibers. The conversion of these cells into tumor cells varied but in essentially all of them the protoplasm became dense and contained spherules like those described by Littau & Black (63). The cells divided by abnormally oriented walls, to form spindle shaped cell clusters. These cell masses increased in size by additional divisions of tumor cells, with some of the derivatives becoming necrotic and others undergoing hypertrophy. Some cells partially differentiated into radially arranged groups of tracheary elements and others into "pseudo phloem." In early stages, the tumors tended to be spherical in shape but as growth continued they became nodular. As this occurred the tumor was composed of spherule-containing meristematic cells between the poorly developed xylem and phloem areas. The cortex and epidermis, which expanded to accommodate the enlarging tumor, appeared to be normal. These tissues were not derived from the tumorous cells. Shikata & Maramorosch (98) observed virions in tumors but not in epidermis. The virions were 60 nm in diameter and hexagonal in shape like those from purified preparations. Sometimes they were enclosed in membranes. Perhaps these were the spherules observed by Littau & Black (63).

Infection spreads out of the phloem; meristems adjacent to phloem are invaded locally and disorganized. This kind of degeneration occurs with the common California strain of tristeza in Mexican lime trees, and possibly also the stem-pitting diseases of peach and of apple, but an associated virus particle has not been found for the latter two diseases.

Many nearby tissues are invaded. Virus-induced yellows of sugar beet appears initially to cause necrosis in the phloem, then pathology appears in adjacent tissues (26). The disease has not been studied developmentally. The virus is transmitted by several species of phloem-feeding aphids (11); translocation probably occurs in the phloem. Masses of long flexuous virus particles occur in the sieve elements and sieve-plate pores as well as in parenchyma cells (32).

Diseases with more than one degenerative sequence. By varying the host variety and the virus strain (43, 116), different symptoms syndromes may be produced in Tristeza disease (87). The yellows type of symptoms (II-A-1-a) are obtained in lemon and sour orange with the seedling-yellow-tristeza strain of virus. Vein clearing and wood pitting are obtained in Mexican lime and sweet orange infected with the common California virus strain. The wood pitting becomes stem pitting if interference with normal cambial activity is severe and prolonged enough to result in sunken areas in the stem (II-B-1). A third symptom syndrome occurs when a pitting type of host is grafted onto the yellows type host; necrosis of sieve tubes occurs immediately below the union. This remote injury may result from hypothetical toxins produced in pathological cells far up the tree in the young shoots. Following sieve-tube necrosis, reserve starch is depleted from the rootstock, fibrous roots die, and the trees either wilt or decline in vigor due to the necrosis of sieve tubes (84).

Virus-like particles associated with tristeza are long flexuous rods (10–12 × 2000 nm). They have been partially purified (6, 58), but the disease has not been reproduced by reinoculating with them.

In Mexican lime, the earliest pathological aberrations occur in new shoot growth about 2 weeks after graft inoculation when chromatic cell formation is prolific (86, 87). The cytoplasm of parenchyma cells next to sieve tubes both in primary and secondary phloem begins to thicken, assumes a clear nonvesicular condition, and then masses, strands, or needle-like bodies that stain darkly with hematoxylin appear within it. The nucleus remains intact. After applying Bald's (3) Giemsa Stain, Schneider (86, 87) observed differential staining within the chromatic cells. The masses and strands were purplish red and other structures were blue. A positive test for argenine was also obtained in affected cells. Both tests are indicative of virus, but they need to be repeated with longitudinal sections of newly-inoculated, young stems in order to be more certain which areas of the protoplasts—the cytoplasm or the masses and strands—give the color reaction. Electron microscopy reveals that part or most of the clear cytoplasm is composed of fine flexuous rods (10–12 nm) without an orderly arrangement (92), while the structures that stain darkly with hematoxylin are composed of aggregates of tubules arranged in parallel. Opinion differs as to what constitutes the virus—the aggregates of tubules or the flexuous randomly arranged rods (92). Shikata & Sasaki (99) regarded the tubes as P-protein, as described by Esau & Cronshaw (30), and the fine flexuous rods as the virus. Others have regarded the tubules as virus (57, 75). In addition to resembling P-protein, the tubules have some resemblance to the proteinaceous filaments associated with tobacco mosaic.

In localized areas of stems with secondary growth, the chromatic condition "spreads" to the cambium; and in newly forming stems, to the ground meristem. In locales where the cambium is affected, xylem and phloem mother cells are not initiated but only an unorganized parenchmatous tissue. When this tissue adheres to the bark upon removal of bark from the wood, pits result in the wood.

In seedling-yellows tristeza of lemon seedlings, a yellows type host, phloem parenchyma cells next to sieve tubes become chromatic as in the pitting type host described above. In this host, however, sieve tubes near chromatic cells become necrotic. This is especially true in young developing roots where phloem forms rapidly to replace the phloem in which the sieve tubes became necrotic. When this occurs, xylem formation is greatly suppressed, and the root system fails to develop. Sieve-tube necrosis and cambial hyperactivity occur with both the common strain and seedling-yellows strain of tristeza; they are mild and inconsequential for the common strain, but drastic and devastating for the seedling-yellows strain (87).

In grafted trees, when the sweet orange (a pitting type host) is used as a top with sour orange rootstocks, one may find chromatic cells in the vascular bundles of stems, leaves, and fruits. Occasionally wood pitting occurs. In small developing roots of the sour orange, chromatic cells and sieve-tube necrosis occur. In the trunk immediately below the bud union, the sour orange sieve tubes become necrotic (84). This essentially girdles the tree, and a sequence of reactions follows in the bud-union phloem. Callose forms on the older sieve tubes above the bud union, the cambium becomes hyperactive producing smaller than normal phloem- and phloem-ray cells. Phloem rays below the bud union may become hyperplastic and woody. After the newly formed sieve tubes below the bud union have functioned for a short time, they too become necrotic. As a result of the girdling, starch being translocated from the tops accumulates above the bud union. Eventually, when the reserve starch in the rootstock is used up, the starved roots decay and the tree wilts or declines.

In *Afraegle* and *Aeglopsis,* two close relatives of *Citrus,* McClean (67) found that inoculations with tristeza virus could be effected by using *Toxoptera citricidis* (Kirk) as a vector, but not by using infected tip grafts from Citrus, which grew poorly and showed signs of incompatibility. Schneider (89) studied the graft unions and reported that sieve tubes were not continuous across the graft unions, although parenchyma cells were, indicating that longitudinal movement of the virus occurs in the sieve tubes. In Egypt, Nour-Eldin and colleagues (74) may have transmitted tristeza virus from the cambium of Mexican lime to that of Aeglopsis by approach grafting diseased to healthy plants.

DISEASES WITH ASSOCIATED MYCOPLASMALIKE BODIES

These diseases are characterized by yellowing and stunting, and may also exhibit witches' broom, phyllody, virescence, and abnormal fruits and seeds. Mycoplasmas in some host plants are sieve-tube limited, while in other hosts they occur in both sieve tubes and phloem parenchyma cells (cf 46, 56)

Diseases with associated Mycoplasma agents have been reviewed recently (15, 44, 55, 66, 119). Morphology of plant Mycoplasmas is briefly considered here. Plant inhabiting Mycoplasmas are pleomorphic. Small (ca 80 nm) forms (called elementary bodies) tend to be spherical and intensely staining. Larger forms to 800 nm stain less intensely and are less regularly shaped (polymorphic) but are often ovoid or spherical. Sometimes they give rise to fine filaments that break up into elementary bodies. Mycoplasmas associated with some diseased plants tend to be elongated parallel to the long axis of sieve tubes; thus their filamentous nature is best seen in longitudinal section (64). Several characteristics are used to identify mycoplasmas: Suspect bodies must have a unit membrane, ribosomes, and DNA-like strands. Cell walls and vacuoles should be wanting.

Degenerative pathways for the few mycoplasma diseases studied to date involve sieve-tube necrosis followed by formation of excessive phloem as in certain of the virus diseases (II-A-1-a in Table 2). Little is known about the way mycoplasmalike bodies affect the structure of the plant, or their distribution in the host, etc.

ASTER YELLOWS   Although the yellows group of diseases was named after peach yellows (113), some pathologists have considered aster yellows the type disease (55). Mycoplasmas have been observed and studied in plants affected by aster yellows (17, 55, 66). Although the pathogenies for aster-yellows diseased tomato and flax plants were reported to be similar to those of curly-top diseased plants (39, 78), a different interpretation seems in order. In curly-top diseased plants, pathology was similar to that in sugar beets and tobacco (18, 21); phloem parenchyma cells adjacent to sieve tubes showed increased stainability and hypertrophy. After hyperplasia of other phloem parenchyma occurred, this was followed by differentiation of many daughter cells into sieve tubes. This abnormal, sieve-tube containing tissue died prematurely. The pathogenies were like those for the II-A-2-a type virus diseases.

If one studies the photomicrographs and text of the papers on tomatoes and flax (39, 78), it becomes apparent that the pathology induced by the aster yellows agent does not follow this degenerative sequence, but is like that for II-A-1-a virus diseases. It seems apparent that sieve-tube necrosis rather than hyperplasia of adjacent parenchyma cells is primary and is followed by excessive procambial or cambial activity. For instance, with regard to phloem poles of tomato roots at 300, 470, and 1200μ from the root apex the legend for figure 17 states: "Progressive degeneration of sieve elements and adjacent cells is seen at each pole. . . ." Neither hyperplasia nor hypertrophy is mentioned or visible in the figures at these early stages (78). That the hyperplastic condition noted (actually excessive phloem formation) in later stages of the disease arose from hyperactivity of either the procambium or cambium in both roots and leaves and not from disorderly primary hyperplasia of phloem parenchyma, is indicated by the following summary statement in the paper of Rasa & Esau (78), page 513: ". . . the abnormal sieve elements differentiated in a more orderly sequence (than curly top), a few cells at a time and were obliterated more gradually than in curly top plants." On page 501, it is stated that the excessive phloem produced in stems appears like "products of a vascular cambium." In other words, primary hyperplasia did not occur but hyperactivity of the cambium did. Girolami (39), in a similar study of curly top and aster yellows in flax, noted similar differences in the two diseases, yet he concluded that the diseases were similar. It appears then that in the phloem of stems of flax plants affected by the aster-yellows agent, sieve-tube necrosis is primary and is followed by hyperactivity of the procambium or cambium to form excessive phloem as occurs in pathogenies classified II-A-1-a for virus diseased plants.

THE BUCKSKIN DISEASE OF CHERRIES AND PEACHES   X-disease and Western X-disease are similar (81, 105). Strains of the buckskin causal agent produce different diseases in different species of *Prunus*. The leaves on sweet cherry trees (*Prunus avium L.*) with sweet cherry rootstocks infected with buckskin disease usually are mildly affected, showing at worst only reddening along the midveins and adjacent lateral veins in autumn. The fruits, on the other hand, may show a combination of several symptoms that vary with the strain of the causal agent: short pedicels, a conical shape, shrinking before ripening, a dull buckskin effect on the styler end, an insipid taste, under size, and failure to mature (80). When sweet cherry is grown on *P. mahleb* rootstocks, trees tend to escape infection but in case of infection they either decline in vigor or suddenly wilt and die (79). Another disease incited by the buckskin agent is a leaf-casting yellows of peach in which irregular areas of the leaf become discolored, then turn brown and drop out (110). Other leaf symptoms are rolling, reddening, and abscission. In the greenhouse, veins become swollen on leaves of chronically diseased trees.

Two types of mycoplasmalike bodies have been found associated with X-diseased plants and their vector. For Western X-diseased celery and also the leaf hopper vector, *Collodonus montanus*, the bodies contained prominent "nuclear" strands and "nuclear" masses (73). Some bodies were electron-transparent and 200–400 nm in diameter; others were interpreted as being cylindrical, and still others were thought to be large spheres without a membrane but with associated small spherical transparent bodies. Mycoplasmalike bodies found in trees are somewhat different. In X-diseased choke cherries and sweet cherries in New York, Granett & Gilmer (42) found filamentous mycoplasma to 3μ in length and 200 nm in diameter with material externally attached to the membrane. In California, straight or undulating filamentous tubules were found associated with X-diseased peach trees; they were to 5.4μ long and 120 to 360 nm in diameter (64).

Sieve-tube necrosis occurs in the stems and leaves of buckskin-infected peach trees. Newly forming shoots are slow to become invaded by the causal agent (83); but eventually when they are invaded even the youngest protophloem sieve tubes in the leaf primordia become necrotic. In stems and leaf veins with secondary phloem, necrosis begins with the older sieve tubes and gradually involves the middle-aged and sometimes the younger ones near the cambium. Then the cambium becomes hyperactive, but only in the production of phloem; and the ring of secondary phloem becomes very wide. As sieve tubes become necrotic, their walls and protoplasts become impregnated with wound gum, and they become red when treated with phloroglucinol in hydrochloric acid. Stem cankers associated with the Merced sources of buckskin disease (83) are no longer believed to be part of the buckskin syndrome.

In trunks of buckskin-infected sweet cherry with Mahleb cherry root-stocks, sieve-tube necrosis occurs below the bud union. Trunk girdling apparently blocks translocation and root deterioration, and tree decline or wilting results.

The symptoms produced in the various hosts by the buckskin agent may be attributed to girdling brought about by sieve-tube necrosis. They are leaf rolling, vein swelling, leaf abscission, arrested fruit development, tree decline and wilting.

PEAR DECLINE   In the orchard, this disease is typified by either wilt or decline of French pear (*Pyrus communis*) varieties when grafted on one of the Oriental pear rootstocks (usually *P. serotina* or *P. ussuriensis*) (91). Quince (*Cydonia oblongata*) is another unsatisfactory rootstock, and decline also occurs on some individuals of "French pear rootstocks." The latter is a designation for heterogeneous seedlings formerly purchased in France. One source of the seeds for their propagation was pear trees in hedge rows. Trees on rootstocks of domestic Bartlett and Winter Nelis (seeds from orchards in the USA) have been little affected. The disease is spread by the pear psylla, *Psylla pyricola* Fors.

Pear decline, as a disease of grafted trees, occurs in two stages. The first occurs on any rootstock and involves the infection of the tree with Mycoplasmalike organisms. This stage is usually symptomless, but a curl and purple discoloration of leaves may result, especially in the autumn (71), and especially with highly susceptible varieties such as Comice. The second stage is necrosis of sieve tubes below the bud unions (88). Although electron microscope studies have not been made of bud union phloem, mycoplasmas are not believed to be directly associated with this killing of sieve tubes. It is suspected that a substance toxic to the rootstock sieve tubes but not to the scion sieve tubes is produced in the leaves and translocated down the trunk, although movement of mycoplasma from a tolerant scion across the union into hypersensitive sieve tubes cannot be ruled out.

Leaves infected with Mycoplasma have been intensively studied in a cultivar of pear called Variolosa that may be either an undescribed species of *Pyrus* or a hybrid. Bud-union pathology is not involved; and trees on their own roots (trees from cuttings) will either decline severely or die. Trees of Variolosa on tolerant rootstocks such as seedlings of Bartlett have a better chance to survive and are less subject to decline, but leaf symptoms occur in the Variolosa tops. In the greenhouse, the major lateral veins, which form secondary phloem, take on a brown discoloration. *Pyrus serotina* is similarly affected. The French pear varieties are for the most part tolerant, and most varieties in the greenhouse show an occasional browning of minor veinlets and occasionally leaf curling late in the season but no decline or wilt.

The first pathology observed in a developmental study of the major lateral veins of Variolosa was necrosis of secondary sieve tubes. This was followed by hyperactivity of the cambium and excessive and abnormal phloem formation (102). Necrosis of sieve tubes continued to occur in the newly forming secondary phloem; and, in conjunction with it, hypertrophy and hyperplasia of phloem parenchyma was common. Mycoplasmas were found in occasional sieve tubes after the pathosis had progressed (46).

With regard to the bud-union pathology in infected orchard trees, a new annual ring of phloem forms each year in pear tree trunks and functions during that growing season. In infected trees, the annual phloem ring functions normally for a time after formation and then the sieve tubes just below the bud union become necrotic. The cambium becomes hyperactive and produces excessive phloem which is abnormal in that cells are smaller than normal and ray cells do not enlarge and differentiate properly. The replacement sieve tubes soon become necrotic (7, 88).

RUBBERY WOOD OF APPLE   This disease causes Lord Lambourne apple trees to be stunted. The scaffold branches and trunks when less than one inch in diameter are flexuous and the weight of the fruit often bends them into a semicircle. Vesicles thought to be mycoplasmalike are associated with the sieve tubes of diseased trees (10). Heat sensitivity of the causal agent also indicates a mycoplasmalike organism as causal (118). Histochemical tests (9) and biochemical assays (94, 103) indicate a deficiency in lignification of the xylem. Although some anatomical observations have been made on the xylem (9, 94), the phloem apparently has not been studied except for the observation of mycoplasmalike bodies. A relationship between possible myco-plasma in the sieve tubes and deficiency of lignin in the xylem has not been established.

ELM PHLOEM NECROSIS   The disease of the American elm, *Ulmus ameri-cana* L., and the winged elm, *Ulmus alata* L., has been under investigation since 1938 when Swingle (106) showed that it was graft transmissible. The vector is a leaf hopper, *Scophoideus luteolus*. Progressive death of roots be-ginning with the fibrous roots and extending into the larger roots occurs and trees finally die. The phloem of the roots and lower trunks is discolored brown. In 1944, McClean (68) described the pathological phloem, which apparently consisted of sieve-tube necrosis accompanied by considerable hy-pertrophy and hyperplasia of phloem and ray parenchyma cells. Recently, mycoplasmalike bodies were reported in the roots of elm, but the kind of phloem elements in which they occurred were not identified nor was any rela-tionship between the infected cells and the phloem pathology established (120).

SUMMARY FOR DISEASES WITH ASSOCIATED MYCOPLASMALIKE ORGANISMS

For three of the diseases reviewed (buckskin, pear decline, and elm phloem necrosis), sieve-tube necrosis accompanied by hypertrophy and hy-perplasia of adjacent parenchyma is primary. The same appears to be true for Aster Yellows of flax and tomato. The phloem of rubbery-wood-diseased apples has not been studied; but the disturbances that occur in the xylem may not be primary; the Mycoplasmalike bodies associated with the disease need further study.

DISEASES WITH ASSOCIATED RICKETTSIALIKE ORGANISMS

The Rickettsias belong to a group of submicroscopic, highly-modified bacte-ria that have rigid, wavy walls. They are parasites of arthropods, mammals, and man, and were suggested as possible causal agents of plant diseases (15).

Recently Rickettsialike organisms were found associated with Pierce's disease of the grape vine—a disease long known to have a xylem limited causal agent (22, 25, 40). Phoney peach also belongs here.

PIERCE'S DISEASE OF THE VINE AND ALFALFA DWARF   The symptoms of Pierce's disease result at least in part from a water deficit within the plant. Typical symptoms of diseased vines are: delayed foliation in the spring, leaf mottling, leaf "burning" or "scalding," dwarfing of the vine, wilting and drying up of fruit, failure of canes to mature evenly, and finally death of the vine (45).

The causal agent was once presumed to be a virus because it is graft and insect transmissible, but a recent report states that diseased vines treated with tetracycline show some improvement (53), and a Rickettsialike organism has been found associated with the xylem (40).

Esau (24) has studied the pathology of this disease by using infectious leaf hoppers (*Draeculacephala minerva* Ball) to inoculate one leaf of each of several grape seedlings. In one experiment, sampling began 71 days after inoculation. At this time, all of the inoculated leaves showed symptoms, and gum deposits and tyloses were present in the vessels. These pathological aberrations were sparsely distributed in the internodes above and below the inoculated leaf. In two other experiments, collections were initially at 3-day and then longer intervals after inoculations. Gum deposits and tyloses in petioles of inoculated leaves were first observed at 12 days in one experiment and at 24 days in the other. Pathosis was observed erratically in the petioles but not at all in stems above and below the inoculated leaf during the 66–74 days that collections were made.

Studies on long-diseased plants showed extensive development of tyloses and gum in lumens of vessels. Other phenomena included failure of phellogens to form and to produce cork in stems, and abnormally small mesophyll cells of leaves. These possibly secondary aberrations could have resulted from toxic effects induced by the presence of the Rickettsialike organisms.

In similar studies on the dwarf disease of alfalfa, the same primary host responses as in the grape were found in the xylem (24) and a similar Rickettsialike organism (40). This disease is probably xylem limited because the xylem-feeding vector must reach the xylem to produce infection (54). The pathology begins in the xylem, and an associated Rickettsialike agent is found there.

DISEASES WITH ASSOCIATED FLAGELLATES

On certain soils in Surinam, *Coffea liberica* is affected by phloem necrosis, a disease with decline or wilt symptoms. The sieve tubes of affected roots, trunks, and scaffold branches contain a flagellate, *Phytomonas leptavasorum* Stahel (104, 111, 112). At the onset of the disease, abnormal divisions are noted near the cambium, probably in the phloem mother cells; the phloem produced thereafter shows characteristic "multiple divisions" of small, disorderly-arranged phloem "cells." The parenchyma cells contain large starch grains. After the population of flagellates increases, the bark no longer slips at the cambium, and the hyperplastic phloem eventually dies. The causal agent is graft transmissible by root scions, but the flagellate has not been mechanically transmitted or grown in culture.

TRANSMISSIBLE DISEASES WITH ASSOCIATED AGENT UNDETERMINED

This temporary class of diseases is analogous to the "Fungi Imperfecti" of fungal classification. Placing diseases wrongly with the Viruses, Mycoplasmas, or Rickettsias might delay or inhibit determination of their causal agents.

There are many more tree diseases for which anatomical studies have been made than space permits mentioning. For most of them several causal agents may be involved and/or the anatomical study was not developmental.

To bring some semblance of organization to the presentation, the outline that was used for virus diseases will be paralleled here. Some diseases will fit readily into the scheme; for others, placement is arbitrary pending further information about the pathogenies of the diseases.

Pathology is Ubiquitous

MOSAICS   Peach mosaic is a graft- and eriophyid-mite transmitted disease (121). It may be caused by a virus, but virus-like particles have not been reported despite several attempts with the electron microscope. The disease resembles tobacco mosaic. In the yellow areas of the leaf I have observed cells that failed to differentiate into palisade and spongy parenchyma cells. The undifferentiated areas may become necrotic and drop out. Results from graft transmission studies indicate that the causal agent occurs in leaves, fruit, bark, and wood. Leaves that were mature when infection occurred did not show symptoms (14).

TRUNK AND STEM DISEASES   The rough bark diseases. Psorosis (scaly bark) of citrus is a presumed virus disease. It is especially severe on sweet orange where its effect on the trunks and larger branches are profound. Small pieces of bark flake off, usually starting in small areas and gradually spreading around the trunk or branch; one rampant type is preceded by gumming and breaks out all at once up and down one side of the tree (114). In addition to a water-soluble gum found in pockets, water insoluble gum is deposited in the vessels (115). Vein clearing occurs along the minor veinlets of newly differentiating leaves when they are about half expanded, and disappears as the leaves mature. No virus particles have been found in such leaves (76). Occasionally, circular chlorotic or necrotic spots occur on mature leaves and also on fruit. The disease is disseminated with infected buds during propagation of trees.

Price (76) believes that translocation of the causal agent occurs rapidly in the phloem, but experiments of longer duration might have shown movement through xylem parenchyma as well.

Webber & Fawcett (117) give some insight as to how bits of bark become affected by psorosis and are then sloughed off by wound-periderm formation. When psorosis lesions first appear, unstained sections reveal groups of brown parencyma cells in the phelloderm and other parenchymatous tissue underlying the young lesions. After the brown cells appear, a wound periderm forms under them that is continuous with the adjacent normal periderm. This causes the affected tissues to be exfoliated. As the disease progresses, cells deeper in the bark turn brown and these in turn are exfoliated by periderm formation. This process is repeated until the bark is very thin. In addition, phloem parenchyma cells and sieve tubes also degenerate (90).

In trunks and branches, the xylem also is affected severely. Gum pockets form that involve rings of immature xylem and adjacent cells of the cambial zone. The cytoplasm may thicken, and the vacuolar space becomes reduced; then the middle lamella and the cell walls dissolve. When this occurs, a portion of the axial system becomes a cavity filled with gum that apparently forms from the dissolved walls and other cellular material. Sometimes cells float freely within the cavity (90). Staining of the wood spreads outward from the pith. Water insoluble wound gum fills the vessels, and the walls of fibers and xylem parenchyma (115).

Diamond Canker of French Prune is another graft transmissible disease that causes bark shelling in limited cankerous areas (101).

Flat limb of apple and finger marks of citrus (65) are examples of *Trunk and limb distortion diseases*. This category of anatomical disturbance is tentative because further anatomical studies are needed on the diseases involved.

Flat limb of apple, a graft transmissible disease of the Gravenstein and other apple varieties, has world wide distribution. The cambium becomes disorganized or malfunctional, causing limbs to become flattened or distorted in areas where normal radial growth does not occur (109). When scions from flat limb trees are used to inoculate Pyracantha, the branches of that plant develop a scaling type of disease that is distinct from that in the apple (109).

Corky bark of Vitis is a disease characterized by strikingly excessive phloem formation and suppressed xylem production (13). The rays are dilated and blocks of phloem that occur in patches between rays apparently are functional. The genesis of the pathological anatomy has not been studied and the disease may not belong here.

Phloem Affinity Diseases

PHLOEM LIMITED PATHOLOGY   *Sieve-tube necrosis.* Grapevine leaf roll is a graft transmissible disease characterized by leaf rolling, early red coloration in the fall, and starch accumulation. Sieve-tube necrosis that is accompanied by hypertrophy and hyperplasia of adjacent parenchyma cells occurs throughout the plant but is more prevalent in young vascular bundles than in secondary phloem of older canes (49). The pathogeny of this disease is strikingly similar to the buckskin disease in peach and the pear decline of the Variolosa pear, both of which have mycoplasma associated with them.

Hyperplasia is the primary anatomical aberration. Tumors arise from one or a few tumor cells. On lime leaves affected by the aphid vectored, vein-enation disease of *Citrus aurantifolia,* tumors develop on thorns and major lateral leaf veins; galls develop on stems, branches, trunks, and roots. The leaf enations are initiated from procambial cells (fiber primordial cells) that lie adjacent to the first formed protophloem sieve tubes in differentiating veins, and for this reason the disease may be regarded as of phloem origin. Normally the procambial cells would differentiate into protophloem fibers, but when one or a cluster of cells become tumor initials, their cytoplasm thickens and becomes darkly staining (52). Repeated cell divisions then occur until a mass of darkly staining cells has been produced abaxially (outwardly) from the protophloem. Meanwhile the metaphloem and xylem develop normally. As the tumorous mass becomes large, the epidermis and underlying parenchyma cells that cover the vein stretch and undergo some divisions to accommodate the tumorous mass. Phytoferratin particles occur in the plastids of this and other diseases; they are sometimes mistaken for virus particles (51).

Stem tumors or galls are initiated from tumor initial cells that form in procambial tissue that lies between the metaxylem and metaphloem and subsequently are incorporated into the cambium. The tumorous cambial cells give rise to excessive xylem tissue which composes the woody galls.

PRIMARY INFECTION OF CELLS APPARENTLY OCCURS SYSTEMICALLY IN THE PHLOEM, BUT LOCALLY PATHOSIS OCCURS IN OTHER TISSUES *Cambium is disorganized.* Stem pitting of apple is a devastating graft transmissible disease of crab apples. The Virginia crab apple is used as a body stock for edible apple varieties, and the causal agent of stem pitting may be latent in the latter. When bark is lifted from trunks, branches, or stems of trees affected by the stem pitting disease, pits may be seen in the wood and there is a wedge of tissue on the bark's cambial face opposite them. They apparently result from localized disorganization of the cambium (48). Other phenomena observed were: multinucleate phloem mother cells, sieve-tube necrosis, and rays that were wider than normal. How these phenomena fit into a degenerative sequence was not determined.

DISCUSSION

For each disease caused by an intracellular pathogen, a characteristic pathogeny (chain of cyto- and histopathological aberrations) occurs following infection. In this review an attempt has been made at a tentative grouping of diseases with similar pathogenesis. The main obstacle in this attempt has been the dearth of developmental studies of diseases. Hopefully, if one is acquainted with the several typical types of pathogenesis, he will have a clearer idea of what is happening in the diseased plants with which he experiments.

Dividing diseases into those caused by Viruses, Mycoplasmas, Rickettsias, and Flagellates simplifies the classification. The Rickettsias produce diseases with pathology unlike those caused by other agents. Until the discovery of these Rickettsialike organisms in the xylem of Pierce's diseased plants, the agents were placed in a group called "xylem-limited viruses" (22). On the other hand, the Mycoplasma diseases and one Flagellate disease are not so different from some of the viral diseases with phloem affinity. Virus diseases with ubiquitous distribution of virions (e.g. mosaics and bark diseases) appear to form a group of transmissible diseases with but one kind of causal agent, the viruses.

It is interesting that diverse organisms may cause similar pathology. Sieve-tube necrosis accompanied by hypertrophy and some hyperplasia of parenchyma cells may be found in diseases caused by Viruses, Mycoplasmalike organisms, and Flagellates. Hyperactive cambial activity and excessive phloem formation may accompany the necrosis, while xylem formation is usually suppressed. The bud-union pathology of tristeza virus infected sweet orange on sour orange rootstocks is very similar to that of pear decline diseased French pear on certain Oriental pear rootstocks. The latter disease appears to be induced by a mycoplasmalike organism.

Perhaps the questions should be raised: Do all intracellular, transmissible, disease-causing agents affect the anatomy of plants? Could some of them cause a release of toxic materials into the translocation or transpiration stream that interferes with biochemical pathways and/or slows down growth? Do they necessarily need to destroy normal cellular structure (necrosis) or cause abnormalities in development (hyperplasia, hypertrophy, and hypoplasia)? Certain diseases that lack anatomical aberrations may fall into this class (16).

Mundry (72) considered virus multiplication to be dependent on the interaction of virus and host genomes. One may extrapolate that disease development is also dependent on the genome of both the host and virus. Changing the species of host changes the genome of the host and hence the nature of the disease. Thus, in a sense, when plants of two species are infected with the same virus there are two diseases with different genomes even though the diseases may be very similar or very different. Such is the case of tristeza virus where it affects two host types—the yellows type and the pitting type.

A most important conclusion from this review is that the earliest pathological aberration in a sequence of degenerative aberrations is usually more specific for diagnosing a disease than those that follow. Two degenerative pathways that begin with the same organism and the same pathological aberration may diverge as the disease progresses in two different hosts; conversely, diseases caused by two causal agents that induce different primary aberrations may become similar as the diseases progress. For instance decline and/or wilt of trees may have several beginnings—sieve-tube necrosis, gopher damage, fungal root rot, etc.

Literature Cited

1. Artschwager, E. F. 1918. Anatomy of the potato plant, with special reference to the ontogeny of the vascular system. *J. Agr. Res.* 14: 221–52
2. Artschwager, E. F. 1923. Occurrence and significance of phloem necrosis in the Irish potato. *J. Agr. Res.* 24:237–45
3. Bald, J. G. 1949. A method for the selective staining of viruses in infected tissues. *Phytopathology* 39:395–402
4. Bald, J. G. 1964. Cytological evidence for the production of plant virus ribonucleic acid in the nucleus. *Virology* 22:377–87
5. Bald, J. G. 1966. Cytology of plant virus infections. *Advan. Virus Res.* 12:103–25
6. Bar-Joseph, M., Loebenstein, G., Cohen, J. 1970. Partial purification of viruslike particles associated with the citrus tristeza disease. *Phytopathology* 60:75–78
7. Batjer, L. P., Schneider, H. 1960. Relation of pear decline to rootstocks and sieve-tube necrosis. *Proc. Am. Soc. Hort. Sci.* 76:85–97
8. Bawden, F. C. 1932. A study on the histological changes resulting from certain virus infections of the potato. *Proc. Roy. Soc. Sect. B,* 111:74–85
9. Beakbane, A. B., Thompson, E. C. 1945. Recognition of "rubbery" condition in Lord Lambourne and some other apple varieties. *Ann. Rep. E. Malling Res. Sta.,* 1944. A28:108–9
10. Beakbane, A. B., Mishra, M. D., Posnette, A. F., Slater, C. H. W. 1971. Mycoplasma-like organisms associated with chat fruit and rubbery wood diseases of apple, *Malus domestica* Borkh., compared with those in strawberry with green petal disease. *J. Gen. Microbiol.* 66:55–62
11. Bennett, C. W. 1960. Sugar beet yellows disease in the United States. *U. S. Dep. Agr. Tech. Bull.* 1218:1–63
12. Bennett, C. W. 1971. The curly top disease of sugarbeet and other plants. *Monograph No. 7, Am. Phytopath. Soc.*
13. Beukman, E. F., Gifford, E. M., Jr. 1969. Anatomic effects of corky bark virus in *Vitus. Hilgardia* 40: 73–103
14. Cochran, L. C., Rue, J. L. 1944. Some host-tissue relationships of the peach mosaic virus. *Phytopathology* 34:934
15. Davis, R. E., Whitcomb, R. F. 1971. Mycoplasmas, rickettsiae, and chlamydiae: possible relation to yellows diseases and other disorders of plants and insects. *Ann. Rev. Phytopathol.* 9:119–54
16. Diener, T. O. 1963. Physiology of virus-infected plants. *Ann. Rev. Phytopathol.* 1:197–218
17. Doi, Y., Teranaka, M., Yora, K., Asuyama, H. 1967. Mycoplasma-or PLT group-like microorganisms found in the phloem elements of plants infected with mulberry dwarf, potato witches' broom, aster yellows, or paulownia witches' broom. *Ann. Phytopath. Soc. Japan* 33:259–66
18. Esau, K. 1935. Ontogeny of the phloem in sugar beets affected by the curly-top disease. *Am. J. Bot.* 22:149–63
19. Esau, K. 1938. Some anatomical aspects of plant virus disease problems. *Bot. Rev.* 4:548–79
20. Esau, K. 1941. Inclusions in guard cells of tobacco affected with mosaic. *Hilgardia* 13:427–34
21. Esau, K. 1941. Phloem anatomy of tobacco affected with curly top and mosaic. *Hilgardia* 13:437–90
22. Esau, K. 1948. Some anatomical aspects of plant virus disease problems. II. *Bot. Rev.* 14:413–49
23. Esau, K. 1948. Phloem structure in the grapevine and its seasonal changes. *Hilgardia* 18:217–96
24. Esau, K. 1948. Anatomic effects of the viruses of Pierce's disease and phony peach. *Hilgardia* 18:423–82
25. Esau, K. 1956. An anatomist's view of virus diseases. *Am. J. Bot.* 43: 739–48
26. Esau, K. 1960. Cytologic and histologic symptoms of beet yellows. *Virology* 10:73–85
27. Esau, K. 1964. Structure and development of the bark in dicotyledons. *Formation of Wood in For-*

est Trees, ed. Zimmerman, Martin, Huldrych, 37–50. New York: Academic 562 pp.

28. Esau, K. 1967. Anatomy of plant virus infections. *Ann. Rev. Phytopathol.* 5:45–76

29. Esau, K. 1968. *Viruses in Plant Hosts—Form, Distribution, and Pathologic Effects.* Madison: Univ. of Wisconsin Press. 225 pp.

30. Esau, K., Cronshaw, J. 1967. Tubular components in cells of healthy and tobacco mosaic virus-infected *Nicotiana. Virology* 33:26–35

31. Esau, K., Cronshaw, J. 1967. Relation of tobacco mosaic virus to the host cells. *J. Cell Biol.* 33:665–78

32. Esau, K., Cronshaw, J., Hoefert, L. L. 1967. Relation of beet yellows virus to the phloem and to movement in the sieve tube. *J. Cell Biol.* 32:71–87

33. Esau, K., Gill, R. H. 1969. Tobacco mosaic virus in dividing mesophyll cells of *Nicotiana. Virology* 38:464–72

34. Esmarch, F. 1932. *Die Blattrollkrankheit der Kartoffel.* Monographien zum pflanzenschutz, Vol. 8. Berlin: Verlag von Julius Springer, 91 pp.

35. Fudl-Allah, A. E. A., Calavan, E. C., Desjardins, P. R. 1971. Comparative anatomy of healthy and exocortis virus-infected citron plants. *Phytopathology* 61:990–93

36. Fulton, R. W. 1966. Mechanical transmission of viruses of woody plants. *Ann. Rev. Phytopathol.* 4:79–102

37. Garnsey, S. M., Jones, J. W. 1967. Mechanical transmission of exocortis virus with contaminated budding tools. *Plant Dis. Reptr.* 51:410–13

38. Gibbs, A. 1969. Plant virus classification. *Advan. Virus Res.* 14:263–328

39. Girolami, G. 1955. Comparative anatomical effects of the curly-top and aster-yellows viruses on the flax plant. *Bot. Gazette* 116:305–22

40. Goheen, A. C., Nyland, G., Lowe, S. K. 1973. Association of a rickettsialike organism with Pierce's disease of grapevines and alfalfa dwarf and heat therapy of the disease in grapevines. *Phytopathol-*

ogy 63:341–45

41. Goldstein, B. 1926. A cytological study of the leaves and growing points of healthy and mosaic diseased tobacco plants. *Bull. Torrey Bot. Club* 53:499–599, 12 plates

42. Granett, A. L., Gilmer, R. M. 1971. Mycoplasmas associated with X-disease in various Prunus species. *Phytopathology* 61:1036–37

43. Grant, T. J. 1959. Tristeza virus strains in relation to different citrus species used as test plants. *Phytopathology* 49:823–27

44. Hampton, R. O. 1972. Mycoplasmas as plant pathogens: perspectives and principles. *Ann. Rev. Plant Physiol.* 23:389–418

45. Hewitt, W. B., Frazier, N. W., Jacob, H. E., Freitag, J. H. 1942. Pierce's disease of grapevines. *Univ. of Calif., Agr. Exp. Sta. Circular 353.* 32 pp.

46. Hibino, H., Schneider, H. 1970. Mycoplasmalike bodies in sieve tubes of pear trees affected with pear decline. *Phytopathology* 60:499–501

47. Hibino, H., Schneider, H. 1971. Virus-like flexuous rods associated with pear vein yellows. *Archiv für die gesamte Virusforschung* 33:347–55

48. Hilborn, M. T., Hyland, F., McCrum, R. C. 1965. Pathological anatomy of apple trees affected by the stem-pitting virus. *Phytopathology* 55:34–39

49. Hoefert, L. L., Gifford, E. M., Jr. 1967. Grapevine leafroll virus—history and anatomic effects. *Hilgardia* 38:403–26

50. Holmes, F. O. 1939. Handbook of phytopathogenic viruses. Minneapolis: Burgess 221 pp.

51. Hooper, G. R. 1968. *Anatomical and cytological aspects of citrus vein-enation virus induced tumors.* PhD thesis, Univ. Calif., Riverside. 155 pp.

52. Hooper, G. R., Schneider, H. 1969. The anatomy of tumors induced on citrus by citrus vein-enation virus. *Am. J. Bot.* 56:238–47

53. Hopkins, D. L., Mortensen, J. A. 1971. Suppression of Pierce's disease symptoms by tetracycline antibiotics. *Plant Dis. Reptr.* 55:610–12

54. Houston, B. R., Esau, K., Hewitt, W. B. 1947. The mode of vector

feeding and the tissues involved in the transmission of Pierce's disease virus in grape and alfalfa. *Phytopathology* 37:247–53

55. Hull, R. 1972. Mycoplasma and plant diseases. *PANS* 18:154–64
56. Kaloostian, G. H., Hibino, H., Schneider, H. 1971. Mycoplasmalike bodies in periwinkle: their cytology and transmission by pear psylla from pear trees affected with pear decline. *Phytopathology* 61:1177–79
57. Kitajima, E. W., Costa, A. S. 1968. Electron microscopy of the tristeza virus in citrus leaf tissues. In *Proc. 4th Conf. Int. Organ. Citrus Virol.*, ed. J. F. L. Childs, 59–64. Gainesville: Univ. Florida Press Florida. 404 pp.
58. Kitajima, E. W., Silva, D. M., Oliveira, A. R., Müller, G. W., Costa, A. S. 1965. Electron microscopical investigations on tristeza. In *Proc. 3rd Conf. Int. Organ. Citrus Virol.*, ed. W. C. Price, 1–9. Gainesville: Univ. Florida Press 319 pp.
59. Kojima, M., Shikata, E., Sugawara, M., Murayama, D. 1969. Purification and electron microscopy of potato leafroll virus. *Virology* 39:162–74
60. Kolehmainen, L., Zech, H., von Wettstein, D. 1965. The structure of cells during tobacco mosaic virus reproduction. *J. Cell Biol.* 25:77–97
61. Küster, E. 1925. *Pathologische pflanzenanatomie.* Ed. 3. Jena: Gustav Fisher 558 pp.
62. Lee, C. L. 1955. Anatomical changes in sweet clover shoots infected with wound-tumor virus. *Am. J. Bot.* 42:693–98
63. Littau, V. C., Black, L. M. 1952. Spherical inclusions in plant tumors caused by a virus. *Am. J. Bot.* 39:87–95
64. MacBeath, J. H., Nyland, G., Spurr, A. R. 1972. Morphology of mycoplasmalike bodies associated with Peach X-disease in *Prunus persica. Phytopathology* 64:935–37
65. Madaluni, A. L. 1968. Studies on finger mark disorder of citrus in Italy. In *Proc. 4th Conf. Int. Organ. Citrus Virol.*, ed. J. F. L. Childs, 10–13. Gainesville: Univ. Florida Press 404 pp.
66. Marmorosch, K., Granados, R. R.,

Hirumi, H. 1970. Mycoplasma diseases of plants and insects. *Advan. Virus Res.* 16:135–93
67. McClean, A. P. D. 1961. Transmission of tristeza virus to *Aeglopsis chevalieri* and *Afraegle paniculata. S. Afr. J. Agr. Sci.* 4:83–94
68. McLean, D. M. 1944. Histopathologic changes in the phloem of American elm affected with the virus causing phloem necrosis. *Phytopathology* 34:818–26
69. McKinney, H. H. 1944. Genera of plant viruses. *J. Wash. Acad. Sci.* 34:139–54
70. McWhorter, F. P. 1965. Plant virus inclusions. *Ann. Rev. Phytopathol.* 3:287–312
71. Millecan, A. A., Gotan, S. M., Nichols, C. W. 1963. Red-leaf disorders of pear in California. *Calif. State Dep. Agr. Bul.* 52:166–70
72. Mundry, K. W. 1963. Plant virus-host cell relations. *Ann. Rev. Phytopathol.* 1:173–96
73. Nasu, S., Jensen, D. D., Richardson, J. 1970. Electron microscopy of mycoplasma-like bodies associated with insect and plant hosts of peach western X-disease. *Virology* 41:583–95
74. Nour-Eldin, F., Tolba, M. A., El-Banna, M. T., El-Attar, S. 1965. Ring callus as a path for non-graft transmitted *Aeglopsis chevalieri* vein-clearing virus. In *Proc. 3rd Conf. Int. Organ. Citrus Virol.*, ed. W. C. Price, 280–84. Gainesville: Univ. Florida Press 319 pp.
75. Price, W. C. 1966. Flexuous rods in phloem cells of lime plants infected with citrus tristeza virus. *Virology* 29:285–94
76. Price, W. C. 1968. Translocation of tristeza and psorosis viruses. In *Proc. 4th Conf. Int. Organ. Citrus Virol.*, ed. J. F. L. Childs, 52–58. Gainesville. Univ. Florida Press 404 pp.
77. Quanjer, H. M. 1931. The methods of classification of plant viruses, and an attempt to classify and name potato viruses. *Phytopathology* 21:577–613
78. Rasa, E. A., Esau, K. 1961. Anatomic effects of curly top and aster yellows viruses on tomato. *Hilgardia* 30:469–515
79. Rawlins, T. E., Parker, K. G. 1934. Influence of rootstocks on the susceptibility of sweet cherry

to the buckskin disease. *Phytopathology* 24:1029–31

80. Rawlins, T. E., Thomas, H. E. 1941. The buckskin disease of cherry and other stone fruits. *Phytopathology* 31:916–25

81. Reeves, E. L., Blodgett, E. C., Lott, T. B., Milbrath, J. A., Richards, B. L., Zeller, S. M. 1951. Western X-disease. *USDA Agr. Handbook* 10:43–52

82. Schlegel, D. E., Smith, S. H., de Zoeten, G. A. 1967. Sites of virus synthesis within cells. *Ann. Rev. Phytopathol.* 5:223–46

83. Schneider, H. 1945. Anatomy of buckskin-diseased peach and cherry. *Phytopathology* 35:610–35

84. Schneider, H. 1954. Anatomy of bark of bud union, trunk, and roots of quick-decline-affected sweet orange trees on sour orange rootstock. *Hilgardia* 22:567–601

85. Schneider, H. 1955. Ontogeny of lemon tree bark. *Am. J. Bot.* 42:893–905

86. Schneider, H. 1957. Chromatic parenchyma cells in tristeza-diseased citrus. *Phytopathology* 47:533–34

87. Schneider, H. 1959. The anatomy of tristeza-virus-infected citrus. In *Citrus Virus Diseases*, ed. J. M. Wallace, 73–84. Berkeley: Univ. Calif. Div. Agr. Sci, 343 pp.

88. Schneider, H. 1959. Anatomy of bud-union bark of pear trees affected by decline. *Phytopathology* 49:550

89. Schneider, H. 1961. Anatomical aspects of tristeza-diseased citrus, Aeglopsis, and Afaegle. In *Proc. 2nd Int. Organ. Citrus Virol.*, ed. W. C. Price, 136–40. Gainesville: Univ. Florida Press 265 pp.

90. Schneider, H. 1969. Pathological anatomies of citrus affected by virus diseases and by apparently-inherited disorders and their use in diagnoses. In *Proc. First Int. Citrus Symp.*, ed. H. D. Chapman, vol. 3:1489–94. Riverside: Univ. Calif. pp. 1105–1839

91. Schneider, H. 1970. Graft transmission and host range of the pear decline causal agent. *Phytopathology* 60:204–7

92. Schneider, H., Sasaki, P. J. 1972. Ultrastructural studies of chromatic cells in tristeza-diseased lime. In *Proc. 5th Conf. Int. Organ.*

Citrus Virol., ed. W. C. Price, 222–28. Gainesville: Univ. Florida Press 301 pp.

93. Schneider, I. R. 1965. Introduction, translocation, and distribution of viruses in plants. *Advan. Virus Res.* 11:163–221

94. Scurfield, G., Bland, D. E. 1963. The anatomy and chemistry of "rubbery" wood in apple var. Lord Lambourne. *J. Hort. Sci.* 38:297–306

95. Semancik, J. S., Weathers, L. G. 1970. Properties of the infectious forms of exocortis virus of citrus. *Phytopathology* 60:732–36

96. Shalla, T. A. 1964. Assembly and aggregation of tobacco mosaic virus in tomato leaflets. *J. Cell Biol.* 21:253–64

97. Sheffield, F. M. L. 1943. Value of phloem necrosis in the diagnosis of potato leaf-roll. *Ann. Appl. Biol.* 30:131–36

98. Shikata, E., Maramorosch, K. 1966. An electron microscope study of plant neoplasia induced by wound tumor virus. *J. Nat. Cancer Inst.* 36:97–116

99. Shikata, E., Sasaki, A. 1969. Long flexuous threads associated with Hassaku dwarf disease of citrus trees. *J. Fac. Agr.*, Hokkaido Univ., 56:219–24

100. Siegel, A., Zaitlin, M. 1964. Infection process in plant virus disease. *Ann. Rev. Phytopathol.* 2:179–202

101. Smith, R. E. 1941. Transmission of diamond canker of the French prune. *Phytopathology* 31:886–95

102. Soma, K., Schneider, H. 1971. Developmental anatomy of major lateral leaf veins of healthy and of pear-decline diseased pear trees. *Hilgardia* 13:471–504

103. Sondheimer, E., Simpson, W. G. 1962. Lignin abnormalities of "rubbery apple wood." *Can. J. Biochem. Physiol.* 40:841–46

104. Stahel, G. 1933. Zur Kenntnis der Siebröhrenkrankheit (Phloëm nekrose) des Kaffebaumes in Surinam. *Phytopathol. Z.* 6:335–57

105. Stoddard, E. M., Hilderbrand, E. M., Palmiter, D. H., Parker, K. G. 1951. X-disease. *USDA Agr. Handbook* 10:37–42

106. Swingle, R. U. 1938. A phloem necrosis of elm. *Phytopathology* 28:757–59

107. Tepfer, S. S., Chessin, M. 1959.

Effects of tobacco mosaic virus on early leaf development in tobacco. *Am. J. Bot.* 46:496–509

108. Thomas, H. E. 1942. Transmissible rough-bark diseases of fruit trees. *Phytopathology* 32:435–36

109. Thomas, H. E. 1961. Virus diseases of apple. *Hilgardia* 31:435–57

110. Thomas, H. E., Rawlins, T. E., Parker, K. G. 1940. A transmissible leaf-casting yellows of peach. *Phytopathology* 30:322–28

111. van Emden, J. H. 1962. On flagellates associated with a wilt of *Coffea liberica. Meded Landbouwhogesch. Opzoekingssta. Staat Gent.* 27:776–84

112. Vermeulen, H. 1968. Investigations into the cause of the phloem necrosis disease of *Coffea liberica* in Surinam, South America. *Neth. J. Plant Pathol.* 74:202–18

113. Walker, J. C. 1950. *Plant Pathology.* New York: McGraw-Hill. 699 pp.

114. Wallace, J. M. 1957. Virus-strain interference in relation to symptoms of psorosis disease of citrus. *Hilgardia* 27:223–46

115. Wallace, J. M. 1959. A half century of research on psorosis. In *Citrus Virus Diseases,* ed. J. M. Wallace, 5–21. Berkeley: Univ. Calif. Div. Agr. Sci. 243 pp.

116. Wallace, J. M., Drake, R. J. 1961. Seedling yellows in California. In *Proc. 2nd Conf. Int. Organ. Citrus Virol.,* ed. W. C. Price, 141–49. Gainesville: Univ. Florida Press 265 pp.

117. Webber, I. E., Fawcett, H. S. 1935. Comparative histology of healthy and psorosis-affected tissues of *Citrus sinensis. Hilgardia* 9:71–109

118. Welsh, M. F., Nyland, G. 1965. Elimination and separation of viruses in apple clones by exposure to dry heat. *Can. J. Plant Sci.* 45:443–54

119. Whitcomb, R. F., Davis, R. E. 1970. Mycoplasma and phytarboviruses as plant pathogens persistently transmitted by insects. *Ann. Rev. Entomol.* 15:405–64

120. Wilson, C. L., Seliskar, C. E., Krause, C. R. 1972. Mycoplasmalike bodies associated with elm phloem necrosis. *Phytopathology* 62:140–43

121. Wilson, N. S., Jones, L. S., Cochran, L. C. 1955. An eriophyid mite vector of the peach-mosaic virus. *Plant Dis. Reptr.* 39:889–92

INDUCTION OF DISEASE BY VIRUSES, ❖ 3569
WITH SPECIAL REFERENCE TO
TURNIP YELLOW MOSAIC VIRUS

R. E. F. Matthews
Department of Cell Biology, University of Auckland, Auckland, New Zealand

INTRODUCTION

Historically, we can distinguish three phases in the study of disease symptoms induced by viruses. The first, from about 1900 to 1935, might be called "The Symptom Era" or "The Era of Chaos." Hundreds of diseases were described in detail and given names. There was widespread confusion about the diseases and their unknown causative agents. Goldstein (30) was the first to attempt to relate disease patterns to the developmental processes of the plant. She found that the distinctive mosaic patterns in successive leaves of tobacco plants infected with tobacco mosaic virus (TMV) were related to the age of the leaf at the time it became infected. She also noted that dark green tissue appeared cytologically normal

The discovery, about 1936, that several plant viruses were ribonucleoproteins focussed attention on the virus particles themselves. With the development of molecular biology after World War II, the study of virus symptoms fell into general disrepute, as being rather second-grade purely descriptive science. This might be called "The Era of Prejudice."

Since about 1965 a new trend has begun to emerge. The knowledge about the structure and replication of viruses provided by molecular biology allows us to think much more constructively about possible ways in which viruses might induce symptoms; very sensitive and specific analytical procedures have been developed; and the widespread availability of high resolution electron microscopes has helped to bridge the gap between biochemical knowledge and descriptions of a disease based on macroscopic observation and the light microscope.

To understand how a virus causes disease we need to be able to explain, in terms of molecular biology and biochemistry, how a single virus particle containing a piece of genetic material with about 2000 base triplets can infect and cause disease in a growing plant, organized into a variety of tissues and organs with specific structures and functions and containing when mature

147

about 10^9 cells. This is a very large and diffuse topic. In this chapter I shall confine myself almost entirely to the problem of the induction of mosaic diseases with special reference to turnip yellow mosaic virus (TYMV) grown in Chinese cabbage (*Brassica pekinensis*) under average glasshouse conditions. There are several major related areas that I have not touched upon: (*a*) The necrotic reaction to viruses; (*b*) the cytology of viral infections other than TYMV; (*c*) abnormalities of growth and differentiation induced by viruses; (*d*) factors influencing symptom expression. The reader is referred to recent general surveys of the kinds of disease induced by plant viruses (9, 47).

THE VIRUS AND ITS GENE PRODUCTS

It might at first thought be assumed that disease symptoms would be a direct consequence of the virus diverting supplies of raw materials into virus production. Where the host plant is already somewhat deficient in some nutrient element before infection, the virus may induce symptoms of nutrient deficiency or make such symptoms more severe. For example, in mildly nitrogen-deficient Chinese cabbage plants the local lesions produced by TYMV have a purple halo, the purple coloration being characteristic of nitrogen starvation (23). On the other hand, in well-nourished Chinese cabbage plants fully infected with TYMV, about 20% of the phosphorus in a leaf may be contained in viral RNA (49) and yet such plants show no symptoms of phosphorus deficiency.

A few examples have been reported where there is an association between periods of more severe disease and rapid virus production [e.g., bean pod mottle virus in soy bean (28) and alfalfa mosaic virus in tobacco (38)]. However, in well-nourished plants there is no general correlation between amount of virus produced and severity of disease. The following considerations suggest that the actual sequestration of amino acids and other materials into virus particles may have no direct connection with the induction of symptoms:

(*a*) Even with viruses, such as TYMV, that reach a relatively high concentration in the diseased tissue ($\simeq 1.0$ mg/gm fresh wt) the amount of virus formed may be quite small relative to the reduction in other macromolecules caused by infection. For example, in Chinese cabbage plants infected for seven weeks with TYMV, normal macromolecular constituents made up of protein and RNA (soluble proteins and ribosomes) had been reduced from 1020 mg/plant to 320 mg. Only 36 mg/plant of virus had been produced in the same period (E. S. Crosbie & R. E. F. Matthews, unpublished).

(*b*) Closely related strains of the same virus may multiply in a particular host to give the same final concentration of virus, and yet have markedly different effects on host cell constituents (e.g., strains of TYMV in the stock culture). However, as discussed later, the rate at which a particular strain of virus replicates in a cell may influence the severity of disease.

(*c*) The type strain of tobacco mosaic virus (TMV) multiplying in White

Burley tobacco produced chlorotic lesions at 35° but none at 20°C. About one tenth as much virus was made at 35° as at 20°C (35).

(d) A single gene change in the tobacco plant may result in a change from the typical mosaic disease produced by TMV to the hypersensitive reaction (50). F_1 hybrids between the two genetic types may respond to TMV with a lethal systemic necrotic disease with greatly reduced virus production (47).

(e) Use of mutants of TMV induced by nitrous acid has shown that changes in disease symptoms can be independent of changes in the coat protein. Mutants of type TMV are known that have defective coat proteins but produce necrotic local lesions like those of the parent strain on appropriate hosts (63). Siegel (62) found mutants of a masked strain of TMV that caused death of cells but had similar serological properties to the parent.

(f) With multiparticle viruses, it is possible to identify and locate certain specific virus functions on particular classes of particle. Thus it is possible to demonstrate that particular symptoms of disease are not associated with coat protein production. For example, the coat protein cistron for both long and short rods of tobacco rattle virus (TRV) is on the RNA of the long rods. Lister (44) and Sänger (61) showed that local-lesion-symptom type in certain hosts is controlled by a function of the long rod RNA. On the other hand, symptoms in *Nicotiana clevelandii* are controlled by a function of the short rod RNA (45).

As far as the relationship of the viral RNA to symptom production is concerned, we know that a single base change, say a cytosine converted to a uracil residue by nitrous acid, is sufficient to produce a mutant virus, giving changed symptoms. It is most improbable that such a change in one out of, say, 7000 bases in a viral RNA could directly bring about altered disease symptoms. Thus we must look to some activities of the virus, other than direct utilization of materials in virus production, for the events that initiate the disease process. As an introduction to this problem let us consider very briefly the current state of knowledge concerning two viruses attacking the bacterium *Escherichia coli*.

The small phage R17 contains one single-stranded piece of RNA about 3500 nucleotides long (mol. wt. of 1.1×10^6). The nature of this genetic material is almost fully understood (e.g., 52, 66). Only three proteins are specified by this RNA—A protein, coat protein, and the RNA polymerase. The first two of these, while appearing as part of the structure of the virus particle, also have other functions during infection and virus replication. The RNA polymerase, responsible for viral RNA synthesis, may have other functions as well. This small piece of RNA, which specifies only three proteins, may thus specify six or more functions.

The phage T7, also attacking *E. coli,* is substantially larger and more complex than R17. T7 contains a single piece of double-stranded DNA of molecular weight 25×10^6. After infection this DNA specifies about 30 proteins. Twenty-three of these have been identified on polyacrylamide gel electrophoresis (69). Many of these proteins form part of the structure of the virus particle, but others appear and function only in the infected cell.

A typical small plant virus such as TYMV (RNA molecular weight 1.9×10^6) could code for five proteins of average size equal to those found in R17 or T7 infections. Single functions for two of these are known—the coat protein and the RNA polymerase. Therefore about three additional proteins of unknown structure and function could be coded for. By analogy with R17, each of these may have two or more roles to play in the infected cell. Thus there are probably 5–10 TYMV-specified functions for which we must search in the infected cell.

It is quite possible that structural proteins in the virus particles may play a direct and specific role in the disease process (as opposed to the nonspecific sequestration of materials discussed above). For example, a separable structural protein of adenovirus (the penton) has been shown to cause rounding up of cells. The mechanisms of cytopathic effects induced by animal viruses has been reviewed by Bablanian (4). Jockusch & Jockusch (33) found that, among four closely related strains of TMV, there was an inverse correlation between the amount of intact virus produced and symptom severity. Disease symptoms were most severe when greatest amounts of defective insoluble coat protein were produced. They suggested that the accumulated coat protein may cause the destruction of cell organelles, but no direct evidence for this is available. Some plant viral coat proteins play a role in control of host range (2).

It seems certain that some of the viral proteins other than those forming part of the virus particle will be found to be involved in the initiation of the disease process, but they may be very difficult to isolate and study for several reasons: (a) They may be present in very low concentration, as a very few molecules per cell of a virus-specific protein could block or derepress some host-cell function. (b) It is quite possible that such proteins would be present in the infected cell for a short period relative to that required for the completion of virus synthesis. (c) An example is known where the virus ($Q\beta$ infecting E. coli) specifies only one subunit in an enzyme made up of four polypeptides, the others being provided by the host cell (34, 37). If several such factors were compounded for a particular virus-specified protein, it might be very difficult to detect and characterize.

In summary, it is possible that some symptoms may be induced by direct action of structural proteins of the virus particles. Nonstructural viral proteins are probably much more commonly involved in the initial events that lead to disease. At present we have no established examples for either kind of protein with plant viruses.

TYMV IN CHINESE CABBAGE

The disease caused by TYMV was first described in 1949 (46). The virus has since been studied by many workers, and it might be supposed that by now a fairly comprehensive description of the macroscopic features of the disease in experimental hosts would be available. We have recently become aware that the full description of this disease is really a very complex problem, and that the gaps in our knowledge are numerous. Recently we initiated experiments to give more detailed and comprehensive information on the disease process. Some of these results are included in the outline that follows. Unless otherwise stated, the data below apply to plants grown in pots in a glasshouse at 20–25°.

The Development of Macroscopic Leaf Symptoms

Following mechanical inoculation of the first well-expanded leaves (about 15 cm long) of a Chinese cabbage plant, virus can first be detected in a small proportion of cells at about 48 hours (73). Virus moves from the inoculated leaves to young leaves 3–4 days after inoculation (48). We can distinguish four successive groups of leaves (in order of appearance on the plant) with respect to the manner in which symptoms develop:

(*a*) Inoculated leaves. Chlorotic or whitish local lesions (depending on strain of virus) appear after 6–8 days, and increase in diameter over a period of about a week.

(*b*) Subsystemic leaves. A group of 2–6 expanded leaves above the inoculated leaves may not become affected by the virus for a period of weeks and may be free of virus at senescence. Alternatively, they may become invaded slowly by the virus from the base towards the tip. In this group, no mosaic symptoms appear, the leaves becoming generally yellowed. These leaves are longer than about 4 cm at the time they are invaded by virus.

(*c*) Mosaic leaves. First visible signs of systemic infection appear about 6 –10 days after inoculation as vein clearing or vein yellowing in young leaves about 3–6 cm long. These leaves would have been less than about 4 cm long at the time of infection. Leaves 2–4 cm long at time of infection then develop numerous small islands of dark green tissue on a background of pale green or yellowish green. Successively younger leaves show a tendency to produce fewer and larger islands of dark green tissue. If the virus isolate is predominantly of one strain, say white, then the dark green islands appear against a background of white tissue. If the plant is infected by a standard stock culture containing a mixture of strains (17), then islands of various shades— pale green, yellowish green, and white—will be intermingled in the mosaic with the dark green islands. In any particular leaf the pattern of the mosaic remains remarkably constant (except for dark green tissue) from the earliest stage of leaf development at which it can be recorded photographically, until the leaves are senescent. In expanded leaves the islands of dark green tissue frequently develop numerous, evenly spaced, small local-lesion-like, yellowish spots, which grow and coalesce over a period of days. White areas in the mosaic tend to become necrotic before the rest of the leaf withers.

(*d*) Leaves of the flowering stalk. These are much smaller than the fully vegetative leaves. Some leaves of this group, particularly those near the base of the stalk, may develop a very clear mosaic pattern. This group is usually characterized by a very high proportion of dark green tissue. Many leaves may be entirely dark green.

Over most of the phase of vegetative growth, infection by TYMV has no effect on the rate at which leaves are initiated at the stem apex. However, in the second week after inoculation, leaf initiation is markedly slowed. It then resumes at about the same rate as healthy plants (E. S. Crosbie & R. E. F. Matthews, unpublished results). A similar brief slowing or halting of leaf initiation is evident in the results of Takahashi (71) for TMV in tobacco, although he did not interpret his results in this way.

Cytological Effects

Cytological abnormalities induced by TYMV appear to be confined to the chloroplasts (17, 27, 51). The different shades of green, yellow, and near white that characterize islands of tissue in the macroscopic mosaic, reflect differing effects of various virus strains on the chloroplasts. Many of these effects can be distinguished by light microscopy of fresh leaf sections (19).

(a) Color. The color of chloroplasts ranges from green in dark green islands to colorless with severe white strains. (b) Clumping and changes in shape. The earliest change after infection seen by light microscopy with all strains of the virus is a rounding up of the chloroplasts. With mild strains of the virus this rounded appearance may persist, with the chloroplasts becoming lined up in rows. With somewhat more severe strains they become arranged in an irregular clump in which the outlines of the chloroplasts are barely discernable. In severely affected chloroplasts (white strains) the clumps of chloroplasts become a spherical amorphous mass. (c) Vesicles. As noted below, small vesicles are seen in all diseased chloroplasts by electron microscopy. With the light microscope only larger vesicles are seen. Different islands of tissue may vary widely in the frequency with which these larger vesicles occur. (d) Fragmentation. In some isolates the chloroplasts undergo fragmentation. The process is synchronous in any cell, giving rise to very numerous small chloroplast bodies in the cell. These smaller bodies may then undergo a further round of fragmentation.

In any block of tissue infected by a single strain of the virus, chloroplasts will show some combination of the above abnormalities. Some change in shape, and some degree of clumping are always present. The chloroplasts appear normal in dark green tissue that has not reached the stage of breakdown. The junctions between different islands of tissue in the mosaic appear remarkably sharp with respect to changes in the state of the chloroplast, when viewed by light microscopy.

The most consistent cytological change in the chloroplasts is the development of numerous small vesicles at the periphery of the organelles. These vesicles, which can be visualized only by electron microscopy, are discussed in detail in a later section.

Besides the small peripheral vesicles, larger vesicles appear in diseased chloroplasts. These are variable in size up to 1-2 μ in diameter, usually have no electron dense contents, are not confined to the periphery of the chloroplast, and are not present at very early times after infection.

TYMV infection may cause several other ultrastructural changes in the chloroplasts. The number of grana per chloroplast, and average granal stack size may be reduced. In pale green strains these effects are not seen or are minimal. They are most marked in white strains. There are increasing effects on the grana as infection proceeds (73). There is a fairly close correspondence between reduction in number of grana and chlorophyll content. Osmiophilic globules about 100 nm dia are seen occasionally in normal mature chloroplasts. Virus infection had no effect on these except in the white strain in old local lesions when many large osmiophilic globules were seen.

Similarly there were no major differences in the size and frequency of starch grains seen in thin sections of leaf harvested near the middle of the day, except in plants systemically infected with a white strain. Regular arrays of phytoferritin molecules are seen occasionally in normal chloroplasts. Their frequency was increased in chloroplasts of leaves systemically infected with more severe strains of the virus. We have never been able to identify mature TYMV particles in the body of the chloroplasts or in the vesicles they contain.

The Fine Structure of the Mosaic

As described above, different strains of the virus affect chloroplasts in distinctive ways, and chloroplasts in dark green tissue appear normal. Areas in systemically infected leaves that macroscopically appear to consist of uniformly infected tissue may on microscopic examination be found to consist of mixed tissue in which chloroplasts in horizontal layers of cells are affected differently. For example, a single upper layer of palisade cells and the lowest layer of spongy mesophyll cells may contain normal (dark green) chloroplasts, while the central zone of cells contains diseased chloroplasts (19). This situation may be reversed, with the central layer containing dark green chloroplasts and the upper and lower layers containing diseased cells. Alternatively, such layering may involve two different strains of the virus (as judged by cytological effects on chloroplasts), with one strain infecting cells in the central layer and the other infecting the outer layers.

In many leaves showing mosaic there are areas consisting of a background of diseased tissue interspersed with very numerous small islands of dark green cells. Many of the small islands in this "stipple" tissue may be visible to the naked eye. In general, the dark green islands appear only near the upper surface of the leaf. They in fact consist of small groups of palisade cells. These are usually dark green cells, but may consist of a group infected with a pale green strain of virus situated in an area of leaf infected predominantly by a more severe strain.

As seen in sections, these areas of horizontal layering may extend for several mm, or they may be quite small, grading down to islands of a few cells or even one cell of a different type. The junction between dark green cells and abnormal cells in the microscopic mosaic is usually very sharp, the chloroplasts in any one cell being affected in the same way.

Origin of the Mosaic

The following statement concerning the origin of the mosaic pattern is based on the hypothesis originally put forward (18, 19, 60): "In cells still undergoing cell division at time of infection, the first virus particle to establish itself in a cell preempts that cell and its progeny, giving rise in the mature leaf to a macroscopic or microscopic island of tissue occupied by that particular strain of virus. Alternatively a cell is converted into the dark green state and gives rise to a dark green island in the mature leaf." Considering this statement in the light of more recent experiments, it will be convenient to consider separately two groups of leaves—those that were approximately 2–30 mm long at the time they were infected, and those below 2 mm.

In leaves in the approximate length range 2–30 mm at the time they were infected, two considerations show that virus or some viral product must move from an initially infected cell into surrounding cells to form an infected cluster that ultimately gives rise to the macroscopic islands (3, 25). First, calculations based on cell numbers, and numbers of cell divisions to maturity, show that a single cell would not be sufficient to give rise to islands of the observed size (25). Statistical considerations show that such groups of cells could not occur together by chance. Therefore some factor must move from an initially infected cell to neighboring cells. Second, in leaves longer than about 2 mm all the histological layers are already formed. Thus if, for example, a single palisade cell became infected with a particular strain of virus it could not form an island of infected tissue all through the leaf other than by cell-to-cell spread.

In leaves that were less than about 2 mm long at the time of infection, we have no need to involve cell-to-cell spread to account for the observed islands of the mosaic. Dark green islands (or islands containing a particular strain of virus) in such leaves could well represent the progeny of single cells. We have found that virus is present in pieces of tissue containing the apical dome and one or two leaf primordia (25). The distribution of the virus in this region has not yet been established. It is possible that a group of cells in the true apical zone remains virus-free, and that successive cells produced by this zone continually become infected by movement of infectious material from the cells below them.

Stewart & Derwen (68), from a study of periclinal chimeras with mericlinal sectors in several species, concluded that all primary growth originates from a group of 1–3 apical initial cells in each of the apical layers. Occasionally we have observed a Chinese cabbage leaf divided about the midrib into two islands of tissue—one half leaf dark green and one half containing a uniform virus infection. This suggests that the apical initials themselves must have been infected with virus.

The following simple experiment demonstrates that the mosaic pattern is already present in latent form in very small leaves (R. E. F. Matthews, unpublished). In the apical region of an actively growing Chinese cabbage plant there are 12–20 leaves in the size range 1–10 mm that form a tight cluster shielded from light and in which virtually no chlorophyll has developed. When excised from the plant these leaves appear uniformly pale yellow, both in healthy plants and plants infected with a white strain of TYMV. When such excised leaves are held in the light under moist conditions for 24–48 hours the healthy leaves become uniformly green, while those from virus infected plants develop a clear mosaic pattern of dark green islands.

The overall pattern of the macroscopic mosaic and the fine structure of the mosaic as revealed by light microscopy show close parallels with patterns seen in chloroplast mutants. These latter must have arisen from clones of cells during leaf ontogeny (15, 19, 67, 68).

The islands of cells seen in stipple tissue, and the layering of cell types seen by microscopy in mature leaves, would be most likely to arise from single cells or very small groups of cells, confined to one (or only some) of the cell layers in the developing leaf. We would not expect to see this microscopic mosaic in leaves in which cell-to-cell spread is a major factor in the formation of islands in the mosaic. Observations fit approximately with this expectation. Stipple tissue and the microscopic mosaic are seen only in leaves that were less than about 1–2 mm long at time of infection.

Subepidermal clusters of normal green cells are seen in leaves with a G W W structure in periclinal chimeras of tobacco carrying the plastogene DP_1. (This structure gives a normal epidermis overlaying mutant mesophyll.) Stewart & Burk (67) suggest that the clusters of green cells arise by occasional periclinal divisions of cells in LI, the inner cell replacing a mutant LII cell and giving rise to a cluster of normal mesophyll cells. The similarities in the clusters illustrated by Stewart & Burk (67) and those of the TYMV stipple tissue (19) are remarkable. If the dark green islands in stipple tissue of TYMV infected leaves arise by a similar mechanism, we would expect the epidermis above patches of stipple tissue to be "dark green" and have cytologically normal chloroplasts. We have recently confirmed this expectation using electron microscopy to detect the presence or absence of peripheral vesicles in the epidermal chloroplasts (T. Hatta & R. E. F. Matthews, unpublished).

In summary, the mosaic induced by TYMV in Chinese cabbage appears to depend on the same basic processes of leaf ontogeny as are involved in the development of periclinal chimeras. In addition to the contributions of the LI, LII, and LIII layers, and the phenomena of replacement and displacement as described by Stewart & Derwen (68), the pattern of the mosaic disease is made even more complex by at least three additional factors: For each of these factors, parallels can be drawn with certain other plants containing mutations affecting the plastids. (*a*) The ability of the virus to spread from an initially infected cell in the apical region to neighboring cells in the early stages of systemic infection. Diffusible factors have also been implicated in some types of plants showing variegated leaves (36). (*b*) The much greater frequency with which TYMV mutates compared with the plastogenes of the tobacco mutant. Not all viruses produce such frequent mutants affecting chloroplasts in different ways. For example, in the mosaic induced by standard TMV, only occasional small yellow areas develop. On the other hand several variegated plants (e.g. *Acer pseudoplanatus*) have highly mutable plastids that sort out in cell lineages in the leaf but do not affect the growing apex (22). (*c*) The breakdown of dark green tissue in the TYMV mosaic after one or a few weeks. However some variegated plants are known in which initially green tissue may turn white, or white tissue green, and in which the environmental conditions may have a marked effect on the pattern produced (21).

Genetic Control of the Mosaic

The description of the macroscopic mosaic given earlier applies to our standard experimental host—Chinese cabbage (cv Wong Bok). In some varieties of the turnip (*Brassica rapa*), e.g. Senator, the stock culture of TYMV gives a very mild diffuse mosaic disease. Furthermore, the clearcut white, yellow-green, and pale green strains isolated from islands of tissue in the Chinese cabbage mosaic do not give distinguishable symptoms on this turnip variety. All show a mild diffuse mottle. However, on back-inoculation to Chinese cabbage these strains give their characteristic symptoms again. (An exception to this statement is that if very small turnip plants are inoculated with the white strain they show distinctive symptoms for a period.)

In an attempt to determine whether the differential response to TYMV strains was controlled by nuclear or cytoplasmic host genes, reciprocal crosses were carried out between the turnip and Chinese cabbage varieties (R. Ushiyama & R. E. F. Matthews, unpublished). The F_1 progeny of both crosses had the major leaf characteristics of the turnip parent except that they had variable and scattered leaf hairs (Turnip was glabrous; Chinese cabbage had scattered hairs). Twelve plants of each cross, together with parental controls, were inoculated with white, yellow-green, or pale green isolates of TYMV from Chinese cabbage.

The F_1 progeny gave clearcut symptoms typical of the Chinese cabbage reaction, whether this species was used as the male or the female parent. This result is strong evidence that the factor controlling the "clearcut" nature of the mosaic in Chinese cabbage is under nuclear control. Unfortunately attempts to carry this experiment through to the F_2 stage failed, as all the F_1 plants were sterile.

Effects on Growth

Provided mineral nutrition does not become limiting, healthy Chinese cabbage plants grow at an exponential rate during the vegetative phase (Fig. 1). Systemic symptoms first appeared in the youngest leaves at about one week after inoculation, but effects on overall growth rate did not become apparent until about three weeks, when virus production was well under way. Growth rate during this time was approximately linear. The effect of a pale green strain of TYMV on overall growth was very similar to that caused by the white strain. For both strains we noticed that infection caused some increase in the rate at which older leaves became senescent in the later stages of the trial.

The average fresh weight of the root system per healthy plant increased in an approximately linear fashion for the duration of the experiment. Roots of plants inoculated with either a white or a pale green strain increased very little after systemic infection and were 30% of the weight of healthy plants at 7.5 weeks.

In the experiment summarized in Figure 1 we excised the subsystemic leaves from groups of virus-infected and healthy plants 3 weeks after inoculation to test the importance of this group of leaves for continued growth of the diseased plant as a whole. Average aerial fresh weights per plant were: Healthy with subsystemic leaves at 7 weeks, 332 gms; healthy without at 7.5 weeks, 361 gms; infected with (at 7 weeks), 105 gms; infected without (at 7.5 weeks), 28 gms. Although this last set of plants made some young growth

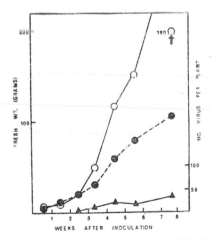

Figure 1 Effect of infection with a white strain of TYMV on total aerial fresh weight of Chinese cabbage plants grown in pots in the glasshouse (E. S. Crosbie & R. E. F. Matthews, unpublished). Each point represents the average for 7–10 plants. Virus was estimated in extracts using analytical ultracentrifugation. O——O Healthy, ●– – –● TYMV infected, ▲——▲ mg virus.

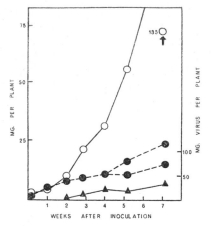

Figure 2   Effects of infection with a white strain of TYMV on the chlorophyll content of Chinese cabbage plants. Plants are of the same experiment as Figure 1. (E. S. Crosbie & R. E. F. Matthews, unpublished.) O —— O Healthy, ●----● TYMV infected, ▲ —— ▲ mg virus. The lower points for chlorophyll A in infected plants at weeks 5 and 7 were estimated, excluding measurable islands of dark green tissue. The upper curve includes the chlorophyll found in dark green islands.

during the period, there was a net loss of fresh weight due to accelerated senescence of older systemically infected leaves.

Infected plants with subsystemic leaves removed at week 3 had root systems averaging 1.8 gms fresh weight, compared with 12.8 gms for equivalent healthy plants. These results, together with those in the next section, suggest that the stunting of the plant as a whole is a consequence of the reduction in chlorophyll caused by the virus. The diseased plant can respond to a limited extent. We have found that when the subsystemic leaves were removed from plants infected for three weeks with a white strain of TYMV, subsequent development of dark green tissue was substantially increased.

Leaf Pigments

We have established (E. S. Crosbie & R. E. F. Matthews, unpublished) that the relative proportions of chlorophylls A and B, and the carotenoids, β-carotene, lutein, violaxanthin, and neoxanthin in healthy Chinese cabbage are very close to those published for spinach (31).

As chlorophyll A is the major photosynthetic pigment, we followed the amount per plant in the same experiment as that of Figure 1. Figure 2 shows that chlorophyll A content rose exponentially in healthy plants. In plants infected with a white strain, net chlorophyll production fell almost to zero by week four. After week four, dark green islands formed an increasing proportion of new leaf tissue, which gave a renewed increase in chlorophyll per plant.

We have studied the effects of TYMV on pigments in more detail for the systemically infected leaf that shows clear vein-clearing and is about 4 cm long when infected (E. S. Crosbie & R. E. F. Matthews, unpublished). All six pigments followed the same pattern of changes. Calculated as amount of pigment per unit fresh weight, concentrations rose in both healthy and infected leaves until 10 days after inoculation. From day 12, pigment concentration in healthy leaves remained constant. Concentrations in infected leaves fell to 10–20% of healthy values by day 21. This drop was due to a cessation of net pigment synthesis, because weight of pigment per leaf continued to rise until the end of the experiment (day 21) in healthy leaves, but reached a plateau by day 12–14 in infected leaves.

A similar situation developed in plants infected with a pale green strain, but the effects were less marked. All six pigments fell to about 50% of the healthy value by about day 12. Net pigment production did not stop with this strain, but continued at a reduced rate. Virus synthesis was well established for both strains (about half final concentration) before significant effects on pigments were seen.

Infection with a white strain or with a pale green strain led to a fall in inoculated leaves in the concentration of both chlorophylls A and B and the four carotenoids to about 65–85% (at day 8) of the corresponding healthy tissue. Because the inoculated leaves were not undergoing expansion, this fall represented a net loss of these pigments. The effect became apparent after virus synthesis was well established. Thus the effects on leaf pigments, although an important feature of the disease process, appear to be a secondary consequence of infection and not essential for virus replication.

Effects on Macromolecules

The yellow-green areas in leaves of Chinese cabbage showing mosaic symptoms had reduced concentration of 68S (chloroplast) ribosomes and Fraction I protein (60). Dark green areas appeared to have normal concentrations of these components. Concentrations of Fraction II proteins (mainly cytoplasmic) and 83S (cytoplasmic) ribosomes were the same in yellow-green and dark green areas.

In a more detailed study, we have shown that in young leaves that were about 4 cm long when they showed systemic vein-clearing, concentration of Fraction I protein fell, relative to healthy leaves, only after virus production had passed the peak rate (6). Ribulose 1,5,diphosphate carboxylase activity followed a similar pattern (7).

Considered on a per plant basis in the same experiment as that of Figures 1 and 2, the macromolecules fell into two classes. Those found in the chloroplast (Fraction I and 68S ribosomes) followed chlorophyll A, and began to fall rapidly compared with healthy plants between two and three weeks after inoculation. These components also showed an actual decrease in concentration in infected leaves. The cytoplasmic components showed delayed and less marked reductions on a per plant basis, which were not due to a reduced concentration in the infected leaves but were a consequence of the overall stunting of growth that follows infection.

A decline in the concentration of Fraction I protein and/or chloroplast ribosomes, which begins during the period of rapid virus synthesis, appears to be a common feature of infections by viruses of widely differing types. It has been found for TMV (53, 54), tomato spotted wilt virus (53), and lettuce necrotic yellows virus (59).

Effects on Photosynthesis

Goffeau & Bové (29) studied the effects of TYMV on certain photosynthetic processes, using chloroplasts isolated from infected and healthy Chinese cabbage leaves. They found that there was an increase in the Hill reaction and a faster rate of ATP formation by both cyclic and noncyclic phosphorylation. Their most interesting experiments were carried out with chloroplasts from inoculated leaves, where they were able to show that the increases occurred during the period of rapid virus replication (about 7–12 days after inoculation).

Using ^{14}C-labelled carbon dioxide we have recently studied the effect of TYMV infection on the early products of photosynthetic carbon fixation (6, 7). We used the systemically infected leaf, which is about 4 cm long at the time it becomes systemically infected. The distribution of labelled carbon in various compounds was altered by infection. Radioactivity entering sugars and sugar phosphates was reduced, while the amounts entering organic acids and amino acids were increased. The extent of these changes closely followed the time course for rate of virus replication in the leaf studied.

The activities of two enzymes involved in the interconversions of early formed products of photosynthesis (phosphoenolpyruvate carboxylase and aspartate aminotransferase) increased compared with healthy leaves during virus replication, and then fell again. The time course of these increases closely followed that for virus replication. The changes in the activity of these two enzymes contrasts with some other effects of infection. The following proteins or activities related to the carbon fixation process remained unaltered until about the time virus concentration had reached a maximum, when they fell substantially and remained lower than in equivalent tissue: 68S ribosomes; Fraction I protein; ribulose diphosphate carboxylase activity; and overall rate of carbon fixation.

These results suggest that, during the period of rapid virus synthesis, TYMV diverts the flow of carbon fixed by the Calvin pathway from sugars to acids by increasing the rate of carboxylation of PEP to form oxaloacetate. From this compound malate and aspartate are formed. These compounds could then readily give rise to many of the components needed for virus synthesis.

As far as we have been able to test the question, the changes outlined above are specific for TYMV. The rises in PEP carboxylase and aspartate aminotransferase activity were found in *Malcomia* and *Cheiranthus* infected with TYMV, but not in Chinese cabbage infected with a virus tentatively identified as cauliflower mosaic virus, and not in tobacco infected with TMV.

Occurrence, Fine Structure, and Activity of Small Vesicles in the Chloroplasts

There is now substantial evidence to suggest that the small vesicles found at the periphery of diseased chloroplasts are intimately involved in virus replication and the induction of disease symptoms. Current knowledge of their occurrence, fine structure, and function will be discussed here in detail.

Seventy-two hours after inoculation with TYMV 14% of the cells in Chinese cabbage leaves observed by light microscopy showed rounding up and some clumping of chloroplasts (73). Examined by electron microscopy these chloroplasts showed peripheral vesicles. To identify cells in which virus had been produced, a wilting procedure was used to induce TYMV particles to form crystalline arrays. Such arrays were seen in cells in which the only abnormalities observed in chloroplasts were slight swelling and the presence of peripheral vesicles. More recently we have examined the state of the chloroplasts in a series of small samples of tissue taken from a developing local lesion, on a radial line from the center out to uninfected tissue (T. Hatta & R. E. F. Matthews, unpublished). Scattered small peripheral vesicles were seen in chloroplasts of normal shape and structure in infected cells most distant from the center of the lesion.

The number of small vesicles per chloroplast varies with strain of virus, as well as with age of infection. They were least numerous in a pale green strain, more numerous for a yellow-green strain, and most numerous for a white strain (73).

The vesicles are found in the chloroplasts of the midribs of infected leaves (73), in leucoplasts of roots, and chromoplasts of petals (41, 43). They are found in all the host plants of TYMV that have been examined [*Brassica rapa*, (73); *Cleome* sp. (43); *Iberis umbellata, Malcomia maritima, M. chia, M. flexuosa, Cheiranthus maritimus, C. allionii, Reseda odorata*, and *Lepidium sativium* (T. Hatta & R. E. F. Matthews, unpublished)].

They are also found in infections caused by other members of the TYMV group, e.g., wild cucumber mosaic virus in *Marah oreganus* (1), belladonna mottle virus in tobacco and other Solanaceae (43). They are not produced by cauliflower mosaic virus in Chinese cabbage (20). They have not been noted in *Brassica perviridis* infected with turnip mosaic virus (24). Thus, they appear to be characteristic for the TYMV group in any host species. Similar peripheral vesicles in diseased chloroplasts have been described for one quite unrelated virus—barley stripe mosaic virus in barley (16). Most of the vesicles described by Carroll were somewhat different in structure from those caused by TYMV, being bounded by a single membrane and enclosed in an expanded part of the periplastidal space. A peripheral reticulum is normal in the chloroplasts of species that have a C4 type of carbon fixation (39), and is also a normal feature of some C3 plants. This reticulum may give the appearance of marginal vesicles in thin sections. Some authors may have mistaken such normal structures for virus-induced vesicles (26). Invaginations from the inner membrane of the plastid are also common in etioplasts of healthy tobacco leaves (8).

The outer membrane of the vesicles induced by TYMV is a continuation of the inner membrane of the chloroplast (41, 73). As observed by Laflèche & Bové (41), some vesicles appear to be definitely open to the cytoplasm, and they suggested that the inner membrane of the vesicles arises from an invagination of the outer membrane of the chloroplasts. This is by no means established. We think that the type of open vesicle illustrated by Laflèche & Bové (41) may possibly represent a degenerating stage of the small peripheral vesicles. For many vesicles the inner membrane appeared to form a continuous sac, as seen in thin sections. For most vesicles it was not possible to observe clearly the details of membranes at the necks, as the sections used were 60–90 nm thick, while commonly observed diameters of the vesicles were 50–80 nm. The necks of the vesicles are no more than 10–30 nm dia as seen in section. We have recently made a further study of the membranes of small vesicles (32), using freeze-fracture methods (13, 14).

We have examined chloroplasts, both in intact leaf tissue and when isolated by standard procedures. Face-on views of the chloroplast membrane show the distribution of the small vesicles over the chloroplast surface. In many chloroplasts they are clustered into groups, with the necks in an approximately hexagonal array (Fig. 3). In other chloroplasts they appear more or less randomly distributed over the surface.

Freeze-fracturing confirms the observations from thin sections, which showed that the outer vesicle membrane is an invagination of the inner chloroplast membrane. Vesicle necks can be seen extending towards the outer membrane; this suggests that there is a connection between the interior of the vesicles and the cytoplasmic space outside of the chloroplast (Fig. 4). However, we have not yet unambiguously established such a connection.

Most biological membranes reveal intramembraneous particles on freeze-fracturing, and these are thought to consist of proteins intercalated in the lipid bilayer (12, 64). Such particles are present in both inner and outer chloroplast membranes and in the outer vesicle membrane. No such particles are present in the inner membranes of the vesicle. Thus, TYMV infection leads to the formation of a membrane with unusual structural properties (32).

When observed in thin sections, many of the vesicles are seen to contain stranded material with the staining properties of double-stranded nucleic acids. The proportion of vesicles showing such strands is higher at early stages after infection (T. Hatta & R. E. F. Matthews unpublished).

The presence of these strands, together with the observation that the small vesicles were the only cytological modification seen under all circumstances where virus replication was occurring, led us to suggest that they might in fact be the site of viral RNA synthesis (73). There are several observations that would be consistent with this role:

(*a*) The double-stranded form of TYMV RNA, which can be isolated from infected plants (58), is associated with the chloroplast fraction rather than with the nuclei or the soluble fraction of leaf extracts (10, 55–57).

(*b*) In their elegant autoradiographic studies, Laflèche and Bové (40) showed that in TYMV-infected tissue treated with Actinomycin D, silver grains accumulated over the spaces between clumped chloroplasts and also over the nucleolus. However, the resolution of the method is insufficient to

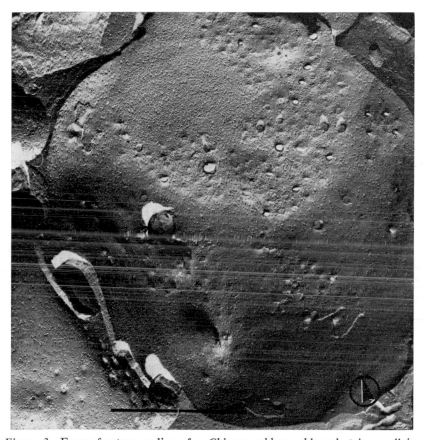

Figure 3   Freeze-fracture replica of a Chinese cabbage chloroplast in a cell infected with TYMV. The fracture has passed through the outer chloroplast membrane and shows a clustered distribution of the depressions about 25 nm dia in the chloroplast surface, representing the sites of the underlying vesicles. (T. Hatta, S. Bullivant, R. E. F. Matthews, unpublished.) Magnification bar = 1 μm. Arrow indicates direction of shadow.

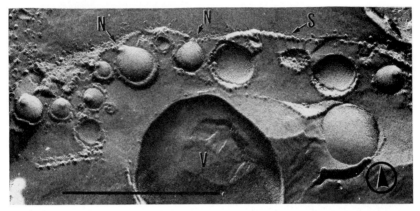

Figure 4 Freeze-fracture replica of a Chinese cabbage chloroplast isolated from leaf tissue infected with TYMV. The fracture has passed through the body of the organelle revealing the membranes of the peripheral vesicles. The necks (N) of some vesicles can be seen to approach very closely to the chloroplast surface (S). V = a large internal vesicle (T. Hatta, S. Bullivant, R. E. F. Matthews, unpublished). Magnification bar = 0.5 μm. Arrow indicates direction of shadow.

distinguish between tracks originating just inside or just outside the chloroplasts. It would be very difficult with radioautographic studies alone to determine whether the labelled material is actually synthesized or merely accumulates at the sites where it is found.

(*c*) Bové and his colleagues (11, 42, 43) have established in various studies that most of the template TYMV RNA ("minus" strands) in the cell and the viral specific replicase are firmly bound to cell structures or organelles that remain in the supernatant after centrifugation of leaf homogenates at 100 g for 5 min, but that sediment at 10,000 g for 15 min. Two lines of evidence strongly suggest that the enzyme-template complex is bound to chloroplast membranes. First, it could be made soluble with the membrane-dispersing detergent Lubrol W. Second, when Actinomycin D-resistant virus-specific RNA synthesis was allowed to proceed in this fraction, using a tritium labelled RNA precursor, and the fraction subjected to centrifugation, most of the incorporated radioactivity was recovered in the pellet. Electron microscopy showed that the pellet consisted largely of the diseased chloroplasts' outer membrane system, including the small peripheral vesicles (11).

SUMMARY AND CONCLUSIONS

TYMV appears to be a virus that has adapted to parasitize primarily one organelle—the chloroplast—rather than the cell as a whole. Viral RNA by itself can initiate infection. It is rather unlikely that an incoming RNA molecule itself could have any function other than to act as a template for the synthesis of viral specific proteins. At an early stage after entry of the RNA into the cytoplasm, the following virus specific proteins are probably synthe-sized: (*a*) The viral specific RNA polymerase. Bové and his colleagues, and others, have provided positive evidence for a specific viral-induced RNA polymerase. (*b*) A "membrane" protein (or proteins) that modifies the chlo-roplast membranes to allow the inner membrane to invaginate, and produces a second membrane to form an ultrastructurally smooth "flask." The coat protein could be involved in this process. Evidence for such a protein is indi-rect. The small peripheral vesicles appear to be specific for TYMV and re-lated viruses, rather than for the host species. (*c*) A protein (or proteins) that causes a higher rate of photophosphorylation and a higher proportion than normal of newly fixed carbon to be converted to amino acids and other acids rather than sugars. These might be virus-specified enzymes for particu-lar metabolic steps, or regulatory proteins affecting host enzyme synthesis or activity. Again, the evidence is indirect. The changed pattern of flow of car-bon fixed in photosynthesis appears to be specific for TYMV, rather than for the host plant.

It may be that following formation of the first peripheral vesicles, one (or possibly a few) RNA complements, each with one or a few virus specific RNA polymerase molecules, becomes established in each vesicle. At about the same time, the virus-induced flow of newly fixed carbon into organic ac-ids and amino acids may begin, along with increased rate of photophosphoryla-tion. Viral RNA strands are probably produced in the vesicles, and at early stages in the infected cell are released into the cytoplasm, where they are tran-scribed. The number of peripheral vesicles per chloroplast increases rapidly as virus production begins.

Virus coat protein is probably synthesized on 83S ribosomes in the cyto-plasm. I wish to suggest that the actual site of assembly of the virus particles is at the neck of the peripheral vesicles. The diameter of the depression in the surface of the chloroplast above a vesicle (e.g., in Fig. 3) is about 25 nm, corresponding quite closely to that of the virus shell. RNA complements that are synthesized in the vesicles may be assembled at the neck with viral coat protein subunits synthesized in the cytoplasm. Completed virus particles (or partially filled or empty shells) would then be released to accumulate in the cytoplasm.

Virus infection causes degenerative processes to begin in mature chloro-plasts, or if cells in a developing leaf are infected, development of normal chloroplast structures may be partially or completely inhibited. Various strains of the virus found in a leaf showing mosaic disease reach about the same concentration in the tissue. As noted earlier, considering the plant as a whole, the amount of TYMV produced is small in relation to the reduction in normal proteins and ribosomes caused by infection. However, the number of peripheral vesicles increases with increasing macroscopic symptom severity (73). Preliminary experiments (E. S. Crosbie & R. E. F. Matthews, unpub-lished) indicated that a white strain increased in inoculated leaves more rap-idly than a pale green strain. Thus, the severity of the pathological state in individual cells caused by different strains of the virus may reflect the rate at which they subvert the chloroplast's normal functions (production of ATP and carbon skeletons), rather than the final quantity of virus produced.

A reduction in chlorophyll content in chloroplast mutants is not necessarily correlated with a reduction in other major plastid constituents or functions. For example, mutants of *Arabidopsis thaliana* had normal levels of ribulose 1,5-diphosphate carboxylase activity, but reduced chlorophyll content and CO_2 assimilation rate (70). Fraction I protein in the tobacco chloroplast mutant (15) is greatly reduced in white tissue, but chloroplast ribosomes are present in normal amounts (R. E. F. Matthews, unpublished). Thus, at this stage there is no obvious biochemical connection between the reduced amounts of pigments, 68S ribosomes, Fraction I protein, and reduced rates of CO_2 assimilation brought about by TYMV infection.

We have no information as to how viruses cause the deficiency in leaf pigments which is a major feature of mosaic diseases. Production of all leaf pigments appears to be reduced to about the same extent by a given virus strain. TYMV infection thus has a similar effect on the two independent pathways of pigment synthesis.

Loss of photosynthetic pigments or a reduced rate of their production is not essential for virus replication. However, as far as the effect of TYMV infection on growth of the plant as a whole is concerned, the prime event may be the reduction in total chlorophyll. Effects on growth (e.g., reduction in size of root system) may follow cumulatively from the reduced carbon assimilation.

The nature of the dark green tissue remains an enigma. For a time at least, this tissue appears able to resist the production of detectable vegetative virus and the accompanying pathological effects. The mechanism of this resistance is unknown. It must involve a factor that can spread from cell to cell in the young leaf. The factor appears to confer resistance on all chloroplasts in a cell, or on none.

We suggested (60) that dark green cells have become resistant to infection by some process like lysogeny in bacteria. This would involve the incorporation of some DNA related to the virus into the host DNA. Certain RNA tumor viruses have been shown to possess an enzyme transcribing the viral RNA into DNA form (5, 72, reviewed in 65). This reverse transcriptase activity in these viruses is found in the virus particle. There is no reason in principle why TYMV (and other viruses causing mosaic diseases) should not possess information for a reverse transcriptase, which is produced only in the infected cell and could form a DNA copy of at least part of the viral RNA.

An alternative explanation is that mutants defective in productive capacity occupy dark green cells and prevent superinfection by related strains of the virus. It would require inspection of 50–100 serial sections to demonstrate the presence or absence of one small peripheral vesicle per chloroplast. It would require many more to find one peripheral vesicle in a cell.

Looked at from the point of view of the plant as a whole, dark green tissue may have some survival value for the host plant and the virus. Plants infected with severe (white) strains produce more dark green islands that those infected with pale green strains. Removal of the subsystemic leaves from plants infected with a white strain leads to a substantial increase in the amount of dark green tissue produced. We believe that there may be a more rapid rate of cell division of dark green cells, compared with cells supporting virus replication, at very early stages in the ontogeny of infected leaves. Now that we can manipulate the amount of dark green tissue it will be possible to test this hypothesis.

In summary, as far as the induction of disease symptoms by TYMV is concerned, there appear to be five major areas worthy of intensive study at the present time: (a) The fine structure, macromolecular composition, and activities of the small peripheral vesicles. (b) The mechanism by which TYMV induces a flow of carbon compounds away from sugars and towards acids. (c) The mechanism by which chlorophyll synthesis is reduced or chlorophyll destroyed during infection. (d) The molecular biological basis for the existence of islands of dark green tissue in the mosaic, including a search for possible "reverse transcriptase" activity induced by the virus. (e) A study of the cytological and biochemical basis for other types of symptom such as vein clearing and the diffuse mottles produced in some hosts other than *Brassica pekinensis.*

Literature Cited

1. Allen, T. C. 1972. Subcellular responses of mesophyll cells to wild cucumber mosaic virus. *Virology* 47:467–74
2. Atabekov, J. 1971. Some properties and functions of the coat protein of plant viruses, including the function of host range control. *Acta Phytopathol. Acad. Sci. Hungaricae* 6:57–60
3. Atkinson, P. H., Matthews, R. E. F. 1970. On the origin of dark green tissue in tobacco leaves infected with tobacco mosaic virus. *Virology* 40:344–56
4. Bablanian, R. 1972. Mechanisms of virus cytopathic effects. *Symp. Soc. Gen. Microbiol. Proc.* 22:359–81
5. Baltimore, D. 1970. RNA dependent DNA polymerase in virions of RNA tumour viruses. *Nature, London* 226:1209–11
6. Bedbrook, J. R., Matthews, R. E. F. 1972. Changes in the proportions of early products of photosynthetic carbon fixation induced by TYMV infection. *Virology* 48:255–58
7. Bedbrook, J. R., Matthews, R. E. F.

1973. Changes in the flow of early products of photosynthetic carbon fixation associated with the replication of TYMV. *Virology* (in press)
8. Boasson, R., Laetsch, W. M., Price, I. 1972. The etioplast-chloroplast transformation in tobacco: correlation of ultrastructure, replication and chlorophyll synthesis. *Am. J. Bot.* 59:217–23
9. Bos, L. 1970. *Symptoms of Virus Disease in Plants.* 2nd Ed., Centre for Agricultural Publication and Documentation. Wageningen: 206 p.
10. Bové, J. M., Bové, C., Rondot, M. J., Morel, G. 1967. Chloroplasts and virus RNA synthesis. *Biochemistry of Chloroplasts.* Ed. T. W. Goodwin. *Proc. NATO Adv. Stud. Inst. Aberystwyth* 1965. Vol. II. New York: Academic, 329–39
11. Bové, C., Mocquot, B., Bové, J. M. 1972. Turnip yellow mosaic virus-RNA synthesis in the plastids: Partial purification of a virus-specific, DNA-independent, enzyme-template complex. (In press)
12. Branton, D. 1971. Freeze-etching studies of membrane structure. *Phi-*

los. Trans. Roy. Soc. Ser. B 261: 133–38

13. Bullivant, S. 1969. Freeze-fracturing of biological materials. *Micron* 1:46–51

14. Bullivant, S., Ames, A. 1966. A simple freeze-fracture replication method for electron microscopy. *J. Cell Biol.* 29:435–47

15. Burk, L. G., Stewart, R. N., Derwen, H. 1964. Histogenesis and genetics of a plastid-controlled chlorophyll variegation in tobacco. *Am. J. Bot.* 51:713–24

16. Carroll, T. W. 1970. Relation of barley stripe mosaic virus to plastids. *Virology* 42:1015–22

17. Chalcroft, J. P., Matthews, R. E. F. 1966. Cytological changes induced by turnip yellow mosaic virus in Chinese cabbage leaves. *Virology* 28:555–62

18. Chalcroft, J. P., Matthews, R. E. F. 1967. Virus strains and leaf ontogeny as factors in the production of leaf mosaic patterns by turnip yellow mosaic virus. *Virology* 33: 167–71

19. Chalcroft, J. P., Matthews, R. E. F. 1967. Role of virus strains and leaf ontogeny in the production of mosaic patterns by turnip yellow mosaic virus. *Virology* 33:659–73

20. Conti, G. G., Vegetti, G., Bassi, M., Favali, M. A. 1972. Some ultrastructural and cytochemical observations on Chinese cabbage leaves infected with cauliflower mosaic virus. *Virology* 47:694–700

21. Darlington, C. D. 1929. Variegation and albinism in *Vicia faba. J. Genet.* 21:161–68

22. Darlington, C. D. 1970. Nucleus, cytoplasm and cell. *Symp. Soc. Exp. Biol.* 24:1–12

23. Diener, T. O., Jenifer, F. G. 1964. A dependable local lesion assay for turnip yellow mosaic virus. *Phytopathology* 54:1258–60

24. Edwardson, J. R., Purcifull, D. E. 1970. Turnip mosaic virus-induced inclusions. *Phytopathology* 60:85–88

25. Faed, E. M., Matthews, R. E. F. 1972. Leaf ontogeny and virus replication in *Brassica pekinensis* infected with turnip yellow mosaic virus. *Virology* 48:546–54

26. Gardner, W. S. 1969. Ultrastructure of *Zea mays* leaf cells infected with Johnson-grass strain of sugarcane mosaic virus. *Phytopathology* 59:1903–07

27. Gerola, F. M., Bassi, M., Giussani, G. 1966. Some observations on the shape and localization of different viruses in experimentally infected plants and on fine structure of host cells. III. Turnip yellow mosaic virus in *Brassica chinensis* L. *Caryologica* 19:457–79

28. Gillaspie, A. G., Bancroft, J. B. 1965. The rate of accumulation, specific infectivity, and electrophoretic characteristics of bean pod mottle virus in bean and soybean. *Phytopathology* 55:906–08

29. Goeffeau, A., Bové, J. M. 1965. Virus infection and photosynthesis. 1. Increased photophosphorylation by chloroplasts from Chinese cabbage infected with turnip yellow mosaic virus. *Virology* 27:243–52

30. Goldstein, B. 1926. A cytological study of the leaves and growing points of healthy and mosaic diseased plants. *Bull. Torrey Bot. Club* 53:499–599

31. Gregory, R. P. F. 1971. *Biochemistry of Photosynthesis.* New York: Wiley-Interscience, 202 pp.

32. Hatta, T., Bullivant, S., Matthews, R. E. F. 1973. Fine structure of vesicles induced in chloroplasts of *Brassica pekinensis* by turnip yellow mosaic virus. *J. Gen. Virol.* (in press)

33. Jockusch, H., Jockusch, B. 1968. Early cell death caused by TMV mutants with defective coat proteins. *Mol. Gen. Genet.* 102:204–09

34. Kamen, R. 1970. Characterisation of the subunits of $Q\beta$ replicase. *Nature* 228:527–33

35. Kassanis, B., Bastow, C. 1971. The relative concentration of infective intact virus and RNA of four strains of tobacco mosaic virus as influenced by temperature. *J. Gen. Virol.* 11:157–70

36. Kirk, J. T. O., Tilney-Bassett, R. A. E. 1967. *The Plastids.* San Francisco: W. H. Freeman and Company, 608 pp.

37. Kondo, M., Gallerani, R., Weissmann, C. 1970. Subunit structure of $Q\beta$ replicase. *Nature* 228:525–27

38. Kuhn, C. W., Bancroft, J. B. 1961. Concentration and specific infectivity changes of alfalfa mosaic vi-

rus during systemic infection. *Virology* 15:281-88

39. Laetsch, M. W., Price, I. 1969. Development of the dimorphic chloroplasts of sugar cane. *Am. J. Bot.* 56:77-87

40. Laflèche, D., Bové, J. M. 1968. Sites d'incorporation de l'uridine tritiée dans les cellules du parenchyme foliare de *Brassica chinensis* saines au infectée par le virus de la mosaique jaune du Navet. *C.R. Acad. Sci.* 266:1839-41

41. Laflèche, D., Bové, J. M. 1969. Development of double membrane vesicles in chloroplasts from turnip yellow mosaic virus infected cells. *Prog. Photosyn. Res.* 1:74-83

42. Laflèche, D., Bové, J. M. 1971. Virus de la mosaique jaune du Navet. Site cellulaire de la replication du RNA viral. *Physiol. Vég.* 9:487-503

43. Laflèche, D., Bové, C., Dupont, G., Mauches, C., Astier, T., Garnier, M., Bové, J. M. 1972. Site of viral RNA replication in the cells of higher plants: TYMV-RNA synthesis on the chloroplast outer membrane system. *Symp. RNA Viruses: Replication and Structure Proc. Fed. Eur. Biochem. Soc. 8th meet., Amsterdam*

44. Lister, R. M. 1968. Functional relationships between virus-specific products of infection by viruses of the tobacco rattle type. *J. Gen. Virol.* 2:43-58

45. Lister, R. M., Bracker, C. E. 1969. Defectiveness and dependence in three related strains of tobacco rattle virus. *Virology* 37:262-75

46. Markham, R., Smith, K. M. 1949. Studies on the virus of turnip yellow mosaic. *Parasitology* 39:330-42

47. Matthews, R. E. F. 1970. *Plant Virology*, New York: Academic, 778 pp.

48. Matthews, R. E. F., Ralph, R. K. 1966. Turnip yellow mosaic virus. *Advan. Virus Res.* 12:273-328

49. Matthews, R. E. F., Bolton, E. T., Thompson, H. R. 1963. Kinetics of labelling of turnip yellow mosaic virus with P^{32} and S^{35}. *Virology* 19:179-89

50. Melchers, G., Jockusch, H., Sengbusch, P. V. 1966. A tobacco mutant with a dominant allele for hypersensitivity against some TMV strains. *Phytopathol. Z.* 55:86-88

51. Miličič, D., Štefanac, Z. 1967. Plastidenveränderungen unter dem Einfluss der Wasserübengelbmosaikvirus (turnip yellow mosaic virus). *Phytopathol. Z.* 59:285-96

52. Min Jou W., Hageman, M., Ysebaert, M., Fiers, W., 1972. Nucleotide sequence of the gene coding for the bacteriophage MS2 coat protein. *Nature* 237:82-88

53. Mohamed, N. A., Randles, J. W. 1972. Effect of tomato spotted wilt virus on ribosomes, ribonucleic acids and fraction I protein in *Nicotiana tabacum* leaves. *Physiol. Plant Pathol.* 2:235-45

54. Oxelfelt, P. 1971. Development of systemic tobacco mosaic virus infection. II. RNA metabolism in systemically infected leaves. *Phytopathol. Z.* 71:247-56

55. Ralph, R. K., Clark, M. F. 1966. Intracellular location of double-stranded plant viral ribonucleic acid. *Biochim. Biophys. Acta* 119:29-36

56. Ralph, R. K., Wojcik, S. J. 1966. Synthesis of double-stranded viral RNA by cell-free extracts from turnip yellow mosaic virus-infected leaves. *Biochim. Biophys. Acta* 119:347-61

57. Ralph, R. K., Bullivant, S., Wojcik, S. J. 1971. Evidence for the intracellular site of double stranded turnip yellow mosaic virus RNA. *Virology* 44:473-79

58. Ralph, R. K., Matthews, R. E. F., Matus, A. I., Mandel, H. G. 1965. Isolation and properties of double-stranded viral RNA from virus-infected plants. *J. Mol. Biol.* 11:202-12

59. Randles, J. W., Coleman, D. F. 1970. Loss of ribosomes in *N. glutinosa* L. infected with lettuce necrotic yellow virus. *Virology* 41:459-64

60. Reid, M. S., Matthews, R. E. F. 1966. On the origin of the mosaic induced by turnip yellow mosaic virus. *Virology* 28:563-70

61. Sänger, H. L. 1969. Functions of the two particles of tobacco rattle virus. *J. Virol.* 3:304-12

62. Siegel, A. 1960. Studies on the induction of tobacco mosaic virus mutants with nitrous acid. *Virology* 11:156-67

63. Siegel, A., Zaitlin, M., Sehgal, O. P.

1962. The isolation of defective tobacco mosaic virus strains. *Proc. Nat. Acad. Sci. U.S.* 48:1845–51

64. Singer, S. J., Nicholson, G. L. 1972. The fluid mosaic model of the structure of cell membranes. *Science* 175:720–31

65. Spiegelman, S., Schlom, J. 1972. Reverse transcriptase in oncogenic RNA viruses. *Fed. Eur. Biochem. Soc. Symp.* 22:115–33

66. Steitz, J. A. 1969. Polypeptide chain initiation nucleotide sequences of the three ribosomal binding sites in bacteriophage R17 RNA. *Nature* 224:957–64

67. Stewart, R. N., Burk, L. G. 1970. Independence of tissues derived from apical layers in ontogeny of the tobacco leaf and ovary. *Am. J. Bot.* 57:1010–16

68. Stewart, R. N., Derwen, H. 1970. Determination of number and mitotic activity of shoot apical initial cells by analysis of mericlinal chimeras. *Am. J. Bot.* 57:816–26

69. Studier, F. W. 1972. Bacteriophage T7. *Science* 176:367–76

70. Švachulová, J. 1971. $^{14}CO_2$ fixation, ribulose 1,5-diphosphate carboxylase activity and free sugar content of two chlorophyll mutants of *Arabidopsis thaliana* L (Heynh.) *Photosynthetica* 5:249–57

71. Takahashi, T. 1972. Studies on viral pathogenesis in plant hosts. II. Changes in developmental morphology of tobacco plants infected systemically with tobacco mosaic virus. *Phytopath. Z.* 74:37–47

72. Temin, H. M., Mizutani, S. 1970. RNA-dependent DNA polymerase in virions of Rous sarcoma virus. *Nature London* 226:1211–13

73. Ushiyama, R., Matthews, R. E. F. 1970. The significance of chloroplast abnormalities associated with infection by turnip yellow mosaic virus. *Virology* 42:293–303

ADVANCES IN THE STUDY OF VESICULAR-ARBUSCULAR MYCORRHIZA[1]

❖ 3570

B. Mosse
Department of Soil Microbiology, Rothamstead Experimental Station, Harpenden, Herts., England

INTRODUCTION

The study of vesicular-arbuscular (VA) mycorrhiza is expanding rapidly. Since Gerdemann's review in 1968, over a hundred papers have been published. During previous five year periods numbers were: 14 (1930–4), 22 1935–9), 17 (1948–52), 43 (1953–7), 56 (1958–62) and 40 (1963–7). These are small numbers considering how long it has been known that VA mycorrhiza are probably the most widespread root infections of plants. With some justification they have been described as the "mal aimée des microbiologistes" (22).

The increase in publications has been accompanied by a shift in subject matter. Most papers, until recently, described the anatomy and recorded the occurrence of VA mycorrhiza, and many efforts were made to culture the fungi; since 1968, 37 papers have dealt with effects of the infection on plant growth.

Several factors probably account for the increased popularity of the subject. The long-standing speculation about the identity of VA endophytes (47, 56) has largely been resolved in favor of one or another species of *Endogone* (32, 46, 95). Very impure inocula consisting of infected roots or of soil containing a normal population of other soil micro-organisms, have been replaced by *Endogone* spores, sporocarps, or "sterilized" soil inoculated with them in the presence of a host plant. Such inocula now regularly produce typical VA infections in experimental plants. With improved techniques, very striking effects of inoculation on plant growth and phosphate uptake have been demonstrated beyond doubt, and this has led to studies of the uptake mechanism and the source of the extra phosphate. These results have also stimulated interest in the fungi themselves, their ecology and taxonomy. The technique of clearing roots in KOH before staining (24, 77, 120, 121) is now

[1] I thank my colleague Dr. D. S. Hayman for many helpful discussions.

171

widely used and has led to more general awareness of the frequency of VA mycorrhiza in many crop plants. Finally, three very good reviews (47, 57, 115) have drawn attention to the potentialities of VA mycorrhiza and helped to raise their study to the level of "a reputable pursuit" (57).

EFFECTS OF VA MYCORRHIZA ON PLANT NUTRITION

Improved growth of plants with VA mycorrhiza has been obtained in soil (5, 12, 13, 19, 31, 32, 45, 46, 64, 93), in peat and sand (25), in sand (34, 90, 153) and in water culture (119). Gerdemann (47) has described the gradual improvement in techniques and discussed some of the technical difficulties that arise from the need to remove indigenous mycorrhizal fungi from most soils, and the inability to culture *Endogone* species on synthetic media. Whereas soils used to be "sterilized" by autoclaving or steaming, they are now usually "sterilized" by γ irradiation (60, 61, 101, 102, 104, 108, 123, 131), fumigated with methyl bromide (32, 41), chloropicrin and methyl bromide (73, 127, 128) treated with aerated steam (73), or steamed between electrodes (16–18). Effects of inoculation in nonsterilized soil have also been examined (40, 45, 66, 71–73, 108, 110, 136). Most experiments have been in pots, but a few (71–73, 127, 136) have also been in field plots. Spores of a species of *Endogone* are now generally used as inoculum. They are multiplied in association with a host plant in "sterilized" soil or sand in greenhouse pots, and are often applied together with particles retained after wet sieving on sieves of selected mesh sizes (17, 34, 45, 51, 60). Ross & Harper (128) raised their inoculum monoxenically in closed jars, and Jackson, Franklin & Miller (66) used lyophilized, ground, infected roots. Filtered washings from the inoculum, containing many of the contaminating organisms, are usually added to the controls; Clark (32) and Baylis (17) added crushed spores.

Baylis (12), Gerdemann (45), and Holevas (64) first suggested that the improved growth of plants inoculated with VA mycorrhiza might be due to increased phosphate uptake. More recently the uptake of phosphate from P-deficient soils, from different sources and amounts of added phosphate, and from isotopically labelled soils and solutions has been studied in some detail.

Phosphorus

UPTAKE FROM UNAMENDED SOIL   Many plants take up more phosphate and grow better in soils containing little available phosphate when inoculated with VA endophytes. This occurred with onions and *Coprosma* grown in 10 different soils (60), with *Griselinia* (12), maize (45, 71), strawberries (64), soyabeans (in one of two soils) (128), wheat (72) and with grape vine (123). Other plants responsive to inoculation have been the tropical grasses, *Paspalum notatum* (101, 102) and *Melinis minutiflora* (109), the tropical legume, *Centrosema pubescens* (109), *Citrus* spp. (73, 86), *Liquidambar styraciflua* (52), *Agathis australis* (14, 92), *Liriodendron tulipifera* (30, 31, 46,

52), *Podocarpus* (14, 19) and *Araucaria sp.* (25, 26), walnut (136), tobacco (9, 119), tomato (34), barley (21) and oats (90). There are also records of negative (80, 90), variable (152), or no response (60), especially after adding phosphate (64, 71, 72, 104, 127).

It is well known that different plant species (8, 42, 67, 146, 154), and even varieties of the same species (81, 137), differ in their ability to extract phosphate from the same soil; some species will show phosphate deficiency symptoms while others grow normally. The reasons are not fully understood, although root distribution and geometry may play some part (10). Even in solution culture, plant species can differ in their ability to take up phosphate at low concentrations (7, 126). It is therefore not surprising that they also differ in their response to VA infection. In an agricultural soil I found that *Liquidambar,* onions, and *Coprosma* responded to inoculation with up to tenfold growth increases, whereas *Nardus stricta* and *Fuchsia* were little improved. The responsive species also grew better with added phosphate, but nonresponsive ones did not. In a much less fertile beach sand, however, growth of *Nardus* was also much improved by inoculation (T. H. Nicolson, private communication). Baylis (16) reported a similar lack of response to mycorrhizal infection in six plant species grown in a soil where *Coprosma robusta* grew much better with mycorrhiza than without. Four of the nonresponsive species, however, grew better with added phosphate. He related mycorrhizal response to root hair development, and suggested that plants with poorly developed root hairs may be obligatory mycotrophs in P-deficient soils. Later, however, he found that in another soil containing only half as much available P, the "nonresponsive" *Leptospermum scoparium* also responded to inoculation. In a final experiment, Baylis (18) studied the growth responses of five plant species to three levels of added phosphate. The results suggested that the species fell into three groups according to their requirements for a minimum threshold value of available P, below which they grew very little. He thought this threshold value might relate to the root:soil interface. An interaction between root geometry and the amount of available soil P would then determine the likely response to mycorrhizal inoculation. Plant species and levels of available phosphate in the soil or growth medium are thus likely to determine plant response to mycorrhiza.

Responses to inoculation were small (45, 136) or not significant (127) in nonsterile soil if the inoculum was mixed with the soil at planting. Responses were as large or even larger than in "sterilized" soil, however, with rough lemon and Troyer citrange (73) and with onions, already mycorrhizal at planting or sown on a cushion of inoculum (108, 110). Khan (71, 72) found that mycorrhizal maize and wheat seedlings planted in a field grew much better and yielded more grain than uninoculated seedlings.

Increasing the weight of inoculum ten-fold (128), or varying spore numbers from 3 to 225 (36), had no significant effect on dry weight of soyabeans, or on dry weight, percent of infection, and P uptake in twelve-week-old tomato plants. Even one spore may constitute an "effective inoculum" (36).

UPTAKE FROM SOILS OR MEDIA GIVEN EXTRA PHOSPHATE   Three aspects have been studied in some detail: (a) the effects of soluble phosphate on infection, phosphate uptake, and growth of mycorrhizal and nonmycorrhizal plants, (b) the effects of different sources of phosphate, and (c) the placement of inoculum and rock phosphate. In general, adding soluble phosphate improves growth of mycorrhizal plants less than that of nonmycorrhizal ones (13, 35, 64, 71, 72, 104, 113, 127); it sometimes even reduces the growth of the mycorrhizal plants. It also tends to reduce infection in both pot (13, 26, 35, 104) and field (71, 72, 74–76, 127) experiments. This may be due to high phosphate concentrations in the host (104, 112).

Hayman & Mosse (60) compared the responses of onions to mycorrhiza and to phosphate in ten irradiated P-deficient soils in a pot experiment. Monocalcium phosphate was given at the rate of 0.4 g/kg soil (approx. equivalent to 221 kg P/ha). All plants were also fed with a complete nutrient solution lacking phosphate. Mycorrhizal plants grew as well in three soils, in one they grew better, and in six worse than those given phosphate. Clearly the mycorrhiza extracted different amounts of phosphate from the different soils, and in six soils it was insufficient for optimum growth.

That responses to inoculation and to phosphate can differ with the soil was also shown in another pot experiment in which 0, 0.2, 0.5, 1.0, and 1.5 g monocalcium phosphate/kg soil were added to both mycorrhizal and nonmycorrhizal plants in four irradiated soils (104). Mycorrhizal plants in all soils without added phosphate were several times larger than nonmycorrhizal. Plants in all soils, whether mycorrhizal or not, reached a maximum weight with 0.2, 0.5, or 1.0 g of monocalcium phosphate, and weighed less when 1.5 g was given. In agricultural and grassland soils, plant P concentrations rose relatively slowly, and mycorrhizal plants weighed more than nonmycorrhizal at all levels of added phosphate. In soils from a forest plantation and a heath, plant P concentrations rose much higher, and with optimum amounts of added phosphate nonmycorrhizal plants grew much better than mycorrhizal. These results were explained on the assumption, suggested by a previous experiment with nonmycorrhizal plants, that plant P concentrations in onions can rise to levels that adversely affect growth. With the larger additions of phosphate, mycorrhizal infections decreased and, after very large phosphate additions, their appearance was changed; arbuscules first became fewer, then all intracellular infection disappeared, and eventually infection was restricted to certain specialized cells in the sub-epidermal layer. With 1.5 g added phosphate infection virtually disappeared from the roots, although the soil remained infective.

Similar results occurred in pot experiments with *Coprosma* (13), where much phosphate reduced infection and caused growth of the infected plants to fall below that of the uninfected. Adding monocalcium phosphate and superphosphate increased the P concentration, but not the weight of mycorrhizal, as compared to nonmycorrhizal maize (113). Adding phosphate to a sand:soil mixture made the growth and P uptake of mycorrhizal and nonmycorrhizal strawberry plants equal (64).

An unusual response to nitrogen and phosphate occurred in *Araucaria cunninghamii* grown in a peat:sand mixture (25). Neither mycorrhizal nor nonmycorrhizal plants responded to phosphate unless a minimum amount of nitrogen was given. Mycorrhizal plants then responded greatly to additions of 5 ppm P, but grew no better with 50 ppm P. Nonmycorrhizal plants responded only slightly to 50 ppm P, and not at all to 5 ppm.

Daft & Nicolson (35) examined the effect of different amounts and times of application of potassium dihydrogen phosphate on the growth of mycorrhizal maize in sand. Plants were fed with a nutrient solution lacking phosphate and were also given a basic dressing of the relatively insoluble tricalcium phosphate. With all additions of potassium phosphate, nonmycorrhizal plants grew as well as mycorrhizal, but without it they grew worse. The added phosphate increased plant weight, but decreased mycorrhizal infection by up to 50% and spore production up to 80%, in direct relationship to the amount of phosphate given but irrespective of the time of application. If the highest dosage was given in one application, these effects were more marked the earlier the application.

Pot experiments are unnatural in providing a limited volume of soil in which the phosphate is so evenly distributed that roots cannot grow away from it. Similar results have, however, also been obtained with soyabeans grown in 1.5 m² bins and given 0, 44, and 176 kg P/ha (127). Yield and weight of nonmycorrhizal plants increased steadily with increasing P additions, while that of mycorrhizal plants, and the number of extra-matrical spores attached to their roots, were progressively reduced. The mycorrhizal plants weighed more at harvest and yielded more seed than nonmycorrhizal plants given 44 kg P/ha. Khan (71) obtained similar results with inoculated maize seedlings planted in a field containing natural infection. At all times during this experiment, plant weight, percent of P, and ear size were greater in unfertilized inoculated maize than in control plants given 250 lb/acre triple superphosphate (40 kg P/ha). Inoculated plants with fertilizer again grew worse than those without, and the fertilizer greatly reduced infection in all plants. Similar responses occurred in wheat (72).

Nitrogen (59, 79), complete fertilizer (2, 78), and bacterial fertilizer (135) can also reduce mycorrhizal infection in the field. Small additions of nutrients to very nutrient-deficient soils or media can, however, increase infection both in field (2, 139) and pot experiments (26, 119, 136), probably by improving plant growth. Whether or not such a stimulation occurs probably depends on the nutrient level in the soil, and the amount of growth a particular plant can make in it.

The formation of VA mycorrhiza in clover seedlings in agar cultures also depended on the nutrient content of the medium (95, 103, 112). With 0.2 g/l K_2HPO_4 and 1 g/l $CaHPO_4$, infection occurred only when there was no nitrogen and when fungal entry was assisted by a culture of bacteria, a cell-free extract from a bacterial culture, or by EDTA (95). If, however, the calcium phosphate content was reduced to 0.5 g/1 and no potassium phosphate was given, infection readily occurred in the presence of potassium nitrate (112). Some infection occurred also without any added phosphate, but both seedling growth and infection were improved by small phosphate additions. As in field and pot experiments, large amounts of phosphate decreased infection. High phosphate concentrations in the plant appeared to make it immune to infection (103). Experiments with transplanted seedlings of different P content confirmed that inability to infect in media containing much phosphate is a result of changed plant metabolism or structure rather than a direct effect of phosphate on the fungus.

UPTAKE FROM DIFFERENT SOURCES   Daft & Nicolson (34) found that mycorrhizal tomatoes responded greatly to small additions of bonemeal, but their relative advantage over nonmycorrhizal plants decreased when 16 and 32 times as much bonemeal was given. Relative improvements in growth were greatest with tricalcium phosphate, less with finely ground apatite, and least with the more soluble dicalcium phosphate. With the relatively insoluble tricalcium phosphate, mycorrhizal maize grew much better than nonmycorrhizal (113). The difference was less with rock phosphate and, given the more soluble monocalcium- and superphosphate, plants with and without mycorrhiza grew equally well. It was inferred from these experiments that, like other micro-organisms (1, 54, 144), the mycorrhizal fungus might be able to utilize insoluble phosphates.

Utilization of rock phosphate was also studied in a placement experiment with maize, soyabeans, and sorghum grown in nonsterile soil in 4-liter containers (66). Soyabeans did not respond to inoculation, but grew better when the phosphate was distributed through the soil than when it was concentrated in bands. Sorghum responded slightly to inoculation, and maize markedly, if either the rock phosphate was distributed throughout the soil and the inoculum added as a concentrated layer, or if both were distributed through a 5 cm layer of soil. With rock phosphate placed in narrow bands, both inoculated and noninoculated maize grew badly.

It has also been suggested that mycorrhiza might utilize organic forms of phosphate. In sand in open pots, where some organic phosphates might quickly be broken down by bacteria, most of the organic phosphates tested were equally available to mycorrhizal and nonmycorrhizal onions (62). Although phosphate concentrations of the mycorrhizal plants were quite often higher than the nonmycorrhizal plants, their growth was not appreciably better, except with inorganic Gafsa rock phosphate; it was considerably worse with glucose-6-phosphate and RNA. Mycorrhiza did, however, greatly improve growth of plants in sand amended with humus, soil from the A horizon, or, to a slightly smaller extent, leaf litter.

UPTAKE FROM ISOTOPICALLY LABELLED SOLUTIONS   The uptake of ^{32}P has been studied from solutions (28, 52, 92), from soil injected with a labelled solution of phosphate (53), and after adding a fungitoxic substance (53). After exposure times of 30 min (28, 92) to 4 days (52) mycorrhizal clover and *Liriodendron* roots, and detached mycorrhizal nodules of *Agathis australis* contained about twice as much activity as nonmycorrhizal. Uptake was halved at 1°C but the difference between mycorrhizal and nonmycorrhizal nodules remained (92). After 40 days growth in a soil injected with labelled phosphate, leaves of mycorrhizal *Liquidambar* contained 15 times more activity than nonmycorrhizal (52). The greatest difference occurred in onions previously starved of phosphate. Roots and shoots of plants exposed for 90 hours to a labelled phosphate solution contained 160 and 45 times more activity, respectively, when the plants were mycorrhizal (53). Mycorrhizal root segments contained 25 times more activity than nonmycorrhizal segments on the same plant, but the latter contained 20 times more activity than segments from nonmycorrhizal roots. Adding parachloronitrobenzene reduced uptake into mycorrhizal roots to the level of nonmycorrhizal roots.

These results confirm that mycorrhizal roots take up more phosphate than nonmycorrhizal, and that this is translocated to the shoots. The effects of the fungitoxicant strongly suggest that it is the fungus that absorbs the extra phosphate. Autoradiography of sections of mycorrhizal roots confirms this (3, 28). Labelling was strongest in hyphae, both within and outside the roots, and was particularly concentrated in the arbuscules. In both studies, however, the labelling of the fungus was uneven.

UPTAKE FROM ISOTOPICALLY LABELLED SOIL   By adding carrier-free ^{32}P to soils it is possible, after thorough mixing and a period of equilibration, to label the "labile" pool of soil phosphate, which contains the phosphate utilized by plants. Insoluble and organic forms of phosphate are not labelled. If a plant, by virtue of its mycorrhiza, was able to utilize such generally unavailable sources of phosphate, the activity per unit of phosphate taken up would be less than in a plant drawing only on the labelled pool. It is thus possible to determine whether mycorrhizal and nonmycorrhizal plants utilize different fractions of the soil phosphate. This was done in three experiments (21, 61, 131).

The specific activity of phosphorus in ten-week-old mycorrhizal and nonmycorrhizal plants in each of eight soils was very similar, though it differed from soil to soil (61). Using one of these soils, Sanders & Tinker (131) found that, throughout the period of their experiment, the specific activity of phosphorus in mycorrhizal and nonmycorrhizal onions was the same, although inflow rates were 3–16 times greater in the mycorrhizal plants. These experiments show that mycorrhizal and nonmycorrhizal plants utilized the same or similarly labelled fractions of soil phosphate, and that mycorrhizal plants did not utilize sources of phosphate unavailable to plants without mycorrhiza. Sanders & Tinker (131) concluded that the greater inflow rates could not be attributed to increased activity of the mycorrhizal roots, but that entry into and subsequent transport through external hyphae must occur. The simplest theory to account for the extra P uptake, therefore, is that the mycelium outside the root constitutes an additional, better-distributed surface for absorbing P from the soil solution.[2]

In a very unusual soil in which plant species differed in their ability to utilize the slowly available P, Benians & Barber (21) found that mycorrhizal barley grew better than nonmycorrhizal, and that the specific activity of the phosphorus in mycorrhizal plants tended to be lower.

Because of the widespread view that mycorrhiza can solubilize unavailable phosphate, we (109) repeated the experiment in some Brazilian soils containing extremely small amounts of $NaHCO_3$- and $CaCl_2$-soluble P. Again there was no indication that mycorrhizal plants used different sources of phosphate. However, some species (e.g., *Centrosema pubescens* and *Paspalum notatum,* but not *Melinis minutiflora*) took up no phosphate at all from these soils unless plants were mycorrhizal. It appears that some plants have a threshold concentration of P in the soil solution below which they do not take up phosphate. This concentration is lower for the mycorrhizal fungus or root, or there may be no threshold value at all. Experiments of Beslow, Hacskaylo & Melhuish (23) support this conclusion. In perlite sub-irrigated with a nutrient solution containing 1 ppm P, red maple seedlings weighed no more than those irrigated with distilled water only. Their weight increased progressively with 5, 10, and 20 ppm P. The beaded roots of red maple are commonly mycorrhizal (VA).

Other Nutrients

The uptake of other nutrients can also be affected by mycorrhiza. Peach seedlings in Yolo loam fed with a nutrient solution containing N, P, K, Ca, Mg, B, and Fe EDTA grew very little and showed severe symptoms of zinc deficiency unless they were inoculated with an *Endogone sp.* and became mycorrhizal (51). Three-year-old inoculated seedlings showed no symptoms of zinc deficiency, weighed nearly twice as much, and contained 2–3 times as much zinc as noninoculated seedlings.[3]

---

[2] In a recent review of phosphate pools, phosphate transport and phosphate availability (*Ann. Rev. Plant Physiol.* 1973, 24:), R. L. Bieleski calculated that with four hyphal connections per mm root and hyphae extending 20 mm from the root surface, P uptake would increase about $60\times$ if diffusion were limiting and $10\times$ if uptake were proportional to surface area.

[3] Mycorrhizal soyabean plants took up more ^{90}Sr than nonmycorrhizal after 1, 3, and 7 days contact with ^{90}Sr amended soil, whether sterilized or not. Lyophilized mycorrhizal roots were used as inoculum. (Jackson, N. E., Miller, R. H., Franklin, R. E. 1973. The influence of vesicular-arbuscular mycorrhiza on uptake of ^{90}Sr from soil by soybeans. *Soil Biol. Biochem.* 5:205–12)

Differences in N, K, Ca, Na, Mg, Fe, Mn, Cu, B, Zn, and Al concentrations of mycorrhizal and nonmycorrhizal plants have been reported (12, 40, 45, 64, 66, 73, 93, 127, 128). The results are, however, inconsistent, with the possible exception of copper, which generally reaches greater concentrations in mycorrhizal plants, and manganese which is generally lower. All these results are subject to the criticism that plants were not really comparable because controls were often very small and suffering from acute phosphate deficiency. Mycorrhizal plants were compared in three experiments with nonmycorrhizal plants given phosphate, and results agree quite well. Ross (127) found concentrations of nitrogen, calcium, and copper higher in mycorrhizal plants, and Holevas (64) found potassium concentrations lower and magnesium higher. In a soil containing very little available P, I found that potassium concentrations of mycorrhizal plants were lower, and magnesium, manganese, copper, and zinc concentrations were higher than in the nonmycorrhizal plants.

Safir, Boyer & Gerdemann (129, 130) measured the rate of water uptake into soyabeans recovering from temporary wilting, and found that mycorrhizal plants recovered more quickly than nonmycorrhizal plants. This was attributable to their better phosphorus nutrition (130), rather than to any specific effect of the mycorrhiza.

Meloh (90) showed that uptake of a much more complex substance (dihydrostreptomycin) was also greater in mycorrhizal than nonmycorrhizal maize.

TAXONOMY OF ENDOGONE

The anatomical similarity of VA infections has in the past led to the widespread assumption that most, if not all, these infections were caused by the same fungus. A variant with fine hyphae, few vesicles, and a tendency towards stranding of the intracellular mycelium was described by Greenall (55) and named *Rhizophagus tenuis*. A somewhat similar endophyte has been described in the roots of oats (142) and in some populations of *Nardus stricta* (4). Pandey & Misra (118) recently described and named another endophyte occurring in *Litchi chinensis*.

Inoculation experiments with different *Endogone* types (105, 116) have shown that VA mycorrhiza are produced by several different *Endogone* species, probably even by different genera of fungi. According to Thaxter's (145) classification, all members of the genus *Endogone* form sporocarps, and species are distinguished by the structure of the sporocarps and the spores they contain. The discovery of different ectocarpic resting spores, attached singly to the extra-matrical mycelium of mycorrhizal roots has made some revision necessary. A definitive classification must probably await the culturing of these fungi, but for purposes of convenience some method of nomenclature is needed.

Gerdemann & Trappe (49) recently completed a monograph on the Endogonaceae in the Pacific Northwest. According to them, four genera, *Glomus, Sclerocystis, Gigaspora,* and *Acaulospora* commonly form (VA) endomycorrhiza. *Glomus* species may be sporocarpic or not, with chlamydospores generally formed terminally on a single undifferentiated hypha. Chlamydospores of *Sclerocystis* borne in sporocarps, are arranged in a single layer around a central plexus of sterile hyphae. *Gigaspora* species bear the spores (probably azygospores) at the tip of a single, large, suspensor-like cell from which a slender hypha usually projects to the spore; they produce, in addition to the resting spores, smaller soil-borne vesicles of distinctive structure. The genus *Acaulospora* does not form sporocarps, and resting spores are borne laterally on a hypha terminating nearby in a large thin-walled vesicle. The name *Endogone* is confined to zygosporic species. At least one produces ectotrophic mycorrhiza, but several others are probably free-living.

Mosse & Bowen (105) described nine types with apparently distinctive spores obtained from Australian and New Zealand soils by wet sieving. The spores differed in their cytoplasmic and wall structure, color, shape, and size, form of the subtending hypha, and in method of germination. We suggested a key for their practical identification, and gave descriptive names to the different spore types. An advantage of such a system is that any new spore type found can easily be incorporated. The name *Endogone,* now quite well established, is retained in this review for all VA endophytes; individual species, strains, or spore types are referred to under the name given in the original publication.

Some of the described types differ in only relatively minor characteristics, and are undoubtedly more closely related than others. Several differ consistently at the fine structure level. The spores are multinucleate. Germ tubes frequently anastomose (94), and there is, therefore, much opportunity for hybridization. Germ tubes of two types tested (laminate, and yellow vacuolate) did not anastomose on cellophane discs. Such observations could be extended, and the genetic homogeneity of a strain could also be checked by studying the variability of the progeny derived from inoculation with a single spore. The best evidence for the distinctness of species is that spores, thought to belong to a distinct spore type, reproduce similar progeny in "sterilized" soil containing a host plant.

The life history of honey-colored sessile spores (105) is so distinctive (98) that they almost certainly belong to the genus *Acaulospora* of Gerdemann & Trappe (49). The resting spore forms as a lateral outgrowth from the swollen subtending hypha of another short-lived spore, whose contents migrate into the rapidly expanding resting spore. The fine structure of this spore exhibits several unusual features. It contains an organism, possibly an actinomycete, that divides by fission and appears to have an independent life cycle (99). The spore wall is very complex, consisting of up to seven layers, one with a very unusual periodic structure (100). Germ tubes arise from peripheral compartments formed within the split wall (98). Three other types [white and bulbous reticulate (105), and a very large reddish-black spore type first found in Nigeria (124)] germinate in a similar way and contain a similar periodic wall layer (125). The bulbous reticulate spore, and the Nigerian spore types have a characteristic swollen, pear-shaped, subtending hypha. The generic name *Gigaspora* (49) is proposed for this group. In all these spores, oil droplets, the characteristic storage material in spores of these fungi, are uniform, small, polygonal, and do not coalesce; they are separated by a fine cytoplasmic network (105). In another group now assigned to the genera *Glomus* and *Sclerocystis* (49), the oil droplets are spherical, coalesce readily, and differ greatly in size within a single spore. Old spores often contain a large central oil vacuole and only a thin peripheral layer of cytoplasm. To a third group, with small round spores formed in closely coherent masses of more than a hundred spores, but lacking a peridium, belong *E. fasciculata* (46) and several related forms described by Gilmore (50).

The Endogonaceae have taxonomic affinities with several other mucoraceous families (49). Their cytology, particularly that of the zygosporic and azygosporic species, is interesting but imperfectly understood. Bucholtz (29) described the delayed fusion of two nuclei in the zygospore, and the disintegration of smaller peripheral nuclei in the suspensor. In the yellow-vacuolate and honey colored types nuclei divide by ingrowth of the inner nuclear membrane while the outer one remains intact, a method also found in some other Phycomycetes. Many nuclei disintegrate during the onset of dormancy.

SPECIFICITY

With the discovery of different *Endogone* endophytes, the question of specificity in VA mycorrhiza has been reinvestigated, both with regard to host range and to specific interactions between particular endophytes and their hosts. Earlier inoculation experiments with infected roots indicated that most VA endophytes could be transferred from one host to another. Only Tolle (147), using surface sterilized root sections, found some evidence of host specificity. The fungi from oat and barley roots could not be transmitted to or from other cereals, but those in wheat and rye were interchangeable. Inoculation experiments with surface-sterilized spores have confirmed that the host range of most endophytes is very wide. For instance, in my experiments, the yellow-vacuolate spore type (105) [*Endogone mosseae* (116)] formed VA mycorrhiza with onions, *Coprosma, Liquidambar, Coleus, Fuchsia,* clover, strawberry, apple, tomato, bean, pea, cucumber, *Nardus* and rye-grass. This type also infected tobacco (9), soyabean (129), citrus (73, 86), maize (71), *Aesculus indica* (70) and barley (21). Other *Endogone* spp. will also infect many of these hosts.

Studies of the effect of different spore types on growth of the host are only beginning. Meloh (90) found that one endophyte (Vesikelmycorrhiza) improved growth of maize and oats in sand irrigated with a nutrient solution, while another (Sporokarpienmycorrhiza) reduced it. Daft & Nicolson (34) found that three *Endogone* spore types differed little in their effects on maize, tomato, and tobacco in sand culture. The first indication that spore types might differ greatly in their effects was obtained when onions, preinoculated with the yellow-vacuolate spore type or sown on a cushion of inoculum (108), were planted in a nonsterile soil. Noninoculated control seedlings became uniformly infected with indigenous mycorrhizal fungi in two of the soils. In one this infection appreciably improved plant growth, in the other it did not, but in both soils the plants infected with the yellow-vacuolate spore type grew much better than noninoculated plants. A similar advantage of inoculated over indigenous strains was shown in two Brazilian soils with *Paspalum notatum* (101) as host plant, and with citrus (73).

Onions inoculated with seven different spore types showed a range of growth responses (102). Two strains produced little infection and slightly improved growth. Another strain caused abundant infection and a ten-fold increase in dry weight; the best strain increased weight fifteen-fold. In a different soil this strain was only fairly good, and four other strains surpassed it. Similar strain effects occurred with *Paspalum*.

Liming in general greatly increased the benefit of inoculation in two acid soils containing little available P (101). One spore type was more beneficial in the unlimed soil. Similar reactions to liming occurred in the same soil with another host plant. Two spore types did not get established in the unlimed soils. Such pH effect on the establishment of certain spore types (yellow-vacuolate, laminate, and RI) has now been observed in several soils (102, 107). It is independent of the host plant. Another spore type (honey-colored) infected rye-grass in only two out of five soils, irrespective of pH (102).

These results indicate that specificity exists in VA mycorrhiza, and that some *Endogone* strains are more beneficial than others in phosphate-deficient soils. The effectivity of a strain appears to depend more on its interaction with a particular soil than with a particular host.

Endogone strains also differed in their ability to improve zinc uptake of mycorrhizal peach seedlings (51).

Even the bacteria associated with inocula of different endophytes had different effects on the growth of nonmycorrhizal onions and rye-grass in phosphate-deficient soils (102). Compared with the effects of the endophytes themselves, these effects were small, but nevertheless statistically significant in onions; in rye-grass, washings from different spore inocula had quite large effects on plant weight. The outer layer of the spore wall of some *Endogone* species contains chitin, and is characteristically free from adhering soil debris. The outer wall layer of others is strongly attacked by bacteria, and may be lost altogether. If such differences in wall structure extend to the hyphae, different endophytes may well have different hyphosphere organisms characteristically associated with them.

ECOLOGY OF ENDOGONE

The relatively simple technique of wet sieving and decanting (43) has been widely used to recover *Endogone* spores from soil. Modifications of the technique have been described (111, 117, 128). Sutton & Barron (143) recently described a technique depending on the adhesion of spores to a glass surface at the meniscus; this gave 94–98% recovery. Counting can be made easier by spreading the spore-containing fraction on "Dicel" cloth (59, 111).

Spore populations have been studied in different continents (49, 69, 106, 124), different ecological environments (69, 91, 124, 141) and in association with particular hosts (27, 122, 133). Effects of manurial treatments (26, 35, 59, 71, 106, 122), of season (59, 87, 91, 124), light (124), and water regime (15, 124) have also been examined in both pot and field experiments.

THE OCCURRENCE OF EXTRA-MATRICAL SPORES AND SPORE TYPES   Most spore types seem to have a world-wide distribution. For instance those with honey-colored sessile spores, easily identifiable because of their unusual origin, have been found in Scotland (48), England (D. S. Hayman, private communication), Australia (27, 105), New Zealand (105), Pakistan (69), South Africa (58), Oregon (49), Florida (133), and in Brazil. Two other spore types (yellow-vacuolate and bulbous-reticulate) are as widespread. Even the crenulate spore type that we found in only one locality in South Australia (105), has now been reported from a restricted area of southeastern Queensland (27), from Nigeria (124), and Germany (H. Kruckelmann, private communication). It is interesting to speculate how such a uniform distribution has come about, as spores are probably too large to be carried into the upper atmosphere. They could, however, very easily be spread on plant roots.

Spores were generally more numerous and varied in cultivated than in noncultivated Australian and New Zealand soils (106). In West Pakistan certain spore types were more common in northern and mid-northern areas where soils contain substantial amounts of clay, than in arid southern regions where they are sandy (69). In six Nigerian areas with differing vegetation types, spores tended to be more numerous in the arid savannahs than in the moist forests, but some fluctuations and unexpected variations occurred in closely adjacent sites (124). Redhead concluded that at least 12 samples should be pooled to obtain a representative spore count. Most of the spores occurred in the top 15 cm.

Although the host range of most spore types is wide, Bevege & Richards (27) found one spore type predominantly associated with *Araucaria cunninghamii* in some 50 rain forest sites, but it was absent from this host in 16 plantations and 5 forest nursery sites. Porter & Beute (122) found *E. gigantea* to be common in peanut, but rare in fescue soils. Adjacent fields on the Rothamsted farm, often carrying the same crop, have quite distinct spore populations. One field contains only a single spore type that is rare in a nearby field with a mixed population. This difference is not attributable to climate, and probably not to soil or any recent crop.

There are marked seasonal fluctuations in spore numbers in temperate climates. Hayman (59) found that the number of spores changed little from December to June, increased greatly in July, and decreased again slowly from September onwards. Mejstrik (91) found fewest spores in April and most in September in a *Molinietum coeruleae* association. Considerable seasonal fluctuations also occurred in Nigerian soils (124). Sutton & Barron (143) counted spores in four Canadian soils growing wheat, maize, strawberries, and tomatoes, and in some comparable soils from England and Florida. They recovered from 20–90 spores/g air-dried soil, a much higher figure than that obtained by most other workers. Spores were least numerous in June and most in October, (strawberries were most in April), but seasonal fluctuations generally were small. More spores occurred within than between rows, and numbers decreased below 28 cm.

The correlation between amounts of mycorrhizal infection and numbers of extra-matrical spores was studied in field plots (59) in wheat grown in field given various formalin and nitrogen treatments, and by Daft & Nicolson (37) in maize grown in pots in sand culture with different phosphate treatments. Both found that numbers of external spores were closely related to root infection, and Daft & Nicolson (37) concluded that, in their experiments, spore numbers were the measure of infection most closely related to plant weight. On the other hand, Redhead (124) thought that the number of spores in soil was not significantly correlated with root infection. Neither he nor I recovered any spores from some Nigerian rain forest and New Zealand bush soils where many plant species were strongly mycorrhizal.

EFFECTS OF LIGHT, WATER REGIME, AND MANURIAL TREATMENTS ON SPORE NUMBERS AND POPULATIONS   Vesicles are mainly storage organs of the endophyte (47, 96) and extra-matrical spores probably have the same function. They are also reproductive stages relatively resistant to adverse conditions. Spore numbers are likely to reflect the nutritional status of the host and/or soil, and the onset of adverse conditions.

It has been inferred from various surveys that spores increase as a result of intermittent root growth (106), during seasons of slow root growth (87), and at sites where many rootlets die annually (124). Baylis (15) did not increase spore numbers by intermittent watering, but he halved them by cutting plants off at ground level. Sporocarps were fewer in logged than in undisturbed stands of sugar maple (68), but site variation was the largest variable. Spores associated with *Khaya* grown outdoors in large containers were most numerous after watering daily, and were reduced to 10% and 25% respectively by watering weekly and twice daily (waterlogging) (124). Heavy shading reduced spore formation by 80%. Neither treatment affected root infection to the same extent. Bevege (26) reported similar effects of shading on spore numbers and infection.

There is good evidence that large applications of nitrogen reduce spore formation in field plots (59, 111, 122) and in pot tests (26). External spores increased with added phosphate in field plots with noninoculated maize, presumably because the plants grew better, but if the maize was already mycorrhizal when planted, they decreased (71). Sporocarp formation was also greatly reduced by added phosphate (64, 97).

Prolonged fertilizer applications may cause changes in numbers and composition of the spore population (59, 106) in some soils. H. Kruckelmann (private communication) found that, in a sandy soil that had grown rye and potatoes for 50 years, unfertilized plots contained few spores, and numbers increased progressively in plots fertilized with NPK, farmyard manure, leaf mold, and farmyard manure plus NPK. However, on a heavy clay soil where wheat had been grown for over a hundred years, fertilizers had the opposite effect. Spore numbers, generally higher in this soil, were greatest in the unfertilized plot, and decreased progressively with additions of farmyard manure, NPK, and farmyard manure + N.

INTERACTIONS WITH OTHER MICROORGANISMS   Exudates from mycorrhizal roots are likely to differ from those of nonmycorrhizal roots because the nutritional status of the host plant may be altered and because the large volume of fungus tissue in mycorrhizal roots could affect root exudates more directly. Changes in root exudates would affect the rhizosphere microorganisms, and might make mycorrhizal roots easier to invade or less susceptible to attack by other fungi. Such effects of ectotrophic mycorrhiza on root pathogens (85) and on nonpathogenic rhizosphere organisms (39) have been reported. Voigt (149) discusses the possibility that mycorrhiza might influence nitrogen fixation by free-living microorganisms.

Only four instances of interactions of VA mycorrhiza with other microorganisms have yet been reported, but other investigations are in progress. Asai (6) found that mycorrhiza seemed to be a necessary precondition for effective nodulation in many legumes. Successful nodulation is known to depend on an adequate phosphate supply, and many volcanic soils of Japan are very deficient in phosphate and some minor elements. In these experiments, however, the mycorrhizae were established from a 5–10 g inoculum of garden soil, which may itself have supplied necessary nutrients. If confirmed, Asai's results would be of considerable interest as many tropical soils are so phosphorus deficient that it is growth limiting, and plants do not respond to nitrogen until phosphate deficiency is corrected (88, 101).

Better nutrition of the host plant was also considered the reason for the larger number of TMV lesions in tobacco plants infected with *Endogone* (134). Daft & Okusanya (38) found that amounts of extractable tomato aucuba, potato X, and Arabis mosaic virus were greater in mycorrhizal plants, and again attributed this to better host nutrition.

Mycorrhizal tobacco plants were less damaged by *Thielaviopsis basicola* than nonmycorrhizal (9). Weight of the latter was reduced by 64%, but that of mycorrhizal plants by only 28%. Fewer chlamydospores of *Thiela-viopsis* formed on mycorrhizal roots, and an extract of such roots inhibited chlamydospore formation by 80–100% in vitro.

SOIL STERILIZATION

Various methods of soil sterilization are used increasingly to kill pathogens, particularly in nurseries. Plant growth is generally improved by these treatments, but occasionally plants grow poorly after sterilization. For instance, zinc deficiency occurred in cotton after soil fumigation (151), and citrus grew badly in several nurseries for unexplained reasons (83, 84). Fumigation of some forest nurseries resulted in excessively vigorous unsuitable nursery trees, and this was corrected by inoculation with ectotrophic mycorrhizal fungi (65).

Kleinschmidt & Gerdemann (73) investigated the growth of citrus seedlings in a nursery that had been fumigated with 400 lb/acre of a 3:1 mixture of methyl bromide and chloropicrin. The seedlings grew unevenly in fumigated soil; the healthy ones were mycorrhizal, whereas stunted chlorotic ones were not. The stunted seedlings, transplanted into sterilized soil that had been inoculated with *E. mosseae,* became healthy, while the uninoculated seedlings remained stunted. Similar responses to inoculation were obtained in a fumigated field containing 63 lb/acre available P and 91 lb/acre acid-soluble P. These are not particularly small amounts of phosphate. In a fumigated nursery in California heavy fertilization only partly overcame the poor growth of some citrus varieties, whereas inoculation with *E. mosseae* greatly improved growth (73). Possibly the mycorrhiza also helped in the uptake of some minor nutrient present at deficiency levels. Experiments with slash pine (148) showed in one of three soils, that adding 200 ppm P reduced plant growth, reduced mycorrhizal infection, and led to copper deficiency in the plants. If, as the frequently higher copper content of mycorrhizal plants suggests, mycorrhiza increase copper uptake, any treatment reducing mycorrhizal infection may lead, for biological rather than chemical reasons, to such unexpected deficiencies.

Fumigating soil cores with methyl bromide (32), or injecting soil to a depth of 0.9 m with methyl bromide and chloropicrin (128) killed the indigenous VA fungi in the soil. In other experiments, however, mycorrhizal infection of sweet gum seedlings was not greatly reduced in a coarse sieved forest soil fumigated with methyl bromide (41).

Effects of formalin, applied as a drench to field soil in situ were studied by Hayman (59). Following autumn application of the sterilant, mycorrhizal infection and spore numbers in the soil were very low until June, but then rose sharply and remained high throughout the following year. After repeated applications of formalin, infection was unexpectedly high, suggesting that strains resistant or tolerant to formalin may have been selected. Wilhelm (150) also reported heavy mycorrhizal infection in strawberries two years after soil fumigation with chlorobromopropene, and I have found the same after treatment with the nematocide, DD.

Nesheim & Linn (114) examined effects of different concentrations of eight fungitoxicants, mixed into a steamed soil inoculated with *E. fasciculata*. All fungitoxicants at all concentrations considerably reduced infection of corn plants. Terraclor, Mylone, and Vapam suppressed it almost completely, Captan was least toxic.

TRENDS

The section headings of this paper show the different directions that research on VA mycorrhiza is taking. Most important from the practical point of view are the studies of VA mycorrhiza in relation to plant nutrition.

Much attention has been paid to the movement of phosphate and carbohydrates between fungus and host in ectotrophic mycorrhiza (57), and to phosphatase and phytase activity of the fungi in vitro. With vesicular–arbuscular mycorrhiza, perhaps because there is no sheath and the fungi cannot be grown in culture, most work has been done with the intact mycorrhizal system, and interest has centered around uptake mechanisms and the source of the extra nutrients absorbed from the soil. As a result it now seems very improbable that VA mycorrhiza utilize insoluble soil phosphate. They greatly increase P uptake by a better exploration of the soil beyond the depletion zone and by hyphal proliferation in favorable micro-environments. Mycorrhizal roots also take up phosphate present in the soil solution at such low concentrations that it is practically unavailable to nonmycorrhizal roots of some plants. Finally, mycorrhizal fungi may help phosphate uptake of the host by competing with other microorganisms for soluble P present in plant debris.

If there is no net increase in available phosphate, it may be questioned whether mycorrhiza do not simply bring about a more rapid exhaustion of limited supplies. While this may be so in the short term, more phosphate will gradually become available from soil reserves by purely chemical processes of desorption and/or solution as phosphate is taken out of the soil solution by mycorrhiza. Thus more phosphate will gradually come into circulation. The most important function of VA mycorrhiza in agriculture may be a better utilization of applied phosphate. In general 25% of added phosphate is used by the crop in the first year (33). The rest reverts to unavailable forms unless a reserve of potentially available P is built up over the years by repeated fertilizer applications. If, because of their improved uptake, mycorrhizal plants can grow as well in a soil solution containing less available P than would be required for equal growth of nonmycorrhizal plants, less fertilizer will be needed for the mycorrhizal plants, especially if they are infected with efficient strains of *Endogone*.

It is probable that the increased use of biocides may pinpoint some other nutrients affected by uptake through mycorrhiza. Others may be found by plant analysis. Such investigations and reported changes in enzyme activities and respiration rates (139) must take into account that a phosphate-deficient plant is likely to differ from a normal plant in many respects. Only mycorrhizal and nonmycorrhizal plants with approximately similar phosphate content should be used for such comparisons.

While it is becoming clearer what advantages the plant may derive from VA mycorrhiza, the role of the plant in fungal metabolism remains quite unexplained. That it is crucial is evident because the fungus is an obligate symbiont. The autotroph in many symbiotic associations releases carbohydrate to the heterotroph that stores it, often in the form of polyols or complex sugars that cannot be metabolized by the autotroph (138). The fungi in ectotrophic mycorrhiza mainly accumulate mannitol and trehalose (57). Many observers have noted the disappearance of starch from infected host cells in VA mycorrhiza. There are reports that mycorrhizal roots contain more soluble sugars (2, 76), but others have found less (139). In two-membered agar cultures, growth of the external mycelium in the medium was stimulated by inositol and glycerol (103), suggesting that these substances might be metabolized by the fungus. D. I. Bevege, G. D. Bowen & M. Skinner (private communication), D. S. Hayman (private communication), and Hepper & Mosse (63) did not find significantly more trehalose or mannitol in roots of various plants infected with VA mycorrhiza, nor any other striking differences in the carbohydrate pattern of infected and noninfected roots. Bevege et al found 74% of photosynthate in VA roots was in the form of soluble carbohydrate, and of the ^{14}C transferred to the external mycelium, 52% was recovered as protein and organic acid, and 30% in structural (cell wall) material. They concluded that the carbohydrate metabolism of VA mycorrhiza differs basically from that of the ectotrophic type.

Present evidence favors the view that mass flow, probably bidirectional, effectively connects the soil mycelium with the intracellular arbuscules. Fine structure (20, 132) clearly shows a complete collapse of arbuscules, leaving behind only remains of fungus wall. My pictures of young arbuscules show them to be full of cytoplasm and organelles. Both active and collapsed arbuscule branches can coexist in the same cell and fat globules apparently get from the fungus into the host cytoplasm (confirming light microscope observations) (89). Transfer of material from fungus to host cell also appears to occur by means of numerous exocytotic vesicles. Various barriers at the host/fungus interface could thus be by-passed. The technique of autoradiography would be very suitable to demonstrate such transfer of large molecules.

The feasibility of inoculation on a field scale is likely to receive further attention, both with the object of reinfecting sterilized soils and of introducing better strains of endophytes into natural soils. Both have already been done. Mycorrhizal plants can be grown in large containers or limited field plots to obtain enough heavily infected soil to inoculate a larger area (73), or preinoculated seedlings can be planted in the field (71, 72). The successful use of lyophilized, ground mycorrhizal roots as storable inoculum (66) raises interesting possibilities.

The need to obtain inoculum of different *Endogone* strains may renew interest in the culture of these fungi. Several workers (82, 140, 147) have obtained limited hyphal growth and new vegetative spores from surface-sterilized mycorrhizal root pieces. At certain stages of root or plant development, such growth regularly occurs in tap-water agar, or other media of low nutrient content. Tolle (147) made the interesting observation that root exudates, but not root extracts, improved such growth. The subculture of detached hyphae has, however, never succeeded, and growth of the attached hyphae stops when the root pieces die. Failure has in the past been attributed to bacterial contamination, but this is incorrect. Extensive hyphal growth can readily be obtained from surface-sterilized, germinated spores. Germ tubes branch, produce vegetative spores, and grow several centimeters, but growth immediately stops if they are severed from the parent spore. Cut hyphae often produce a new growing point and make limited growth from the cut end by the method of wound healing described by Gerdemann (44). Unless they rejoin the hypha from which they were cut, or anastomose with another attached to a spore, no further growth will be made. Axenically germinated spores of the yellow-vacuolate type can easily be obtained (94), and probably offer the best starting point for attempts to culture *Endogone*. Barrett's (11) technique for producing cultures of *"Rhizophagus"* and their probable authenticity as VA endophytes, have been discussed by Gerdemann (47). Although the culture of VA endophytes presents a challenge, the reward is great for practical purposes and for comparative studies with other obligate and pathogenic fungi.

Literature Cited

1. Agnihotri, V. P. 1970. Solubilization of insoluble phosphates by some soil fungi isolated from nursery seedbeds. *Can. J. Microbiol.* 16:877–80
2. Alexandrova, E. I. 1968. Mycorrhiza formation of barley in relation to the action of mineral fertilizers. [Russian] *Uch. Zap. Perm. Ped. Inst.* 64:274
3. Ali, B. 1969. Cytochemical and autoradiographic studies of mycorrhizal roots of *Nardus. Arch. Mikrobiol.* 68:236–45
4. Ali, B. 1969. Occurrence and characteristics of the vesicular-arbuscular endophyte of *Nardus stricta. Nova Hedwigia* 17:409–25
5. Asai, T. 1943. Die Bedeutung der Mykorrhiza für das Pflanzenleben. *Jap. J. Bot.* 12:359–436
6. Asai, T. 1948. Über die Mykorrhizenbildung der leguminosen Pflanzen. *Jap. J. Bot.* 13:463–85
7. Asher, C. J., Loneragan, J. F. 1967. Responses of plants to phosphate concentration in solution culture. I. Growth and phosphorus content. *Soil Sci.* 103:225–33
8. Ballard, S. S., Dean, L. A. 1941. Soil studies with radioactive phosphorus: Significance of biological measurements of the retention of applied phosphorus by soils. *Soil Sci.* 52:173–82
9. Baltruschat, H., Schönbeck, F. 1972. Untersuchungen über den Einfluss der endotrophen Mycorrhiza auf die Chlamydosporenbildung von *Thielaviopsis basicola* in Tabakwurzeln. *Phytopathol. Z.* 74:358–61
10. Barley, K. P., Rovira, A. D. 1970. The influence of root hairs on the uptake of phosphate. *Soil Sci. Plant Anal.* 1:287–92
11. Barrett, J. T. 1961. Isolation, culture and host relation of the phycomycetoid vesicular-arbuscular endophyte *Rhizophagus. Rec. Advan. Bot.* 2:1725 Univ. Toronto Press
12. Baylis, G. T. S. 1959. Effect of vesicular-arbuscular mycorrhizas on growth of *Griselinia littoralis* (Cornaceae). *New Phytol.* 58:274–80
13. Baylis, G. T. S. 1967. Experiments on the ecological significance of phycomycetous mycorrhizas. *New Phytol.* 66:231–43
14. Baylis, G. T. S. 1969. Synthesis of mycorrhiza of *Podocarpus* and *Agathis* with Endogone spores. *Nature, London* 221:1267–68
15. Baylis, G. T. S. 1969. Host treatment and spore production by Endogone. *N.Z. J. Bot.* 7:173–74
16. Baylis, G. T. S. 1970. Root hairs and phycomycetous mycorrhizas in phosphorus-deficient soil. *Plant Soil* 33:713–16
17. Baylis, G. T. S. 1971. Endogonaceous mycorrhizas synthesized in *Leptospermum* (Myrtaceae). *N.Z. J. Bot.* 9:293–96
18. Baylis, G. T. S. 1972. Minimum levels of phosphorus for nonmycorrhizal plants. *Plant Soil* 36:233–34
19. Baylis, G. T. S., McNabb, R. F. R., Morrison, T. M. 1963. The mycorrhizal nodules of podocarps. *Trans. Brit. Mycol. Soc.* 46:378–84
20. Becking, J. H. 1965. Nitrogen fixation and mycorrhiza in Podocarpus root nodules. *Plant Soil* 23:213–26
21. Benians, G. J., Barber, D. A. 1972. Influence of infection with *Endogone* mycorrhiza on the absorption of phosphate by barley plants. *Letcombe Lab. Ann. Rept.* 1971: 7–9
22. Bertrand, D. 1972. Interactions entre éléments minereaux et microorganisms du sol. *Rev. Écol. Biol. Sol.* 9:349–96
23. Beslow, D. T., Hacskaylo, E., Melhuish, J. H. 1970. Effects of environment on beaded root development in red maple. *Bull. Torrey Bot. Club* 97:248–52
24. Bevege, D. I. 1968. A rapid technique for clearing tannins and staining intact roots for detection of mycorrhizas caused by *Endogone* spp., and some records of infection in Australasian plants. *Trans. Brit. Mycol. Soc.* 51:808–10
25. Bevege, D. I. 1970. Vesicular-arbuscular mycorrhizas of *Araucaria cunninghamii* and their role in nitrogen and phosphorus nutrition.

Paper given at *42nd Congr. Austr. N.Z. Assoc. Adv. Sci.* Sect. 12 (microfiche)

26. Bevege, D. I. 1972. *Vesicular-arbuscular mycorrhizas of Araucaria: Aspects of their ecology and physiology and role in nitrogen fixation.* Ph.D. Thesis, Univ. of New England, Armidale, N.S.W.

27. Bevege, D. I., Richards, B. N. 1971. Some aspects of *Endogone* forming mycorrhizas with Hoop Pine (*Araucaria cunninghamii* Ait.). *XV IUFRO* Congr. Sect. 24

28. Bowen, G. D., Rovira, A. D. 1968. The influence of micro-organisms on growth and metabolism of plant roots. In: *Root Growth*, 170–99. Ed. W. J. Whittington, Butterworths: London

29. Bucholtz, F. 1912. Beiträge zur Kenntniss der Gattung Endogone *Link. Beih. Bot. Centralbl.* Ser. II 29:147–224

30. Clark, F. B. 1963. Endotrophic mycorrhizae influence yellow-poplar seedling growth. *Science* 140:1220–21

31. Clark, F. B. 1964. Micro-organisms and soil structure affect yellow-poplar growth. *U.S. Forest Serv. Res. Paper,* CS-9:12

32. Clark, F. B. 1969. Endotrophic mycorrhizal infection of tree seedlings with *Endogone* spores. *Forest Sci.* 15:134–37

33. Cooke, G. W. 1965. The responses of crops to phosphate fertilizers in relation to soluble phosphorus in soils. In: *Soil Phosphorus*, Min. Ag. Fish. Tech. Bull. 13:64–74

34. Daft, M. J., Nicolson, T. H. 1966. Effect of *Endogone* mycorrhiza on plant growth. *New Phytol.* 65: 343–50

35. Daft, M. J. Nicolson, T. H. 1969. Effect of *Endogone* mycorrhiza on plant growth. II. Influence of soluble phosphate on endophyte and host in maize. *New Phytol.* 68: 945–52

36. Daft, M. J., Nicolson, T. H. 1969. Effect of *Endogone* mycorrhiza on plant growth. III. Influence of inoculum concentration on growth and infection in tomato. *New Phytol.* 68:953–61

37. Daft, M. J., Nicolson, T. H. 1972. Effect of *Endogone* mycorrhiza on plant growth. IV. Quantitative relationships between the growth of the host and the development of the endophyte in tomato and maize. *New Phytol.* 71:287–95

38. Daft, M. J., Okusanya, B. O. 1973. Effect of *Endogone* mycorrhiza on plant growth. V. Influence of infection on the multiplication of virus in tomato, petunia and strawberry. *New Phytol.* 72: In press

39. Davey, C. B. 1971. Nonpathogenic organisms associated with mycorrhizae. In: *Mycorrhizae,* ed. E. Hacskaylo., US Dep. Ag. Misc. Publ. 1189:114–21

40. Deal, D. R., Boothroyd, C. W., Mai, W. F. 1972. Replanting of vineyards and its relationship to vesicular-arbuscular mycorrhiza. *Phytopathology* 62:172–75

41. Filer, T. H., Toole, E. R. 1968. Effect of methyl bromide on mycorrhizae and growth of sweet gum seedlings. *Plant Dis. Reptr.* 52:483–85

42. Fried, M. 1953. The feeding power of plants for phosphates. *Proc. Soil Sci. Soc. Am.* 17:357–59

43. Gerdemann, J. W. 1955. Relation of a large soil-borne spore to phycomycetous mycorrhizal infections. *Mycologia* 47:619–32

44. Gerdemann, J. W. 1955. Wound healing of hyphae in a phycomycetous mycorrhizal fungus. *Mycologia* 47:916–18

45. Gerdemann, J. W. 1964. The effect of mycorrhiza on the growth of maize. *Mycologia* 56: 342–49

46. Gerdemann, J. W. 1965. Vesicular-arbuscular mycorrhizae formed on maize and tulip-tree by *Endogone fasciculata. Mycologia* 57: 562–75

47. Gerdemann, J. W. 1968. Vesicular-arbuscular mycorrhiza and plant growth. *Ann. Rev. Phytopathol.* 6:397–418

48. Gerdemann, J. W., Nicolson, T. H. 1963. Spores of mycorrhizal *Endogone* species extracted from soil by wet sieving and decanting. *Trans. Brit. Mycol. Soc.* 46:235–44

49. Gerdemann, J. W., Trappe, J. M. 1973. The Endogonaceae in the

Pacific Northwest. *Mycologia Memoirs.* In press

50. Gilmore, A. E. 1968. Phycomycetous mycorrhizal organisms collected by open-pot culture methods. *Hilgardia* 39:87–105

51. Gilmore, A. E. 1971. The influence of endotrophic mycorrhizae on the growth of peach seedlings. *J. Am. Soc. Hort. Sci.* 96:35–38

52. Gray, L. E., Gerdemann, J. W. 1967. Influence of vesicular-arbuscular mycorrhiza on the uptake of phosphorus-32 by *Liriodendron tulipifera* and *Liquidambar styraciflua. Nature, London* 213:106–7

53. Gray, L. E., Gerdemann, J. W. 1969. Uptake of phosphorus-32 by vesicular-arbuscular mycorrhizae. *Plant Soil* 30:415–22

54. Greaves, M. P., Wilson, M. J. 1970. The degradation of nucleic acid and montmorillonite-nucleic acid complexes by soil micro-organisms. *Soil Biol. Biochem.* 2:257–68

55. Greenall, J. M. 1963. The mycorrhizal endophytes of *Griselinia littoralis* (Cornaceae). *N.Z. J. Bot.* 1:389–400

56. Harley, J. L. 1950. Recent progress in the study of endotrophic mycorrhiza. *New Phytol.* 49:213–47

57. Harley, J. L. 1969. *The biology of mycorrhiza.* 2nd ed., Leonard Hill: London

58. Hattingh, M. J. 1972. A note on the fungus *Endogone. J. S. Afr. Bot.* 38:29–31

59. Hayman, D. S. 1970. *Endogone* spore numbers in soil and vesicular-arbuscular mycorrhiza in wheat as influenced by season and soil treatment. *Trans. Brit. Mycol. Soc.* 54:53–63

60. Hayman, D. S., Mosse, B. 1971. Plant growth responses to vesicular-arbuscular mycorrhiza. I. Growth of *Endogone*-inoculated plants in phosphate-deficient soils. *New Phytol.* 70:19–27

61. Hayman, D. S., Mosse, B. 1972. Plant growth responses to vesicular-arbuscular mycorrhiza. III. Increased uptake of labile P from soil. *New Phytol.* 71:41–47

62. Hayman, D. S., Mosse, B. 1972. The role of vesicular-arbuscular

mycorrhiza in the removal of phosphorus from soil by plant roots. *Rev. Écol. Biol. Sol.* 9:463–70

63. Hepper, C., Mosse, B. 1973. Trehalose and mannitol content of roots with *Endogone* mycorrhiza. *Rep. Rothamsted Exp. Sta. 1972*

64. Holevas, C. D. 1966. The effect of a vesicular-arbuscular mycorrhiza on the uptake of soil phosphorus by strawberry (*Fragaria* sp. var. Cambridge Favourite). *J. Hort. Sci.* 41:57–64

65. Iyer, J. G., Lipas, E., Chesters, G. 1969. Correction of mycotrophic deficiencies of tree nursery stock produced on biocide-treated soils. In: *Mycorrhizae,* ed. E. Hacskaylo, U S Dep. Agr. Misc. Publ. 1189:233–38

66. Jackson, N. E., Franklin, R. E., Miller, R. H. 1972. Effects of VA mycorrhizae on growth and phosphorus content of three agronomic crops. *Proc. Soil Sci. Soc. Am.* 36:64–67

67. Kalra, Y. P. 1971. Different behaviour of crop species in phosphate absorption. *Plant Soil* 34:535–39

68. Kessler, K. J., Blank, R. W. 1972. *Endogone* sporocarps associated with sugar maple. *Mycologia* 64:634–38

69. Khan, A. G. 1971. Occurrence of *Endogone* species in West Pakistan soils. *Trans. Brit. Mycol. Soc.* 56:217–24

70. Khan, A. G. 1972. Podocarp-like mycorrhizal nodules in *Aesculus indica. Ann. Bot.* 36:229–38

71. Khan, A. G. 1972. The effect of vesicular-arbuscular mycorrhizal associations on growth of cereals. I. Effects on maize growth. *New Phytol.* 71:613–19

72. Khan, A. G. 1973. The effect of vesicular-arbuscular mycorrhizal associations on growth of cereals. II. Effects on wheat growth. *J. Appl. Biol.* In press

73. Kleinschmidt, G. D., Gerdemann, J. W. 1972. Stunting of citrus seedlings in fumigated nursery soils related to the absence of endomycorrhiza. *Phytopathology* 62:1447–53

74. Kirillova, V. P. 1968. Effect of application of mineral fertilizers on the formation of mycorrhiza

in *Agrostis tenuis, Alopecurus pratensis* and other plants which are components of a grass meadow community of the Karelian isthmus. [Russian] *Uch. Zap. Perm. Ped. Inst.* 64:279–82. (*Chem. Abstr.* 1970. 73:3112 f)

75. Khrushcheva, E. R. 1960. Conditions favourable for the formation of mycorrhiza of maize. [Russian] *Agrobiologiya* 4:588–93

76. Khrushcheva, E. R. 1960. The mycorrhiza of wheat and its importance for the growth and development of the plant. [Russian] *Izv. Akad. Nauk. Ser. Biol.* 2: 220–39. *Soils Fert.* 1960. 23: abstr. 1220

77. Kryuger, L. V., et al. 1968. Method for determining the amount of infection in endotrophic mycorrhiza and for quantitative characterisation of mycosymbiotrophism in plant associations. [Russian] *Uch. Zap. Perm. Ped. Inst.* 64:260

78. Lanowska, J. 1962. An investigation on the influence of fertilizers on the occurrence of mycorrhiza in *Cucurbita pepo.* [Polish] *Acta Microbiol. Pol.* 11:349–58

79. Lanowska, J. 1966. Effect of different sources of nitrogen on the development of mycorrhiza in *Pisum sativum.* [Polish] *Pan Pulawski* 21:365–86

80. Laycock, D. H. 1945. Preliminary investigations into the function of the endotrophic mycorrhiza of *Theobroma cacao* L. *Trop. Agr.* 22:77–80

81. Lyness, A. S. 1936. Varietal differences in the phosphorus feeding capacity of plants. *Plant Physiol.* 11·665–88

82. Magrou, J. 1946. Sur la culture de quelques champignons de micorrhizes à arbuscules et à vésicules. *Rev. Gén. Bot.* 53:49–77

83. Martin, J. P., et al. 1953. Effect of soil fumigation on growth and chemical composition of citrus plants. *Soil Sci.* 75:137–51

84. Martin, J. P., Baines, R. C., Page, A. L. 1963. Observations on the occasional temporary growth inhibition of citrus seedlings following heat or fumigation treatment of soil. *Soil Sci.* 95:175–85

85. Marx, D. H. 1971. Ectomycorrhizae as biological deterrents to pathogenic root infections. In: *Mycorrhizae,* ed. E. Hacskaylo, U S Dep. Agr. Misc. Publ. 1189: 81–96

86. Marx, D. H., Bryan, W. C., Campbell, W. A. 1971. Effect of endomycorrhiza formed by *Endogone mosseae* on growth of citrus. *Mycologia* 63:1222–26

87. Mason, D. T. 1964. A survey of numbers of *Endogone* spores in soil cropped with barley, raspberry and strawberry. *Hort. Res.* 4:98–103

88. McClung, A. C., et al. 1957. Preliminary fertility studies on 'campos cerrados' soils in Brazil. *IBEC Res. Inst. Bull.* 13:5–17

89. McLennan, E. I. 1926. The endophytic fungus of Lolium. II. The mycorrhiza on the roots of *Lolium temulentum* L. with a discussion of the physiological relationships of the organisms concerned. *Ann. Bot.* 40:43–68

90. Meloh, K. A. 1963. Untersuchungen zur Biologie der endotrophen Mycorrhiza bei *Zea mays* L. und *Avena sativa* L. *Arch. Mikrobiol.* 46:369 81

91. Mejstrick, V. K. 1972. Vesicular-arbuscular mycorrhizas of the species of a *Molinietum coeruleae* L. I. association: the ecology. *New Phytol.* 71:883 90

92. Morrison, T. M., English, D. A. 1967. The significance of mycorrhizal nodules of *Agathis australis.* *New Phytol.* 66:245–50

93. Mosse, B. 1957. Growth and chemical composition of mycorrhizal and non-mycorrhizal apples. *Nature, London* 179:922–24

94. Mosse, B. 1959. The regular germination of resting spores and some observations on the growth requirements of an *Endogone* sp. causing vesicular-arbuscular mycorrhiza. *Trans. Brit. Mycol. Soc.* 42:274–86

95. Mosse, B. 1962. The establishment of vesicular-arbuscular mycorrhiza under aseptic conditions. *J. Gen. Microbiol.* 27:509–20

96. Mosse, B. 1963. Vesicular-arbuscular mycorrhiza: an extreme form of fungal adaptation. In: *Symbiotic Associations,* ed. P. S. Nutman, B. Mosse, Symp. Soc.

Gen. Microbiol. 13:146–70. Cambridge Univ. Press

97. Mosse, B. 1967. Effects of host nu-. trient status on mycorrhizal infection. *Rep. Rothamsted Exp. Sta. 1966*, p. 79

98. Mosse, B. 1970. Honey-coloured, sessile *Endogone* spores. I. Life history. *Arch. Mikrobiol.* 70:167–75

99. Mosse, B. 1970. Honey-coloured, sessile *Endogone* spores. II. Changes in fine structure during spore development. *Arch. Mikrobiol.* 74:129–45

100. Mosse, B. 1970. Honey-coloured, sessile *Endogone* spores. III. Wall structure. *Arch. Mikrobiol.* 74: 149–59

101. Mosse, B. 1972. Effects of different *Endogone* strains on the growth of *Paspalum notatum*. *Nature, London* 239:221–23

102. Mosse, B. 1972. The influence of soil type and *Endogone* strain on the growth of mycorrhizal plants in phosphate deficient soils. *Rev. Écol. Biol. Sol.* 9:529–37

103. Mosse, B. 1972. Growth of *Endogone* mycorrhiza in agar medium. *Rep. Rothamsted Exp. Sta. 1971*, p. 93

104. Mosse, B. 1973. Plant growth responses to vesicular-arbuscular mycorrhiza. IV. In soil given additional phosphate. *New Phytol.* 72: 127–36

105. Mosse, B., Bowen, G. D. 1968. A key to the recognition of some *Endogone* spore types. *Trans. Brit. Mycol. Soc.* 51:469–83

106. Mosse, B., Bowen, G. D. 1968. The distribution of *Endogone* spores in some Australian and New Zealand soils and in an experimental field soil at Rothamsted. *Trans. Brit. Mycol. Soc.* 51:485–92

107. Mosse, B., Hayman, D. S. 1970. Effect of *Endogone* mycorrhiza on plant growth. *Rep. Rothamsted Exp. Sta. 1969*, p. 95–96

108. Mosse, B., Hayman, D. S. 1971. Plant growth responses to vesicular-arbuscular mycorrhiza. II. In unsterilised field soils. *New Phytol.* 70:29–34

109. Mosse, B., Hayman, D. S., Arnold, D. 1973. Plant growth responses to vesicular-arbuscular mycor-

rhiza. V. Phosphate uptake from ^{32}P labelled soil solution by three plant species. *New Phytol.* 72: In press

110. Mosse, B., Hayman, D. S., Ide, G. J. 1969. Growth responses of plants in unsterilized soil to inoculation with vesicular-arbuscular mycorrhiza. *Nature, London* 224: 1031–32

111. Mosse, B., Jones, G. W. 1968. Separation of *Endogone* spores from organic soil debris by differential sedimentation on gelatine columns. *Trans. Brit. Mycol. Soc.* 51:604–8

112. Mosse, B., Phillips, J. M. 1971. The influence of phosphate and other nutrients on the development of vesicular-arbuscular mycorrhiza in culture. *J. Gen. Microbiol.* 69:157–66

113. Murdoch, C. L., Jackobs, J. A., Gerdemann, J. W. 1967. Utilization of phosphorus sources of different availability by mycorrhizal and non-mycorrhizal maize. *Plant Soil* 27:329–34

114. Nesheim, O. N., Linn, M. B. 1969. Deleterious effects of certain fungitoxicants on the formation of mycorrhiza on corn by *Endogone fasciculata* and on corn root development. *Phytopathology* 59: 297–300

115. Nicolson, T. H. 1967. Vesicular-arbuscular mycorrhiza—a universal plant symbiosis. *Sci. Progr. Oxf.* 55:561–68

116. Nicolson, T. H., Gerdemann, J. W. 1968. Mycorrhizal *Endogone* species. *Mycologia* 60:313–25

117. Ohms, R. E. 1957. A flotation method for collecting spores of a phycomycetous mycorrhizal parasite from soil. *Phytopathology* 47:751–52

118. Pandey, S., Misra, A. P. 1971. *Rhizophagus* in mycorrhizal association with *Litchi chinensis* Sonn. *Mycopath. Mycol. Appl.* 45:337–54

119. Peuss, H. 1958. Untersuchungen zur Ökologie und Bedeutung der Tabakmycorrhiza. *Arch. Mikrobiol.* 29:112–42

120. Peyronel, B. 1940. Prime osservazioni sui rapporti tra luce e simbiosi micorrizica. *Annuar. Lab. Chanousia* Giardino Botanico

dell'Ordine Mauriziana al Piccolo San Bernardo 4:3–19

121. Phillips, J. M., Hayman, D. S. 1970. Improved procedures for clearing roots and staining parasitic and vesicular-arbuscular mycorrhizal fungi for rapid assessment of infection. *Trans. Brit. Mycol. Soc.* 55:158–61

122. Porter, D. M., Beute, M. K. 1972. *Endogone* species in roots of Virginia type peanuts. *Phytopathology* 62:783 (abstr.)

123. Possingham, J. V., Obbink, J. G. 1971. Endotrophic mycorrhiza and the nutrition of grape vines. *Vitis* 10:120–30

124. Redhead, J. F. 1971. *Endogone* and endotrophic mycorrhizae in Nigeria. *XV IUFRO Congr.*, Sec. 24

125. Redhead, J. F., Old, K. M., Nicolson, T. H. 1973. A new species of mycorrhizal *Endogone* from Nigeria with a distinctive spore wall. *New Phytol.* 72: In press

126. Rorison, I. H. 1968. The response to phosphorus of some ecologically distinct plant species. I. Growth rates and phosphorus absorption. *New Phytol.* 67:913–23

127. Ross, J. P. 1971. Effect of phosphate fertilization on yield of mycorrhizal and non-mycorrhizal soybeans. *Phytopathology* 61: 1400–03

128. Ross, J. P., Harper, J. A. 1970. Effect of *Endogone* mycorrhiza on soybean yields. *Phytopathology* 60.1552–56

129. Safir, G. R., Boyer, J. S., Gerdemann, J. W. 1971. Mycorrhizal enhancement of water transport in soybean. *Science* 172:581–83

130. Safir, G. R., Boyer, J. S., Gerdemann, J. W. 1972. Nutrient status and mycorrhizal enhancement of water transport in soybean. *Plant Physiol.* 49:700–03

131. Sanders, F. E., Tinker, P. B. 1971. Mechanism of absorption of phosphate from soil by *Endogone* mycorrhizas. *Nature, London* 233: 278–79

132. Scannerini, S., Bellando, M. 1968. Sull' ultrastruttura delle micorrize endotrofiche di *Ornithogalum umbellatum* L. in attivita vegetativa. *Atti. Accad. Sci. Torino* 102: 795–809

133. Schenck, N. C., Hinson, K. 1971. Endotrophic vesicular-arbuscular mycorrhizae on soybean in Florida. *Mycologia* 63:672–75

134. Schonbeck, F., Schinzer, U. 1972. Untersuchungen über den Einfluss der endotrophen Mycorrhiza auf die TMV-Läsionenbildung in *Nicotiana tabacum* L. (var. Xanthinc). *Phytopathol. Z.* 73:78–80

135. Shelonina, I. M. 1959. Effect of bacterial fertilisers on mycorrhiza formation in maize. [Russian] *Kukuruza* 4:42–43; (*Rev. Appl. Mycol.*, 39:306 1960.)

136. Shemakhanova, N. M., Mazur, O. P. 1968. Mycorrhiza in walnut (*Juglans regia* L.) and conditions of its formation. [Russian] *Izv. Akad. Nauk SSSR Ser. Biol.* 41: 517–29

137. Smith, S. N. 1934. Responses of inbred lines and crosses in maize to variations of nitrogen and phosphorus supplied as nutrients. *J. Am. Soc. Agron.* 26:785–804

138. Smith, D., Muscatine, L., Lewis, D. 1969. Carbohydrate movement from autotrophs to heterotrophs in parasitic and mutualistic symbiosis. *Biol. Rev.* 44:17–90

139. Sogina, I. I. 1970. *Mycorrhiza of maize and its effects on some physiological processes* [Russian]. Autoreferat Diss.

140. Stahl, M. 1949. Die Mycorrhiza der Lebermoose mit besonderer Berücksichtigung der thallosen Formen. *Planta* 37:103–48

141. Stelz, T. 1968. Mycorrhizes et végétation des pelouses calcaires, *Rev. Socs. sav. haute-Normandie* 50:69–85

142. Strzemska, J. 1955. Investigations on the mycorrhiza in corn plants. [Polish] *Acta Microbiol. Pol.* 4: 191–204

143. Sutton, J. C., Barron, G. L. 1972. Population dynamics of *Endogone* spores in soil. *Can. J. Bot.* 50: 1909–14

144. Tardieux-Roche, A., Tardieux, P. 1970. La biosynthèse des phosphates condensés par la microflore du sol et son rôle dans la nutrition des vegetaux. *Ann. Agron.* 21:305–14

145. Thaxter, R. 1922. A revision of the Endogoneae. *Proc. Am. Acad. Arts. Sci.* 57:291–350

146. Thomas, W. 1930. The feeding power of plants. *Plant Physiol.* 5: 443–89

147. Tolle, R. 1958. Untersuchungen über die Pseudomycorrhiza von Gramineen. *Arch. Mikrobiol.* 30: 285–303

148. van Lear, D. H., Smith, W. H. 1972. Relationships between macro- and micro-nutrient nutrition of slash pine on three coastal plain soils. *Plant Soil* 36:331–47

149. Voigt, G. K. 1971. Mycorrhizae and nutrient mobilization. In: *Mycorrhizae* ed. E. Hacskaylo, U S Dep. Agr. Misc. Publ. 1189: 122–31

150. Wilhelm, S. 1959. Parasitism and pathogenesis of root-disease fungi. In: *Plant Pathology* 1908–58, 356–66 ed. C. S. Holton et al., Univ. Wisconsin Press

151. Wilhelm, S., George, A., Pendery, W. 1967. Zinc deficiency in cotton induced by chloropicrin-methyl bromide soil fumigation to control Verticillium wilt. *Phytopathology* 57:103 (abstr.)

152. Winter, A. G., Birgel, G. 1953. Untersuchungen über die Verbreitung, Ökologie und funktionelle Bedeutung der endotrophen Mykorrhiza bei gärtnerischen Kulturpflanzen. *Naturwissenschaften* 40:393–94

153. Winter, A. G., Meloh, K. A. 1958. Untersuchungen über den Einfluss der endotrophen Mykorrhiza auf die Entwicklung von *Zea mays* L. *Naturwissenschaften* 45:319

154. Wuenscher, M. L., Gerloff, G. C. 1971. Growth of *Andropogon scoparius* (Little Bluestem) in phosphorus deficient soils. *New Phytol.* 70:1035–42

HEARTWOOD, DISCOLORED WOOD, AND MICROORGANISMS IN LIVING TREES

❖ 3571

Alex L. Shigo and W. E. Hillis

Northeastern Forest Experiment Station, U.S.D.A. Forest Service, Durham, New Hampshire, and Forest Products Laboratory, Division of Applied Chemistry, Commonwealth Scientific and Industrial Research Organization, South Melbourne, Victoria, Australia

The state of our knowledge of plant pathology is reflected by the terms we use (120). Consider some of the terms used to describe wood altered by processes associated with aging and injury of living trees: heartwood, wound heartwood, pathological heartwood, traumatic heartwood, false heartwood, precocious heartwood, blackheart, brownheart, red heart, blue butt, mineral streak, mineral stain, woundwood, discolored wood, wound-initiated discolored wood, wetwood, ripewood, reaction zone, protection wood, and even true wood. Indeed, there is confusion!

The main visible change observed in wood of trees is change of color. This can be the result of processes associated with aging (heartwood), injury (discolored wood), or both. However there are more important characteristics than color for the sapwood altered by these processes. The factors that initiate the formation of heartwood, discolored wood, and extractives (which are largely responsible for color) are different.

This is the major reason for the confusion in understanding a situation when color alone is the basis for distinguishing the type of tissue under study. When injury-altered tissues are considered as age-altered tissues, and the role of microorganisms in the processes are not considered, it is impossible to interpret the situation accurately. The confusion is further compounded when the injury processes occur in tissues already altered by aging. Clarification of these processes obviously is needed.

We will discuss in this review those processes in living trees that are associated with colored wood, in the hope of clarifying them so that future research will be more accurately oriented and the opportunity to bring these changes under our control will be improved. We will consider and contrast two types of wood, which we will refer to as "heartwood" and "discolored wood."

197

Heartwood occurs in mature specimens of most, but not all, tree species. It is formed in a more or less regular manner in individual trees, and is not normally associated with detectable injury. Discolored wood occurs in many species, but it is distributed irregularly among and within trees, often not being prominent in most individuals of the species, and it is usually associated clearly with mechanical or biological damage.

In view of the complexity of the situation and the volume of the literature, we will consider only the *major* points, and only a portion of the published work is cited because of space limitations.

Formation of Wood

Herbaceous and woody plants have points of both similarity and dissimilarity in their life cycles (167). In almost all cases, trees are larger than herbaceous plants, and can live longer and grow larger in mass than any other organisms on earth. It is remarkable that they can achieve this despite their inability to move away from destructive forces as other organisms can. It would be rare indeed for trees to live even a short time without receiving wounds or damage from external agencies, in addition to those received from the natural shedding of branches.

Because wounds were probably common to trees as they evolved, the survival of trees as tall perennial plants depended in part on their development, through natural selection, of effective systems for protection (as with heartwood) and repair of wounds (resulting in many cases in discolored wood). Despite the different habits or appearance, and the wood anatomy of the Gymnosperms (cone-bearing trees, mostly evergreens) and the Angiosperms (flowering broad-leaved hardwood trees with spreading branches, often deciduous) the wood-formation processes are similar.

A series of biological processes results in the formation of wood. The thin-walled early-wood portion of a growth ring is produced during the period of terminal growth of the tree. The fiber (or tracheid) diameter is regulated by hormones produced by the foliage and transmitted to the developing fibers in the cambium. The fiber wall thickness appears to be genetically controlled, but also is dependent on the amount of available photosynthate. The interaction of the two physiological processes results in the type of fiber produced, and also ultimately the amount of primary metabolites in the sapwood and the amount of extractives in the heartwood or injured wood.

When the lignification of the fiber wall (composed largely of cellulose microfibrils packed in a matrix) and the middle lamella is completed, the cell dies. Capillaries existing in the secondary wall occupy a significant volume, estimated to be 25% in *Pinus resinosa* green sapwood (9). These capillaries probably have a diameter between 16 and 60×10^{-10} meters, and collapse when the wood dries (75).

In contrast to fibers and tracheids, the transverse and longitudinal parenchyma (about 7% in conifers and 17% or more in deciduous hardwoods) can remain viable for many years. In *Tamarix aphylla* sapwood, however, the fibers retain their living protoplasts for the same period as the parenchyma and ray cells (36). The shape of the nuclei of living ray parenchyma cells in the sapwood of several species of Gymnosperms and Angiosperms changes and they eventually disintegrate (45, 64). They lose organelles (36), their vitality (115), nitrogen-containing compounds (106), starch (69), and their ability to consume oxygen (65) as their distance from the cambial region increases. The amount of sugars and biotin and pyridoxine can decrease abruptly at the heartwood boundary (27, 181).

The sapwood vessels in Angiosperms occasionally contain tyloses; but as heartwood and discolored wood form, they usually appear in much greater numbers in many species, so that movement of liquids and perhaps microorganisms is blocked. Aspiration of the pits in conifer tracheids as heartwood forms has a similar effect on liquid movement.

To meet different physiological needs throughout the year, primary metabolites are stored in the sapwood in the form of starch or fats, according to the type of tree (68). The amounts vary according to prior needs and current demands, and are not uniform across the sapwood. The distribution of photosynthates within a tree can be considered a system of competing metabolic sinks, which are constantly changing in size according to the needs of a particular zone at a particular time. There is little information about the dynamic translocation of carbohydrate through the rays of the sapwood.

Formation of Heartwood

The sapwood of the trunk, branches, and roots of many—but not all—uninjured trees changes abruptly in appearance and function after a certain age. This interior core is "heartwood," of which one definition is: "The inner layers of wood which, in the growing tree, have ceased to contain living cells and in which the reserve materials (e.g. starch) have been removed or converted into heartwood substances. It is generally darker in color than sapwood, though not always clearly differentiated" (84).

The proportion, and even the existence, of heartwood in a mature tree varies within the family, genus, and even species (25, 76). Within a species, under normal circumstances, the amount and rate of heartwood formation varies to a lesser extent with tree age (66), growth rate, environment, and silviculture practice (154). In some genera (such as *Eucalyptus*) and in some species, the age of the sapwood transformed to heartwood is remarkably constant (25), and possibly is mainly genetically determined. It has been observed with some species that heartwood formation commences at some distance above ground level (25, 166) and that the proportion of heartwood in some species remains greatest at this level. Sometimes, however, heartwood may never form, as in *Alstonia scholaris* (19). Living cells 115 years old have been found in *Acer saccharum* (50).

The periphery of the approximately conical-shaped central core of heart-wood often undulates vertically and horizontally and can cut across parts of annual rings as abrupt tongues.

Transition of sapwood to normal heartwood is initiated by internal processes rather than by external conditions. Once it begins, the process continues so that frequently the number of sapwood rings remain more or less constant during the life of the tree. There are also species in which heartwood formation is initiated after many years, but thereafter more than one ring of sapwood is transformed annually to heartwood.

The little available information indicates that the area of sapwood relative to heartwood is greater in the trunk than in the roots and branches of similar size, but the ratio can vary with species.

Usually heartwood contains less moisture than sapwood, and the decrease can be abrupt and considerable, as in *Picea glauca* which has 136–162% moisture in the sapwood and 47–48% in the heartwood (22). Some *Pinus* species show similar decreases (25). On the other hand, the heartwood of some hardwood species contains more moisture than the sapwood (74, 177, 178). Analyses of the gas in the heartwood of a number of trees show, in addition to nitrogen, a large proportion of carbon dioxide and a small amount of oxygen (25, 85, 110, 111).

Freshly cut cross-sections of many trees reveal a transition, intermediate, or white zone surrounding the heartwood. Usually the transition zone is less than 1 cm wide, and the width can increase with tree height and with season. There are also intermediate zones of up to 15 cm, as reported in *Sloanea woollsii* (19). The zone is paler in color, clearly distinguishable from the sapwood, with a moisture content sometimes even lower than that of the heartwood (25, 179), increased amounts of nicotinic acid amide, biotin, pyridoxine (181, 182), and in some cases protein nitrogen (182). In pine species the amount of extractives is low (52).

In contrast to reports of the absence of organelles at the heartwood boundary mentioned previously, the amount of organelles in the transition zone of *Cryptomeria japonica* has been found to be only decreased (118). Furthermore, respiration and the activity of the malate and glucose-6-phosphate dehydrogenases in this zone in *Pinus radiata* is higher than in the surrounding sapwood during the dormant season (L. Shain & J. F. G. Mackay, unpublished data). When slowly dried, the boundary of the sapwood-transition zone of some species (e.g. eucalypts) can become very dark colored (W. E. Hillis, unpublished data), and this is probably due to localized increases of phenol oxidases. Increases in peroxidase activity in the "pre-heartwood" zone of *Larix europea* and *P. sylvestris* have been reported (95).

In some species the transition zone has not been recognized or does not exist, so some properties cannot be allocated precisely to the transition zone or to the heartwood periphery. Some of the recent results contradict those of earlier workers (45, 65). Cytological studies of ring-porous hardwoods have shown an increase in vitality in the parenchymous cells (80), and in respiration (180) at the heartwood periphery. Extractives are formed mainly in the radial parenchyma, but the longitudinal parenchyma can also form them. Peroxidase activity increased markedly at the periphery of the heartwood (27, 95, 172), as did the activities of amylase (78), phenol oxidases, malate dehydrogenase, etc. (182). Recent studies showed that tyloses form before extractives in *Eucalyptus* and *Nothofagus* species (V. Nečesaný, unpublished), and that calcium oxalate crystals are not found in the cells containing polyphenols (G. Scurfield, unpublished). The chemical reactivity of cellulose in Douglas-fir trees, as shown by the accessibility of hydroxyl groups, revealed a marked maximum at the sapwood-heartwood boundary. The carbonyl index, moisture content, and extractives content also showed maxima at the same point, indicating increased biochemical activity in this region (20).

The transformation of sapwood to heartwood is accompanied by necrosis of the xylem parenchyma, although some enzymic activity may be found in the heartwood. Phenol-oxidizing enzymes have been reported in the heartwood of *Pinus lambertiana* (159), and the two found in *P. radiata* heartwood were probably of host origin (137). Other major differences from sapwood can include aspiration of the pits in Gymnosperms (53), formation of tyloses in Angiosperms, or gum when the pit aperture is less than 10μ (18). Although starch is absent from heartwood, small amounts of free sugar may be found (74). The nitrogen content is lower (106), and the pH higher than in sapwood (74). Fatty and resinous materials that are stored in sapwood of some trees instead of starch, are changed in composition as heartwood forms (62, 109). The most noticeable change is the formation of nonstructural material, (extractives), sometimes in amounts exceeding 30% of the total wood, which increases the density, color, durability, and many other properties of the wood (68, 74, 76).

Extractives accumulate in the lumen, or occlude or encrust pits and walls (7, 38, 91, 92). In some species, phenolic substances can diffuse from the ray parenchyma cells into the cell walls and into fiber lumens (38). The capillaries of the cell wall are wide enough to accommodate the molecules of some extractives (75). There is a good deal of indirect evidence for the presence of extractives in cell walls of the heartwood of different species (162, 170) (W. E. Hillis, unpublished data). New techniques, such as the use of gas-liquid chromatography and microspectrophotometry, make it possible for very small amounts of most extractives to be estimated or detected in small parts of tissues. These and other techniques will enable the relative amounts of extractives in the cell wall and lumen to be determined (10, 92). Toxic components probably convey greater durability if present in the cell wall than in the lumen.

The amounts of polyphenols formed in cultures of *Juniperus communis* (23), rose (26), and tea (40) were dependent on the sugar concentration.

These findings support an earlier proposal (69), that the amount of polyphenols in heartwood is related to the amount of carbohydrate reaching the boundary.

A large number of secondary compounds has been identified in heartwood (68). The composition varies according to the family, genus, and species to such an extent that sometimes species can be identified from their heartwood extractives, although there are cases where there is variation in composition within a species (71). The composition of extractives in the sapwood often differs from that in the heartwood of the same tree (74). The existence of extractives in heartwood (and in discolored wood) is possibly concerned with the evolution of species that can resist insect predators and microorganisms (42).

A large portion of published data supports the view that the extractives are formed at the heartwood periphery from carbohydrate (76). Other views on heartwood formation have been expressed (158). Recent data give further evidence in support of *in situ* formation (L. Shain & W. E. Hillis, unpublished information), although it has been proposed that the factors controlling the amount and composition of heartwood extractives are incorporated in the ray cells during the early stages of their development (61).

The wide range of colors seen in heartwood is due largely to extractives. The amount in heartwoods of similar age of some species is partly genetically controlled (44), but a fast growth rate can lower the amount normally formed (76), and even the soil type can affect the color (117).

Usually the amount of extractives increases from the pith to the heartwood periphery, and this is generally considered to be due to the larger amount formed at the time of heartwood formation. There are cases of a uniform distribution of extractives across the heartwood of Scots pine (30). The work of Anderson et al (3), Rudman (128), and others has shown that the toxicity of extractives in the interior of heartwood to fungi and insects can decrease on aging (93, 132). This could be due to enzymic oxidation (99, 137), free radical reactions, polymerization, hydrolysis, and changes caused by acid and microbiological degradation. Extractives have widely differing toxic or repellent properties to different wood-destroying microorganisms (107, 132). Even mildly toxic components present in adequate amounts can confer durability (58).

Mechanism of the Formation of Heartwood and Extractives

The abrupt change of sapwood to heartwood reveals the existence of an active situation. It is generally considered that the change is a DNA-coded aging effect that can be influenced by environment. However, the erratic and undulating heartwood boundaries that cross over several growth rings in many trees indicate that more study is required to define the trigger that sets off the changes.

The differences between sapwood and heartwood already pointed out pinpoint the narrow transition zone (when present) or the heartwood periphery as dynamic zones in the living tree. Chattaway (19) suggested that heartwood formation must be preceded by a period of increased metabolic activity. Other workers suggested that both the Krebs cycle and pentose shunt enzymes were affected (66, 70, 72, 74, 182).

Although some direct evidence of increased metabolic activity has been found at the heartwood periphery (95, 105, 180), most studies have been unsuccessful in this regard. Hirai (77) produced evidence that heartwood formation takes place mainly when cambial growth ceases, so that the frequently reported lack of evidence of activity at the heartwood periphery could be due to collection of samples at inappropriate periods. The recent work of L. Shain & J. F. G. Mackay (unpublished data) has shown that increases in respiration and the activity of malate and glucose-6-phosphate dehydrogenases in the transition zone of *P. radiata* are seasonal, maximum amounts occurring in the dormant period of tree growth. The factor that initiates these increases in activity requires consideration.

Many studies have shown that ethylene acts as a regulatory hormone (14, 119) in a variety of physiological changes occurring at many stages in the ontogeny of plants. It can play an important role in the regulation of cellular metabolism, which is related not only to morphological changes, but also to basic cell processes. Very small amounts (1–5 ppm and smaller) of ethylene effectively trigger a wide range of events according to the tissue involved (12, 14, 119). Because of the ready production of ethylene on injury of many tissues and their sensitivity to it, experimental work in this area is fraught with difficulties (119). The considerable data collected on ethylene in studies of vegetative tissues point to a probable pattern of events that lead to heartwood and extractives formation.

Ethylene is produced by the transition zone surrounding the heartwood of *Pinus radiata* (the peak of production taking place in the dormant period) and in larger amounts than the adjacent sapwood (L. Shain & W. E. Hillis, unpublished data). In *P. radiata,* the transition zone contains very small amounts of polyphenols and a lower moisture content than the heartwood. The transition zone of *Eucalyptus tereticornis* also produces more ethylene than the sapwood, but in this species the transition zone contains more polyphenols than the sapwood, and has a moisture content similar to that of sapwood and heartwood (W. E. Hillis, unpublished data). Cell suspension or callus cultures of different plants, including sycamore, also release ethylene; and a very sharp peak of production occurs in the latter after 10–14 days of culture (113) or toward the end of the growing phase of cell cultures (96).

The factor initiating ethylene production by injured, diseased, or senescing plant tissues has not been established (119). There is an absolute need for oxygen (102) and, at low oxygen concentration, sensitivity of the tissue to ethylene is decreased. Once the threshold value is exceeded, the system can produce ethylene autocatalytically. However, the system does not seem to be fully activated until a certain physiological age is reached (119). Carbon dioxide is competitive at the receptor site of the ethylene, particularly at low oxygen levels (13, 103).

It has been found that polyphenol formation increases in *Sirex-Amylostereum* lesions within 24 hours of the increases of ethylene and respiration (135). Heartwood polyphenols have been formed in fresh blocks of *P. radiata* sapwood stored in a container continuously ventilated with humidified air containing ethylene (5 ppm), but not in air containing carbon dioxide (27.5%) (L. Shain & W. E. Hillis, unpublished data). The composition of polyphenols, however, more closely resembled that from injured wood than that from heartwood of this species.

The factors initiating dehydration of the transition zone or heartwood in many species are unknown. However, moisture stress may be important for the stimulation of ethylene production (113, 114). It is noteworthy that, in *P. radiata* blocks used for in vitro polyphenol synthesis, the latter were formed predominantly in the partially desiccated zone near the surface. The formation of pinosylvins in *P. resinosa* cultures (88) or of heartwood in stem sections of *Fagus sylvatica* (183) (which has been attributed to dehydration) may have been caused directly by ethylene.

Ethylene increases, in a short period, RNA and protein synthesis (104) and the activity or the *de novo* synthesis of a number of enzymes. Those so far reported include phenylalanine ammonia lyase (PAL) (82, 122), polyphenol oxidase (156), α-amylase (86), cellulase (1), and particularly peroxidase (83, 123). Several cases are known in which ethylene increases the rate of respiration (81, 121). The increase in respiration of the transition zone (relative to the surrounding sapwood) in *Pinus radiata* increases (L. Shain & J. F. G. Mackay, unpublished data) over a similar period to the increase in ethylene content (L. Shain & W. E. Hillis, unpublished data). So far, there are no specific data showing that ethylene enhances the activity of enzymes of the Krebs cycle and pentose shunt, such as those found to increase in the transition zone in the dormant period of *P. radiata* (L. Shain & J. F. G. Mackay, unpublished data).

The amount of ethylene present has been related to the concentrations of a polyphenol formed in carrot tissues (17). The amount of polyphenol formed was reduced by carbon dioxide. Also, treatment with natural and synthetic auxins produced the characteristic polyphenol in carrot (17), and other polyphenols in the cell and callus suspension cultures of other plants (8, 26, 108).

An association between auxin and formation of heartwood and extractives has been considered (74, 182). Hormones, metabolites, etc. could be more readily available for heartwood formation after periods of active growth in the cambium. Several workers have shown that natural and synthetic auxins initiate, stimulate, and prolong ethylene production in higher plants (46, 79, 119, 131). Recent work indicates, however, that the balance between auxin-ethylene is more important than absolute amounts of either (14, 103). **The** overall effect of auxin in heartwood formation requires further study. A very high concentration of carbon dioxide has also been reported to be conducive to the formation of heartwood polyphenols in *Acacia mearnsii* (15).

Although the most detailed studies on ethylene were concerned with the production of polyphenols, it is well established that ethylene promotes an increased production of rubber in *Hevea braziliensis* (2) and carbohydrate gum in *Prunus* spp. (W. E. Hillis, unpublished data).

In summary, evidence indicates that ethylene plays a key role in the formation of extractives, which are largely responsible for the color of heartwood. Whether the initiation of ethylene formation is triggered by water stress, which has been suggested as a key factor in heartwood formation (129, 183), remains to be determined. Heartwood extractives are formed at the heartwood periphery or in the transition zone during the dormant season, from translocated or stored carbohydrate. Peroxidase, whose activity increases at the heartwood periphery (95, 172), and the phenol oxidases in the heartwood (27, 89), can cause darkening of the tissues after exposure to air.

The role played by ethylene in the formation of discolored wood requires determination. It is notable that different trees of *P. radiata* respond differently to *Sirex-Amylostereum* damage in the formation of ethylene, polyphenols, and discolored wood. The ethylene appears to result from host-parasite interaction. It is known that some fungi produce ethylene; and whether these produce discolored wood—in contrast to those that do not form ethylene—requires further study. It should be noted, however, that discolored wood can have a different composition from that of heartwood in the same tree.

Formation of Discolored Wood

The major conditions in heartwood formation—cell death, depletion of nutrients, deposits in cells with darkening of tissues—also occur in formation of discolored wood, but there are other processes too.

Though discoloration is a condition of the wood, the color is a poor indicator of the changes that have occurred (144). Attention should be focussed on the events that follow wounding, rather than on one minor condition—discoloration—of these important events. Although this minor condition has served as the focus for so many studies, it will be treated here within the broad context of the events that follow wounding.

The many events that occur from wounding to total decomposition of tissues are continuous over time, and actually it is not possible to separate them. But, for the sake of clarifying the events that follow wounding and putting discoloration and decay in proper perspective, the sequence of events in a model system are separated into three major stages (143) (See Table 1).

Stage I includes all processes associated with host response to wounding, in which both the tree and the environment are considered. Slight discoloration may occur in the xylem as a result of chemical processes, including those involving formation of phenols and other components, and oxidation resulting from exposure to air (37, 47, 90, 98, 160).

Stage II includes those events that occur when microorganisms surmount the chemical protection barriers and invade the xylem. These pioneer invaders are usually, but not always, bacteria and nonhymenomycetous fungi (139, 140, 141, 142, 144, 146). The discoloration of the wood is intensified as a result of interactions between invading microorganisms and living xylem cells (144). There is now also a host response to invasion (133).

Stage III includes the events that occur when decay microorganisms, especially hymenomycetes, invade and degrade the cell-wall substances. The microorganisms compete among themselves; all cells in the xylem are now dead. After the pioneer decay microorganisms invade the wood, many other microorganisms—phycomycetes, actinomycetes, myxomycetes, and nematodes—follow and compete for the remaining portions of the tissues (51, 140, 146).

Table 1   The Sequence of Events in the Decomposition of Wood

	Stage I	Stage II	Stage III
WOUND >	Host response to wounding	Infection & invasion by pioneer microorganisms	Decay processes
			──────────────> DECOMPOSITION
	Tree and abiotic environment	Tree, abiotic environment, and microorganisms	Interactions among microorganisms as they digest cell-wall substances

When trees are vigorous and the wounds are not severe, the processes stop in Stage I. As the vigor of the tree decreases, or as more severe wounds are inflicted, the processes may go on to Stage II before they stop. When the vigor of the tree is low, the wounds are severe, and the aggressiveness of the microorganisms is strong, the processes may go to Stage III.

In summary, there are several stages following the many wounds inflicted on trees before they reach maturity. A host response to wounding always occurs—Stage I; infection and invasion of xylem by microorganisms that surmount the protective barriers of the tree and a host response to invasion occur most of the time—Stage II; and some of the time decay follows—Stage III. While these events are occurring, the tree is growing, branches are dying, and other wounds are occurring.

TYPES OF WOUNDS   There are two basic types of wounds: those that expose primarily the axis of the stem or root of trees via broken branches, broken tops, and broken roots; and those that expose primarily the xylem immediately under the bark by means of mechanical wounds, animal wounds, fire wounds, etc.

The most common type is the branch wound. Branches die for many reasons. When they die or when they are broken off, air and microorganisms may quickly or gradually enter the center of the tree and the growth layers in the trunk that extend into the branch. A broken top or broken root exposes similar tissues. Most branch wounds heal and prevent exposure of the trunk to air and microorganisms, but healing may be complete only after some of the trunk tissues have been altered. One of the minor conditions of the altered tissues in Stage I is discoloration. More discoloration is associated with branch wounds than with any other type of wound, and it is usually in the center of the tree.

The severity of the wound and the vigor of the host affect the rate and effectiveness of the tree's response to the wound (116, 161). Wounds that break the bark, but injure the cambium and xylem only slightly, usually heal rapidly (21, 143), although it has been found that the "wound heartwood" of *Pinus sylvestris* forms with the beginning of cambial activity and terminates in winter (100). The processes can stop in Stage I, and some discoloration may be associated with the wound.

It is noteworthy that when the sapwood is deeply penetrated by a wound, a pale-colored transition zone (similar to that around heartwood) surrounds the discolored wood in a number of species (133, 144). A similar zone has been observed around the discolored wood of lesions resulting from attack on *P. radiata* by the *Sirex-Amylostereum* complex (136), and on *Picea abies* by *Fomes annosus* (134). Thus, with some species at least, there is a visual similarity between the formation of discolored wood and heartwood, and further examination may show the existence of the transition zone to be more widespread.

HOST RESPONSE TO WOUNDING   The response to wounding in herbaceous and woody plants is similar in principle: a chemical protective response occurs, and tissues darken (127). Most woody plants survive after wounding because the protective response is effective most of the time. However, in some cases, the tree may be so low in vigor, or the conditions for invasion by microorganisms may be so favorable because of inoculum quantity, environmental conditions, and severity of wound, that invasion occurs rapidly. Between the extremes of no invasion and rapid invasion are all degrees of effectiveness of host protection and aggressiveness of microorganisms. Also, between the extremes there are all degrees and gradations of color changes in the wood.

In general, the living sapwood cells show a dynamic response, and discolored wood containing extractives is formed in a zone several millimeters wide around the area containing microorganisms (133). Heartwood shows a passive response. When the protection processes in Stage I function effectively, the xylem altered by host response to wounding is indeed a protective wood that resists invasion by microorganisms, and "protection wood" (43, 63, 87) is an accurate term for these tissues.

The extractives formed in "protection wood" or "reaction zone," as in the lesions from *Fomes annosus* (133, 135), *Sirex noctilio-Amylostereum areolatum* (73), etc., can be different from those of the heartwood and even those of lesions in different trees of the same species. The extractives can play a significant part in enabling the living tree to hinder the extension of the damaged zone.

INFECTION AND INVASION PROCESSES   Propagules of many microorganisms are carried to moist fresh wound surfaces by wind, rain, snow, animals, man, and insects (4, 143). Saprophytes infecting wound surfaces may prevent the establishment of other more aggressive microorganisms that can invade (11, 33, 34, 49, 107, 125).

When severe wounds are inflicted and conditions are favorable, invasion by microorganisms occurs. The principal pioneer microorganisms that invade wounds are usually bacteria and nonhymenomycetous fungi (5, 6, 31, 97, 101, 140, 143, 151). There is some indication that they can utilize or detoxify the chemicals formed in Stage I (149, 163, 164). As the pioneers enter, the living cells distal to the wound at the margins of the lesion continue to react (133, 144), but the response is more to the invasion of microorganisms than to the wound (133). Here the term "reaction zone" is an accurate one (133), describing the margin of the lesion where the interactions are occurring between tree and microorganism. Shain (133, 134) has given details on the events occurring in the reaction zone.

The pioneer invaders in some cases—such as root rots—are hymenomycetes (150, 169, 173). Bacteria and yeasts are intimately associated with some of the pioneer nonhymenomycetous filamentous fungi (101, 140, 143). The bacteria may be aerobes (24), facultative anaerobes (24), or obligate anaerobes (145, 157). The ray parenchyma cells are usually the first to be invaded by pioneer invaders. These cells contain a high concentration of nutrients, and they are also the cells that show the first signs of darkening as their contents break away from the cell wall and begin to degenerate (138, 160). Some microorganisms produce polyphenol oxidases (150) that increase oxidation of the phenols (90, 150), leading to further darkening of the tissues.

As the tissues die and discolor in Stage II, pH increases, minerals accumulate, and moisture increases (50, 54, 142). Although these changes are accompanied by intensification of discoloration in most species, the changes in some species occur without a color change, a condition commonly called "wetwood" (32, 59, 94, 168, 174). Bacteria are commonly associated with wetwood (16, 59).

DECAY PROCESSES   The processes that result in decay follow discoloration (4, 140, 141, 143), but there are great differences between discoloration processes and decay processes, especially in rate. A large column of discoloration does not indicate that a large column of decay will follow (64, 142). The rate of decay depends on the aggressiveness of microorganisms.

The columns or lesions of discoloration and decay advance most rapidly above and below the wound, but towards the center of the tree in species of *Acer* that have no colored core of age-altered wood. In species of *Quercus* that have a colored core of age-altered wood, the columns of discoloration and decay advance most rapidly above and below the wound along the sap-wood-heartwood boundary that was present at the time of wounding (146). When severe wounds occur, however, the entire column of wood present at the time of wounding may discolor and decay (64).

SUCCESSION OF MICROORGANISMS   Microorganisms that inhabit wood in living trees have the greatest survival advantage when they attack wounds in a sequential manner. Each invading microorganism exerts its specific force against the dynamic protective barriers formed by the wounded tree. The pioneer microorganisms first alter the substrate to their advantage and then digest the cell contents. As the pioneers advance, the substrate is altered further to the advantage of other organisms that follow—a *succession* (5, 6, 31, 34, 51, 97, 101, 107, 140, 163).

A good account of microorganisms associated with heartwood in *Thuja plicata* was given by Eades & Alexander (28) and by Findlay & Pettifor (39). Dark and light heartwood occur in this species; the dark heartwood contained nondecay fungi, but no organisms were found in the light heartwood. Findlay & Pettifor concluded that the fungi were responsible for the dark heartwood and its reduced strength and specific gravity. The toxicity of dark heartwood is also low (48). Consequently it would be more correctly defined as discolored wood. The hyphae they observed in the cells was shown by Roff (126) to be due to nonhymenomycetous fungi, and thus the conclusion of Findlay & Pettifor was supported that the fungus in the "dark heartwood" probably was not a decay fungus.

Findlay & Pettifor also reported (39) results from laboratory tests showing that test blocks of dark heartwood were susceptible to attack by certain hymenomycetous fungi such as *Coniophora cerebella*, whereas it was with great difficulty that fungi could be induced to grow at all over the light-colored heartwood. The results indicated that the pioneer nonhymenomycetous fungi altered the wood to the advantage of *C. cerebella*. Similar results with *Acer saccharum* suggested that the pioneer microorganisms attack wood altered as a result of host response to injury, and the alterations are to the advantage of hymenomycetes (142). Tissues in Stage I may have evolved as effective deterrents to invasion by hymenomycetes. The bacteria and nonhymenomycetous fungi then probably adapted to the new substrate.

COMPARTMENTALIZATION   As resistant as the tissues may be in Stage I, under certain conditions some microorganisms are able to surmount the chemical protective barriers and invade. At this time the tree forms a second line of defense and restricts the path of the invaders.

One of the first mechanical barriers to form in tissues after wounding is plugged vessels (124, 138). In those species capable of doing so, tyloses are formed; and in the other species the vessels are plugged with a gummy material (138, 160). These plugs begin to form in *Acer rubrum* a few days after wounding (124). The most dramatic anatomical response to wounding occurs when the uninjured cambium around the wound begins to form cells that are different from those formed normally (138, 147). The severity of the wound determines the extent of the reaction. When wounds occur during the time when the cambium is not forming cells, the barrier wall begins to form as soon as cambial activity resumes (147).

After injury, the active living cambium, even well away from the wound, forms cells with thicker walls (138, 147). These cells in *Acer* are similar to late wood (138). The ray cells have thicker walls, and the cells are more rounded. Vessel production is retarded (147). These and other changes account for a barrier wall separating tissues formed after wounding from those present at the time of wounding. As subsequent tissues and annual growth layers form, they remain separated from the compartmentalized injured tissues. The living ray parenchyma acts as a barrier wall tangential to the sides of the injury.

These changes also occur in tropical species that have no distinctive growth rings. A different behavior is shown by eucalypts (152), in which the cambium shortly after injury in the growing season goes from the induction of multiple cambial divisions to the production of anomalous parenchyma, which give rise to roundish patches of thin-walled cells that develop into lacunae with layers of tissue. Eventually the cambium resumes normal xylem formation, but the tissues in the lacunae break down, and the region is filled with polyphenols ("kino"). In most eucalypts kino has a markedly different composition from xylem polyphenols (68, 70).

When the invading microorganisms spread, they do so along the path of least resistance, vertically through the compartmentalized tissues (143, 146). Apparently this is the weakest link to the compartment. The rims or margins of the compartments remain intact throughout the life of the tree. At the portion of the column distal to the wounds—the reaction zone—the pioneer fungi are still present in most cases (133, 142, 144).

When a tree is wounded another time, another barrier wall or compartment begins to form and to envelop the inner compartments (143, 146). This process is similar to having a pipe slide over other smaller pipes. The wood between the rim of the last compartment and the new compartment includes the tissues invaded by microorganisms. The degree of tissue alteration or the rate of invasion of each new compartment may be different, and different microorganisms may be involved (140, 142). This explains the ring rots commonly found in trees; they are compartments between the barrier walls (140, 142). This also explains the darker streaks of tissues within altered heartwood.

The barrier wall formed by the cambium after wounding functions effectively most of the time in compartmentalizing the injury and microorganisms. The wall acts as a partition between wood formed before and after wounding. The wall is the major site of partitioning of growth rings, termed "shake," which occurs when other pressures are exerted, such as drying processes, growth of microorganisms (112, 147, 171), etc. Shakes are associated with wounds, but not all wounds form shakes (147).

If part of the cambium of eucalypts and some other trees containing heartwood is killed, the region between that part and the heartwood remains as "included sapwood" when the adjacent sapwood is converted to heartwood during subsequent growth of the tree (68). These areas can subsequently discolor and decay. Included sapwood is frequently seen in trees that have been scarred by fire (146).

Differences Between Heartwood and Discolored Wood, and Means of Recognition

When living sapwood cells encounter different stimuli, they usually respond differently when forming colored wood. The latter can be heartwood when the oldest tissues are affected first, or different types of discolored wood when often the youngest tissues are affected first. The differences between sapwood-heartwood and sapwood-discolored wood are mainly in the amount and distribution of inorganic elements, the pH, the amount and composition of extractives, and moisture content.

EXTRACTIVES Normal heartwood has a similar color throughout the cross-section of a log, and a chemical composition that is in almost all cases constant for a particular species. In injured and discolored wood, the amount of extractives is higher than in the sapwood, amorphous deposits of melanistic substances are more abundant than in heartwood (55), and the extractable materials in these tissues frequently differ qualitatively from each other (74, 76, 135). Discolored woods of the same species, and apparently resulting from the same cause, can contain different ratios of components (73).

Components in discolored wood can be different from those found in heartwood, as in Prunus species affected by Trametes versicolor (60) and Stereum purpureum (67). The cellular inclusions in histological examinations have been defined as tannins, deposits, etc.—in most cases without consideration of their variable composition, which can be different even a few cells apart (92). Aside from the confusion caused by the theories regarding their biogenesis, such loose terminology overlooks their difference in properties to invading organisms, either by presenting physical barriers such as gums or by forming toxic components. Certain types of wounds in certain species show a stimulation of the synthesis and accumulation of materials inhibitory to decay fungi; in other species the discolored wood surrounding the wounds is no more decay-resistant than the sapwood of that particular species (57). The age of the sapwood that has given rise to discolored wood may also influence the ability of the latter to resist decay fungi. It should be noted, however, that the ability of the sapwood to form discolored wood containing toxic components may be linked with an ability to produce heartwood resistant to decay fungi (57).

When heartwood is wounded, it can discolor, although in dark heartwood the changes are difficult to see. Kondo (89) pointed out that heartwood contains enzyme systems that can function after injury, and oxidase enzymes have been reported in the heartwood of *Pinus* spp. (137). The brown stain of hemlock heartwood can be controlled with an enzyme inhibitor (35). Discolored heartwood has been associated also with galleries of insects (155) and with microorganisms (175, 176).

The pioneer invaders of wood—bacteria and nonhymenomycetous fungi—are often ignored in studies of heartwood. The heartwood altered by microorganisms is then considered—on the basis of color—to be another type of heartwood.

When heartwood is wounded, microorganisms can invade (142, 146); but when discolored tissues in Stage I that have resulted from host response to wounding are wounded again, even when they are in heartwood, they seldom discolor further (142). This indicates that tissues in Stage I are more a protective tissue than unaltered heartwood.

Much useful information has been gained from in vitro tests about the relative durability of sapwood, discolored wood, and heartwood from the same tree, and between different trees and species. Some of these conclusions may need to be modified when more becomes known about the complexities of function of specific substrate molecules and their relationship to gene and enzyme activation. These aspects are not usually considered in agar plate and wood-block tests.

INORGANIC ELEMENTS   The main mineral constituents of wood are salts of calcium, potassium, and magnesium, but many other elements are present in minor amounts. The acid radicals are carbonates, phosphates, silicates, sulfates, and oxalates—and probably the acidic groups of components of the cell wall. The inorganic material occurs scattered throughout the cell wall or as accumulations in the form of crystals, either deposits in the cells or intracellular large lumps.

The relative amounts of ash in the sapwood and heartwood show great variation between species, sometimes even in the same species. Usually the amount of ash is higher in the sapwood than in the heartwood, but the amount of certain elements is selectively different. During the transformation of the sapwood of *Robinia pseudoacacia* and *Maclura pomifera* to heartwood, the ash fell about 30%, but the amount of phosphorus dropped about 95%, and calcium 24% and 41% respectively, but the other elements remained constant (55). Magnesium and manganese tended to be concentrated in the heartwood of pine, unlike calcium which steadily decreased with the succeeding growth rings (41).

An increase in inorganic components in discolored tissues of sapwood has been reported by many workers. These accumulations are commonly calcium carbonate, but other components can be present. The significant aspect of discolored wood is that the amount of inorganic material is higher than that in the surrounding tissue. Discolored wood of sugar maple had 6 times more ash than normal sapwood; however, there was a 9-fold increase in calcium but a 56% reduction in potassium (E. L. Ellis, unpublished data). There can also be an increase in manganese with this species (148). The differences in these three aspects in shagbark hickory are much greater (29).

The situation can be more complex. Hart (55) found in the discolored wood of *Robinia pseudoacacia* a 136% increase in ash as compared with normal sapwood, when potassium increased 61%, calcium 100%, and magnesium 168%, but phosphorus decreased 35%. A similar pattern was observed with *Maclura pomifera*. The minerals in the stained wood of *Acer saccharum* are not removed by water, and this behavior may be due to combination with the polyphenols (N. Levitin, unpublished). It is interesting that, whereas the accumulation of calcium salts in aging cells of plants is well documented (167), the increase observed in discolored wood has not so far been observed in heartwood.

The discolored woods of *Quercus alba, Maclura pomifera, Robinia pseudoacacia, Juglans nigra,* and *Acer saccharum* had higher pH, moisture, and ash contents than uninjured sapwood (56). The deeper the stain of discolored wood of *A. saccharum*, the higher the pH and mineral content (50; N. Levitin, unpublished). As the wound that resulted in the surrounding discolored wood of sugar maple is approached, there is a general increase in moisture (144), pH, and ash (144, 163, 164). The pH of the discolored wood of *Picea abies* (134) and *Quercus rubra* (130, 142, 171) is also higher near the wound. The pH of discolored wood of many species is above 6, even as high as 9 (50, 144), whereas that of sapwood and, in particular, heartwood in the same stem is below the pH of discolored wood, usually below pH 5.5.

Valuable use has recently been made of the content of inorganic materials and of developments in electrical techniques to detect discoloration and decay in living trees. The technique enables a quantitative and objective assessment of discolored wood and heartwood.

The changes in concentration of ions in wood are in direct relation to resistance to a pulsed electric current (153, 165). As tissues die, discolor, and decay, the resistance to a pulsed electric current decreases, as long as the moisture content of the wood being measured remains above the fiber saturation point (165). The decrease in resistance is related to the increase in mobile ions (such as potassium) in dying tissues, leading to discolored wood (165). Resistance to a pulsed electric current throughout unaltered and uninfected heartwood in *Quercus* spp. is higher than that of sapwood (165). The electrical apparatus indicates the differences between age-altered high-resistance (12–60 thousand ohms) tissues and injury-altered low-resistance (1–20 thousand ohms) tissues. The measurements indicate accurately the degree of degradation of injury-altered tissues (153, 165) in living trees. With this method, numbers can be put on the model system to show that, as the tissues go from Stages I to II to III, the resistance to a pulsed electric current decreases steadily.

Commercial Importance of Distinguishing Between Heartwood and Discolored Wood

The need for understanding processes associated with color changes in wood has increased in the last decade; it will increase much more in the future. The demand for high-quality wood—indeed all wood—is increasing throughout the world.

Properties associated with change in color are important economically because they affect wood quality and utilization. Darkened age-altered wood often yields high-value products. The adoption of multiple-use forestry practices will lead to a higher incidence of injury in trees, but in this case the darkened wood resulting from injury, and inhabited by microorganisms, has little or no value for quality products.

Techniques are required (such as the above electrical technique) that will make possible the ready detection and quantification of the biggest loss of wood in the forests of today and of the future, and ultimately to assist in control of such losses. Investigations into the formation of colored wood—heartwood or discolored wood—must be made so that the maximum amount of the most suitable wood can be obtained to meet the world's expanding needs.

Literature Cited

1. Abeles, F. B., Leather, G. R. 1971. Control of cellulase secretion by ethylene. *Planta* 97: 87–91
2. Abraham, P. D., Wycherley, P. R., Pakianathan, S. W. 1968. Stimulation of latex flow in *Hevea brasiliensis* by 4-amino-3,5,6-trichloropicolinic acid and 2-chloroethane-phosphoric acid. *J. Rubber Res. Inst. Malaya* 20:291–305
3. Anderson, A. B., Scheffer, T. C., Duncan, C. G. 1963. The chemistry of decay resistance and its decrease with heartwood aging in incense cedar (*Libocedrus decurrens*). *Holzforschung* 17:1–5
4. Bakshi, B. K., Singh, S. 1970. Heart rots in trees. *Int. Rev. Forest. Res.* 3:197–251
5. Basham, J. T. 1958. Decay of trembling aspen. *Can. J. Bot.* 36: 491–505
6. Basham, J. T. 1966. Heart rot of jack pine in Ontario. *Can. J. Bot.* 44:275–95
7. Bauch, J., Liese, W., Scholz, F. 1968. Über die Entwicklung und staffliche Zusammensetzung der Hoftüpfelmembranen von Längatracheiden im Coniferen. *Holzforschung* 22:144–53
8. Berlin, J., Barz, W. 1971. Metabolism of isoflavones and coumestanes in cell and callus suspension cultures of *Phaseolus aureus*. *Planta* 98:300–14
9. Berlyn, G. P. 1969. Microspectrophotometric investigation of free space in plant cell walls. *Am. J. Bot.* 56:498–506
10. Bland, D. E., Hillis, W. E. 1969. Microspectrophotometric investigation of lignin and polyphenol distribution in wood sections. *Appita* 23:204–10
11. Brooks, F. T., Moore, W. C. 1926. Silver leaf disease—V. *J. Pomol. Hort. Sci.* 5:61–97
12. Burg, S. P., Burg, E. A. 1965. Ethylene action and the ripening of fruits. *Science* 148:1190–96
13. Burg, S. P., Burg, E. A. 1967. Molecular requirements for the biological activity of ethylene. *Plant Physiol.* 42:144–52
14. Burg, S. P. 1968. Ethylene, plant senescence and abscission. *Plant Physiol.* 43:1503–11
15. Carrodus, B. B. 1970. Carbon dioxide and the formation of heartwood. *New Phytol.* 70:939–43
16. Carter, J. C. 1945. Wetwood of elms. *Ill. Nat. Hist. Surv. Bull.* 23:407–48
17. Chalutz, E., De Vay, J. E., Maxie, E. C. 1969. Ethylene-induced isocoumarin formation in carrot root tissue. *Plant Physiol.* 44:235–41
18. Chattaway, M. M. 1949. The development of tyloses and secretion of gum in heartwood formation. *Aust. J. Biol. Sci.* 2B:227–40
19. Chattaway, M. M. 1952. The sapwood-heartwood transition. *Aust. Forest.* 16:25–34
20. Chow, S.-Z. 1972. Hydroxyl accessibility, moisture content, and biochemical activity in cell walls of Douglas-fir trees. *Tappi* 55: 539–44
21. Chudnoff, M. 1971. Tissue regeneration of debarked eucalypts. *Forest Sci.* 17:300–05
22. Clark, J., Gibbs, R. D. 1957. Studies in tree physiology. Part IV. *Can. J. Bot.* 35:219–53
23. Constabel, F. 1968. Gerbstoffproduktion der Calluskulturen von *Juniperus communis*. *Planta* 79: 58–64
24. Cosenza, B. J., McCreary, M., Buck, J. D., Shigo, A. L. 1970. Bacteria associated with discolored and decayed tissues in beech, birch and maple. *Phytopathology* 60:1547–51
25. Dadswell, H. E., Hillis, W. E. 1962. Wood. In: *Wood extractives and their significance to the pulp and paper industry.* Ed. W. E. Hillis 3–55 New York: Academic 513 p.
26. Davies, M. E. 1972. Effects of auxin on polyphenol accumulation and the development of phenylalanine ammonia-lyase activity in darkgrown suspension cultures of Paul's scarlet rose. *Planta* 104: 66–77
27. Dietrichs, H. H. 1964. Chemischphysiologische Untersuchungen über die Split-Kern-Umwandlung

der Rotbuche (*Fagus sylvatica*). *Bundesforsch. Anst. für Forst u. Holzwirtschaft. Mitt.* 58:1–141. Reinbek

28. Eades, H. W., Alexander, J. B. 1934. Western red cedar: significance of its heartwood colorations. *Can. Dep. Int. Forest Serv. Circ. 41.* 15 p.

29. Ellis, E. L. 1965. Inorganic elements in wood. In: *Cellular Ultrastructure of Woody Plants* 181–89. Ed. W. A. Côté, Jr. Syracuse Univ. Press. 603 p.

30. Erdtman, H., Rennerfelt, E. 1944. Der Gerhalt des Kiefernkernholzes an Pinosylvin-Phenolen. Ihre quantitative Bestimmung und ihre hemmende Wirkung gegen Angriff verschiedener Fäulpilze. *Svensk Papperstid.* 47:45–56

31. Etheridge, D. E. 1961. Factors affecting branch infection in aspen. *Can. J. Bot.* 39:799–816

32. Etheridge, D. E., Morin, L. A. 1962. Wetwood formation in balsam fir. *Can. J. Bot.* 40:1335–45

33. Etheridge, D. E. 1969. Factors affecting infection of balsam fir (*Abies balsamea*) by *Stereum sanguinolentum* in Quebec. *Can. J. Bot.* 47:457–79

34. Etheridge, D. E. 1970. *Ascocoryne sarcoides* (Jacq. ex Gray) Groves and Wilson and its association with decay of conifers. In: *Interaction of Organisms in the Process of Decay of Forest Trees.* Univ. Laval Bull. 13:19–26. Quebec

35. Evans, R. S., Halvorson, H. N. 1962. Cause and control of brown stain in western hemlock. *Forest Prod. J.* 12:367–73

36. Fahn, A., Arnon, N. 1963. The living wood fibres of *Tamarix aphylla* and the changes occurring in the transition of sapwood to heartwood. *New Phytol.* 62:99–104

37. Farkas, G. L., Király, Z. 1962. Role of phenolic compounds in the physiology of plant diseases and disease resistance. *Phytopathol. Z.* 44:105–50.

38. Fengel, D. 1970. Ultrastructural changes during aging of wood cells. *Wood Sci. Technol.* 4:176–88

39. Findlay, W. P. K., Pettifor, C. B. 1941. Dark coloration in western

red cedar in relation to certain mechanical properties. *Empire Forest. J.* 20:64–72

40. Forrest, G. I. 1969. Studies of the polyphenol metabolism of tissue cultures derived from the tea plant (*Camellia sinensis*). *Biochem. J.* 113:765–72

41. Fossum, T., Hartler, N., Libert, J. 1972. The inorganic content of wood. *Svensk Papperstid.* 75:305–09

42. Fraenkel, G. S. 1959. The raison d'etre of secondary plant substances. *Science* 129:1466–72

43. Frank, A. B. 1895. Die Krankheiten der Planzen. 2nd ed., 344 p. Verlag Trewendt, Breslav

44. Franklin, E. C., Taras, M. A., Volkman, D. A. 1970. Genetic gains in yields of oleoresin, wood extractives and tall oil. *Tappi* 53:2302–05

45. Frey-Wyssling, A., Bosshard, H. H. 1959. Cytology of the ray cells in sapwood and heartwood. *Holzforschung* 13:129–37

46. Fuchs, Y., Lieberman, M. 1968. Effects of kinetin, IAA and gibberellin on ethylene production and their interactions in growth of seedlings. *Plant Physiol.* 43:2029–36

47. Gagnon, C. 1967. Polyphenols and discoloration in the elm disease investigated by histochemical techniques. *Can. J. Bot.* 45:2119–24

48. Gardner, J. A. F. 1963. The chemistry and utilization of western red cedar. *Can. Forest. Dep. Publ.* 1023. 26 pp.

49. Ginns, J. H., Driver, C. H. 1970. The mycobiota of slash pine stumps and its influences on the occurrence of *Fomes annosus* root rot. In: *Interaction of Organisms in the Process of Decay of Forest Trees.* Univ. Laval Bull. 13:11–18. Quebec

50. Good, H. M., Murray, P. M., Dale, H. M. 1955. Studies on heartwood formation and staining in sugar maple *Acer saccharum* Marsh. *Can. J. Bot.* 33:31–41

51. Good, H. M., Nelson, J. I. 1962. Fungi associated with *Fomes igniarius* var. *populinus* in living poplar trees and their probable

significance in decay. *Can. J. Bot.* 40:615–24

52. Harris, J. M. 1954. Heartwood formation in *Pinus radiata*. *New Zealand Forest Serv. Forest Res. Inst. Tech. Pap. 1*

53. Harris, J. M. 1954. Heartwood formation in *Pinus radiata*. *New Phytol.* 53:517–24

54. Hart, J. H. 1965. Formation of discolored sapwood in three species of hardwoods. *Mich. Agr. Exp. Sta. Quart. Bull.* 48:101–16. East Lansing

55. Hart, J. H. 1968. Morphological and chemical differences between sapwood, discolored sapwood and heartwood in black locust and osage orange. *Forest Sci.* 24:334–38

56. Hart, J. H., Wardell, J. F., Johnson, K. C. 1969. Abstr. *XI Int. Bot. Congr. 11:* Seattle

57. Hart, J. H., Johnson, K. C. 1970. Production of decay-resistant sapwood in response to injury. *Wood Sci. Technol.* 4:267–72

58. Hart, J. H., Hillis, W. E. 1972. Inhibition of wood-rotting fungi by ellagitannins in the heartwood of *Quercus alba*. *Phytopathology* 62:620–26

59. Hartley, C., Davidson, R. W., Crandall, B. S. 1961. Wetwood, bacteria, and increased pH in trees. *U. S. Dept. Agr. Forest Serv. Forest Prod. Lab. Rept.* 2215. 34 p.

60. Hasegawa, M., Shirato, T. 1959. Abnormal constituents of *Prunus* wood. Isoolivil from *P. jamasakura* wood. *J. Jap. Forest. Soc.* 41:1–4

61. Hemingway, R. W., Hillis, W. E. 1970. Heartwood formation in living stumps of Douglas-fir. *Wood Sci. Technol.* 4:246–54

62. Hemingway, R. W., Hillis, W. E. 1971. Changes in fats and resins of *Pinus radiata* associated with heartwood formation. *Appita* 24:439–43

63. Hepting, G. H., Blaisdell, D. J. 1936. A protective zone in red gum fire scars. *Phytopathology* 26:62–67

64. Hepting, G. H., Shigo, A. L. 1972. Difference in decay rate following fire between oaks in North Caro-

lina and Maine. *Plant Dis. Reptr.* 56:406–07

65. Higuchi, T., Fukazawa, K., Shimada, M. 1967. Biochemical studies on the heartwood formation. *Hokkaido Univ. Coll. Exp. Forest. Res. Bull.* 25:167–94

66. Higuchi, T., Shimada, M., Watanabe, K. 1967. Studies on the mechanism of heartwood formation. Pt V. *J. Jap. Wood Res. Soc.* 13:269–73

67. Hillis, W. E., Swain, T. 1959. Phenolic constituents of *Prunus domestica*. III. *J. Sci. Food Agr.* 10:533–37

68. Hillis, W. E. 1962. The distribution and formation of polyphenols within the tree. In: *Wood Extractives and their Significance to the Pulp anl Paper Industry:* 59–131 Ed. W. E. Hillis, Academic: New York. 513 pp.

69. Hillis, W. E., Humphreys, F. R., Bamber, R. K., Carle, A. 1962. Factors influencing the formation of phloem and heartwood polyphenols. *Holzforschung* 16:114–21

70. Hillis, W. E. 1964. The formation of polyphenols in trees. II. The polyphenols of *Eucalyptus sieberiana* kino. *Biochem. J.* 92:516–21

71. Hillis, W. E. 1966. Variation in polyphenol composition within species of *Eucalyptus*. *Phytochemistry* 5:541–56

72. Hillis, W. E., Inoue, T. 1966. The formation of polyphenols in trees. III. The effect of enzyme inhibitors. *Phytochemistry* 5:483–90

73. Hillis, W. E., Inoue, T. 1968. The formation of polyphenols in trees. IV. The polyphenols formed in *Pinus radiata* after *Sirex* attack. *Phytochemistry* 7:13–22

74. Hillis, W. E. 1968. Chemical aspects of heartwood formation. *Wood Sci. Technol.* 2:241–59

75. Hillis, W. E. 1969. The contribution of polyphenolic wood extractives to pulp colour. *Appita* 23:89–101

76. Hillis, W. E. 1971. Distribution, properties and formation of some wood extractives. *Wood Sci. Technol.* 5:272–89

77. Hirai, S. 1951. Study on the pro-

cess of heartwood growth in the Japanese Larch stem. *Jap. Forest. Soc. Trans.* 59:231–34

78. Höll, W. 1972. Stärke und Stärkeenzyme im Holz von *Robinia pseudoacacia* L. *Holzforschung* 26:41–45

79. Holm, R. E., Abeles, F. B. 1967. 2,4-Dichlorophenoxyacetic acid induced ethylene evolution: its role in soybean hypocotyl swellings. *Plant Physiol.* 42:30–31

80. Hugentobler, U. H. 1965. Zur Cytologie der Kernholzbildung. *Naturforsch. Gesell. Zurich Vierteljahrsschrift* 110:321–42

81. Hulme, A. C., Rhodes, M. J. C., Wooltorton, L. S. C. 1971. The effect of ethylene on the respiration, ethylene production, RNA and protein synthesis for apples stored in low oxygen and in air. *Phytochemistry* 10:1315–23

82. Hyodo, H., Yang, S. F. 1971. Ethylene enhanced synthesis of phenylalanine ammonia-lyase in pea seedlings. *Plant Physiol.* 47:765–70

83. Imaseki, H. 1970. Induction of peroxidase activity by ethylene in sweet potato. *Plant Physiol.* 46:172–74

84. International Association of Wood Anatomists. 1957. International glossary of terms used in wood anatomy. *Tropical Woods* 107:1–35

85. Jensen, K. F. 1967. Measuring oxygen and carbon dioxide in red oak trees. *U. S. Dept. Agr. Forest Res. Note NE-74.* 4 pp.

86. Jones, R. L. 1968. Ethylene enhanced release of α-amylase from barley aleurone cells. *Plant Physiol.* 43:442–44

87. Jorgensen, E. 1962. Observations on the formation of protection wood. *Forest. Chron.* 38:292–94

88. Jorgensen, E., Balsillie, D. 1969. Formation of heartwood phenols in callus tissue cultures of red pine (*Pinus resinosa*). *Can. J. Bot.* 47:1015–16

89. Kondo, T. 1964. On the wood enzyme. *J. Jap. Wood Res. Soc.* 10:43–48

90. Kosuge, T. 1969. The role of phenolics in host response to infection. *Ann. Rev. Phytopathol.* 7:195–222

91. Krahmer, R. L., Côté, W. A. 1963. Changes in coniferous wood cells associated with heartwood formation. *Tappi* 46:42–49

92. Krahmer, R. L., Hemingway, R. W., Hillis, W. E. 1970. The cellular distribution of lignans in *Tsuga heterophylla* wood. *Wood Sci. Technol.* 4:122–39

93. Kumar, S. 1971. Causes of natural durability of timber. *J. Timber Develop. Assoc. India* 18:1–21

94. Lagerberg, T. 1935. Barrträdens vattved. Vattvedens Natur praktiska Betydelse. *Svenska Skogsvårdsför. Tidskr.* 33:177–264

95. Lairand, D. B. 1963. Contribution to the cytochemistry of wood. *Drevarsky Vyskum* 1:1–11

96. La Rue, T. A. G., Gamborg, D. L. 1971. Ethylene production in plant cell cultures, variations in production during growing cycle and in different plant species. *Plant Physiol.* 48:394–98

97. Lavallee, A. 1970. Observations in inoculations of hardwood species with *Pholiota aurivella* (Batsch ex. Fr.) Kummer. In: *Interaction of Organisms in the Process of Decay of Forest Trees. Univ. Laval Bull.* 13:27–37. Quebec

98. Lorenz, R. C. 1944. Discolorations and decay resulting from increment borings in hardwoods. *J. Forest.* 42:37–43

99. Lyr, H. 1962. Enzymatische Detoxifikation der Kernholztoxine. *Flora* 152:570–79

100. Lyr, H. 1967. Über den jahreszeitlichen Verlauf der Schutzkernbildung bei *Pinus sylvestris* nach Verwandungen. *Archiv Forstwesen* 16:51–57

101. Maloy, O. C., Robinson, V. S. 1968. Microorganisms associated with heart rot in young grand fir. *Can. J. Bot.* 46:306–09

102. Mapson, L. W. 1970. Biosynthesis of ethylene and the ripening of fruit. *Endeavour* 29:29–33

103. Mapson, L. W., Hulme, A. C. 1970. The biosynthesis, physiological effects and mode of action of ethylene. In: *Progress in Phytochemistry* 2:343–84. Ed. L. Reinhold, Y. Linschitz

104. Marei, N., Romani, R. 1971. Ethylene-stimulated synthesis of ribosomes, ribonucleic acid and

protein in developing fig fruits. *Plant Physiol.* 48:806–08

105. Matsukuma, N., Kawano, H., Shibata, Y., Kondo, T. 1965. Studies on the intermediate zone of sugi wood, some physiological activities of the xylem tissue. *J. Jap. Wood Res. Soc.* 11:227–31

106. Merrill, W., Cowling, E. B. 1966. Role of nitrogen in wood deterioration: Amounts and distribution of nitrogen in tree stems. *Can. J. Bot.* 44:1555–80

107. Merrill, W. 1970. Spore germination and host penetration by heartrotting hymenomycetes. *Ann. Rev. Phytopathol.* 8:281–300

108. Miller, C. O. 1969. Control of deoxyisoflavone synthesis in soy bean. *Planta* 87:26–35

109. Mutton, D. B. 1962. Wood resin. In: *Wood Extractives and their Significance to the Pulp and Paper Industry:* 337–63. Ed. W. E. Hillis, Academic: New York. 513 pp.

110. MacDougal, D. T. 1927. *Composition of gases in trunks of trees.* Carnegie Inst. Wash. 26:162–63

111. MacDougal, D. T., Overton, J. B., Smith, G. M. 1929. *The hydrostatic-pneumatic system of certain trees.* Carnegie Inst. Wash. 1–98

112. McGinnes, E. A., Jr., Chang, C. I. J., Wu, K. Y. T. 1971. Ringshake in some hardwood species: The individual tree approach. *J. Polymer Sci.* 36:153–76

113. MacKenzie, I. A., Street, H. E. 1970. Studies on the growth in culture of plant cells. VIII. The production of ethylene by suspension cultures of *Acer pseudoplatanus. J. Exp. Bot.* 21:824–34

114. McMichael, B. I., Jordan, W. R., Powell, R. D. 1972. An effect of water stress on ethylene production by intact cotton petioles. *Plant Physiol.* 49:658–60

115. Nečesaný, V. 1966. Die Vitälitarsveranderung der Parenchymzellen als physiologische Grundlage der Kernholzbildung. *Holzforschung Holzverwert.* 18:61–65

116. Neely, D. 1970. Healing of wounds on trees. *J. Am. Hort. Sci.* 95:536–40

117. Nelson, N. D., Maeglin, R. R., Wahlgren, H. E. 1969. Relationship of black walnut wood color

to soil properties and site. *Wood & Fiber* 1:29–37

118. Nobuchi, T., Harada, H. 1968. Electron microscopy of the cytological structure of the ray parenchyma cells associated with heartwood formation of sugi (*Cryptomeria japonica*). *Jap. Wood Res. Soc.* 14:197–202

119. Pratt, H. K. Goeschl, J. D. 1969. Physiological roles of ethylene in plants. *Ann. Rev. Plant Physiol.* 20:541–84

120. Pringle, R. B., Scheffer, R. P. 1964. Host-specific plant toxins. *Ann. Rev. Phytopathol.* 2:133–56

121. Reid, M. S., Pratt, H. K. 1972. Effects of ethylene on potato tuber respiration. *Plant Physiol.* 49:252–55

122. Rhodes, M. J. C., Wooltorton, L. S. C. 1971. The effect of ethylene on the respiration and on the effect of phenylalanine ammonia lyase in swede and parsnip root tissue. *Phytochemistry* 10:1989–97

123. Ridge, I., Osborne, D. J. 1970. Regulation of peroxidase activity by ethylene in *Pisum sativum. J. Exp. Bot.* 21:720–24

124. Rier, J. P., Shigo, A. L. 1972. Some changes in red maple, *Acer rubrum,* tissues within 34 days after wounding in July. *Can. J. Bot.* 50:1783–84

125. Rishbeth, J. 1951. Observations on the biology of *Fomes annosus,* with particular reference to East Anglian pine plantations. *Ann. Bot.* 15:1–21

126. Roff, J. W. 1964. Hyphal characteristics of certain fungi in wood. *Mycologia* 56:799–804

127. Rubin, B. A., Artsikhovskaya, E. V. 1964. Biochemistry of pathological darkening of plant tissues. *Ann. Rev. Phytopathol.* 2:159–78

128. Rudman, P. 1965. The causes of variations in the natural durability of wood: Inherent factors and ageing and their effects on resistance to biological attack. *Holz u. Organismen* 1:151–62 Ed. G. Becker, W. Liese. Duncker & Humblot: Berlin

129. Rudman, P. 1966. Heartwood formation in trees, *Nature* (London) 210:608–10

130. Sachs, I. B., Ward, J. C., Bulgrin, E. H. 1966. Heartwood stain in red oak. *Holz Roh Werkst.* 24: 489–97

131. Sakai, S., Imaseki, H. 1971. Auxin-induced ethylene production by mungbean hypocotyl segments. *Plant & Cell Physiol.* 12: 439–59

132. Scheffer, T. C., Cowling, E. B. 1966. Natural resistance of wood to microbial deterioration. *Ann. Rev. Phytopathol.* 4:147–70

133. Shain, L. 1967. Resistance of sapwood in stems of loblolly pine to infection by *Fomes annosus*. *Phytopathology* 57:1034–45

134. Shain, L. 1971. The response of sapwood of Norway spruce to infection by *Fomes annosus*. *Phytopathology* 61:301–07

135. Shain, L., Hillis, W. E. 1971. Phenolic extractives in Norway spruce and their effects on *Fomes annosus*. *Phytopathology* 61:841–45

136. Shain, L., Hillis, W. E. 1972. Ethylene production in *Pinus radiata* in response to *Sirex-Amylostereum* attack. *Phytopathology* 62: 1407–09

137. Shain, L., Mackay, J. F. G. 1973. Phenol-oxidizing enzymes in the heartwood of *Pinus radiata*. *Forest Sci.* (In press)

138. Sharon, E. M. 1973. Some histological features of *Acer saccharum* wood formed after wounding. *Can. J. Forest Res.* 3 (In press)

139. Shigo, A. L. 1965. The pattern of decays and discolorations in northern hardwoods. *Phytopathology* 55:648–52

140. Shigo, A. L. 1967. Successions of organisms in discoloration and decay of wood. *Int. Rev. Forest. Res.* 2:237–99

141. Shigo, A. L. 1967. The early stages of discoloration and decay in living hardwoods in northeastern United States: A consideration of wound-initiated discoloration and heartwood. *IUFRO Congr. Proc.* 9:117–33. Munich

142. Shigo, A. L., Sharon, E. M. 1968. Discoloration and decay in hardwoods following inoculations with Hymenomycetes. *Phytopathology* 58:1493–98

143. Shigo, A. L., Larson, E. vH. 1969. A photo guide to the patterns of discoloration and decay in living northern hardwood trees. *U. S. Dept. Agr. Forest Serv. Res. Paper* NE-127, 100 pp.

144. Shigo, A. L., Sharon, E. M. 1970. Mapping columns of discolored and decayed tissue in sugar maple, *Acer saccharum*. *Phytopathology* 60:232–37

145. Shigo, A. L., Stankewich, J., Consenza, B. J. 1971. *Clostridium* sp. associated with discolored tissues in living oaks. *Phytopathology* 61: 122–23

146. Shigo, A. L. 1972. Successions of microorganisms and patterns of discoloration and decay after wounding in red oak and white oak. *Phytopathology* 62:256–59

147. Shigo, A. L. 1972. Ring and ray shakes associated with wounds in trees. *Holzforschung* 26:60–62

148. Shortle, W. C. 1970. Concentration of manganese in discolored and decayed wood of sugar maple *Acer saccharum* Marsh. *Phytopathology* 60:578

149. Shortle, W. C., Tattar, T. A., Rich, A. E. 1971. Effects of some phenolic compounds on the growth of *Phialophora melinii* and *Fomes connatus*. *Phytopathology* 61:552–55

150. Siegle, H. 1967. Microbiological and biochemical aspects of heartwood stain in *Betula papyrifera* Marsh. *Can. J. Bot.* 45:147–54

151. Singh, S., Tewari, R. K. 1970. Role of a precursor fungus in decay in standing teak. *Indian Forest.* 96:874–75

152. Skene, D. S. 1965. The development of kino veins in *Eucalyptus obliqua* L'Herit. *Aust. J. Bot.* 13: 367–78

153. Skutt, H. R., Shigo, A. L., Lessard, R. A. 1972. Detection of discolored and decayed wood in living trees using a pulsed electric current. *Can. J. Forest Res.* 2:54–56

154. Smith, J. H. G., Walters, J., Wellwood, R. W. 1966. Variation in sapwood thickness of Douglas-fir in relation to tree and section characteristics. *Forest Sci.* 12:97–103

155. Solomon, J. D., Toole, E. R. 1971.

Stain and decay around carpenter-worm galleries in southern hardwood trees. *U. S. Dept. Agr. Forest Serv. S. Forest Exp. Sta. Res. Note.* 4 pp.

156. Stahmann, M. A., Clare, B. G., Woodbury, W. 1966. Increased disease resistance and enzyme activity induced by ethylene production by black rot infected sweet potato tissue. *Plant Physiol.* 41: 1505–12

157. Stankewich, J. P., Cosenza, B. J., Shigo, A. L. 1971. *Clostridium quercicolum* sp. n. isolated from discolored tissues in living oak trees. *Anton. van Leeuwenhoek* 37:299–302

158. Stewart, C. M. 1966. Excretion and heartwood formation in living trees. *Science* 153:1068–74

159. Stutz, R. E. 1959. Control of brown stain in sugar pine with sodium azide. *Forest Prod. J.* 9: 459–63

160. Sucoff, E., Ratsch, H. Hook, D. 1967. Early development of wound-initiated discoloration in *Populus tremuloides* Michx. *Can. J. Bot.* 45:649–56

161. Swarbrick, T. 1926. The healing of wounds in woody stems. *J. Pomol. Hort. Sci.* 5:98–114

162. Tarkow, H., Krueger, J. 1961. Distribution of hot-water soluble material in cell walls and cavities of redwood. *Forest Prod. J.* 11: 228–29

163. Tattar, T. A., Shortle, W. C., Rich, A. E. 1971. Sequence of microorganisms and changes in constituents associated with discoloration and decay of sugar maples infected with *Fomes connatus. Phytopathology* 61:556–58

164. Tattar, R. A., Rich, A. E. 1973. Extractable phenols in clear, discolored, and decayed wood tissues and bark of sugar maple. *Phytopathology* 63: (in press)

165. Tattar, T. A., Shigo, A. L., Chase, T. 1972. Relationship between the degree of resistance to pulsed electric current and wood in progressive stages of discoloration and decay in living trees. *Can. J. Forest. Res.* 2:236–43

166. Trendelenburg, R., Mayer-Wegelin, H. 1955. In: *Das Holz als Rohstoff:* 245–62. Hanser Verlag: Munich

167. Varner, J. E. 1961. Biochemistry of senescence. *Ann. Rev. Plant Physiol.* 12:245–64

168. Wallin, W. B. 1954. Wetwood in balsam poplar. *Minn. Forest Note* 28, 2 pp.

169. Wallis, G. W., Reynolds, G. 1965. The initiation and spread of *Poria weirii* root rot of Douglas fir. *Can. J. Bot.* 43:1–9

170. Wangaard, F. F., Granados, L. A. 1967. The effect of extractives on water-vapor sorption by wood. *Wood Sci. Technol.* 1:253–77

171. Ward, J. C., Hann, R. A., Baltes, R. C., Bulgrin, E. H. 1972. Honeycomb and ring failure in bacterially infected red oak lumber after kiln drying. *U. S. Dept. Agr. Forest Serv. Res. Paper FPL-165.* 36 pp.

172. Wardrop, A. B., Cronshaw, J. 1962. Formation of phenolic substances in the ray parenchyma of angiosperms. *Nature* (London) 193:90–92

173. Whitney, R. D. 1961. Root wounds and associated root rots of white spruce. *Forest. Chron.* 37:401–11

174. Wilcox, W. W. 1968. Some physical and mechanical properties of wetwood in white fir. *Forest Prod. J.* 18:27–31

175. Wilson, C. L. 1959. The Columbian timber beetle and associated fungi in white oak. *Forest Sci.* 5: 114–27

176. Wright, E. 1938. Further investigations of brown-staining fungi associated with engraver beetle (*Scolytus*) in white fir. *J. Agr. Res.* 57:759–73

177. Yazawa, K., Ishida, S., Miyajima, H. 1965. On the wet-heartwood of some broad-leaved trees grown in Japan. I. *J. Jap. Wood Res. Soc.* 11:71–76

178. Yazawa, K., Ishida, S. 1965. On the wet-heartwood of some broad-leaved trees grown in Japan. II. *J. Fac. Agr. Hokkaido Univ.* 54: 123–36

179. Yazawa, K., Ishida, S. 1965. On the existence of the intermediate wood in some broad-leaved trees

grown in Hokkaido, Japan. *J. Fac. Agr. Hokkaido Univ.* 54:137–50

180. Zelawski, W. 1960. Respiration intensity of oak wood in particular annual rings of sapwood. *Bull. Acad. Pol. Sci. Ser. Sci. Biol.* 8: 509–76

181. Ziegler, H. 1963. Storage, mobilization and distribution of reserve material in trees. In: *The Forma-tion of Wood in Forest Trees,* 303–20. Ed. M. H. Zimmerman. Academic: New York. 562 pp.

182. Ziegler, H. 1967. Biologische Aspekte der Kernholzbildung. *Holz Roh-Werkstoff* 26:61–68

183. Zycha, H. 1948. Über die Kernbildung und verwandte Vorgänge un Holz der Rotbuche. *Forstwiss. Cent.* 67:80–109

BIOLOGICAL ACTIVITIES OF VOLATILE FUNGAL METABOLITES

❖ 3572

S. A. Hutchinson

Department of Botany, The University, Glasgow, Scotland

The authors of the last two reviews (46, 62) of this topic have questioned whether it is worthwhile to base a discussion on "volatility" at all. Outside the cell the distinction between volatility and nonvolatility is only qualitative, in the magnitude of the vapor pressures in particular conditions; within the cell solutions the distinction serves no purpose. They agree, however, that in practice it has become a stimulating focus for thoughts and that this has led to useful new work and knowledge. Six attributes seem to affect this closely:

(*a*) The ability to approach an organism through the gas phase may be particularly relevant for that majority of fungi that develop part of their mycelium and their entire reproductive structure in air above wet or liquid substrates (46).

(*b*) Many lipophilic compounds produced directly into the air and almost insoluble in water may, even at some distance from a donor source, accumulate faster in the plasma membrane of an acceptor cell if the transfer takes place via the gas phase instead of via the liquid phase (46).

(*c*) During movement in the gas phase, metabolites will be exposed only to gaseous and physical inactivating or stimulating factors. Those diffusing in complex liquid solutions are likely to be exposed to more concentrated chemical inactivating or stimulating factors, and their movements will be limited by discontinuities in water films (46, 62).

(*d*) Because the substances are active as gases, they are likely to be relatively simple molecules; with modern methods their identification is probably less of a problem than their measurement and control; this leads to a common experimental approach, particularly for their chemistry (62).

(*e*) The very low concentrations in which some of the identified ones are active suggests a comparison with antibiotics, growth factors, and vitamins, all areas of knowledge in which a loosely defined concept has promoted inquiry and discovery (62).

(*f*) Perhaps the most important practical point; that their volatility leads to impermanence in particular situations, and to risks of escape from observation (62).

223

This discussion will develop around the themes of activities of known volatile primary metabolites, and those of known volatile secondary metabolites or of unidentified factors that appear to be volatile metabolites of some sort. This arrangement is convenient because of the origins of these two groups of substances. Primary metabolites, as used here, refers to those produced by and subsequently involved in the processes of normal energy exchange and synthesis of cell substance. As these processes are believed to follow the same basic pathways in all living matter, many of these primary metabolites are likely to be present in most living environments, to be produced by linked reactions, and to show patterns of activity that may support widely applicable generalizations. The term secondary metabolites is used here to refer to those produced by living cells that have no known function in subsequent metabolism. They may be produced by unusual changes in basic metabolism or by idiosyncratic processes; hence compared with primary metabolites they are less likely to be present in most living environments and less likely to be subject to widely applicable generalizations.

The two groups are not sharply separated; they are, however, well established and convenient concepts in many biological and chemical reports. This usage does not preclude more rigorous definition if a need for it develops, but in this discussion this does not seem to be justified. More significantly, any such rigor might put a misleading emphasis on an arbitrary discontinuity between ideas of these and of other biological interactions. It has been suggested that the need to divide a meal into chewable bites should not be so overemphasized that a cow comes to be regarded as a collection of cuts of meat (60). To extend this analogy, communication will be hindered rather than helped if an undisciplined wish for arbitrary precision makes it impossible to discuss the relative palatability of oxtail stew and grilled steak without rigorous preliminary definition of where the tail finishes and steak begins. In this discussion, for example, there is an obvious paradox in discussing the biological activity of "secondary metabolites," as any "secondary metabolite" that has a biological activity must affect metabolism and hence it is not a secondary metabolite at all! It seems best to leave such paradoxes to be resolved by common sense unless any reasonable possibility of significant confusion is shown to arise from them.

Similarly the term "volatile" is used here colloquially, to refer to any metabolite that is a gas, or has a high vapor pressure, in the usual conditions in which it is liberated from a cell. The adjectives "high" and "usual" are convenient empirical qualifications; in particular they will restrict any discussion of the many metabolites that might volatilize in substantial quantities if they were liberated from a cell in unusual conditions, and those that might affect cells if they were to accumulate in unusual concentrations.

Because most primary metabolites are probably products of metabolic pathways found in all living things, those produced by fungi are also likely to be formed by other living things in any shared environment. In fact, in this context, fungi share the inverse of the concern for Dr. Johnson's dog (16). The interest lies not in their ability to produce the compounds, but in their ability to liberate enough of any one or more of them to have a significant effect on the environment (121). Secondary metabolites may be unusual and intrinsically interesting, and they are more likely to contribute characteristic ecological properties to the producer, but again this depends on quantities and sensitivities.

It is therefore tempting to restrict the review to cases in which a living fungus has been shown to liberate a sufficient quantity of such a metabolite into a gas phase between itself and a sensitive organism to produce a biological effect. This could produce a misleading survey, however, as some interesting records are based on the correlation of observations of biological effects of culture gases with analysis of the potential activity of volatile substances identified in dead mycelial extracts. For example, gases from cultures of *Dipodascus aggregatus* have been found to be biologically active: 11 esters, 9 alcohols, 5 acids, and 3 carbonyl compounds have been identified in extracts from the mycelium. Many have stimulatory or inhibitory effects on other organisms in gas concentrations that may reasonably be expected to develop above the cultures. Hence, while the culture gases have not been analyzed quantitatively, it is likely that these compounds contribute to their effects (94, 95).

It is also appropriate to distinguish between the knowledge of volatile metabolites obtained from studies in controlled laboratory conditions, and that obtained from analysis of field conditions. It has been suggested that studies in controlled laboratory conditions are only examination of potentials. How much such potentials will be expressed in any ecosystem will depend on the balance that develops between production and removal of the metabolite and the sensitivity of the organisms concerned. And, even if the maximum potential is expressed, whether the factor it produces has any local significance will depend on the balances of the other limiting factors of the ecosystem (62).

The discussion of each metabolite or group of metabolites therefore covers:

(*a*) Records of the effects of each on fungi in controlled conditions, with emphasis on, but not complete restriction to, studies of gaseous concentrations known to be produced by fungal cells, and records of the effects of concentrations known to be produced by fungi on other organisms in controlled conditions.

(*b*) Records of the existence or likely existence of conditions that could support such activity in the less controlled environments in which fungi are likely to grow, and of known effects of compounds on the ecology of fungi in these conditions.

Within this framework carbon dioxide is discussed separately because of the disproportionately large amount of work that has been done on it. Other primary metabolites are then reviewed as a whole. The concept of "secondary" metabolites is based on the negative character of absence of any known part in subsequent normal metabolism of the producer. Hence there is no reason to expect them to be other than a heterogeneous group, and each is discussed separately.

PRIMARY METABOLITES

Carbon Dioxide

The enormous amount of work on carbon dioxide obviously relates to its almost ubiquitous formation. The most recent of several reviews of its effect on fungi (132) gives a broad and modern conspectus, and only representative references will therefore be given here. The total number of species discussed in this review is trivially small compared with the total number of fungi, and many effects are greatly modified by other environmental factors (21), by differences shown by different strains of a species (32, 90), or both. It is therefore dangerous to deduce generalizations at this stage; it seems likely, however, that at least the following patterns will be confirmed by more comprehensive tests:

(a) Carbon dioxide fixation by fungal hyphae through nonphotosynthetic processes has been established (41, 71, 111, 112). At low concentrations it commonly stimulates growth, and there are several claims that it is an essential metabolite for normal fungal development (25, 116). It seems unlikely that this hypothesis can be disproved, because of the very low concentrations that have been shown to affect growth in some cases, and the practical difficulty of studying living tissue that is not producing any respiratory CO_2.

(b) At higher concentrations CO_2 is commonly inhibitory. There may be some consistency in the order of sensitivity shown by different species, e.g., a big proportion of the fusaria that have been examined can tolerate high CO_2 pressures in their environment. A large proportion of the other species and genera that have been examined are visibly inhibited under about 10%–20% volume/volume of CO_2 in air (21, 84, 130). There may be great variation in sensitivity of individual cells in a population (42, 43), and between cells in different phases of an organism's life cycle (110).

(c) The gas may affect the type of growth as well as the amount; for example, the formation of sexual organs by members of the Blastocladiales (26), and of sporophores of Basidiomycetes (104), can be modified by changing the CO_2 of the environment. The effects on the induction of yeast-like growth by normally filamentous fungi seems particularly interesting (7).

(d) The effects may be greatly limited by the balance of other factors in the environment (5, 6, 90), and sensitivity may be greatest when temperature, nutrient status or both are near critical limits (21).

(e) Tolerance and reaction to CO_2 seem likely to be two of the common factors controlling sporulation, and contributing to the well established principle that conditions for sporulation of most species are more narrowly defined than those for vegetative growth (73, 82, 114, 115).

(*f*) The effects on spore germination and initial development of germ tubes fall within the same pattern of reactions as those on mycelial growth. Spore germination may be increased by the presence of low levels of CO_2, inhibited by higher ones (102), and at least some spores may be unable to germinate in CO_2-free environments (120). Effects of high concentrations are not generally lethal, spores exposed to inhibitory concentrations being capable of normal germination on return to environments with lower concentrations (73).

Other Primary Metabolites

The enormous literature on the production of ethanol, acetone, acetaldehyde, and other primary metabolites by yeasts can be traced easily in many textbooks and reviews (57), and it cannot reasonably be reviewed here. The biological effects of the concentrations of these substances in the gases above yeast cultures have been examined only rarely. In one recent study (49) the gases from cultures of *Saccharomyces cereviseae* were shown to have a variety of inhibitory effects on the growth and sporulation of cultures of *Aspergillus niger* and on the germination of seeds of *Lepidium sativum*. Acetaldehyde, ethyl acetate, ethanol, n-propanol, iso-butanol, and a mixture of isopentanols were identified in the gases. The effects could be reproduced by various mixtures of authentic samples of these identified constituents, in the concentration in which they were present in the tests with the cultures of *S. cereviseae*. Attention was drawn to the likelihood of slight changes in the environment having a big effect on the relative concentration of any constituent of the mixture of gases in the culture head spaces, and so to its contribution to any biological action of the total mixture.

Ethanol and acetaldehyde have been studied in other fungi more closely than any volatile primary metabolites other than CO_2. Ethanol is probably a common product from many species (40, 100, 125). Earlier examinations of its effects in controlled conditions were mostly incidental to searches for useful disinfectants and decontaminants; this resulted in study being concentrated on the immediately inhibitory effects of highly concentrated aqueous solutions in direct exposures of minutes or less, and in the general opinion that it and other alcohols are only weakly inhibitory to fungi compared with other common agents (88, 109).

The effects of the presence of substantially lower concentrations, presented either in solution or as a gas, have been found to be more varied. Rhizomorphs were not formed by cultures of *Armillaria mellea* in defined media with a variety of nitrogen and carbon sources unless an additional growth factor was present. This could be ethanol at 50 ppm, and optimal response was obtained with 500 ppm (140). This could explain the stimulus to rhizomorph formation associated with the presence of a volatile factor from cultures of *Aureobasidium pullulans* (99). Low concentrations of ethanol and of amyl alcohol can induce an endogenous rhythm in coremial formation in a species of *Penicillium* (37); in minute quantities it increases the efficiency of utilization of glucose in certain fungi, and in others it makes good growth possible with carbon sources that otherwise are utilized only with difficulty (46).

Many of these reports deal with amounts in media that might reasonably be expected to develop through transfer of ethanol vapor in the concentrations found in the head space gases of cultures. There are fewer direct tests of the effects of exposure to identified concentrations of the vapor in such culture gases, or to air mixtures made with authentic ethanol. The rates of increase of colony diameters of *Rhizopus nigricans, Thielaviopsis paradoxa, Gloeosporium musarum, Botrytis cinerea,* and *Trichoderma viride* were lower in atmospheres equilibrated with aqueous solutions of 1%–3% vol/vol ethanol than in normal air. The effects became visible very rapidly on exposure to the ethanol mixture and, once equilibrium was established, the lower growth rates characteristic of each concentration were constant for the duration of the experiments (134). The rates of elongation of cress seedlings, and the rates of increase of diameter of colonies of *Aspergillus niger,* were reduced by the presence of 1 mg ethanol per liter of air. This was the concentration found in culture gases of *Saccharomyces cereviseae,* which had been found to produce similar inhibition in the same conditions (49). Cultures of *Trichoderma harzianum* produced vapor concentrations equal to that given by volatilizing 0.1 ml to 0.5 ml ethanol in 1 liter of air. These culture gases, and mixtures with 0.5 ml authentic ethanol per liter of air, produce similar reduction in pigmentation of lettuce seedlings. Sporulation by colonies of *Pestalotia rhododendri* was completely inhibited by these culture gases and by atmospheres with 0.25 ml authentic ethanol per liter of air; sporulation of *Aspergillus niger* and rate of increase of colony diameters of *P. rhododendri* and *A. niger* were each inhibited by culture gases but they were apparently unaffected by this concentration of ethanol vapor. Spore germination of *A. niger* and of *P. rhododendri* was apparently unaffected by increased ethanol vapor concentration up to that produced by 0.5 ml volatilized in 1 liter of air (63). Some of the morphogenic effects of gases from fungal cultures on fern prothalli have been attributed to ethanol formation (61, 123). Ethanol was found to constitute 90% or more of the total organic content of most collections of volatile metabolites from cultures of *Fusarium oxysporum* that inhibited the germination of sporangiospores of *Rhizopus stolonifer;* from investigations with air mixed with the same concentrations of authentic ethanol it was concluded that this was not the sporostatic factor (115).

Records of production of other common volatile primary metabolites are overt or implicit in many studies of metabolic pathways, but there are relatively few direct demonstrations of biological effects by the amounts extruded from the mycelium. Gaseous acetaldehyde at 50–150 ppm in air progressively delayed germination of spores of *Rhizopus nigricans, Thielaviopsis paradoxa, Gloeosporium musarum, Botrytis cinerea,* and *Trichoderma viride;* when colonies were exposed to gas mixtures after germination, the growth rates were initially reduced, then increased to a similar maximum rate and limit for each concentration. This contrasts with the effects of alcohol discussed above. Acetone and chloroform had similar effects of those of alcohol; H_2S, HCN, and NH_3 had similar effects to those of acetaldehyde (134). In low concentrations acetaldehyde extracted from cultures of *Fusarium oxysporum* delayed the development of two-dimensional growth by fern prothalli, but this delay was not as great as that produced by similar extracts of the ethanol from the cultures (123). It was found to be the main factor controlling volatile sporostasis of sporangiospores of *Rhizopus stolonifer* by *Fusarium oxysporum*, though other possible products may also be implicated (114, 115). A wide range of compounds liberated in low gaseous concentration from solutions were found to stimulate sporulation of *Pestalotia rhododendri*, in the order alcohols > esters > acids > aldehydes > ketones. The production of many of these in cells of *Dipodascus aggregatus* was affected by nutrition of the culture (95). Isovaleric acid has been shown to be produced in cultures of *Agaricus bisporus* and to stimulate spore germination in a similar way to gases from such cultures (83).

This review cannot reasonably include references to every relevant record of all these ubiquitous substances; even if this were done the total number of species recorded would be only a small sample of the whole range of fungi. Within this sample great differences have been seen between the amounts of the substances produced by a fungus in different conditions, between different strains of a species, and between species and higher groups (32, 106). Hence while it can reasonably and usefully be deduced that such volatile metabolites may have biological effects not directly part of the main pathways of primary metabolism itself, there do not yet appear to be any other sufficiently clear patterns in the results to justify wider inductive generalizations.

ECOLOGICAL EFFECTS OF VOLATILE PRIMARY METABOLITES

It is difficult to distinguish between causation and correlation when studying the relationship of changes in biological activity of mixtures to multiple changes in their components, and more so when the components are mostly all produced by linked reactions. This is particularly relevant to the interpretation of the many reports of the effects of carbon dioxide in natural environments. These are not all based on unequivocal information (18, 55, 62), and in many cases there seems to be a need to consider possible correlations between CO_2 output and the amount of ethanol, and of acetaldehyde liberated into the substrate and hence indirectly into the gas phase (12, 29, 40, 114). Environmental factors will affect the proportions of constituents of such gas mixtures as they are liberated from cells; they may also have a differential effect on sorption or other inactivation of individual constituents during movement of the mixture from site of origin to site of reaction. The production of such mixtures in which more than one constituent may be present in active or nearly active concentration has been demonstrated (49). This general concern applies of course to the study of any complex interactions, and it is impracticable here to analyze all existing reports in relation to it. The remainder of the summary in this section is therefore of the author's claims. These are presented in the themes of ecology of saprophytes, ecology of parasites and of pathogens, and of applied ecology associated with the prevention of deterioration of, or with the production of materials used by man.

The ecology of saprophytes has been studied most extensively in soil. The marked reduction in numbers in lower layers (65) has been correlated with sensitivity to CO_2 and to the differences in the amount in soil atmospheres at various depths (1, 24, 36). In a detailed study (24) the linear growth of *Penicillium nigricans,* a typical inhabitant of surface layers, was greatly reduced in atmospheres with more than 5% vol/vol CO_2; *Zygorhyncus vuillemenii,* a typical inhabitant of deep layers, was little affected by up to 20% vol/vol; that of *Gliomastix convoluta,* present throughout the soil, was reduced under 5% vol/vol, but growth was reduced by a much smaller amount than with *P. nigricans.* Because fungi from the deeper layers were facultative anaerobes it was concluded that CO_2 rather than O_2 tolerance or requirement may determine the vertical distribution of these fungi. Differential effects on different nutrient groups have been recorded; e.g., it has been suggested that conditions of high CO_2 and low O_2 resulting from the initial rapid decomposition of organic material may be a feed-back control system enabling rhizosphere fungi to establish themselves in competition with faster growing but CO_2 intolerant primary decomposers (126, 127). In the relatively few cases that have been examined, high CO_2 is commonly inhibitory but not toxic, spores or vegetative cells reverting to normal growth if the CO_2 levels fall appropriately. If this were found to be general by more comprehensive survey it would be a particularly important feed-back device, dormant cells so produced being available to maintain the potential and flexibility of a complex population in a fluctuating environment. The way in which the addition of CO_2 to the soil atmosphere can overcome soil fungistasis has not been explained (34). It has been suggested that microbial action is implicated in the failure of *Pythium mammilatum* to invade wooden blocks containing glucose, which have been buried in soil before inoculation, but that this is unlikely to be associated with gaseous metabolites (8). The justification for this deduction is not clear, as the situation seems to be one in which such metabolites could well be effective.

Effects on the ecology of pathogens in the soil have been reported frequently (3, 119, 132). Evidence from extensive study of the ecology of *Gaeumannomyces graminis* (*Ophiobolus graminis*) has supported the hypothesis that its spread on host roots is directly related to CO_2 concentration in the environment, and control has been achieved by measures designed to increase the level of this gas in the soil (47). From physical principles, however, it has been deduced that O_2 availability is more likely to be a factor (122), and examination of the CO_2 tolerance of the pathogen in vitro led earlier to the proposition that concentrations likely to affect it significantly are not likely to build up in normal soils (38). The significance of this culture evidence is reduced by knowledge of the great effects that difference in nutrient supply and other environmental conditions may have, and by uncertainties about the concentration of CO_2 that may build up in local microclimates. The control of *Rhizoctonia solani* in soil by the addition of organic matter has similarly been attributed to the build-up of respiratory CO_2 by stimulation of the microflora (14). In studies of the effects of this fungus on radish and beet it was found that the pathogenic phase was more sensitive that the saprophytic one, and more seedlings emerged from infested soils under a high CO_2 mixture than from soils under normal aeration (97). More complex interactions have been proposed to account for the control of several diseases by soil flooding techniques. For example up to 85% of indigenous fusaria in banana plantation soils were eradicated within 40 days of flooding (131), and during this period the fusaria in dead stems and roots were destroyed by the activities of putrifying organisms. This destruction was not uniform throughout the soil, living fusaria being isolated from the top 1″ of soil, which had been under water for 6–10 months, but not from lower layers; it was suggested that the production of such gases as CO_2, methane, and H_2S in toxic concentrations may assist in this destruction. Growth of the *Fusarium* isolates in culture was, however, stimulated by incubation under increased concentrations of CO_2 in air; isotope studies confirmed that this may be associated with CO_2 fixation by the fungi (129). This general tolerance to or stimulation of growing mycelia by CO_2 has led to the development of selective isolation techniques by which fusaria present in small numbers in a soil population can be grown out onto agar culture for isolation (11, 98). It is not clear, however, to what extent this depends on the development of tolerance by the fusaria, or to the enrichment of the substrate (127). The apparent paradox between control and stimulation was resolved by the demonstration of the interacting effects on conidial germination and chlamydospore formation. Under normal air, or under mixtures of nitrogen and oxygen, or under nitrogen alone, conidia germinate to form mycelium and chlamydospores. Conidia do not germinate or germinate only slowly under high CO_2, and chlamydospores are not formed on the mycelium. This supports the hypothesis that the elimination of fusaria from flooded soil is due fundamentally to the suppression of chlamydospore formation; this results in mycelium and ungerminated conidia dying when the immediately available and colonizable material in the soil is exhausted (92). This is specifically a response to CO_2, not to O_2 shortage, though high O_2 availability has also been shown to be needed for chlamydospore production (53).

Flooding technique for the control of *Urocystis tritici* on wheat (66), *Sclerotinia sclerotiorum* on celery (128), and *Phytophthora parasitica* var. *nicotianae* on tobacco (118) have similarly been attributed to microbial activity, but these gaseous relationships have not been analyzed rigorously. The infection of wheat seedlings by *Fusarium culmorum* is increased in soils with a high CO_2 content, suggesting a more direct relationship (84).

There are, as is to be expected, fewer reports of CO_2 build-up affecting fungal attack of aerial parts of plants, but effects on spore germination in liquid culture will obviously be related to effects in infection drops. The penetration of wheat seedlings by *Puccinia graminis* was enhanced in CO_2-free air in light or in dark, but it is not clear whether this is due to effect on the host or on the fungus. *Puccinia recondita,* however, was found to be relatively insensitive to CO_2, and changes in its concentration had only slight effects on disease etiology (141). The growth of fungi attacking maple wood was better in mixtures of 10% CO_2 in air than in normal air, and this concentration was found commonly in attacked wood (133).

Applied Ecology

The effects of CO_2 have been examined frequently in relation to commercial methods for fruit storage. In an early investigation (21) increases up to 60% in air had a negligible effect on the growth of *Penicillium expansum* and other species causing common fruit rots. Subsequently, however, raising concentrations to 30% in air was found to delay the rate of development of decay by *Penicillium expansum* and by *Sclerotinia fructicola* (19). Increased CO_2 was found to be a more useful storage factor when applied in conjunction with lowered temperature, and many reports on this topic have been reviewed (117). In dealing with storage rots of living tissues, changes in CO_2 concentration may affect the host, but the saprophytic rots of other organic material may also be affected. Mold growth on bread was significantly reduced by storage under 17% CO_2 in air. The growth of *Thamnidium* sp. on meat at 1°C was reduced by half under 10% CO_2 in air, and it was almost completely suppressed under 20% in air. At 20°C, however, the growth was very little affected by exposure to 20% CO_2 in air, and it was pointed out that in some conditions raising the environmental CO_2 level may enhance the rate of decay (90). It has been found in studies of grain storage that fungal growth was reduced under 12–21% CO_2 in air (48). In other investigations it was found that in the presence of 21% O_2 differences of CO_2 concentration below 13.8% had little effect on fungal growth. Under 50–70% CO_2 there was no fungal growth and grain germination was normal (101).

Penicillin production in fermenters was found to be optimum in one process when air to the fermenters was supplemented with 0.25% CO_2 vol/vol, but the requirement was enhanced by mineral deficiencies (68).

VOLATILE SECONDARY METABOLITES AND UNIDENTIFIED FACTORS

We have only random and slight knowledge of the possible variety and abundance or rarity of volatile secondary metabolites; the expected variety in their biogenesis suggests that there is no obvious reason to expect future knowledge to reveal any widespread correlation with patterns of other characters. This irregularity is illustrated by the results of a recent survey of the genus *Fomes* prompted by the observation that culture gases from a strain of *Fomes anno-sus* strongly inhibited the mycelial growth of five species of fungi (33). These effects, and others on green plants and bacteria, were subsequently found to be due to the liberation of triacetylene (Hexa 1–3–5 triyne) from the fungus mycelium (50). The conditions leading to the liberation of so highly reactive a compound from a living cell must be unusual, but they have not been inves-tigated. In the wider survey of 37 other species (32) triacetylene was not found in any samples of culture gases excepting those from 3 of 7 strains of *F. annosus*. The production of methyl chloride by six species and of HCN by *F. noxius*, is referred to below. In the remaining 31 species there was great variation in the amounts of primary volatile metabolites recorded by G L chromatography, but no general consistent pattern of production of either primary or secondary metabolites appeared. It was also clear that quite small changes in culture conditions could substantially affect the production of at least some volatile metabolites.

It seems most appropriate in these circumstances to present separate dis-cussions of each of a representative and reasonably comprehensive range of compounds, or of unidentified volatile complexes. These appear below in ar-bitrary order, based mostly on the amount known about each.

Hydrogen Cyanide

Hydrogen cyanide is one of the products to which the concept of primary and secondary metabolism has little relevance. The many records of its pro-duction by fungi have been reviewed recently (87). Most of the reports re-viewed have dealt with its production and effects in solution. Locquin (80), after examining 300 species, concluded that it is a normal fungal metabolite, but that in some cases it may be liberated or produced in too small a quantity for regular detection, or it may be an intermediate that is normally reused immediately on formation. Bach (2) felt more strongly that the liberation of free HCN is the result of disturbance of normal metabolism. She found in particular that damaged basidiomycete sporophores produce more than un-damaged ones, and that old cultures with much moribund mycelium produce more than young ones. Robbins et al (113) concluded that the gaseous HCN produced by an unidentified basidiomycete was "probably formed by autoly-sis," as they were unable to demonstrate its production by actively growing mycelium. The demonstration of the production of abundant gaseous HCN by actively growing young cultures of several species of *Fomes* and *Clitocybe* (86) supports Locquin's view. However, it seems unlikely that all this pro-duction could come from the "maladjusted cells of the mycelium" envisaged by Bach. Tomkins (135) reported that the reaction of several species of fungi was similar to gaseous HCN, acetaldehyde, and hydrogen sulphide. In each case spore germination was delayed, the growth rates following germi-nation were reduced initially but they increased as the colony got larger, and the concentration required to inhibit growth was larger than that required to check germination. When appropriate concentrations of any of these substances were introduced into the air above normal growing cultures the hyphal growth was immediately reduced, but it subsequently increased to approximately normal rates. He suggested that this correlates with the known relationships of these substances to common respiratory pathways. Marshall & Hutchinson (87) examined the effects after seven days, hence their measurements would not demonstrate any effects of such earlier inhibitions. They found that the HCN-containing gases from cultures of *Fomes scutellatus* inhibited the germination of lettuce seeds and growth of lettuce seedlings, but that they had no visible effects on eight species of bacteria, or on other colonies of *F. scutellatus*, and only slight inhibitory effects on colonies of *Aspergillus niger*. They draw attention to the known flexibility of the respiratory pathways of many bacteria and fungi, which would suggest that they are likely to be able to overcome inhibition of any one pathway by HCN. This would also provide a mechanism to explain patterns of initial inhibition followed by return to normal growth rates (86). The inhibition of bacteria reported by Locquin (80) is less understandable. The association of cyanide with respiratory pathways, particularly with terminal oxidation, suggests that at least some of the product is a "primary metabolite." Detection outside the cell may be a measure of overproduction, or of the cell's inability to retain what it produces. Ecological effects in field conditions are illustrated mostly from studies of diseases of green plants. Snow mold of alfalfa has been attributed to HCN production by an unidentified basidiomycete mycelium associated with diseased plants (75, 76, 138, 139), and "fairy ring" disease of turf grasses to its production by *Marasmius oreades* (9, 39, 77). These reports are, however, based on the analysis of culture filtrates or mycelial extracts and their effects on the suscepts, not on the analysis of direct effects of culture gases. Lettuce seedlings show similar disease symptoms when grown on mineral agar in pure culture exposed to HCN containing gases from cultures of *Fomes scutellatus*, or on sterile soil irrigated with these gases (86). The characteristic resistance of many basidiomycete mycelia to decay in soil may be affected by their HCN content (67), though it has been pointed out that the fruit bodies of cyanogenic fungi are not necessarily impervious to attack by animals (2). Trione (136) concluded that resistance of flax to *Fusarium lini* is not associated with HCN production.

Ethylene

It seems likely that ethylene is a common fungal product. It has been identified in head space gases from cultures of *Agaricus bisporus* (79), *Penicillium digitatum* (142), *Blastomyces dermatidis* (93), *Mucor hiemalis* (85), and particularly abundantly from *Fusarium oxysporum*, *Verticillium nubilum*, and *Pyronema confluens* (J. Hillman, personal communication). Another survey of 228 species was carried out by removing mycelium or cells from liquid cultures by filtering or centrifuging, placing samples of the isolated fungal material in the barrel of a gas syringe, incubating the assembly for 24 hours, then analyzing samples of the head space gas in the syringe by G L chromatography. In these conditions there was at least some evidence of ethylene being present in samples from 58 species. *Aspergillus clavatus* produced an outstanding 514 ppm vol/vol. *Thamnidium elegans, Aspergillus flavus, Cephalosporium gramineum, Penicillium corylophorum,* and *Ascochyta imperfecti* produced between 10 ppm vol/vol and about 16 ppm vol/vol, the remaining species produced < 4 ppm, mostly < 1 ppm vol/vol (64). It is unfortunate that the experimental procedure in this interesting survey does not enable one to distinguish between products of normal healthy cells and those from damaged cells or cells in unusual starvation conditions. The investigators concluded, however, that ethylene is a common fungal product, which should be considered in any studies of growth disturbances of plants associated with fungal attack.

Mucor hiemalis has been isolated from anaerobic soils in the field containing ethylene at concentrations that could affect the root extension of cereals. This fungus produced ethylene anaerobically, but more ethylene per g of organism was formed with increasing oxygen (85). The authors suggest that the apparent conflict with the observation that ethylene is not accumulated in soil having concentrations of oxygen above 1% is probably due to the fact that anaerobic conditions are necessary for the prior mobilization of the required substrates, or to increased ethylene breakdown in aerobic conditions.

Ethylene has often been found in host tissues invaded by fungi, but in many cases it has not been possible to determine how much of this has been formed by host cells and how much by the pathogens. For example, it has been looked for but not found in cultures of *Ceratocystis fimbriata,* but it has been formed in sweet potato tissues invaded by this fungus. The investigators suggested that the fungus stimulates the host cells to produce the gas; the gas in turn then stimulates the host cells to produce peroxidases and polyphenol oxidases, so inducing an acquired resistance to further fungal invasion (124). Other investigators subsequently found much but variable amounts of ethylene in head space gases of cultures of this species. The amount in invaded tissue does not correlate with the amounts produced by different strains of the pathogen in culture, however, and they conclude that the relative participation of host and pathogen in ethylene production remains undetermined (28).

Odorous and Related Compounds in Ecology and Taxonomy

The odors produced by fungi have an obvious effect as attractants and repellants. It is impracticable here to review the enormous literature on their effects on the palatability of mankind's food and drink. The effects on relationships of fungi with insects has been examined more fully than with any other animals. Attraction to truffles and other edible fungi is well known in folk lore (107), but identification of the substances involved was not possible until modern analytical techniques became available. Several aldehydes, including phenyl crotonaldehyde, were identified in the strong-smelling gases liberated from ripe sporophores of *Ithyphallus impudicus,* which appear to attract flies to the exposed sticky spore masses and so to foster dispersal (78). The attraction of house flies to sporophores of *Amanita muscaria* has been attributed to the formation of partially characterized esters with carbonyl groups (91), but it is not clear whether these are identical with the volatile substances found in the sporophores, which have a differential toxic effect on the two sexes of fly (81, 91). There have been speculations about the probability of smell affecting beetle movements in the development of many of their close associations with fungal sporophores and mycelial mats in colonized wood (10, 30, 35). These have also been extended by the identification of many characteristic odorous carbonyl compounds, esters, etc. in extracts from species of *Ceratocystis, Endoconidiophora,* and related genera of "sap stain" and other wood decaying fungi (20, 31). Six species of *Fomes* (*F. conchatus, F. occidentalis, F. pomaceus, F. ribes, F. rimosus,* and *F. vinosus*) out of 37 examined were found to produce methyl chloride on media containing sodium chloride, and *F. pomaceus* was shown to form methyl bromide on media containing sodium bromide, but the amounts produced in the test conditions had no significant apparent effect on the activity of small numbers of *Diosophila melanogaster, Tenebris molipor,* or *Megoura vicea* (62). The known insecticidal action of methyl bromide, and records of its effects on bacteria (69), *Armillaria mellea* (74), and many other fungi suggests, however, that this product could have an ecological effect in some conditions.

Smell has been used extensively as a taxonomic character or as an aid to recognition, particularly in Hymenomycetes (121). Relatively few of the compounds contributing to the odors have been identified, however (12). Their use is therefore limited by the problem of unequivocal communication of mental concepts of the subjective property of smell; e.g., the interpretation of Fries's description of *Stereum odoratum* (44) as ". . . odor forte peregrinus," demands a knowledge of the personal habits of visitors to Sweden in the early 19th century, which few of us can claim (107). The only consistent pattern of volatile production that emerged from a survey of 37 species of *Fomes* was the formation of methyl chloride by 6 species reported above (32). These species have been allocated to a variety of genera in the various taxonomic treatments proposed for the Polyporaceae. The presence of this character may help to resolve these uncertainties, if it is found to be consistent in wider examination. Only one strain of each species has been tested, however, and the finding that only three out of five strains of *Fomes annosus* produced identifiable amounts of triacetylene in standard tests (unpublished records, Joint Mycological Laboratory, Glasgow University) indicates the possible limitations of such evidence.

Effect on Reproduction of Phycomycetes

Identification of the part played by volatile metabolites in sexual reproduction of Zygomycetes has been helped by considering the processes in the successive phases of: (*a*) initiation of sexual branches (zygophores); (*b*) elongation and orientation of zygophores; (c) differentiation of gametangia and gametes, fusion of gametes, and zygospore formation.

These structures do not all differentiate to the same degree in all genera; e.g., in several species of *Mucor* gametangia develop only on well-defined preformed zygophores, in species of *Absidia* and *Rhizopus* they are formed on aerial branches that are not obviously different from any other aerial branches in the colony, and in species of *Phycomyces* they may apparently form on any cells that come into contact (15).

The third stage has been seen only after zygophores or other gamete-forming cells have come into direct contact. Volatile substances are involved in the scond stage, and there have been claims that they may be involved in the first.

The initiation of zygophores in unmated + or − cultures of heterothallic species can be brought about by the addition of trisporic acid to the culture (51). This substance is apparently identical with the "gamone" reported in earlier work (103). It has been chemically identified in both + and − mycelia in mated cultures of *Mucor mucedo*, *Phycomyces blakesleeanus*, *Blakeslea trispora*, and + mycelia of *Blakeslea trispora* paired with a culture of the homothallic *Zygorhyncus moelleri;* mutual induction of zygophores in many interspecific pairings of cultures suggests that it is a sex hormone common throughout the Mucorales (51). It has not been detected in homothallic species. It has been detected in pure cultures of each mating type of *Blakeslea trispora* growing in contact with cell-free extracts from cultures of the opposite mating type, and in +, but not − cultures of *B. trispora* to which the precursor trisporol C has been added (51). This possibility is supported by observations of effects of volatile stimulants to zygospore fomation in other Zygomycetes. The effects on *Rhizopus sexualis* have not been analyzed by stages, but the number of zygospores formed in cultures was greatly reduced by incubation at low temperature (ca 10°C), and this reduction was overcome by irrigating the cultures at low temperature with gases from cultures of the same species growing at room temperature. The volatile factor is said to be basic, not CO_2, ethanol, or ammonia, and may be methylamine. Gases from other cultures of *R. sexualis* were most effective, those from *Phycomyces blakesleeanus* were less, those from *Sordaria fimicola* were least, and those from *Rhizoctonia solani* were ineffective in comparative tests (58).

In the second phase of elongation and orientation, zygophores of opposite mating types grow steadily towards each other in air from distances of up to 2 mm apart; those of the same mating type bend away from each other. This occurs both with zygophores formed on paired cultures of the two mating types, and with those induced on unpaired cultures by the addition of trisporic acid. Zygotropism of this sort is a property of zygophores that is not dependent on the forming hyphae being able to carry out the trisporic acid synthesis (G. W. Gooday, personal communication). Very slight protrusions appear on mutually facing walls while the zygophores are separated by distance equal to about one third of their width, which suggest that forces drawing the filaments laterally had effected a bulging of the delicate cell walls at the growing points (15). The possibility of this interaction being controlled by development of some sort of polar electric field has been examined by inserting a variety of membranes between mycelia of different mating types. Earlier results supported this hypothesis, but they were later found to be invalid (4, 137). Other experiments with various perforated and impermeable membranes etc. strongly support the hypothesis that volatile substances are implicated in the reaction. The simplest hypothesis that would account for the mutual attraction between zygophores of opposite mating types, and repulsion between those of the same type, is the existence of two volatile metabolites, each providing a specific reaction in one strain only (4, 22). No such metabolites have yet been detected in many chemical investigations with G L chromatography and other techniques (51).

In other studies of the Phycomycetes, the formation of sporangia by *Pilobolus kleinii* has been found to be stimulated by a volatile factor from cultures of *Mucor plumbens,* identified as ammonia (96), growth of cultures of *Phytophthora citrophthora* has been stimulated by an unidentified volatile metabolite from cultures of *Mucor spinosus* (13), and oospore formation by single strains of heterothallic *Phytophthora* spp. has been found to be induced by unidentified volatiles from cultures of *Trichoderma viride* (17).

Ideas of possible biological significance of these factors have been widened by observation that hyphae of the parasitic species *Parasitella simplex* and *Chaetocladium brefeldii* are attracted to those of their hosts (species of *Absidia, Mucor,* etc.) by volatile factors. The factors are "sex linked," + strains of the parasites being attracted to − strains of the hosts, and − strains of parasites to + strains of hosts. This has been interpreted as a development of a "sexual" attractant as a means to more efficient parasitism (23).

There are obviously large technical problems in examining the minute amounts of substances likely to be involved in tropic reactions in this scale. There seem to be even greater credibility problems involved in considering the likely effects of differences in reasonable concentrations of such substances. It has been calculated that if a zygophore 10 μ in diameter lies transverse to the diffusion path 2 mm from a point source, the concentration of molecules hitting the front face of the hyphae would be about 1% more than that hitting the far side. With a broader source (e.g., many neighboring zygophores) or at greater distances, the difference would be less. In test conditions to which these calculations were related, and in which zygophore tropism was seen, it seemed unlikely that the difference would have fallen below 0.3%; it was pointed out that this is a similar order of difference to that calculated for radiant energy falling on the near and far sides of phototropic sporangiophores of *Phycomyces* sp. (4).

It seems likely, though difficult to accept, that differences of this order in radiant energy have a differential effect on growth, and in high concentrations of molecules it may imply large differences in total numbers of molecules hitting a given area in a given time. The very low gaseous concentrations at which some of the identified volatile metabolites are active suggests, however, that the total number of molecules hitting the surface in a given time will be very small. It is even more difficult to conceive that differences of the order of 0.3% in this very small absolute number can produce differences in absorption of energy or material that could account for the large rapid effects seen. This order of difference is that expected for hyphae lying transverse to the diffusion path. The differences on either side of the tip of a 10μ broad hypha that would be involved in keeping it growing approximately parallel to a diffusion path would be infinitesimal.

The proof of the pudding is in the eating, however! Zygophores grow towards each other, or repel each other, when separated by about 2 mm of air or less, and their behavior fits the hypothesis that in these conditions their orientation is controlled by two strain-specific volatile metabolites.

Volatile Compounds of Arsenic, Selenium, and Tellurium

The most recent of several reviews of this subject (27) records many reports that a wide range of bacteria and fungi can tolerate appreciable quantities of these elements in the media. No correlation has been found between this tolerance and the ability to produce volatile metabolites containing these substances. They were not found in careful searches of any of the tolerant bacteria or yeasts, but only in some fungi (70). The fungi that do produce such volatiles are apparently able to carry out biological methylation, producing trimethyl arsine, dimethyl selenide, and/or dimethyl telluride. This ability has been used as a delicate test for the presence of traces of arsenic in many assimilable compounds. Speculations about well-documented cases of arsenical poisoning in the 19th century were confirmed by finding that many molds, particularly *Scopulariopsis brevicaule,* could produce trimethyl arsine from arsenic-containing wallpaper. More recently, in 1931, the deaths of two children were clearly related to the absorption of arsenic liberated by fungal action on arsenic-containing wall plaster (27). With this impressive evidence, it is surprising that the potential significance (18) of these compounds in microbial ecology has not been examined more fully.

Other Records of Secondary and Unidentified Metabolites

The few other isolated records of the activity of unusual secondary metabolites, or of unidentified volatile factors, do not apparently fall into any general pattern, although they add significantly to the general picture that emerges from the above survey (e.g., 45, 54, 59, 72, 108).

CONCLUSIONS

Two recent reviews (46, 132) have discussed the biogenesis of these products, and the biochemistry of the reactions in which they are involved. These present a comprehensive up-to-date account of the present state of knowledge. This review has therefore been restricted to the reports of their occurrence and of their possible significance in ecology. The range and total number of species that have been examined critically is a trivially small sample of the total number of living fungi. Within this sample there is great variety in the sorts, amounts, and biological activities of the volatile metabolites produced, but the total information is rarely sufficient for more than speculative generalizations. Such generalizations may, however, be helpful if used as guide lines for new investigations rather than as principles to be invoked in explanations. For example, the suggestion that volatiles are likely to be most active where other limiting factors are near critical levels (21) is perceptive and stimulating, but it is based on observation of a relatively small number of precise tests. It also seems sensible to examine the degree to which patterns seen in this small range of observations reflect a bias in the interests or methods of the investigators, rather than a generally distributed phenomenon. For example, the use of sealed vessels for many quantitative tests by the author and his colleagues will have favored the frequency with which they have found CO_2 and ethanol to be produced in inhibitory levels by a variety of fungi. These tests have, however, also drawn attention to the great differences between the amounts of these metabolites that may be liberated by different fungi in a standard environment; in many cases examined these amounts seem too small to have any significant growth-promoting or inhibiting action. Similarly, most of the work reviewed deals with the effects of selected substances on selected organisms. This is a reasonable first step in the analysis of complex situations, but it may give little information of likely effects of higher-order interactions [e.g., the response of rhizomorphs of *Armillariella elegans* to atmospheres containing various amounts of CO_2 and O_2 could not have been anticipated from study of the response to changes in the partial pressure of each gas applied separately (52)]. In another case, the effect of the volatile products from the mycelium of *Agaricus bisporus* on its fruiting in compost beds is indirect; the products stimulate soil bacteria, which in turn produce an unidentified volatile that stimulates sporophore production (56).

With these qualifications it is likely that interactions with gaseous metabolites may be a significant factor in many ecological situations. The characteristic properties of, and effects of volatility in diffusion in air and liquids have been referred to in the introduction. Rapid diffusion in the gas phase could have particular adaptive value in cyclic or rapid *ad hoc* reactions; it could result in the removal of residual material and so reduce the need for any other inactivating system (105). It is appropriate to be particularly alert to possible interactions with these metabolites in controlled aeration systems in fermenters, food stores, etc. A rapid flow of air through such a system may affect not only CO_2, O_2, and humidity, but also the amounts of other gaseous growth-regulating factors. It may also affect economics by the premature removal of foodstuffs which might be assimilated if left in contact with the fungal cells. The evidence that some vitamins can diffuse from media through the air in sufficient quantity to support the growth of vitamin-requiring fungi on vitamin free media (89) gives further indication of ways in which such interactions might exist.

This review discusses information of the effects of such fungal metabolites. The thought of likely significance that emerges from it is reinforced by appreciating that this is an arbitrarily restricted selection from a continuum of interactions with similar gaseous products from all other living things. A quietly balanced emphasis on such potential implications may result in the facts speaking for themselves.

Literature Cited

1. Abeygunawardena, D. V., Wood, R. K. S. 1957. Factors affecting the germination of sclerotia and mycelial growth of *Sclerotium rolfsii* Sacc. *Trans. Brit. Mycol. Soc.* 40:220–31
2. Bach, E. 1953. The Agaric *Pholiota aurea. Dansk Bot. Arkiv.* 16: 4–220
3. Baker, R. 1968. Mechanisms of biological control of soil-borne pathogens. *Ann. Rev. Phytopathol.* 6:263–94
4. Banbury, G. H. 1954. Physiological studies of the Mucorales. III. The zygotropism of zygophores of *Mucor mucedo* Brefeld. *J. Exp. Biol.* 6:235
5. Barinova, S. A. 1953. Influence of CO_2 on mould growth. *Mikrobiologiya* 22:391–98
6. Barinova, S. A. 1954. Effects of carbon dioxide on respiration in molds. *Mikrobiologiya* 23:521–26
7. Bartnicki-Garcia, S., Nickerson, W. J. 1962. Induction of yeast like development in *Mucor* by carbon dioxide. *J. Bacteriol.* 84:829–40
8. Barton, R. 1960. Antagonism among some sugar fungi. *The Ecology of Soil Fungi*, 160–7. Ed. D. Parkinson, J. S. Waid, *Liverpool University Press.* 324 pp.
9. Bayliss, J. S. 1911. Observations on *Marasmius oreades* and *Clitocybe gigantea* as parasitic fungi. *J. Econ. Biol.* 6:111–32
10. Benick, L. 1952. Pilzkäfer und Käkerpilze. *Acta Zool. Fenn.* 70: 5–246
11. Bergman, H. F. 1959. Oxygen deficiency as a cause of disease in plants. *Bot. Rev.* 25:417–85
12. Birkinshaw, J. H., Chaplen, P., Findlay, W. P. K. 1957. Biochemistry of wood rotting fungi. 9.

Volatile metabolic products of *Stereum subpilatum* Berk. and Cent. *Biochem. J.* 66:188–92
13. Bitancourt, A. A., Rosetti, V. 1951. Stimulation of growth of *Phytophthora citrophthora* by a gas produced by *Mucor spinosus. Science* 113:531
14. Blair, I. D. 1943. Behaviour of the fungus *Rhizoctonia solani* Kühn in the soil. *Ann. Appl. Biol.* 30: 118–27
15. Blakeslee, A. F. 1904. Sexual reproduction in the Mucorineae. *Proc. Am. Acad. Arts Sci.* 40:206
16. Boswell, J. 1791. *The life of Samuel Johnson, Ll.D.*, London
17. Brazier, C. M. 1971. Induction of sexual reproduction in single A^2 isolates of *Phytophthora* species by *Trichoderma viride. Nature New Biol.* 231:283
18. Brian, P. W. 1960. Antagonistic and competitive mechanisms limiting survival and activity of fungi in soil. *The Ecology of Soil Fungi.* pp. 115–29. Ed. D. Parkinson, J. S. Waid, *Liverpool University Press.* 324 pp.
19. Brooks, C., Bratley, C. O., McColloch, L. P. 1936. Transit and storage diseases of fruits and vegetables as affected by initial CO_2 treatments. *U.S. Dep. Agr. Tech. Bull.* 519:1–24
20. Brown, T. S. 1970. *Germination of basidiospores of Fomes applanatus.* MS thesis, Pennsylvania State University, State College. 21 pp.
21. Brown, W. 1922. On the germination and growth of fungi at various temperatures and in various concentrations of oxygen and of carbon dioxide. *Ann. Bot., London* 36:257–300

242   HUTCHINSON

22. Buller, A. H. R. 1933. *Researches on Fungi 5*: London: Longmans Green & Co. 416 pp.
23. Burgeff, H. 1927. Untersuchungen über sexualität und parasitismus bei Mucorineen. 1. *Bot. Abhandl.* 1. (4): 1–135
24. Burges, N. A., Fenton, E. 1953. The effect of carbon dioxide on the growth of certain soil fungi. *Trans. Brit. Mycol. Soc.* 36:104–08
25. Buston, H. W., Moss, M. O., Tyrell, D. 1966. The influence of carbon dioxide on growth and sporulation of *Chaetomium globosum. Trans. Brit. Mycol. Soc.* 49:387–96
26. Cantino, E. C. 1961. The relationship between biochemical and morphological differentiation in non-filamentous aquatic fungi. *Microbial Reactions to the Environment. Soc. Gen. Microbiol. Symp.* 11, 243 pp.
27. Challenger, F. 1945. Biological methylation. *Chem. Rev.* 36:315–61
28. Chalutz, E., DeVay, J. E. 1969. Production of ethylene in vivo and in vitro by *Ceratocystis fimbriata* in relation to disease development. *Phytopathology* 59:750–55
29. Coley-Smoth, J. R., Cooke, R. C. 1971. Survival and germination of fungal sclerotia. *Ann. Rev. Phytopathol.* 9:65–92
30. Collins, R. P., Kalnins, K. 1965. Carbonyl compounds produced by *Ceratocystis fagacearum. Am. J. Bot.* 52:751–54
31. Collins, R. P., Kalnins, K. 1966. Production of carbonyl compounds by several species of endoconidium-forming fungi. *Mycologia* 58:622–28
32. Cowan, M. E., Glen, A. T., Hutchinson, S. A., McCartney, M. E., Mackintosh, J. M., Moss, A. M. 1973. Production of volatile metabolites by species of the genus *Fomes. Trans. Brit. Mycol. Soc.* 60:(In press)
33. Dick, C. M., Hutchinson, S. A. 1966. Biological activity of volatile fungal metabolites. *Nature,* 211:268
34. Dobbs, C. G., Hinson, W. 1953. A widespread fungistasis in soils. *Nature* 172:197–204

35. Dorsey, C. K., Leach, J. G. 1956. The bionomics of certain insects associated with oak wilt, with particular reference to the Nitulidae. *J. Econ. Ent.* 49:219–30
36. Durbin, R. D. 1955. Straight-line function of growth of microorganisms at toxic levels of carbon dioxide. *Science* 121:734–35
37. Faraj, Salman, A.-G. 1970. Induktion einer endogenen Rhythmik der Koremienbildung durch Alkohol bei einer mutante von *Penicillium claviforme* Bainer und einer Varietät davon. *Biochem. Physiol. Pflanzen* 161:42–49
38. Fellows, H. 1928. The influence of oxygen and carbon dioxide on the growth of *Ophiobolus graminis* in pure culture. *J. Agr. Res.* 37:349–55
39. Filer, T. H. 1966. Effects on grass and cereal seedlings of hydrogen cyanide produced by mycelium and sporophores of *Marasmius oreades. Plant Dis. Reptr.* 50:264–66
40. Foster, J. W. 1949. *Chemical Activities of fungi.* New York: Academic Press, 648 pp.
41. Foster, J. W., Ruben, S., Kamen, M. D. 1941. Radioactive carbon as an indication of CO_2 utilization. VII. The assimilation of CO_2 by moulds. *Proc. Nat. Acad. Sci., U.S.* 27:590–96
42. Frankel, C. 1889. Die Einwirkung der Köhlensäure auf die Lebenstätigkeit der Microorganismen. *Z. Hyg.* 5:332–62
43. Frankland, P. F. 1889. Über den Einfluss der Köhlensäure auf die Lebenstätigkeit der Microorganismen. *Z. Hyg.* 6:13–22
44. Fries, E. 1836. *Epicrisis Systematis Mycologici et Synopsis Hymenomyceti.* Upsala p. 553
45. Fries, N. 1961. The growth promoting activity of some aliphatic aldehydes from fungi. *Svensk Bot. Tidskr.* 55:1–16
46. Fries, N. 1973. Effects of volatile organic compounds on the growth and development of fungi. *Trans. Brit. Mycol. Soc.* 60:1–21
47. Garrett, S. D. 1970. *Pathogenic Root-Infecting Fungi.* London: Cambridge University Press. 294 pp.
48. Geddes, W. F., Cuendet, L. S., Christensen, C. M. 1955. Recent

investigations on grain storage. *3rd Int. Bread Congr.*, Hamburg, 1955. 136–39

49. Glen, A. T., Hutchinson, S. A. 1969. Some biological effects of volatile metabolites from cultures of *Saccharomyces cerevisiae* Meyer ex Hansen. *J. Gen. Microbiol.* 55:19–27

50. Glen, A. T., Hutchinson, S. A., McCorkindale. N. J. 1966. Hexa 1-3-5 tri-yne, a metabolite of *Fomes annosus. Tetrahedron Lett.* 4223–25

51. Gooday, G. W. 1973. Differentiation in the Mucorales. *Symp. Soc. Gen. Microbiol.* 23, London. (In press)

52. Griffin, D. M. 1972. *Ecology of Soil Fungi.* London: Chapman & Hall. 195 pp.

53. Griffin, D. M., Nair, N. G. 1968. Growth of *Sclerotium rolfsii* at different concentrations of oxygen and carbon dioxide. *J. Exp. Bot.* 19:812–16

54. Halim, A. F., Collins, R. P. 1971. An analysis of the odourous constituents of *Trametes odorata. Lloydia* 34:451

55. Harley, J. L. 1960. The physiology of soil fungi. *The Ecology of Soil Fungi.* pp. 265–76. Ed. D. Parkinson, J. S. Waid, *Liverpool University Press.* 324 pp.

56. Hayes, W. A., Randle, P. E., Last, F. T. 1969. The nature of the microbial stimulus affecting sporophore formation in *Agaricus bisporus* (Lange) Sing. *Ann. Appl. Biol.* 64:177–87

57. Henrici, A. T. 1947. *Moulds, Yeasts and Actinomycetes.* New York: John Wiley. 2nd Ed. 409 pp.

58. Hepden, P. M., Hawker, L. E. 1961. A volatile substance controlling early stages of zygospore formation in *Rhizopus sexualis. J. Gen. Microbiol.* 24:155–64

59. Hodgkiss, I. J., Harvey, R. 1971. Factors affecting the fruiting of *Pleurage anserina* in culture. *Trans. Brit. Mycol. Soc.* 57:533–35

60. Horsfall, J. G. 1945. *Fungicides and Their Action.* Waltham: Chronica Botanica. 239 pp.

61. Hutchinson, S. A. 1967. Some effects of volatile fungal metabolites on the gametophytes of *Pteri-*

dium aquilinum. Trans. Brit. Mycol. Soc. 50:285–88

62. Hutchinson, S. A. 1971. Biological activity of volatile fungal metabolites. *Trans. Brit. Mycol. Soc.* 57: 185–200

63. Hutchinson, S. A., Cowan, M. E. 1972. Identification and biological effects of volatile metabolites from cultures of *Trichoderma harzianum. Trans. Brit. Mycol. Soc.* 59: 71–77

64. Ilag, L., Curtis, R. W. 1968. Production of ethylene by fungi. *Science* 159:1357–58

65. Jeffreys, E. G., Brian, P. W., Hemming, H. G., Lowe, D. 1953. Antibiotic production by the microfungi of the acid heath soils. *J. Gen. Microbiol.* 9:314–41

66. Jones, G. H., Sief-el-Nasr, A. el-G. 1940. Control of smut diseases in Egypt with special reference to sowing depth and soil moisture. *Egypt. Min. Agr. Bull.* 224. 46 pp.

67. Juillet, A., Jaulmes, P., Susplugas, J. 1943. Recherches sur l'acide cyanohydrique chez *Marasmius oreades* Fries. *Soc. Pharmacol. Montpelier. Séance* 27:100–12

68. Koffler, H., Emerson, R. L., Perlman, D., Burris, H. R. 1945. Chemical changes in submerged penicillin fermentations. *J. Bacteriol.* 50:517–48

69. Kolb, R. W., Schneider, R. 1949. The germicidal and sporicidal efficacy of methyl bromide for *Bacillus anthracis. J. Bacteriol.* 59: 401–11

70. Kotskova-Kratokhvalova, A., Gebauerova, A., Grdinova, M. 1956. The production of volatile arsenical compounds by fungi. *Czeka Mykol.* 10:77–87

71. Krause, A. W. 1930. Untersuchungen über den Einfluss der Ernährung, Belichtung und Temperatur auf die Perithecienproduktion einiger Hypocreaceen. Beitrag zur Kulturmethodik einiger parasitärer und saprophytischer Pilze. *Z. Parasitenkund,* Abt. F, 2:419–76

72. Krupa, S., Fries, N. 1971. Studies of ectomycorrhizae of Pine. I. Production of volatile organic compounds. *Can. J. Bot.* 49: 1425–31

73. Lambert, E. B. 1933. Effect of ex-

cess CO_2 on growing mushrooms. *J. Agr. Res.* 47:599–608

74. LaRue, J., Paulus, A., Wilbur, W., O'Reilly, H., Darley, E. 1962. Armillaria root rot fungus controlled with methyl bromide soil fumigation. *Calif. Agr.* 16:8–9

75. Lebeau, J. B., Cormack, M. W., Moffat, J. E. 1959. Measuring pathogenesis by the amount of toxic substance produced in alfalfa by a snow mould fungus. *Phytopathology* 49:303–5

76. Lebeau, J. B., Dickson, J. G. 1955. Physiology and nature of disease development in winter crown rot of alfalfa. *Phytopathology* 45:667–73

77. Lebeau, J. B., Hawn, E. J. 1963. Formation of HCN by the mycelial stage of a fairy ring fungus. *Phytopathology* 53:1395–96

78. List, P. H., Freund, B. 1966. Phenylcrotonaldehyd und andere Aldehyde als Geruchstoffe der Stinkmorchel, *Phallus impudicus* L. *Naturwiss.* 53:585

79. Lockhard, J. D., Kneebone, L. R. 1962. Effects of mushroom gases studied. *Mushroom Growers Assoc. Bull.* 148:143–47

80. Locquin, M. 1944. Degagement et localization de l'acide cyanohydrique chez les Basidiomycetes et les Ascomycetes. *Bull. Soc. Linnienne, Lyon,* 13:151–57

81. Locquin-Linard, M. 1967. Un problem à eclairer: celuie de la tuemouche. Étude de l'action de l'*Amanita muscaria* sur les mouches. III. *Rev. Mycol.* 32:428–37

82. Lopriore, G. 1895. Ueber die Einwirkung der Köhlensäure auf das Protoplasma der lebenden Pflanzenzelle. *Jahrb. Wiss. Bot.* 28:531–625

83. Losel, D. M. 1964. The stimulation of spore germination in *Agaricus campestris* by living mycelium. *Ann. Bot.* 28:541–54

84. Lundegårdh, H. 1923. Die Bedeutung des Kohlensäureghalts und der Wasserstoffionkonzentration des Bodens für die Entschung der Fusariosen. *Bot. Notiser* 1:25–52

85. Lynch, J. M., Harper, S. H. T. 1972 Requirements for substrates and oxygen in the formation of ethylene by *Mucor hiemalis. Proc.*

Soc. Gen. Microbiol., 65th Meet. p. 23

86. Marshall, A. M. 1970. Studies in the Production and Effects of Volatile Fungal Metabolites. Ph.D. Thesis, University of Glasgow 162 pp.

87. Marshall, A. M., Hutchinson, S. A. 1970. Biological activity of volatile metabolites from cultures of *Fomes scutellatus. Trans. Brit. Mycol. Soc.* 55:239–51

88. Martin, H. 1940. *The Scientific Principles of Plant Protection.* London: Arnold. 3rd Ed. 385 pp.

89. Meissel, M. N., Medredeva, G. A. 1947. The volatility of certain vitamins and the possibility of their utilization from the atmosphere by certain micro-organisms. *Biokhimiya* 12:303–13

90. Moran, T., Smith, E. C., Tomkins, R. G. 1932. The inhibition of mould growth in meat by carbon dioxide. *J. Soc. Chem. Ind.,* London 51:114–16

91. Muto, T., Sugawara, R. 1965. The house fly attractants in mushrooms. Part 1. Extraction and activity of the attractant. *J. Agric. Chem. Soc.* Japan 29:949–54

92. Newcombe, M. 1960. Some effects of water and anaerobic conditions on *Fusarium oxysporum* f. *cubense* in soil. *Trans. Brit. Mycol. Soc.* 43:51–59

93. Nickerson, W. J. 1948. Ethylene as a metabolic product of the pathogenic fungus *Blastomyces dermatitidis. Arch. Biochem.* 17:225–33

94. Norrman, J. 1968. Morphogenetic effects of some volatile organic compounds on *Pestalotia rhododendri. Arch. Mikrobiol.* 61:128–42

95. Norrman, J. 1969. Production of volatile organic compounds by the yeast fungus *Dipodascus aggregatus. Arch. Mikrobiol.* 68:133–49

96. Page, R. M. 1959. Stimulation of sexual reproduction of *Pilobolus* by *Mucor plumbens. Am. J. Bot.* 46:579–85

97. Papavizas, G. C., Davey, C. B. 1962. Activity of *Rhizoctonia* in soil as affected by carbon dioxide. *Phytopathology* 52:759–66

98. Park, D. 1961. Isolation of *Fusarium oxysporum* from soils. *Trans. Brit. Mycol. Soc.* 44:119–22

99. Pentland, J. D. 1967. Ethanol produced by *Aureobasidium pullulans*

and its effect on the growth of *Armillaria mellea*. *Can. J. Microbiol.* 13:1631–39

100. Perlman, D. 1950. Observations on the production of ethanol by fungi. *Am. J. Bot.* 37:237–41

101. Peterson, A., Schlegel, V., Hummel, B., Cuendet, L. S., Geddes, W. F., Christensen, C. M. 1956. Grain storage studies 22. Influence of oxygen and carbon dioxide concentrations on mold growth and grain deterioration. *Cereal Chem.* 33:53–66

102. Platz, J. A., Durrell, L. W., Howe, M. E. 1927. Effect of carbon dioxide upon the germination of chlamydospores of *Ustilago zeae* Beck (Ung.). *J. Agr. Res.* 34:137–47

103. Plempel, M. 1965. Sexualreaktion und Carotin Synthese bei Zygomyceten. *Planta* 65:225

104. Plunkett, B. E. 1954. The aeration complex of factors and fruit body formation in pure culture of Hymenomycetes. *VIII Conf. Int. Bot. Paris*, Rapp. et Comm. Sec. 18–20 pp. 101–02

105. Pratt, H. K., Goeschl, J. D. 1969. Physiological roles of ethylene in plants. *Ann. Rev. Plant Physiol.* 20:541–84

106. Raistrick, H., Birkinshaw, J. H., Charles, J. H. V., Clutterbuck, P. W., Coyne, F. P., Hetherington, A. C., Lilly, C. H., Rintoul, M. L., Rintoul, W., Robinson, R., Stoyle, J. A. R., Thom, C., Young, W. 1931. Studies in the biochemistry of micro-organisms. *Phil. Trans. Roy. Soc. Ser. B.* 220:1–367

107. Ramsbottom, J. 1953. *Mushrooms and Toadstools*. London: Collins 306 pp.

108. Rankine, B. L. 1964. Hydrogen sulphide production by yeasts *I Sci. Food Agr.* 15:872–77

109. Reddish, G. F. 1957. *Antiseptics, Disinfectants, Fungicides and Chemical and Physical Sterilization*. London: Henry Kempton. 975 pp.

110. Reed, H. S., Crabill, C. H. 1915. The cedar rust diseases of apples caused by *Gymnosporangium Juniperae-virginianae* Schw. *Virginia Agr. Exp. Sta. Tech. Bull.* 9. 106 pp.

111. Rippel, A., Bortels, H. 1927. Vorlaufige über die allgemeine Bedeuung der Köhlensäure für die Pflanzenzelle (Versuche an *Aspergillus niger*). *Biochem. Z.* 184:237–44

112. Rippel, A., Heilmann, F. 1930. Action of CO_2 on heterotrophs. *Arch. Mikrobiol.* 1:119–36

113. Robbins, W. J., Rolnick, A., Kavanagh, F. 1960. Production of hydrocyanic acid by cultures of a Basidiomycete. *Mycologia* 42:161–66

114. Robinson, P. M., Park, D. 1966. Volatile inhibitors of spore germination produced by fungi. *Trans. Brit. Mycol. Soc.* 49:639–49

115. Robinson, P. M., Park, D., Garrett, M. K. 1968. Sporostatic products of fungi. *Trans. Brit. Mycol. Soc.* 51:113–24

116. Rockwell, E., Highberger, J. H. 1927. The necessity of CO_2 for the growth of bacteria yeasts and moulds. *J. Infect. Dis.* 40:438–46

117. Schelhorn, M. von. 1951. Control of microorganisms causing spoilage in fruit and vegetable products. *Advan. Food Res.* 3:429–82

118. Schrevan, D. A. van. 1948. Investigations of certain pests and diseases of Vorstandenden tobacco. *Tijdschr. Plantenziekten* 54:149–74

119. Sewell, G. W. F. 1965. The effect of altered physical condition of soil on biological control. *Ecology of Soil-borne Plant Pathogens*, ed. K. F. Baker, W. C. Snyder, 479–94. Berkeley: Univ. Calif. Press. 571 pp.

120. Sibilia, C. 1928. Richerche sulle rigini dei ceriali. *Boll. R. Stat. Patol. Veg.* (Roma). 88:235–47

121. Singer, R. 1968. *The Agaricales in Modern Taxonomy*. Weinheim: Cramer, 2nd ed. 915 pp.

122. Smith, A. M., Noble, D. 1972. Effects of oxygen and carbon dioxide on the growth of two varieties of *Gaeumannomyces graminis*. *Trans. Brit. Mycol. Soc.* 58:499–503

123. Smith, D. L., Robinson, P. M. 1969. The effects of fungi on the morphogenesis of gametophytes of *Polypodium vulgare* L. *New Phytol.* 68:113–22

124. Stahmann, M. A., Clare, B. G., Woodbury, B. G. W. 1966. Increased disease resistance and enzyme activity induced by ethylene, and ethylene production in black

rot infected sweet potato tissue. *Plant Physiol.* 41:1505–12

125. Stevens, R. 1960. Beer flavour. 1. Volatile products of fermentation. A review. *J. Inst. Brew.* 66:453–71

126. Stotzky, G., Goos, R. D. 1965. Effect of high CO_2 and low O_2 tensions on the soil microbiota. *Can. J. Microbiol.* 11:853–68

127. Stotzky, G., Goos, R. D. 1966. Adaptation of the soil microbiota to high CO_2 and low O_2 tensions. *Can. J. Microbiol.* 12:849–61

128. Stoner, W. N., Moore, W. D. 1953. Lowland rice farming as a possible cultural control for *Sclerotinia sclerotiorum* in the Everglades. *Plant Dis. Reptr.* 37:181–86

129. Stover, R. H. 1958. Studies on Fusarium wilt of bananas. II. Some factors influencing survival and saprophytic multiplication of *Fusarium oxysporum* f. *cubense* in soil. *Can. J. Bot.* 36:311–24

130. Stover, R. H., Freiburg, S. R. 1958. Effect of carbon dioxide on multiplication of *Fusarium* in soil. *Nature* 181:788–89

131. Stover, R. H., Waite, B. H. 1953. An improved method of isolating *Fusarium* spp. from plant tissue. *Phytopathology* 43:700–01

132. Tabak, H. H., Cooke, W. B. 1968. The effects of gaseous environments on the growth and metabolism of fungi. *Bot. Rev.* 34:126–252

133. Thacker, D. C., Good, H. M. 1952. The composition of air in trunks of sugar maple in relation to decay. *Can. J. Bot.* 30:475–85

134. Tomkins, R. G. 1930. Volatile substances and the growth of moulds. *Rept. Food Invest. Board,* 1930:48–55

135. Tomkins, R. G. 1932. The action of certain volatile substances on the growth of mould fungi. *Proc. Roy. Soc.,* Ser. B, 111:210–26

136. Trione, E. J. 1960. Extra cellular enzyme and toxin production by *Fusarium oxysporum* f. *lini*. The HCN content of flax in relation to flax wilt resistance. *Phytopathology,* 50:480–82, 482–86

137. Vandendries, R. 1934. Le barrage sexuel chez les *Lenzites betulina*. *C.R. Acad. Sci. Paris.* 198:193

138. Ward, E. W. B. 1964. On the source of hydrogen cyanide in cultures of a snow mould fungus. *Can. J. Bot.* 42:319–27

139. Ward, E. W. B., Thorn, G. W. 1966. Evidence for the formation of HCN from glycine by a snow mould fungus. *Can. J. Bot.* 44:95–104

140. Weinhold, A. R., Garroway, M. O. 1965. Period of exposure to ethanol in relation to rhizomorph production by *Armillaria mellea*. *Phytopathology* 55:1082

141. Yirgou, D., Caldwell, R. M. 1963. Stomatal penetration of wheat seedlings by stem and leaf rust: effect of light and carbon dioxide. *Science,* 141:272–73

142. Young, R. E., Pratt, H. K., Biale, J. B. 1951. Identification of ethylene as a volatile product of the fungus *Penicillium digitatum*. *Plant Physiol.* 26:304–10

A LYSOSOMAL CONCEPT FOR PLANT PATHOLOGY

❖ 3573

Charles L. Wilson[1]
U. S. Department of Agriculture, Agricultural Research Service,
North Central Region, Delaware, Ohio; Department of Plant Pathology,
Ohio State University, Columbus

Introduction

The lysosomal concept plays an important role in explaining numerous pathological processes in animals (27). For some reason plant pathologists have been reluctant to borrow or even consider concepts used in animal pathology. However, evidence is mounting that lysosomes may be just as important in plant pathology as in animal pathology.

A rationale for the existence of lysosomes in both animal and plant cells can be developed without having to examine cells directly. Anabolism and catabolism take place simultaneously in living cells. These two processes are antagonistic if allowed to occur at the same site. The compartmentalization of hydrolytic enzymes into membrane-bound organelles would provide a mechanism whereby enzymes and substrates of biosynthesis in the cell would not be destroyed during anabolism. Such organelles would also provide loci for the hydrolysis of macromolecules.

Since the original discovery of lysosomes in animals by DeDuve & Novikoff the lysosomal concept has been continually expanded (26). Originally DeDuve emphasized the autolytic function of lysosomes, calling them "suicide bags." Subsequent research has shown that lysosomes are important organelles in normal metabolism and that a number of pathological processes can be explained by their malfunction (27).

A lysosome can be defined as a single-membrane-bounded body that contains more than one acid hydrolytic enzyme (34). Because there is now information on the origin and disposition of lysosomes, we can think in terms of a lysosomal system. Lysosomes are involved essentially with catabolic pro-

[1] I am grateful for the help and encouragement of J. R. Aist, V. N. Armentrout, J. E. Baker, J. M. Daly, J. E. DeVay, R. D. Durbin, J. L. Hall, J. Kuć, Ph. Matile, W. J. Page, and R. Prasad. Figure 1 was reproduced through the kind consent of Dr. R. Wattiaux and Figure 4 through the cooperation of Dr. W. J. Page. The author appreciates the opportunity to use this material.

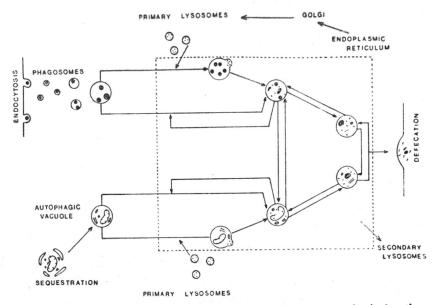

Figure 1 Schematic representation of the heterophagic and autophagic functions of lysosomes. From Wattiaux (118).

cesses with the only anabolic processes being lipid synthesis and possibly hydrolase synthesis.

In animals, lysosomes play a role in digestion and excretion in the cell. Primary lysosomes are pinched off as vesicles from the Golgi apparatus and contain hydrolytic enzymes. These vesicles can fuse with other vacuolar elements in the cell to form autophagic vacuoles or secondary lysosomes. The schematic representation (Figure 1) of the formation of primary and secondary lysosomes in animals was presented by Wattiaux (118).

The schematic figure shows that it is difficult or impossible to identify lysosomes on the basis of morphology alone. Part of the Golgi apparatus itself is technically a lysosome, as are variously shaped vesicles and vacuoles with a variety of contents. To identify lysosomes clearly, information must be obtained through histochemical tests for hydrolytic enzymes with the light and electron microscope or cellular fractionation and enzyme analysis.

Lysosomal Concept in Higher Plants

Our present understanding of higher plant lysosomes is rudimentary. We agree that some of the hydrolytic enzymes in plant cells are compartmentalized, but disagree as to where these compartments are located. Originally plant spherosomes were thought to be the primary sites of lysosomal enzymes (6, 8, 37, 42, 51, 72, 98, 103, 110, 115, 116, 123). However, recent workers contend that the greatest concentration of hydrolytic enzymes is packaged in small vacuoles (9, 44, 71, 95) and/or vesicles produced by the Golgi apparatus (62).

The histochemical and cellular fractionation techniques used to detect lysosomes are variable and the results vary with the investigator, the physiological state and location of the tissue, and the plant species involved. Initial studies have attempted to focus on one particular cellular inclusion as a lysosome rather than considering a lysosomal system in the cell. Therefore, discrepancies are understandable.

In spite of these shortcomings a lysosomal concept is developing for higher plants. Gahan (35) and Matile (63) have recently summarized our knowledge of plant lysosomes and each has come to a different conclusion. Gahan contends that there is enough divergence in the enzyme content and behavior of plant and animal lysosomes to make them distinct. He proposes that the term "spherosome" rather than "lysosome" be used for single-membrane-bounded bodies containing hydrolytic enzymes in plants. This proposal was premature; subsequently a number of similarities have been found between animal and plant lysosomes. Matile emphasizes the importance of vacuoles as plant lysosomes and presents a lysosomal system based on the interaction of vacuoles in the living cell. He also points to the origin of primary lysosomes from the endoplasmic reticulum (ER).

The schematic drawing (Figure 2) is based on our present knowledge of higher plant and fungal lysosomes. It is presented as a working model for further discussions of disease processes in the cell. To give some substance to this model we should look in detail at evidence for its various components.

GOLGI APPARATUS   The Golgi apparatus in plants consists of stacked cisternae called dictyosomes. Dictyosomes produce secretory vesicles that may have diverse function in the translocation of compounds in the cell. In fact, dictyosomes in the same cell may produce vesicles with different contents. The best documentation of the function of the Golgi apparatus in plants is in the accumulation and secretion of polysaccharides, particularly in wall synthesis (81, 83, 121).

Lysosomes have been reported to originate both from the Golgi apparatus and from vesicles produced by the ER. Although there is good evidence of this in animals, very little is known of the origin of lysosomes in plants. Marty & Buvat (62) have recently shown dictyosome-produced vesicles that contain acid phosphatase in *Euphorbia characias* L. These vesicles fuse with autophagic vesicles to form secondary lysosomes, a system quite comparable with that found in animals.

Although acid phosphatase and other hydrolytic enzymes have been found in cisternae of the Golgi apparatus in plants (27, 84, 108), Dauwalder et al (24) warn that this does not necessarily point to the Golgi as the origin of lysosomes. In their study of developing *Zea mays* L. root tips, they were able

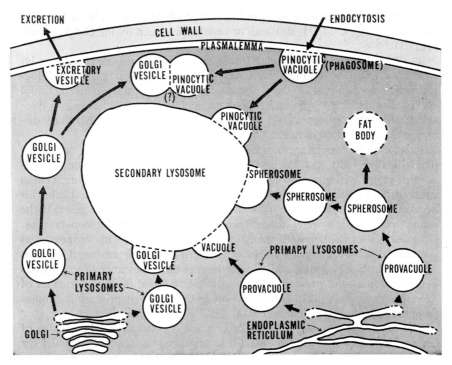

Figure 2 Schematic representation of possible lysosomal behavior in higher plants and fungi.

to find acid phosphatase within the Golgi apparatus in only a few cells near the apex of the root. They were also unable to find structures clearly identifiable as lysosomes in these cells.

VACUOLAR SYSTEM (VACUOME)   The vacuolar system in animals has been demonstrated to be part of the lysosomal system (26) and there is mounting evidence that this is also the case in plants. There is confusion over what constitutes a vacuole in plants, what vacuoles should be called, and how they function. According to Matile (63) "The term vacuole designates an empty space, that is, a particle lacking internal structures." Such a definition would not accommodate the "secondary vacuoles" described by Mahlberg (59, 60), which contain numerous microvesicles. Berjak (9) contends that vacuoles are organelles that show a developmental sequence that establishes the central vacuole of mature cells.

Fixation and embedding procedures may remove the contents of single-membrane-bounded bodies making them appear as empty vacuoles in electron micrographs. With this in mind, let us examine the origin of vacuoles in plants and their fate where it is known.

Berjak (9) has recently presented evidence that vacuoles in cress root cells originate by vesiculation of the ER. Acid phosphatase was also found associated with the membranes that formed provacuoles. Such activity was not found in adjacent ER. Berjak has apparently been able to pinpoint the site of acid phosphatase synthesis and provacuole formation. She observed provacuoles formed by both terminal and intermediate vesiculation of the ER. Hall & Davie (44) found the highest concentration of β-glycerophosphatase to be associated with small vacuoles in *Zea mays*.

Matile (63) also believes that provacuoles and vacuoles are derivatives of the ER. He has found the same ribosome-like particles sculpturing the outer surface of vacuoles that are present in the ER. Matile & Wiemken (71) found that a yeast vacuolar membrane contained enzymes characteristic of the ER.

Using freeze etching techniques, Matile & Moor (69) conclude that provacuoles (primary lysosomes) originate from the ER. Provacuoles fuse with one another and inflate to form small vacuoles that on fusion with provacuoles or other vacuoles increase in volume. The large central vacuole represents the end product of this process. Vacuoles also have a meristematic capability and can multiply by evagination and constriction of separate vacuoles.

Although the central vacuole in plants is thought of as an internally nonstructured organelle, in reality there are generally many small vesicles within the tonoplast and other organelles and membranes apparently in the process of intracellular digestion, synthesis, or deposition. More work is needed to explore the possible segregation of biochemical events within the central vacuole itself.

PINOCYTIC VACUOLES (PHAGOSOMES)   Pinocytic vacuoles are generally not lysosomes themselves as they usually do not contain hydrolytic enzymes. However, they are part of the lysosomal system. Mahlberg (59) has recently shown that pinocytic vacuoles ("secondary vacuoles") are formed in living trichomes of *Tradescantia virginiana*. Such vacuoles are formed by the invagination of the plasma membrane. These vacuoles upon detachment from the plasma membrane are transported throughout the cell by cyclosis. Most of the vacuoles had vesicular inclusions. Inclusions resembling "secondary vacuoles" were also observed in the central vacuole, which indicates a fusion or engulfment of "secondary vacuoles" by the tonoplast.

Mahlberg (60) also found that neutral red accumulated in structures that corresponded to "secondary vacuoles." After uptake by the "secondary vacuoles" the central vacuole also showed accumulation of neutral red. This accumulation indicates that the "secondary vacuoles" had either fused with the vacuole or their contents had been transferred into it.

The presence of a cell wall has restricted observations of pinocytosis in plant cells. Pinocytosis has been proposed as the mechanism for the uptake of macromolecules by plant cells (74), and Mahlberg (5, 60) was the first to show clearly that pinocytosis can occur in intact plant cells. His electron micrographs of "secondary vacuoles" resemble lomosomes described in other plant cells (28).

SPHEROSOMES   There is confusion over the nature of the membrane that surrounds spherosomes, their enzyme content, and the relation of spherosomes to the other lysosomal components in the cell (32, 48, 77, 79, 110). Part of this confusion appears to have resulted from the diversity of spherosomes in different plants and tissues. Other points of confusion have arisen because different criteria have been used to define spherosomes, particularly with the light and electron microscope.

Fine structure studies of spherosomes have shown them with (110) and without (48) membranes. These discrepancies are perhaps explained by Schwarzenbach (100, 101) in his study of spherosomes of *Ricinus communis* endosperm cells. He found that immature spherosomes (prospherosomes) are bound by a unit membrane, the mature spherosomes are not. Spherosomal membranes normally undergo a differentiation process during seed development resulting in the separation of the inner and outer layer of the original unit membrane. The inner layer is left surrounding the mature spherosome.

The enzymatic composition of spherosomes is poorly understood. Although there are numerous histochemical studies with the light microscope reporting concentrations of hydrolytic enzymes in spherosomes, it has been *assumed* that the reactive particles are spherosomes. This assumption is based on the similar size and distribution of reactive particles and spherosomes. Recent reports of the packaging of hydrolytic enzymes in small vacuoles raises the question as to whether early "spherosomal particles" were in fact small vacuoles. Recent histochemical studies at the electron microscope level show acid phosphatase concentration primarily in small vacuoles (9, 44, 71, 95).

Enzymatic analyses of spherosomes in cellular fractions are scarce because of the difficulties in fractionating plant cells. Matile & Spichiger (70) and Balz (6) found acid protease, phosphatase, and esterase activity concentrated in spherosomal fractions from tobacco and corn. Semadeni (103) detected the presence of three particulate fractions carrying hydrolases in corn seedlings. The heaviest fraction contained protease, phosphatase, RNase, and acid esterase. A higher fraction contained the same acid hydrolases, glucose-6-phosphatase, arylsulfatase-C, and small amounts of amylase activity. By staining with fluorochromes the structures containing hydrolases were identified as two kinds of spherosomes and fragments of ER.

Semadeni (103) found that spherosomes in corn seedlings contained all the enzymes necessary for lipid synthesis. He concluded that the high activity of lipid synthesis in the spherosomes points to the conversion of these bodies into lipid bodies. Jacks et al (55) recently reported that in isolated peanut spherosomes the activity of lipase and fatty acid–(Coenzyme A) synthetase is not associated with the isolated spherosomes. They suggest that peanut spherosomes are principal sites of lipid storage but not lipid degradation. Lipase has been found associated with spherosomes in *Ricinus* (86).

From the foregoing discussion we can conclude that much more research is needed to clarify the role of spherosomes in the lysosomal system in plants. My opinion is that a number of biochemically distinct spherosomes will be found, some functioning as lysosomes and others as storage organelles. Once these have been more clearly delineated, we may want to revamp the term "spherosome" and give it more specific meaning.

Undoubtedly some organelles now called spherosomes are involved in the lysosomal system. The question is whether they serve as primary or secondary lysosomes. Because spherosomes apparently arise from the ER, they may be primary lysosomes. No clear fusion of spherosomes with vacuolar components of the cell has been demonstrated. However, spherosome-like bodies within vacuoles are common (38). Spherosomes may serve as primary lysosomes during certain periods of cellular development and as storage organelles during other periods. Also, spherosomes may function differently in different tissues.

In histochemical studies there has been overreliance on the detection of one hydrolytic enzyme, acid phosphatase, to demonstrate lysosomes. Matile (63) states: "Acid phosphatase seems to represent a doubtful marker enzyme of lysosomes; the simplicity of its assay and its easy cytochemical demonstration have raised the popularity of this enzyme out of all proportion to our knowledge about its functional significance." Recently some workers (22, 99) have questioned the validity of certain interpretations using the Gomori reaction.

Function of Lysosomes in Higher Plants

We must consider the normal functioning of plant lysosomes because their malfunction may account for certain plant disease processes. Undoubtedly the major function of plant lysosomes is the compartmentalization of hydrolytic enzymes so that normal catabolic reactions can occur in the same cell with synthetic processes. Very little is known about the detailed function of plant lysosomes in metabolism.

Concentrations of hydrolytic enzymes have been found in sites other than particles in the cytoplasm, such as the cell wall (20), nucleus (43), and ER (98). Heavy concentrations of hydrolytic enzymes have been found in the cell walls, particularly in root tissue (20, 43–46, 104, 105). The outer cells of the roots show most of the activity. Such enzymes probably play important roles in the digestion and movement of nutrients by the plant. Lysosomes probably have transported these enzymes to such sites, but *de novo* synthesis cannot be ruled out.

When seeds germinate there is extensive autodigestion and turnover of metabolites. We can reasonably assume that lysosomes play an important role in this process. In fact, lysosomal behavior may explain the initiation of the germination process. Meyer et al (76) showed evidence for the activation of acid phosphatases in germinating lettuce seed. They were also able to liberate phosphatase activity by treatment of cell homogenates with detergents and trypsin, indicating that these enzymes were compartmentalized in the cell. Flinn & Smith (31) concluded that in cotyledons of *Pisum arvense* the acid phosphatase is localized initially in lysosome-like particles and is released during early stages of germination to aid in some way in the mobilization of the storage reserves.

Lysosomes probably play a key role in the mobilization of reserve nutrients in the cell. Vacuolization is a common event at the initiation of growth and such vacuoles probably serve in the digestion of food reserves. Matile (68) has pointed to the aleurone vacuoles in plants as a lysosome that contains stored protein and hydrolytic enzymes. He has indicated that the pattern of acid hydrolases localized in aleurone vacuoles (proteases, phosphatase, RNase, β-amylase and α-glucosidase) suggests that not only the aleurone grain protein and phosphate reserves but also other cell constituents are broken down in this lysosome.

Secondary lysosomes or autophagic vacuoles are common in higher plants. Such vacuoles contain various organelles, membranous elements or myelin-like bodies (23). These vacuoles may serve in the "recycling" of cellular components during growth and differentiation. Coulomb & Buvat (23) have described such a lytic process in *Cucurbita pepo*.

During mitosis in animal cells there is increased permeability of the lysosomal membrane (27). Gahan (35) found no such behavior in mitotic cells from root and shoot meristems of *Solanum lycopersicum* ($=$ *Lycopersicon esculentum*). However, Olszewska & Gabara (85) suggest a relationship between lysosomes and cytokinesis in a number of other plants.

Sorokin & Sorokin (109) have shown that the physiological state of stomata of *Campanula persicifolia* is related to the enzymatic activity of spherosomes within the guard cells. Under conditions of decreased turgor in closed and partly closed stomata, acid phosphatase is clearly demonstrable in the spherosomes of guard cells. Conversely, under conditions of increased turgor in guard cells, the spherosomes are always unreactive for acid phosphatase. One can speculate that lysosomes play a key role in stomatal opening and closing by regulating osmotic concentrations in the cell.

Cellular differentiation in plant tissues may involve lysosomes. Gahan & Maple (33) observed that the terminal stage of xylem differentiation in root tips is accompanied by a change from a particulate to a diffuse distribution of acid phosphatase. Wardrop (117) found in *Chara* with the development of the secondary wall that the level of acid phosphatase and β-glucosidase increased in the cytoplasm. Before this stage both enzymes were confined to particles. Paralleling these changes in enzyme distribution the plasmalemma developed an irregular scalloped outline and various vacuolar bodies appeared.

Mlodzianowski (78) and Nougarede & Pilet (84) have found acid phosphatase associated with developing chloroplasts and they indicate that hydrolases may be involved in the differentiation process. Nougarede & Pilet were able to demonstrate that the acid phosphatase in plastids of *Lens culinaris* Medic. apparently originated in the dictyosomes.

Senescence of tissue, if it is genetically controlled, can be considered a form of differentiation (or dedifferentiation). Matile & Winkenbach (67) found that the rapid senescence of the *Ipomoea* corolla is characterized by the breakdown of proteins and nucleic acids. At the onset of wilting, the activities of deoxyribonuclease, ribonuclease, and β-glucosidase increase dramatically. Because Actinomycin D inhibited the increase in RNase it was concluded that protein synthesis was a prerequisite for the changes in enzyme activity in the senescing corolla.

Butler (17) outlines the cytological changes that occur in senescing cotyledons of cucumber. He was not able to identify a single initial cytological change that might initiate senescence. However, he suggests that the initial damage to the cell could result from the release of hydrolytic enzymes.

Berjak & Villiers (11) studied senescence of the root cap of *Zea mays* during germination. An association between ER and lysosome-like bodies was seen during germination, together with an intensification of the acid phosphatase reaction within the lysosome. In the outermost cells the lysosomes ruptured and acid phosphatase activity became detectable in the ground cytoplasm. The sequence of events leading to senescence of root cap cells led Berjak & Villiers to conclude that such senescence is genetically controlled.

There are a number of examples in animal cells where lysosomes serve as secretory vesicles in which cellular materials are accumulated and transported (27). Figier (30) found acid phosphatase concentrations in the Golgi, ER, and plasmalemma of *Vicia faba*. He suggests that there is an active transport system for the movement and excretion of sugars involving lysosomal enzymes.

The foregoing isolated reports indicate that lysosomes are important in a number of diverse metabolic functions in plant cells. Although association of lysosomal behavior with cellular events is apparent, direct cause and effect relationships have not been clearly shown. Much more research is needed in this area.

Lysosomal Concept in Plant Pathogens

Hydrolytic enzymes are extremely important in pathogenesis, because they provide the pathogen a means of chemical ingress into the host and a process whereby nutrients can be digested. Many studies have been made on the hydrolytic enzymes produced by plant pathogens, but very little is known about the compartmentalization and deployment of these enzymes within cells of the pathogen.

Lysosomes have been found in fungi (66, 92, 96, 102, 122) and nematodes (27). The lack of elaborated membranes in bacteria and mycoplasma may preclude the existence of lysosomes in these organisms.

Only with the fungal pathogens has enough information been accumulated to allow the synthesis of a concept of lysosomal behavior. A lysosomal system comparable with that in animals and higher plants may exist in fungi. However, the same unanswered questions exist concerning the nature of spherosomes and the origin of primary lysosomes. Because a Golgi apparatus has not been found in some fungi (13), the origin of primary lysosomes in these fungi needs clarification.

SPHEROSOMES   As with higher plants it is thought that spherosomes are primary sites of hydrolytic enzymes in fungi (122). Matile & Wiemken (71) introduced the idea that the only lysosome in yeast is the central vacuole and that particulate concentrations of hydrolase in filamentous fungi may be small vacuoles rather than spherosomes. Other workers (41) have shown that there are small particulate concentrations of hydrolytic enzymes in yeast in addition to the central vacuole.

Smith & Marchant (107) illustrate the fusion of spherosomes in *Saccharomyces cerevisiae* Hansen with the central vacuole. They present the framework of a lysosomal system that starts with spherosomes (primary lysosomes) that originate from the ER. The spherosomes migrate and release their contents into the central vacuole forming a secondary lysosome.

GOLGI APPARATUS   The Golgi apparatus in fungi is poorly understood. Recognizable dictyosomes have been found in the Phycomycetes (13, 40). There are only two examples of a Golgi apparatus in higher fungi. One is in the Ascomycete *Neobulgaria pura* (Fr.) Petri (80) and the other in a yeast (112). Bracker (13) states: "it is apparent that most plant pathogens are in groups that do not possess this organelle." Although there is not a recognizable Golgi apparatus in a number of fungi, it can be assumed that there are specialized secretory areas of the ER. Such rudimentary dictyosomes appear to exist in the yeast *Saccharomyces fragilis* Jorg. (112) and perhaps in other fungi.

Grove et al (40) have shown a membrane conversion in the Golgi of *Pythium ultimum* Trow. They demonstrate a transition in membrane morphology across stacks of dictyosome cisternae from endoplasmic reticulum-like at one pole to plasma membrane-like at the opposite pole. They conclude: "a major function of the Golgi apparatus is to elaborate secretory vesicles where limiting membranes can fuse with plasma membranes."

DeDuve (25) has presented a hypothesis on the segregation of the vacuoles in cells that is supported nicely by the work of Grove et al. In order not to distort DeDuve's hypothesis I will quote it directly.

On the bases of present information, we may view the vacuome as a system endowed with the same type of continuity as that existing in a chemical factory comprising a complex network of interconnected vats and pipes, controlled by valves that are intermittently opened and shut in a non-synchronous manner. The space delimited by the system is always discontinuous at any given time, but can nevertheless be traversed in its entirety thanks to the shifting of the boundaries that divide it. That the vacuome possesses this kind of continuity is proven by the flow pattern of the biosynthetic products made in the innermost cisternae of rough-surfaced endoplasmic reticulum. Some of them are channelled through the secretory line right up to the extracellular space. Others, such as the acid hydrolases, follow an initially similar line, but are then divided into the phagosome-lysosome system, from which they may eventually be ejected also into the extracellular medium.

However, this functional continuity exists only in the outward direction. Objects that enter the vacuome by endocytosis can invade an appreciable part of it, as has been clearly indicated with a number of endocytic markers; but they never reach the innermost parts of the system. To my knowledge at least, there is no image or record of an endocytized substance ending up in the lumen of a rough-surfaced cisterna of the endoplasmic reticulum. Therefore, the vacuome consists of two distinct spaces interconnected by a *one-way lock* which allows materials to move from the inner to the outer-space, but not in the reverse direction. The inner space corresponds roughly to that bounded by the endoplasmic reticulum and may appropriately be called the *endoplasmic space*. I propose to designate the outer space as the *exoplasmic* space.

DeDuve feels that the membrane conversion that occurs in the Golgi apparatus is the locus for the "one-way lock." He also believes that the thickening of ER-like membranes to become plasmamembrane-like is through the addition of glycoproteins. Such an addition prevents these membranes from fusing with the ER, thus effecting the one-way conversion of membranes.

VACUOME   Although definitive evidence is scarce, there is probably a vacuolar system in fungi comparable with that in animals and higher plants. Matile & Wiemken (71) have been able to isolate the central vacuole of *Saccharomyces cerevisiae* and analyze its enzyme content. They found two acid endopeptidases, p-nitrophenylacetate-esterase, and RNase. Activity of these enzymes was 20–30 times greater in the vacuoles than in the lysed protoplasts. Because they were unable to detect any other structures containing hydrolases, they assumed that the vacuoles are the only lysosomes of yeast cells. Matile (63) infers that the particulate concentrations of hydrolytic enzymes in fungal hyphae described by Pitt (92) are vacuoles rather than spherosomes.

Repeated photomicrographs of fungi show osmiophilic bodies (spherosome-like) within vacuoles (3, 53). These bodies may appear within the vacuole or in close proximity. As mentioned, Smith & Marchant (107) have shown such bodies fusing with the central vacuole in yeast. Such associations can be interpreted at least two ways. If the osmiophilic bodies contain hydrolytic enzymes, these bodies may be serving as primary lysosomes that deposit their enzymes into the vacuole (or secondary lysosome). If these bodies do not contain hydrolytic enzymes, their fusion with vacuoles may provide a way of mobilizing the lipid and protein that they contain.

Vesicles or a vesicular network may be present in fungal vacuoles (14, 39, 52). Autophagic vacuoles with membrane debris are also common (113). Myelin-like figures that are typical of animal lysosomes are also identifiable (73). The function of the internal compartmentalization of fungal vacuoles is not well understood.

PINOCYTIC VACUOLES (PHAGOSOMES)   Pinocytosis has not been clearly demonstrated in fungi. Carroll (19) suggests that lomasomes serve in micropinocytosis during ascosporogenesis in *Saccobolus keruerni*. Some lomasomes may be pinocytic vacuoles. Bracker (13) has stated that fungal lomasomes "have gained much prominence on little information." We literally don't know whether lomasomes are coming or going. Some workers favor the idea that they serve in the excretion of enzymes and wall materials and others that they function in absorption.

I have been able to follow the apparent pinocytic uptake of neutral red by living spores of *Ceratocystis ulmi* (Fig. 3). Conidia of *C. ulmi* were placed in an aqueous solution of neutral red (0.001 g of neutral red in 200 ml of water) and examined with Normarski differential interference contrast. Spores with extra large vacuoles were observed in order to get a clearer view of the initial formation of neutral-red-containing bodies.

Apparent pinocytic invaginations of the outer spore membrane can be seen to protrude into large vacuoles (Fig. 3, A-E). Such protrusions are red and appear to be connected to the plasmalemma. They enlarge rapidly and pinch off neutral-red-containing bodies from their tip into the vacuole. These bodies continue to enlarge and move freely within the vacuole (Fig. 3, F-J).

The initial invagination into the vacuole can be seen within a minute after the spores come in contact with the neutral red. The entire pinocytic process resulting in neutral-red-bodies in the vacuole generally occurs within 7 minutes. Some neutral-red-containing bodies lose their color in 20–30 minutes after they appear, indicating that the neutral red is transformed or excreted (Fig. 3, J).

Lysosomes in the Normal Physiology of Pathogens

Again our discussion must be limited to the fungi because so little is known about lysosomes in other plant pathogens. Although the information is sketchy, it suggests that lysosomes function in fungal growth and differentiation.

Page & Stock (88) studied the role of lysosomes in the germination of macroconidia of *Microsporum gypseum* (Bodin) Guiart. They found alkaline protease-containing vesicles in the spore coat before macroconidial germination. Hydrolized areas were seen around the vesicles during germination. Other vesicles, believed to contain phosphodiesterase, were also observed in

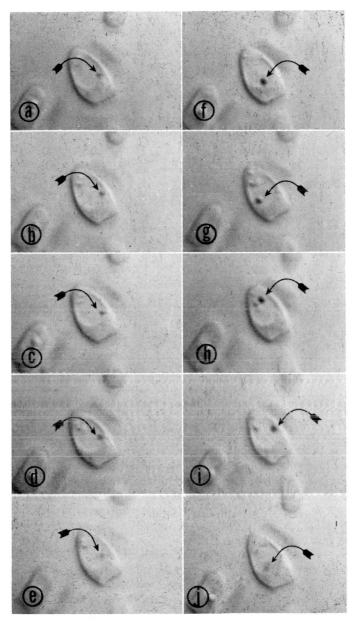

Figure 3 Apparent pinocytic uptake of neutral red into a spore of *Ceratocystis ulmi*. Arrows point to a neutral red body during formation. Seven minutes elapsed between A and I. Twenty minutes elapsed between I and J.

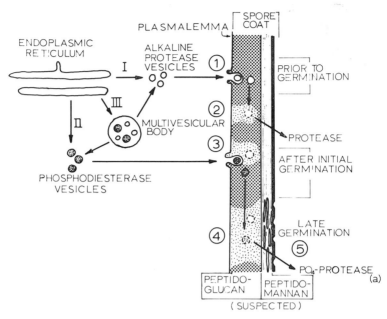

Figure 4 Proposed schematic portrayal of the origins, distribution, and function of germination lysosomes of *Microsporum gypseum*. From Page & Stock (88, 89).

the spore coat during germination. It was postulated that the electron transparent vesicles containing the most alkaline protease activity are inserted into the spore coat before germination and that the electron dense vesicles containing the greatest phosphodiesterase activity are inserted after germination and the release of alkaline protease. This would allow spore coat hydrolysis to precede protease inactivation (89) during germination. Figure 4 is the schematic diagram that Page & Stock used to portray the origin, distribution, and function of lysosomes during the germination of *M. gypseum*.

There are numerous cytological studies of spore germination in fungi (3, 13, 16, 49, 50, 53, 111). The vacuolar system in spores undergoes changes during germination. Vacuoles have been seen to enlarge, increase in number, and have their contents modified.

Spore germination requires the mobilization of numerous stored substances from which new cytoplasm can be synthesized. Because vacuoles appear to be the primary sites of hydrolytic enzymes, we can assume that they are also the primary sites for the mobilization and perhaps synthesis of new metabolites. In *Ceratocystis ulmi* certain lysosomes appear to be membrane bound within the vacuole of spores (unpublished). This compartmentalization would perhaps allow hydrolysis within the lysosome and synthesis in other parts of the vacuole. Vacuoles commonly have internal vesicles that could segregate catabolic and anabolic activities.

Buckley et al (15) found dark-appearing "storage bodies" in the conidia of *Botrytis cinerea* Pers. ex Fr. before germination. These single-membrane-bound bodies are thought to contain lipid or phospholipid. Upon germination they form vacuoles that contain whorled structures, spherules, granules, and membrane loops. It is suggested that these inclusions provide material for the assembly of membranous organelles during germination.

Gay & Greenwood (36) have described "cleft-vesicles" in *Saprolegnia* that have dark-staining contents. These vesicles are numerous at the onset of sporangium development and are fewer in the zoospores. The myelin-like figures that appear in these vesicles are interpreted by Gay & Greenwood as evidence that membrane synthesis is occurring. However, myelin-like figures are interpreted by some workers to indicate autolysis (113).

Remsen et al (97) contend that the compact, membrane-bound, electron dense bodies contained in the vacuoles of *Penicillium megasporum* Orpurt and Fennell probably have a role in conidial germination. They infer that these bodies may be lysosomes that trigger germination through the release of hydrolytic enzymes.

Lysosomes may play a key role in the fructification of certain fungi. The formation of fruiting structures by fungi in some cases requires the mobilization, translocation, and concentration of large quantities of metabolites. There must be considerable autolysis and translocation of materials in wood-rotting basidiomycetous hyphae in trees prior to the production of fruiting bodies. Autolysis within secondary lysosomes is probably involved in part of this process.

Spores of a number of fungi are borne within a fluidal matrix. Such matrices would presumably be formed through the digestion of cells or through excretion. Wilson et al (122) found a relation between the behavior of lysosome-like bodies and the deliquescence of asci in *Ceratosystis fimbriata* Ell. and Halst. At ascospore maturity, lysosome-like bodies appeared to rupture and cause the digestion of the ascus and the formation of a sticky matrix in which the spores are embedded.

Some basidiomycetous fruiting bodies characteristically lyse after basidiospore formation. Iten & Matile (54) found that autolysis of mature fruiting bodies of *Coprinus lagopus* Fr. is caused by the digestion of cell walls. After sporulation, chitinase activity, which is localized in vacuoles, is released and causes cell wall degradation. It was not clear whether the cells actively secrete chitinase and thus induce their own autolysis or whether passive release occurs from cells whose metabolic activity had ceased completely.

During fungal sporulation it can be assumed that there is considerable "recycling" of cellular components. Thornton (113) found that certain endoplasmic cisternae of the sporangiophores of *Phycomyces blakesleeanus* Burgeff are engaged in the isolation and lysis of cytoplasm. The occurrence of nuclei with widely separated inner and outer membranes and the extension of nuclear membranes into the cytoplasm suggested to Thornton that the vesicle membranes originate from nuclear membranes.

It has been demonstrated repeatedly that plant pathogenic fungi can secrete hydrolases (75). Lysosomes probably play a role in this process. Matile (65) has shown that acid proteases in *Neurospora crassa* Shear and B. O. Dodge are stored intracellularly in a special type of lysosome called a "protease particle." Matile feels that protease particles can function in the secretion of protease or they can fuse with the vacuoles. Calonge (18) has indicated that multivesicular bodies in *Sclerotinia fructigena* Aderh. and Ruhl. may function in the secretion of extracellular enzymes. Lysosomes may also function in the secretion of pigments such as the "pigment granules" of *Endogone* (82).

Lysosomes in Host-Parasite Interactions

Lysosomes in the host and parasite may play important roles during host-parasite interactions. Hydrolytic enzymes are of prime importance to a parasite, as they degrade the host into utilizable nutrients. Hydrolytic enzymes within the host could also play important roles in degrading parasite enzymes and toxins. A parasite could decompartmentalize host lysosomes and thus lead to its autolysis and vice versa.

PREINFECTION INTERACTIONS   Hydrolytic enzymes are present in the wall of higher plants (57), fungi (21), and bacteria (7). Presumably these enzymes are excreted by lysosomes because no site for their synthesis has been found in the wall.

When a host and parasite approach one another, a battle between their enzymes may precede their actual contact. The parasite may digest the host cell wall if it excretes the proper enzymes in proper concentrations or, as in the case where chitinase has been found in bean leaves (1), the host may digest the cell wall of the parasite. A battle can also be theoretically waged between the host and parasite enzymes themselves with one perhaps winning by its ability to degrade or inactivate the enzymes of the other.

Before infection the endogenous lysosomal system of the host and parasite may be affected by extracellular enzymes. Such effects can lead to the autolysis of host and/or parasite cells through the decompartmentalization of hydrolases and their subsequent attack on the "endoplasmic space." I could find no explanations of this nature in plant pathology literature, but such a process might explain the action of certain toxins. Lysosomal disruption has been found to occur in animal cells in response to certain bacterial toxins (12).

CELLULAR PENETRATION   Penetration of cell walls may take place through mechanical and/or enzymatic means. Mechanical penetration itself may induce a response in host lysosomes, because cellular injury can cause an increase in hydrolase activity (124).

When the parasite is dependent on enzymatic digestion of the host wall, hydrolases are of primary importance. Webber & Webber (119) found a concentration of "lysosome-like organelles" in the penetration peg of a lichen haustorium of *Parmelia sulcata* Tayl. They postulated that these bodies were important in providing the enzymes for cell wall penetration and perhaps digestion of host cytoplasm. Armentrout & Wilson (5) noticed a migration of lysosome-like bodies into the appressorium of the mycoparasite *Mycotypha microspora* Fenner on *Piptocephalis virginiana* Leadbeater and Mercer. After penetration, the haustorium and sheath showed a concentration of acid phosphatase.

NONBALANCED PARASITISM   Most plant pathogens kill host cells in advance of invasion. Among the chemical weapons (enzymes, growth regulators, and toxins) that cause disease and necrosis, hydrolases are by far the most important. Apparently, the parasite may benefit not only from digestion of host substrates by its own hydrolytic enzymes, but also from hydrolases released by the host in response to invasion (autolysis). A portion of the "action in advance" in host tissue may be explained by the disruption of the host lysosomal system. Hydrolases released from compartmentalized lysosomes in one host cell may act on lysosomal membranes in adjacent cells and thus set in motion a chain of "reactions in advance."

Pitt & Coombes (93, 94) have shown that infection of *Solanum tuberosum* L. tissue by *Phytophthora erythroseptica* Pethyhr., *Phytopthora infestans* (Mont.) d By., and *Fusarium coeruleum* (Lib.) Sacc. caused swelling and disruption of host cytoplasmic particles containing acid phosphatase, esterases, and proteases. They also demonstrated a decrease in acid phosphatase and RNase activities in the particulate fractions of infected cell homogenates and increases in the activity of these enzymes in the supernatant fluid fractions.

We can assume that the release of hydrolases into host cells results in digestion of cellular components and the provision of a wider range of nutrients for the parasite. Direct evidence is needed on the effects of lysosomal disruption on other organelles in plant cells. Weissman et al (120) were able to show that the liberation of lysosomal enzymes in rabbit tissue preceded liberation of malic dehydrogenase from mitochondria of these preparations. As Pitt & Coombes (93) indicate, similar studies are needed in diseased plant cells.

Increased liberation of hydrolases in diseased cells does not necessarily lead to necrosis. Sheikh et al (106) found that there was increased activity of acid β-glycerophosphatase, acid naphthol-AS-BI-phosphatase, and naphthol esterase during the induction of crown gall in *Lycopersicon esculentum* Mill. by *Agrobacterium tumefaciens* (Smith and Townsend) Conn. They suggested that the increased esterase activity may play a role in cell wall development of the maturing gall. An investigation of possible lysosomal involvement in other plant diseases with tissue overgrowth would be interesting.

The relationship of lysosomes to plant virus infection needs investigation. Allison & Mallucci (4, 61) have indicated that certain animal lysosomes may be involved in "uncoating processes" and the initiation of certain pathological responses to viral infection. They suggest that the participation of lysosomes in two other processes initiated by viruses, namely polykaryocytosis and malignant transformation, deserve consideration. Otsuki et al (87) have shown that tobacco mosaic virus can enter tobacco mesophyll protoplasts via pinocytosis.

Kazama & Aldrich (56) studied acid phosphatase localization during the digestion of *Escherichia coli* (Migula) Castellani and Chalmers by the Myxomycete *Physarum flavicomum*. Berk. They found acid phosphatase localized in food vacuoles, dictyosomes, membrane-bound vesicles, and tentatively, in smooth endoplasmic reticulum. Kazama & Aldrich suggest that there may be a digestive process in *P. flavicomum* comparable with that found in various protozoans.

BALANCED PARASITISM   Ellingboe[2] presents the following hypothetical involvement of lysosomes in balanced parasitism:

> Electron microscopy has shown what appear to be invaginations of host plasma membranes surrounding haustoria. Assuming these observations to be correct, a means of bringing host organelles into the parasite can be visualized. Invaginations of host plasma membrane may contain both host cytoplasm and host organelles. These small packets of host material could coalesce with lysosomes of the parasite to form secondary lysosomes. Host organelles could be digested in the lysosomes and constituents of the host cytoplasm and organelle products could go to the parasite through the membrane of the lysosomes. The disposition of parasite lysosomes, if necessary for a growing parasite, may be the reverse of the procedure for bringing host material into the parasite. What has been observed as invaginations of host plasma membrane into the parasite, and interpreted as a means for getting host material into the parasite, may also be involved in getting material out of the parasite. Blebbing on the outside of the haustorial sheath may be a similar process for exchange of cytoplasm and organelles. Proof that these interpretations of function are correct is still lacking, as is any conjecture of function made from still pictures.

CELLULAR INJURY   Plant lysosomes may respond to cellular injury other than that produced by living pathogens. Petrovskaya-Baronova (91) found that the destruction of organelles in wheat-quack grass during freezing was due to an increased activity of hydrolytic enzymes, presumably resulting from damage to lysosome-like organelles.

Wyen et al (124) point out that the number of hydrolytic enzymes, including ribonuclease, phosphodiesterase, phosphatase, and peptidase, increases in responses to leaf excision. They believe that the increase in acid phosphatase found in excised *Avena* leaves is due to an increase in the number of lysosomes. Increased hydrolase production when leaves are excised is generally explained as an "accelerated senescence" of the leaf.

---

[2] Ellingboe, A. H. 1968. Inoculum production and infection by foliage pathogens. *Ann. Rev. Phytopathol.* 6:317–30

Evans & Miller (29) found that acid phosphatase activity increased within mesophyll cells of ponderosa pine during ozone exposure. Lysosomes also may be involved in host responses to other types of pollutant injury.

Berjak & Villiers (10) have found that lysosome-like bodies may have been responsible for the removal of certain organelles in *Zea mays* cells that were responding to age-induced injury.

Lysosomes and Plant Disease Resistance

As indicated previously, enzymes secreted by lysosomes may play a role in host resistance by attacking the pathogen or degrading its enzymes. Because hydrolases appear to be abundant in host and pathogen cell walls, more attention should be given to their interaction as it relates to resistance.

Hypersensitivity is the most universally resistant response in plants to infection by pathogens. The hypersensitive reaction involves pathogen-induced morphological and histological changes which cause necrosis of the infected tissue as well as inactivation of the pathogen. Agrios (2) states, "Although the exact mechanism that triggers the hypersensitive reaction is still unknown, this type of defense against infection is one of the most important in plants."

Rapid cellular changes involving increased membrane permeability and the denaturing of protein with eventual necrosis characterize the hypersensitive reaction. Klement & Goodman (58) have shown that SH-containing compounds can mimic the hypersensitive reaction. They postulate that these compounds do so by denaturing protein. Such changes could account for increased membrane permeability.

Although we have information on the biochemical events that occur in host cells during the hypersensitive reaction, no accepted explanation has been given for the initiation of this process. I submit that the disruption of lysosomal membranes and the subsequent release of their proteases could account for the rapid degradation of protein and necrosis in cells. The release of hydrolases from lysosomes could also affect membranes of lysosomes in adjacent cells and result in a still further release of proteases and protein degradation.

Lysosomes and Fungicides

Originally it was thought that fungitoxicants accumulated on conidial surfaces where they reacted mainly with ligands. Subsequently it has been shown that most fungitoxicants are taken up and concentrated in the cell (114). However, very little is known about the specific mechanisms of uptake and the sites of activity of fungitoxicants.

The possibility that lysosomes may play a role in fungicide uptake and activity should be considered. Pinocytic vacuoles characteristically sequester foreign particles and chemicals which enter the cell. The lysosomal system, at least in animals, provides the easiest avenue for the entrance of foreign material through pinocytosis. Since pinocytosis appears to occur in fungal spores

H — HYDROLASES
F — FUNGITOXICANT

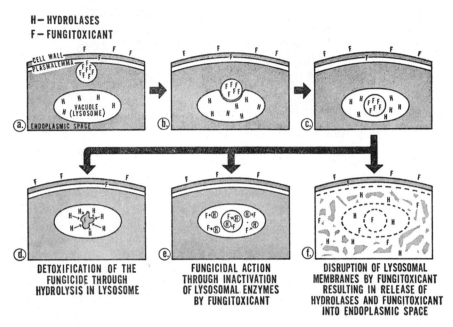

Figure 5 Proposed mechanisms for fungicidal action where fungitoxicants are taken up through pinocytosis.

(Fig. 4), the possible involvement of the lysosomal system in fungicidal activity is hypothesized (Fig. 5).

In animal cells heavy metals such as gold are taken into the cells through pinocytosis and sequestered in lysosomes (90). Because heavy metals are effective fungicides, their possible uptake into the lysosomal system in fungi should be investigated. If heavy metals are taken into the fungal cell through a pinocytic process and concentrated in lysosomes, this concentration could, in turn, cause disruption of lysosomal enzymes or membranes and subsequent death of the cell. Lysosomes could also be important as sites for detoxification of fungicides (Fig. 5).

We should keep in mind that lysosomes themselves within the fungal spore have tremendous fungicidal potential. These "suicide bags" are only 100 Å away from destroying the cell. The massive release of hydrolases into the cell would be expected to cause irreversible changes in the cytoplasm. Means of manipulating lysosomal membranes to effect these changes should be explored. Matile (64) found that when inositol-requiring mutants of *Neurosporia crassa* were grown in a suboptimum amount of exogenous inositol, necrosis occurred. Death was attributed to inositol deficiency in the protease particles (lysosomes), which allowed proteases to escape and degrade cellular protein.

Summary and Conclusions

There is evidence for the existence of lysosomal systems in higher plants and fungi comparable to those described in animals. These systems appear to be a vital part of the normal metabolism of these organisms and could serve specialized functions in such important phenomena as spore and seed germination, fructification, storage, autolysis, senescence, translocation, secretion, morphogenesis, disease resistance, pathogenicity, cellular injury, detoxification, and fungicidal activity.

Extensive research is needed to fill the gaps in our knowledge of higher plant and fungal lysosomes. The potential importance and possible application of research findings in this area should justify a greatly accelerated effort. In human pathology the lysosomal concept has led to an explanation and therapy for a number of major diseases (27). There is reason to believe that this could also happen in plant pathology.

Literature Cited

1. Abeles, F. B., Bosshart, R. P., Forrence, L. E., Habig, W. H. 1970. Preparation and purification of glucanase and chitinase from bean leaves. *Plant Physiol.* 47: 129–34
2. Agrios, G. N. 1969. *Plant Pathology*. Academic: New York & London, p. 629
3. Aitken, W. B., Niederpruem, D. J. 1970. Ultrastructural changes and biochemical events in basidiospore germination of *Schizophyllum commune*. *J. Bacteriol.* 104:981–88
4. Allison, A. C., Mallucci, L. 1965. Histochemical studies of lysosomes and lysosomal enzymes in virus-infected cell cultures. *J. Exp. Med.* 121:463–76
5. Armentrout, V. N., Wilson, C. L. 1969. Haustorium-host interaction during mycoparasitism of *Mycotypha microspora* by *Piptocephalis virginiana*. *Phytopathology* 59: 897–905
6. Balz, H. P. 1966. Intrazelluläre Lokalisation und Funktion von hydrolytischen Enzymen bei Tabak. *Planta* 70:207–36
7. Bayliss, M., Glick, D., Siem, R. A. 1948. Demonstration of phosphatases and lipase in bacteria and true fungi by staining methods and the effect of penicillin on phosphatase activity. *J. Bacteriol.* 55:307–16
8. Benes, K., Zdenek, L., Horavka, B. 1961. A contribution to the histochemical demonstration of some hydrolytic and oxidative enzymes in plants. *Histochemie* 2:313–21
9. Berjak, P. 1972. Lysosomal compartmentation: ultrastructural aspects of the origin, development, and function of vacuoles in root cells of *Lepidium sativum*. *Ann. Bot.* 36:73–81
10. Berjak, P., Villiers, T. A. 1972. Ageing in plant embryos. II. Age-induced damage and its repair during early germination. *New Phytol.* 71:135–44
11. Berjak, P., Villiers, T. A. 1970. Ageing in plant embryos. I. The establishment of the sequence of development and senescence in the root cap during germination. *New Phytol.* 69:929–38

12. Bernheimer, A. W., Schwartz, L. L. 1964. Lysosomal disruption by bacterial toxins. *J. Bacteriol.* 87:1100–04
13. Bracker, C. E. 1967. Ultrastructure of fungi. *Ann. Rev. Phytopathol.* 5:343–74
14. Brushaber, J. A., Jenkins, S. F., Jr., 1971. Lomasomes and vesicles in *Poria monticola*. *Can. J. Bot.* 49: 2075–79
15. Buckley, P. M., Sjaholm, V. E., Sommer, N. F. 1966. Electron microscopy of *Botrytis cinerea* conidia. *J. Bacteriol.* 91:2037–44
16. Buckley, P. M., Sommer, N. F., Matsumoto, T. T. 1968. Ultrastructural details in germinating sporangiospores of *Rhizopus stolonifer* and *Rhizopus arrhizus*. *J. Bacteriol.* 95:2365–73
17. Butler, R. D. 1967. The fine structure of senescing cotyledons. *J. Exp. Bot.* 18:535–43
18. Calonge, F. D. 1969. Multivesicular bodies in *Sclerotinia fructigena* and their possible relation to extracellular enzyme secretion. *J. Gen. Microbiol.* 55:177–84
19. Carroll, G. C. 1966. Doctoral Thesis, Univ. Texas, Austin, Texas
20. Chang, C. W., Bandurski, R. S. 1964. Exocellular enzymes of corn roots. *Plant Physiol.* 39:60–64
21. Cheung, D. S. M., Barber, H. N. 1971. Uredospore wall proteins of wheat stem rust. *Arch. Mikrobiol.* 77:239–46
22. Corbett, J. R., Price, C. A. 1967. Intracellular distribution of p-nitrophenyl-phosphatase in plants. *Plant Physiol.* 42:827–30
23. Coulomb, C., Buvat, R. 1968. Processus de degenerescence cytoplasmique portielle dans les cellules de jeunes racines de *Cucurbita pepo*. *C. R. Acad. Sci.* (Paris) 267:843–44
24. Dauwalder, M., Whaley, W. G., Kephart, J. E. 1969. Phosphatases and differentiation of the Golgi apparatus. *J. Cell. Sci.* 4: 455–97
25. DeDuve, C. 1969. The lysosome in retrospect. In *Lysosomes in Biology and Pathology; Frontiers of Biology* (monogr. ser.) ed. J. T. Dingle, H. B. Fell. North-Holland: Amsterdam & London

26. DeDuve, C., Wattiaux, R. 1966. Function of lysosomes. *Ann. Rev. Physiol.* 28:435–92
27. Dingle, J. L., Fell, H. B., Eds. 1969. *Lysosomes in Biology and Pathology; Frontiers of Biology* (monogr. ser.) North-Holland: Amsterdam & London
28. Ehrlich, M. A., Schafer, J. F., Ehrlich, H. G. 1968. Lomasomes in wheat leaves infected by *Puccinia gramminis* and *P. recondita. Can. J. Bot.* 46:17–20
29. Evans, L. S., Miller, P. R. 1972. Ozone damage to ponderosa pine: A histological and histochemical appraisal. *Am. J. Bot.* 59:297–304
30. Figier, J. 1968. Localisation infrastructurale de la phosphomonoesterase acide dans la stipule de *Vicia faba* L. au niveau du nectaire. *Planta* (Berl.) 83:60–79
31. Flinn, A. M., Smith, D. L. 1967. The localization of enzymes in the cotyledons of *Pisum arvense* L. during germination. *Planta* 75:10–22
32. Frey-Wyssling, A., Grieshaber, E., Mühlethaler, K. 1963. Origin of spherosomes in plant cells. *J. Ultrastruct. Res.* 8:506–16
33. Gahan, P. B., Maple, A. J. 1966. The behavior of lysosome-like particles during cell differentiation. *J. Exp. Bot.* 17:151–55
34. Gahan, P. B. 1967. Histochemistry of lysosomes. *Int. Rev. Cytol.* 21:1–63
35. Gahan, P. B. 1968. Lysosomes. In *Plant Cell Organelles,* ed. J. B. Pridham, 228–37. New York: Academic
36. Gay, J. L., Greenwood, A. D. 1966. Structural aspects of zoospore production in *Saprolegnia ferax* with particular reference to the cell and vacuolar membranes. The Fungus Spore, *Proc. Symp. Colston Res. Soc. 18th,* Bristol, England, p. 95–108
37. Gorska-Brycass, A. 1965. Hydrolases in pollen grains and pollen tubes. *Acta Soc. Bot. Poloniae.* 34:589–604
38. Griffiths, D. A. 1971. The fine structure of *Verticillium dahliae* Kleb. colonizing cellophane. *Can. J. Microbiol.* 17:79–81
39. Grove, S. N., Bracker, C. E., Morré, D. J. 1968. Cytomembrane differentiation in the endoplasmic reticulum-golgi apparatus-vesicle complex. *Science* 161:171–73
40. Grove, S. N., Morré, D. J., Bracker, C. E. 1967. Dictyosomes in vegetative hyphae of *Pythium ultimum. Proc. Indiana Acad. Sci.* 76:210–14
41. Gunther, W. K., Merker, H. J. 1966. Über das Verhalten und die Lokalisation der sauren Phosphatase von Hefezellen bei Repression und Derepression. *Exp. Cell Res.* 45:133–47
42. Hall, J. L. 1969. Histochemical localization of β-glycerophosphatase activity in young root tips. *Ann. Bot.* 33:399–406
43. Hall, J. L. 1969. Localization of cell surface adenosine triphosphatase activity in maize roots. *Planta* (Berl.) 85:105–07
44. Hall, J. L., Davie, C. A. M. 1971. Localization of acid hydrolase activity in *Zea mays* L. root tips. *Ann. Bot.* 35:849–55
45. Hall, J. L., Butt, V. S. 1969. Adenosine triphosphate activity in cell wall preparation and excised roots of barley, *J. Exp. Bot.* 20:751–62
46. Hall, J. L., Butt, V. S. 1968. Localization and kinetic properties of β-glycerophosphatase in barley roots. *J. Exp. Bot.* 19:276–87
47. Halperin, W. 1969. Ultrastructural localization of acid phosphatase in cultured cells of *Daucus carota. Planta* 88:91–102
48. Harwood, J. L., Sodja, A., Stumpf, P. K., Spurr, A. R. 1971. On the origin of oil droplets in maturing castor bean seeds *Ricinus communis. Lipids* 6:851–54
49. Hawker, L. E., Hendy, R. J. 1963. An electron-microscope study of germination of conidia of *Botrytis cinerea. J. Gen. Microbiol.* 33:43–46
50. Heintz, C. E., Niederpruem, D. J. 1971. Ultrastructure of quiescent and germinated basidospores and oidia of *Coprinus lagopus. Mycologia* 63:745–66
51. Holcomb, G. E., Hildebrandt, A. C., Evert, R. F. 1967. Staining and acid phosphatase reaction of spherosomes in plant tissue culture cells. *Am. J. Bot.* 54:1204–09
52. Hughes, G. C., Bisalputra, A.

1970. Ultrastructure of hyphomycetes. Conidium ontogeny in *Peziza ostracoderma*. *Can. J. Bot.* 48: 361–66

53. Hyde, J. M., Walkinshaw, C. H. 1966. Ultrastructure of basidiospores and mycelium of *Lenzites saepiaria*. *J. Bacteriol.* 92:1218–27

54. Iten, W., Matile, P. 1970. Role of chitinase and other lysosomal enzymes of *Coprinus lagopus* in the autolysis of fruiting bodies. *J. Gen. Microbiol.* 61:301–09

55. Jacks, T. J., Yatsu, L. Y., Aitschul, A. M. 1967. Isolation and characterization of peanut spherosomes. *Plant Physiol.* 42:585–97

56. Kazama, F. W., Aldrich, H. C. 1972. Digestion and the distribution of acid phosphatase in the myxoamebae *Physarum flavicomum*. *Mycologia* 64:529–38

57. Kivilaan, A., Beaman, T. C., Bandurski, R. S. 1961. Enzymatic activities associated with cell wall preparations from corn coleoptiles. *Plant Physiol.* (Lancaster) 36:605–10

58. Klement, Z., Goodman, R. N. 1967. The hypersensitive reaction to infection by bacterial plant pathogens. *Ann. Rev. Phytopathol.* 5:17–44

59. Mahlberg, P. 1972. Further observations on the phenomenon of secondary vacuolation in living cells. *Am. J. Bot.* 59:172–79

60. Mahlberg, P. 1972. Localization of neutral red in lysosome structures in hair cells of *Tradescantia virginiana*. *Can. J. Bot.* 50:857–59

61. Mallucci, L., Allison, A. C. 1965. Lysosomal enzymes in cells infected with cytopathic and noncytopathic viruses. *J. Exp. Med.* 121:477–85

62. Marty, F., Buvat, R. 1972. Distributions des activites phosphatasique acides au cours du processus d'autophagie cellulaire dans les cellules du meristeme radiculane d'*Euphorbia characias* L. *C. R. Acad. Sci.* (Paris) 274:206–09

63. Matile, P. 1969. Plant lysosomes in biology. In *Frontiers of Biology* (monogr. ser.) ed. J. T. Dingle, H. B. Fell. North-Holland: Amsterdam & London. 14:406–30

64. Matile, P. 1966. Inositol deficiency resulting in death: An explanation of its occurrence in *Neurospora crassa*. *Science* 151:86–88

65. Matile, P. 1965. Intrazelluläre Lokalisation Proteolytischer Enzyme von *Neurospora crassa*. *Z. Zellforschung* 65:884–96

66. Matile, P. 1964. Die Funktion proteolytischer Enzyme bei der Proteinaufnahme durch *Neurospora crassa*. *Naturwissenschaften* 51:489–90

67. Matile, P., Winkenbach, F. 1971. Function of lysosomes and lysosomal enzymes in the senescing corolla of the morning glory (*Ipomoea purpurea*). *J. Exp. Bot.* 23: 759–71

68. Matile, P. 1968. Aleurone vacuoles as lysosomes. *Z. Pflanzenphysiol.* 58:365–68

69. Matile, P., Moor, H. 1968. Vacuolation: Origin and development of the lysosomal apparatus in root-tip cells. *Planta* 80:159–75

70. Matile, P., Spichiger, J. 1968. Lysosomal enzymes in spherosomes (oil droplets) of tobacco endosperm. *Z. Pflanzenphysiol.* 58:277–80

71. Matile, P., Wiemken, A. 1967. The vacuole as the lysosome of the yeast cell. *Archiv. Mikrobiol.* 56:148–55

72. Matile, P., Balz, J. P., Semadeni, E., Jost, M. 1965. Isolation of spherosomes with lysome characteristics from seedlings. *Z. Naturforschung* 20:693–98

73. McKeen, W. E., Mitchell, N., Smith, R. 1967. The *Erysiphe cichoracearum* conidium. *Can. J. Bot.* 45:1489–96

74. McLaren, A. D., Jensen, W. A., Jacobson, L. 1960. Absorption of enzymes and other proteins by barley roots. *Plant Physiol.* 35: 549–56

75. Meyer, J. A., Garber, E. D., Shaeffer, S. G. 1964. Genetics of phytopathogenic fungi. XII. Detection of esterases and phosphatases in culture filtrates of *Fusarium oxysporum* and *F. xylarioides* by starch-gel zone electrophoresis. *Bot. Gaz.* 125:298–300

76. Meyer, H., Mayer, A. M., Harel, E. 1971. Acid phosphatases in germinating lettuce—Evidence of

partial activation. *Physiol. Plant* 24:95–101

77. Mishra, A. K., Colvin, J. R. 1970. On the variability of spherosome-like bodies in *Phaseolus vulgaris*. *Can. J. Bot.* 48:1477–80

78. Mlodzianowski, F. 1972. The occurrence of acid phosphatase in the thylakoids of developing plastids in lupin cotyledons. *Z. Pflanzenphysiol.* 66:362–65

79. Mollenhauer, H. H., Totten, C. 1971. Studies on seeds. II. Origin and degradation of lipid vesicles in pea and bean cotyledons. *J. Cell Biol.* 48:395–405

80. Moore, R. T., McAlear, J. H. 1963. Fine structure of Mycota. The occurrence of the Golgi dictyosome in the fungus *Neobulgaria pura* (Fr.) Petrak. *J. Cell. Biol.* 16:131–41

81. Morré, D. J., Mollenhauer, H. H., Bracker, C. E. 1971. Origin and Continuity of Golgi Apparatus. In *Results and Problems in Cell Differentiation*, Vol. 2: Origin and Continuity of Cell Organelles, 82–126. Ed. J. Reinert, H. Ursprung. Springer-Verlag

82. Mosse, B. 1970. Honey-colored, sessile *Endogone* spores: II. Changes in fine structure during spore development. *Arch. Mikrobiol.* 74:129–45

83. Northcote, D. H. 1971. The Golgi apparatus. *Endeavour* 30:26–33

84. Nougarede, A., Pilet, P. 1967. Activité et localisation, au niveau des infrastructures, de phosphomonoesterase acide dans la racine et dans les jeunes feuilles du *Lens culinaris* L. *C. R. Acad. Sci.* (Paris) 265:663–66

85. Olszewska, M. J., Gabara, B. 1964. *Protoplasma* 59:163–79

86. Ory, R. L., Yatsu, L. Y., Kircher, H. S. 1968. Association of lipase activity with the spherosomes of *Ricinus communis*. *Arch. Biochem. Biophys.* 264:255–64

87. Otsuki, Y., Takebe, I., Honda, Y., Matsui, C. 1972. Ultrastructure of infection of tobacco mesophyll protoplasts by tobacco mosaic virus. *Virology* 49:188–94

88. Page, W. J., Stock, J. J. 1972. Isolation and characterization of *Microsporum gypseum* lysosomes: role of lysosomes in macroconidia germination. *J. Bacteriol.* 110: 354–64

89. Page, W. J., Stock, J. J. 1971. Regulation and self-inhibition of *Microsporum gypseum* macroconidia germination. *J. Bacteriol.* 108:276–81

90. Persellin, R. H., Ziff, M. 1966. The effect of gold salt on enzymes of the peritoneal macrophage. *Arthritis Rheum.* 9:57–65

91. Petrovskaya-Baronova, T. P. 1971. Nuclei and chloroplasts in leaves of the wheat-quack grass hybrid during freezing. *Fiziol. Rast. Physiol.* 18:941–46

92. Pitt, D. 1968. Histochemical demonstration of certain hydrolytic enzymes within cytoplasmic particles of *Botrytis cinerea* Fr. *J. Gen. Microbiol.* 52:67–75

93. Pitt, D., Coombes, C. 1969. Release of hydrolytic enzymes from cytoplasmic particles of *Solanum tuber* tissues during infection by tuber-rotting fungi. *J. Gen. Microbiol.* 56:321–29

94. Pitt, D., Coombes, C. 1968. The disruption of lysosome-like particles of *Solanum tuberosum* cells during infection by *Phytophthora erythroseptic* Petrybr. *J. Gen. Microbiol.* 53:197–204

95. Poux, N. 1970. Localisation d'activites enzymatiques dans le meristeme radiculaire de *Cucumis rativas* L. III. Activite phosphatasique acide. *J. Microscop.* 9:407–34

96. Reiss, J. 1971. Cytochemischer Nachweis von Hydrolasen in Pilzzellen. II. Aminopeptidase. *Acta Histochem.* 39:277–85

97. Remsen, C. C., Hess, W. M., Sassen, M. M. H. 1967. Fine structure of germinating *Penicillium megasporium* conidia. *Protoplasma* 64:439–51

98. Rodkiewicz, B., Kwiatkowski, M. 1965. Enzymy hodrolityczne W. Rozwijajacym sif Woreczku Zaiazkow m Lilii. *Acta Soc. Bot. Polon.* 34:235–42

99. Rosenthal, A. S., Moses, H. L., Ganote, C. E. 1970. Interpretation of phosphatase cytochemical data. *J. Histochem. Cytochem.* 18:915

100. Schwarzenbach, A. M. 1971. Aleuronvakuolen und Spharosomen im Endosperm von *Ricinus communis* wahrend der Samen-

rcifung und Keimung. *Ber. Schweiz. Bot. Ges.* 81:70–91

101. Schwarzenbach, A. M. 1971. Observations on spherosomal membranes. *Cytobiologie* 4:145–47

102. Scott, W. A., Munkres, K. D., Metzenberg, R. L. 1971. A particulate fraction from *Neurospora crassa* exhibiting aryl sulfatase activity. *Arch. Biochem.* 143:623–32

103. Semadeni, E. G. 1967. Enzymatische Charakterisherung der Lysosomenaquivalente (Sparosomen) von Maiskeimlingen. *Planta* 72:91–118

104. Sexton, R., Sutcliffe, J. F. 1969. The distribution of β-glycerophosphatase in young roots of *Pisum sativum* L. *Ann. Bot.* 33:407–19

105. Sexton, R., Sutcliffe, J. F. 1969. Some observations on the characteristics and distribution of adenosine triphosphatases in young roots of *Pisum sativum,* cultivar Alaska. *Ann. Bot.* 33:683–94

106. Sheikh, K., Gahan, P. B., Stroun, M. 1971. A cytochemical study of acid phosphatases and esterases during the induction of crown gall in the tomato. *Histochem. J.* 3:179–83

107. Smith, D. G., Marchant, R. 1968. Lipid inclusions in the vacuoles of *Saccharomyces cerevisiae. Arch. Mikrobiol.* 60:340–47

108. Sommer, J. R., Blum, J. J. 1965. Cytochemical localization of acid phosphatases in *Euglena gracilis. J. Cell Biol.* 24:235–51

109. Sorokin, H. P., Sorokin, S. 1968. Fluctuations in the acid phosphatase activity of spherosomes in guard cells of *Campanula persicifolia. J. Histochem. Cytochem.* 16:791–802

110. Sorokin, H. P., Sorokin, S. 1966. The spherosomes of *Campanula persicifolia* L. A light and electron microscope study. *Protoplasma* 62:216–36

111. Stocks, D. L., Hess, W. M. 1970. Ultrastructure of dormant and germinated basidiospores of a species of *Psilocybe. Mycologia* 62:176–91

112. Strunk, C. 1970. Golgi-cisternen in *Saccharomyces* Protoplasten. *Cytobiologie* 2:251–58

113. Thornton, R. M. 1968. The fine structure of Phycomyces. I. Autophagic vesicles. *J. Ultrastruct. Res.* 21:269–80

114. Torgeson, D. C. (ed.). 1969. *Fungicides* 2:742. Academic

115. Walek-Czernecka, A. 1963. Note sur la detection d'une esterase non secifrque dans les spherosomes. *Acta Soc. Bot. Polon.* 32:407–08

116. Walek-Czernecka, A. 1965. Histochemical demonstration of some hydrolytic enzymes in the spherosomes of plant cells. *Acta Soc. Bot. Polon.* 34:573–88

117. Wardrop, A. B. 1968. Occurrence of structures with lysosome-like functions in plant cells. *Nature* 218:978–80

118. Wattiaux, R. 1969. Biochemistry and function of lysosomes. In *Handbook of Molecular Cytology; Frontiers of Biology* (monogr. ser.) ed. A. Lima-De-Faria. North-Holland: Amsterdam & London. 15:1159–78

119. Webber, M. M., Webber, P. J. 1970. Ultrastructure of lichen haustoria; Symbiosis in *Parmelia sulcata. Can. J. Bot.* 48:1521–24

120. Weissmann, G., Keiser, H., Bernheimer, A. W. 1963. Studies on lysosomes. III. The effects of streptolysis O and S on the release of acid hydrolases from a granular fraction of rabbit liner. *J. Exp. Med.* 118:205

121. Whaley, W. G., Dauwalder, M., Kephart, J.E. 1972. Golgi apparatus: Influence on cell surfaces. *Science* 175:596–99

122. Wilson, C. L., Stiers, D. L., Smith, G. G. 1970. Fungal lysosomes or spherosomes. *Phytopathology* 60:216–27

123. Wilson, K. S., Cutter, V. M. 1955. Localization of acid phosphatase in the embryo sac and endosperm of *Cocos nucifera. Am. J. Bot.* 42:116–19

124. Wyen, N. V., Udvardy, J., Farkas, G. L. 1971. Changes in the level of acid phosphatases in *Avena* leaves in response to cellular injury. *Phytochemistry* 10:765–70

THE FUNGAL HOST-PARASITE RELATIONSHIP[1]

❖ 3574

H. L. Barnett and F. L. Binder

Division of Plant Sciences, West Virginia University, Morgantown, West Virginia, and Department of Biology, Marshall University, Huntington, West Virginia

Introduction

Knowledge of the host-parasite relationship between economic plants and microorganisms is fundamental to plant pathology, yet the process of obtaining the knowledge has been slow and difficult, which is indicative of the complex nature of this relationship. Much of this study has been aimed at the elucidation of the resistance mechanism in the host rather than the metabolism and nutritional requirements of the parasite. An exception has been the recent surge of interest and expanded investigation into the nutrition of the rust fungi (20, 25, 52, 55, 65, 66).

The study of the nutrition of fungi parasitic on other fungi (mycoparasites) has to a large extent paralleled that of similar studies of parasites on higher plants, except that it has been more recent and more limited. Parasites that can be cultivated easily on common laboratory media have received major attention. Only in recent years has the relatively young science of fungus physiology matured to the extent that it could serve as a firm foundation for the more specialized nutritional studies of the parasites believed to require living hosts for their survival. The success that has been achieved in determining the special nutritional requirements for some of these parasitic fungi is not based so much on improved techniques as on a better understanding of their general basic nutritional needs.

The use of mycoparasites in studies aimed at elucidation of basic principles of parasitism has certain advantages over the use of parasites on higher plants; (*a*) there is a saving of time and space; (*b*) the environment can be controlled rigidly; (*c*) the host nutrition can be controlled.

A great number of fungi have been observed growing on other fungi in nature but most of these must be considered merely as fungicolous fungi until a nutritional relationship has been demonstrated. Lists of keys and discussions of fungicolous fungi have been given by several authors (32, 35, 37, 38,

[1] Published with the approval of the Director of the West Virginia University Agricultural Experiment Station as Scientific Paper No. 1237.

273

46, 58). Almost all taxonomic groups of fungi are included and in some cases both host and parasite belong to the same genus (37, 39, 40).

The mycoparasites may be separated into two major groups based on the mode of parasitism. The *necrotrophic* (destructive) parasite makes contact with its host, excretes a toxic substance, which kills the host cells, and utilizes the nutrients that are released. The *biotrophic* (balanced) parasite is able to obtain nutrients from the living host cell, a relationship that normally exists in nature. It causes little or no harm to the host, at least in the early stages of development. This group would include those that have been called "obligate" parasites, as well as some others that have been cultured on nonliving media. The term obligate, however, is inappropriate and should be abandoned, for basically it indicates (by the usual definition) only that we do not know the nutritional conditions required for growth of the parasite in the absence of a living host. The terms biotrophic and necrotrophic have been generally accepted in literature referring to mycoparasites and they have also been proposed for the parasites of higher plants (57).

Other reviews have treated various aspects of the fungi parasitic on other fungi (6, 7, 18, 26, 46, 63). In this review we propose to discuss the interrelationships between a number of fungal host-parasite combinations, with emphasis on morphology and physiology, and to present recent information on the special nutritional requirements of the biotrophic mycoparasites.

Necrotrophic Mycoparasites

Most of the necrotrophic mycoparasites are capable of indefinite saprophytic existence and are characterized by relatively rapid growth on a wide variety of substrata. They are facultative or opportunistic, having enzymes that enable them to compete strongly with other organisms for space and nutrients. Although these parasites do not require nutrients from other organisms, their growth is often enhanced as they overgrow and kill colonies of susceptible fungi. Relatively few species have received the thorough study that they deserve, and many of our concepts are vague and our information is incomplete.

HOST RANGE   Host ranges of the necrotrophic mycoparasites are characteristically broad, but host ranges have very little meaning because they are only partially known. There is variability due to differences in isolates of parasites and hosts and to the influence of nutrition on parasitism.

Haskins (37) made a special study of the host range of *Pythium* sp. (*P. acanthicum?*) and reported that 69 of the 98 potential host fungi tested were attacked, including *P. mamillatum,* resulting in death of some cells or hyphae. Hosts included representatives of all major groups of fungi. This parasite also caused browning of root tips of seedlings of several plants.

Gliocladium roseum is commonly observed in nature overgrowing many different fungi and its broad host range has been confirmed in the laboratory

(11). Severely attacked were species of *Ceratocystis, Trichothecium roseum, Thamnidium elegans* and others, and no species has been found to be completely immune in all stages of development.

Three unidentified species of *Cephalosporium* were similar in virulence on three species of *Helminthosporium* (41). Conidia, conidiophores, and hyphae of *H. teres* and *H. vagans* were killed by all three species but only a few cells of the mycelium of *H. sativum* were killed. One isolate of *C. acremonium,* causal agent of black bundle disease of corn, was similar in virulence.

We have observed a basidiomycete, identified by cultural characteristics as *Polyporus adustus,* attacking and destroying mycelium of many other fungi, including both Mucorales and the higher fungi. Isolates varied in virulence but both monokaryotic and dikaryotic isolates obtained as contaminants and those obtained from known fruit bodies of *P. adustus* were similar in behavior. A special study comparing several wood rotting basidiomycetes revealed that several additional species were similar to *P. adustus* in their ability to parasitize and destroy other fungi in culture (34). The more virulent species included *Polyporous versicolor* and *Pleurotus ostreatus.*

Rhizoctonia solani, the well-known pathogen of economic crops, is highly successful in competition with other soil fungi. An extensive study showed that a number of isolates of *R. solani* could destroy hyphae of several other fungi under favorable conditions in the laboratory (21). The most susceptible hosts were *Rhizopus nigricans, Mucor* spp., *Helicostylum* sp., *Pythium* spp., and *Amblyosporium botrytis,* the only host outside the Phycomycetes.

Ampelomyces quisqualis (Cicinnobolus cesati) is known in nature only as a parasite of powdery mildew fungi, although it is easily cultured on laboratory media. Recently it has been described as an internal parasite of mycelium, sporangia, and sporangiophores of several Mucorales, causing severe damage (45). Some species of *Mortierella* were resistant.

A number of other fungi are frequently found on aged mushrooms and other fleshy fungi. These include species of *Verticillium, Cephalosporium, Mycogone, Dactylium, Didymocladium,* and *Sepedonium,* some of which may cause serious losses in commercial mushroom beds. *Calcarisporium arbuscula* has been isolated from apparently healthy wild mushrooms (61). Little is known about the host-parasite relationship of these fungi.

MODE OF PARASITISM   The mode of parasitism among necrotrophic mycoparasites is similar in many species but is known to vary with the different host-parasite combinations. The concept of parasitism implies prolonged contact with or without penetration of the host, but experimentally it is often difficult to determine what part diffusible antibiotics might play in the relationship.

Species of *Trichoderma,* principally *T. viride,* have long been known to be antagonistic to other soil and wood-inhabiting fungi. Weindling (62) was one of the first to point out the parasitic nature of this fungus but some of its antagonistic activity has since been attributed to antibiotics (42, 63). We may conclude that the degree of destruction by this parasite is dependent upon the isolate of the species and of the host, and upon nutrition and environment.

Gliocladium roseum and the wood-rotting basidiomycetes usually make contact by means of short branches, which touch or curl around the host hyphae or spores, the contents of which soon begin to disintegrate (11, 34). There is no evidence of any diffusible toxic substance in advance of the growing hyphae of these or other necrotrophic mycoparasites. Some other parasites of this group, *Rhizoctonia solani* and *Papulospora stoveri,* may form extensive loose coils of hyphae with few or no branches (21, 60).

Penetration of host cells may follow their death, as in *Gliocladium roseum* (11) and *Pleurotus ostreatus,* or may occur while the host is alive, as by *Apelomyces quisqualis* in nature (29) and *Rhizoctonia solani* on *Mucor* sp. and *Rhizopus nigricans* (21). The latter host sometimes shows a biological resistance by early digestion of parasite hyphae, or mechanical resistance by deposition of wall-like material surrounding the invading hypha (21), the latter being relatively common among the filamentous fungi.

A cytological study by Emmons (29) revealed that for a short time following the invasion of *Erysiphe cichoracearum* by a hypha of *A. quisqualis,* the appearance of the host cell remained unchanged. Apparently the parasite was obtaining nutrients from the living cell. After a short time the nuclear membrane of the host cell disappeared and the contents showed signs of death and disintegration as the parasite continued its growth and development. This appears to be a parasite that combines early biotrophic existence with later necrotrophic parasitism.

Although the details of the relationship between *Darluca filum* and its rust hosts are not well understood, there is some evidence that the parasite is destructive (1). Aecia may be infected at any stage of development and hyphae of the parasite ramify among but do not penetrate aecial hyphae or spores, some of which show disintegration. The net result is to check the development of the rust pustule without complete destruction. *D. filum,* like the typical necrotrophic parasite, grows on common laboratory media and utilizes a variety of carbon and nitrogen sources (24, 49, 51). Isolates are very variable but there seems to be no nutritional deficiency that would account for the constant association of this parasite with rust fungi only.

Only a few studies of the fine structure of necrotrophic mycoparasites have been made. Although coiling of hyphae of *Trichoderma longibrachiatum* around hyphae of *Pellicularia sasakii* was common, a series of electron micrographs showed that the mycoparasite sometimes penetrated and invaded living cells of the host but caused little disorganization of the contents during early stages of infection (36). The first sign of infection was the formation of an internal infection papilla composed of wall material surrounding the invading hypha. The sheath-like papilla was pierced by the parasite hypha, which then continued its growth inside the host cell.

EFFECT OF NUTRITION AND ENVIRONMENT   Isolates of *Rhizoctonia solani* differed widely in virulence under the same cultural conditions (21). Susceptibility of four fungi to *R. solani* was decidedly greater in darkness than in white light. Susceptibility of *Mucor recurvus* to *R. solani* was high at 25°C and decreased to zero at 15°C. In general, good mycelial growth of both the parasite (*R. solani*) and the host (*M. recurvus* or *R. nigricans*) was required for heavy parasitism, with certain sugars in a complete medium being more favorable than others (21).

On the other hand, a medium giving an imbalance of carbon to nitrogen, favored parasitism by *Polyporus adustus* and other wood-rotting basidiomycetes (34). This emphasizes the fact that the substrate that furnishes the same nutrients for both host and necrotrophic parasite may favor one fungus at the expense of the other.

It is interesting to speculate about the possibility of using the necrotrophic mycoparasites or their metabolic products as an aid in reducing growth or sporulation of plant pathogens. Although the results of most studies have been disappointing, it would seem desirable that some of these parasites should receive continued investigation. Yarwood (67) demonstrated that growth and conidium formation by colonies of clover powdery mildew may be checked following artificial inoculation with *Ampelomyces quisqualis* under controlled conditions.

Biotrophic Mycoparasites

The biotrophic mycoparasites differ basically from the necrotrophic group in their ability to obtain nutrients from the living host cell. Throughout the many years of close association with the host, some of these parasites apparently have lost the ability to synthesize one or more required nutrients, and must depend on a host to furnish a continuous supply. The loss in synthetic ability has not always involved the same nutrient, so that the deficiencies of the parasites may differ. Furthermore, the susceptible host must not only synthesize and contain the required nutrient within its cells, but the parasite must have the ability to obtain the nutrient from the host. Thus, permeability of the host cell membrane may play an important role in determining the susceptibility of the host fungi and the success of the parasite.

Until recent years few of these mycoparasites that we now call biotrophic had been cultured successfully on nonliving media in the laboratory, and there was some justification for believing them to be obligate parasites, similar to the rusts, powdery mildews, and downy mildews on higher plants. However, our present knowledge based on research of the last two decades has indicated that most, and perhaps all, of these mycoparasites have the ability to make continuous near-normal growth on artificial laboratory media, provided that nutrients for which the parasite is deficient are added. Specific nutritional requirements for axenic growth of several biotrophic mycoparasites are discussed in a separate section of this review.

Three distinct types or modes of biotrophic mycoparasitism are known:

(*a*) the *internal* parasites represented by the Chytrids develop within cells of other fungi; (*b*) the *contact* parasites do not produce any haustoria or other internal hyphae; (*c*) the *haustorial* parasites produce distinct haustoria within the host hyphae. The three types are so distinct in morphology and physiology that they must be discussed separately.

Internal Mycoparasites

The internal mycoparasites are tentatively placed in the biotrophic group because they appear to cause little or no harm to the host during early development, although they may destroy the host protoplasm prior to sporulation. Little is known of their nutritional requirements.

Karling (39, 40) has discussed taxonomy and morphology of a number of parasitic chytrids and their hosts, but very little is known about the nutritional relationship of parasite and host. Included here is a brief discussion of two parasitic chytrids which penetrate and grow within the host, completely surrounded by host cytoplasm. *Rozella cladochytrii* has a known host range of three species of *Nowakowskiella* and three of *Cladochytrium* (39). The parasite enters the rhizomycelium of *N. profusum* and develops as a naked protoplast surrounded by and scarcely distinguishable from the host protoplasm. In the early stages there is little change except a local swelling of the host, but later the host protoplasm is destroyed as the parasite forms a surrounding wall and produces zoospores.

Olpidiopsis incrassata is a unicellular internal parasite of species of *Saprolegnia* and *Isoachlya*. These hosts were susceptible only before the mycelium produced sporangia or oogonia, and the period of susceptibility could be prolonged by delaying initiation of reproduction (56). It was proposed that a chemical precursor for reproduction is also essential for infection by the parasite. Such a mechanism for controlling susceptibility would be unique and this host-parasite relationship deserves more study.

Biotrophic Contact Mycoparasites

Although it seems probable that this mode of parasitism may be relatively common in nature, it was not described in detail until 1958 (10). Even now only five mycoparasites of this group are known and have been studied in the laboratory, although some of them have been recognized as being common associates of other fungi in nature. All are imperfect fungi and their hosts are ascomycetes or imperfects.

NATURAL OCCURRENCE AND HOST RANGE   *Calcarisporium parasiticum* is known only from laboratory cultures in West Virginia. It has appeared five times (1954–1956) in isolations of *Dothiorella quercina* from red oak twigs and was described as a new species (5). The known host range is limited to several species of *Physalospora* and closely related fungi.

Gonatobotrys simplex is widely distributed but is not commonly isolated. It is associated with *Alternaria* spp. in nature, and in culture parasitizes species of *Alternaria* and *Cladosporium* (64).

Gonatobotryum fuscum has been reported as a parasite on *Ceratocystis* in England (59) and from West Virginia growing on old mats of *Ceratocystis fagacearum* (53). In the laboratory the host range has been extended to additional species of *Ceratocystis, Graphium,* and *Leptographium* (53).

Gonatorrhodiella highlei is a common parasite on *Nectria coccinea* var. *faginata,* causal agent of the beech bark disease in the New England states, and is not known to be associated with any other fungus in nature. In culture *Tritirachium* sp. and *Cladosporium* sp. are also parasitized (31).

Stephanoma phaeospora is known only from a culture with *Fusarium* sp. isolated from an orchid (22). The host range in culture has recently been extended to nine species of ascomycetes and imperfects and two species of *Ustilago* (50).

SPORE GERMINATION AND TROPISMS   The conidia of *Calcarisporium parasiticum* and *Gonatobotrys simplex* do not germinate in distilled water or on a synthetic medium, but require a substance secreted by the host hyphae or furnished by some natural products such as yeast extract, malt extract, mycelium extract, or orange juice (10, 64). Germination on a yeast extract medium is typically by secondary spores or buds (Figs. 1, 3). At this stage an attractant is apparently secreted by the germinating spores, for there is a positive tropism of the nearby host hyphae directly toward the parasite (Figs. 2, 4). The unusual ability to cause a tropism of the host is evidently a survival mechanism because the parasite at this stage does not produce long term tubes. In contrast, conidia of *G. fuscum, G. highlei* and *S. phaeospora* produce long germ tubes on germination and the host hyphae do not show a positive tropism.

MODE OF CONTACT   Parasitism is by means of specialized branches, often no more than a few microns long, which contact the host hypha, may partially or completely surround it, or touch end-to-end a short branch of the host (Figs. 5–8). The small "buffer cell" at the point of contact is characteristic of *Calcarisporium parasiticum* (Fig. 5). The presence of these special absorptive branches or cells is considered as evidence of parasitism.

There is evidence that a required nutrient is normally held within the cells of most hosts and very little escapes into the substrate before autolysis (10, 64). The contact cells must, therefore, function in some way to increase the permeability of the host cell membrane to this nutrient. However, recently it was shown that several fungi excreted a growth factor into the medium that stimulated growth of *Stephanoma phaeospora* (50).

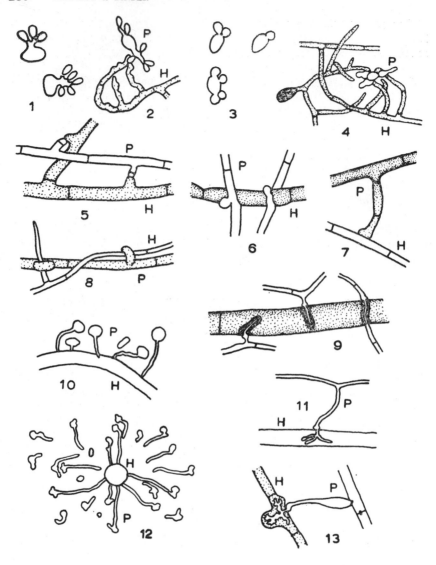

Figures 1–13 Spore germination, tropisms and morphology of host-biotrophic mycoparasite relationships. 1. Spore germination of *Calcarisporium parasiticum* by secondary spores (10). 2. Tropism of *Sphaeropsis malorum* hyphae toward *C. parasiticum* (10). 3. Spore germination of *Gonatobotrys simplex* by budding (64). 4. Tropism of *Alternaria* hyphae toward *G. simplex* (64). 5. Mode of contact between *C. parasiticum* and *S. malorum* (10). 6. Mode of contact between *G. simplex* and *Alternaria* sp. (64). 7. Mode of contact between *Gonatobotryum fuscum*

EFFECTS OF PARASITE ON GROWTH OF HOST   The contact biotrophic parasite generally causes no apparent harm to its hosts except a reduction in rate of growth. Since there is no evidence of toxin production, the slower growth may be due to competition for nutrients. A typical example is that of *Calcarisporium parasiticum* growing in liquid media on *Physalospora obtusa*. Less dry mycelium was produced by the combined host and parasite than by the host alone (10). This phase of the relationship between host and contact biotrophic parasite needs much more investigation.

EFFECTS OF HOST NUTRITION ON DEVELOPMENT OF THE PARASITE   The composition of the host medium, i.e., the nutrients and their concentrations, is known to affect host metabolism which in turn may determine the relative success or failure of the parasite. *Physalospora obtusa*, a highly susceptible species to *Calcarisporium parasiticum*, supported about equal growth of the parasite regardless of the carbon-nitrogen ratio, whereas highly resistant *P. ilicis* was more heavily parasitized on a medium high in available nitrogen (10).

Parasitic growth of *Gonatobotryum fuscum* was greater on a medium with relatively high carbon-nitrogen ratio (17, 53). A complex nitrogen source favored the parasite, as did an excess of microelements, particularly manganese.

Gonatorrhodiella highlei showed distinct inhibition by increased concentration of asparagine before the host, *Nectria coccinea* var. *faginata*, was affected (31). This inhibition was attributed to accumulation of ammonia in the cultures containing high asparagine. Increased KNO_3 in the medium was not inhibitory and did not result in accumulation of ammonia. An excess supply of thiamine and biotin in the medium may be essential to good parasitism by *G. highlei*.

An unusual relationship involving biotin and pyridoxine occurs when *G. fuscum* is cultured with the host *Graphium* sp., which is deficient for pyridoxine (8). Parasitism is heavy in darkness on media containing adequate pyridoxine and thiamine, but in light the pyridoxine is destroyed so rapidly that the host growth is limited and the parasite is strongly inhibited. Biotin must be added when the concentration of pyridoxine is very low (ca 12 μg/

---

and *Graphium* sp. (53). 8. Mode of contact between *Stephanoma phaeospora* and *Tritirachium* sp. 9. Resistance sheaths of *G. fuscum* preventing penetration by *Graphium* sp. 10. Spore germination and tropism of *Piptocephalis virginiana* toward hypha of *Choanephora cucurbitarum* (15). 11. Haustorium of *P. virginiana* in hypha of *C. cucurbitarum* (15). 12. Spore germination and tropism of *Dispira cornuta* toward swollen spore of *Cokeromyces recurvatus* (43). 13. Contact branch and haustorium of *Dispira simplex* in hypha of *Chaetomium* sp. (19). H = host, P = parasite.

liter or less). It is now known that *G. fuscum* is deficient for both thiamine
and biotin (23).

A mutualistic symbiotic relationship can be demonstrated between *G. fuscum* and *Graphium* sp. on a medium containing thiamine and only a trace of
biotin and pyridoxine. After a very slow start *G. fuscum* synthesizes pyridoxine needed by *Graphium* sp., which in turn synthesizes enough biotin for
growth of the parasite (8). The rate of growth increases as the mass of mycelium increases.

Another unusual complex relationship between *G. fuscum* and *Graphium*
sp., which may be related to nutrition, has been observed frequently in aging
cultures. There is attempted penetration of hyphae and conidia of *G. fuscum*
by hyphae of *Graphium* sp. The *Graphium* hypha forms a slight appressorium-like swelling and there is a deposition of dark wall material around the
penetrating hypha. These resistance sheaths may extend partially or completely across the hyphal cavity (Fig. 9). Only occasionally is the sheath penetrated and internal hyphae of *Graphium* sp. become evident. Is this a reversal of parasitism in which the parasite becomes the reluctant host? The existence of resistance sheaths surrounding penetrating hyphae of mycoparasites
has been reported for *Rhizoctonia solani* (21) and for *Trichoderma longibrachiatum* (36) and one may assume that they are common.

TEMPERATURE AND LIGHT   In general, the favorable temperatures for both
host and mycoparasite fall within the same range, and no striking effects of
temperature on parasitism have been reported. The destruction of pyridoxine
by light and the effects on the deficient host, *Graphium* sp., are reported
above. It has also been reported that *Gonatorrhodiella highlei* on *Nectria
coccinea* var. *faginata* and on *Tritirachium* sp. is completely inhibited by
light (200fc) (31).

Biotrophic Haustorial Mycoparasites

The filamentous haustorial mycoparasites belong to the morphological group,
merosporangiferous Mucorales, i.e., those that produce spores in rod-like
sporangia (12, 13). The principal genera are *Syncephalis* and *Piptocephalis,*
in the Piptocephalidaceae, and *Dispira, Dimargaris* and *Tieghemiomyces,* in
the Dimargaritaceae. The species have received intensive taxonomic and morphological study by Benjamin (12, 13).

HOST RANGES   For the most part the hosts of these parasites belong to other
families of the Mucorales. *Piptocephalis virginiana* parasitized 22 species of
Mucorales (15). Most other species of *Piptocephalis* have similar host
ranges, except *P. xenophila,* which can parasitize several species of ascomycetes and imperfects (27).

Three species of *Dispira* vary in their host ranges. *D. cornuta* parasitizes
only species of Mucorales (3, 14). Susceptibility to *D. parvispora* was influ-

enced by the concentration and ratio of carbon to nitrogen in the medium (19). An ascomycete, *Monascus purpureus,* was parasitized only on a high nitrogen medium. *D. simplex* is unique in that it parasitizes only species of *Chaetomium* (13, 19). Limits of the host ranges of other haustorial mycoparasites are poorly known and some have not been determined.

SPORE GERMINATION AND TROPISMS   Spore germination by the haustorial mycoparasites generally requires one or more special nutrients or stimulants which may be furnished by several complex natural products, by excretions of a nearby living host fungus, or by filtrates of host culture. Spores of *Piptocephalis* spp. swelled to several times their volume and germinated by one or more germ tubes on a medium containing yeast extract but no swelling or germination occurred on water agar (15).

In *Dispira cornuta,* two factors (or nutrients) again appeared to be involved, one for swelling of the spore and the other for formation of germ tubes (43). Spores swelled and produced only distorted short germ tubes on yeast extract, but with a living host or host extract there was rapid production of long germ tubes. Germ tubes of *Piptocephalis virginiana* and of *D. cornuta* show strong positive tropism toward living host cells or hyphae (Figs. 10, 12) (Refs 15, 43).

MODE OF PARASITISM AND EFFECT ON HOST   The events following germination and prior to establishing a nutritional relationship with the host appear to be uniform. The hypha of the parasite contacts the host wall and usually forms a conspicuous appressorium-like swelling, followed by penetration of the host wall by a slender infection peg, and formation of a branched haustorium (Figs. 11, 13). Haustoria have been observed only in susceptible fungus hyphae.

Resistance of *Phycomyces blakesleeanus* and *Rhizopus nigricans* to *Piptocephalis virginiana* appears to be principally mechanical. Young hyphae were heavily parasitized, whereas aged hyphae seldom were (30). However, both young and aged mycelium of *Choanephora cucurbitarum* were heavily parasitized. When the toughness of the walls of the two hosts was compared by sonification of mycelium it was found that the hyphal walls at all ages of *C. cucurbitarum* were much more easily broken than those of *P. blakesleeanus* (30). Apparently either the wall composition or thickness is important in mechanical resistance.

The type and degree of response of a highly susceptible host, such as *Choanephora cucurbitarum* to *P. virginiana* depends to a large extent on the number of penetrations. When severely parasitized, a young host hypha frequently responded by a flurry of lateral branching, resembling a witches broom (30). *Dispira simplex* caused an enlargement of the host cells that contain haustoria (Fig. 13). There is little or no apparent inhibition of host growth by the haustorial mycoparasites.

FINE STRUCTURE OF THE HAUSTORIUM   In general the fine structure of the haustorium of *Piptocephalis virginiana,* the only mycoparasite studied with the electron microscope, is similar to that of haustorial parasites of higher plants (2, 47). The haustorium formed within hyphae of *Mycotypha microspora* is bounded by an electron dense sheath with a convoluted surrounding membrane. The haustorial wall is continuous with the wall of the parasite hypha and there is evidence of a collar of host wall material at its base. The infection apparatus formed by *P. virginiana* in hyphae of *Choanephora cucurbitarum,* also a highly susceptible host, varied only in minor points from that described above (47).

Table 1   The relation of amounts of glucose and yeast extract to percentage dry weight of TCA-soluble nitrogen in the mycelium of *Mortierella pusilla* and degree of parasitism by *Piptocephalis virginiana* at the end of 7 days (54).

	g Glucose—g Yeast Extract			
	5-1	25-1	5-5	25-5
Mg dry mycelium	82	185	138	256
% TCA-nitrogen	0.61	0.34	1.46	0.40
Parasite rating	Poor	None	Excellent	Trace

HOST NUTRITION AND GROWTH OF PARASITE   Ayers (3) was a pioneer in investigations of effects of host nutrition on degree of parasitism by the haustorial mycoparasites. He demonstrated that media high in carbon were unfavorable to growth of *Dispira cornuta* on its host, whereas media rich in a usable nitrogen source favored parasitism. These results have been generally confirmed and the experiments extended (17, 43, 44). *Cokeromyces recurvatus* was heavily parasitized by *D. cornuta* on a medium with a low carbon-nitrogen ratio but not at all on a medium high in carbon and low in nitrogen (17).

A more extensive study was made on the effects of nitrogen sources and concentrations on parasitism by *Piptocephalis virginiana* (14). On most hosts alanine, asparagine, glutamic acid, and casein hydrolysate favored the development of the parasite, whereas glycine, urea, ammonium sulfate, and ammonium tartrate were poor nitrogen sources. The host, *Mortierella pusilla,* made about equal growth on media with glutamic acid and ammonium sulfate but there was no growth of the parasite on the host on the latter medium. *M. pusilla* was parasitized on a medium with high concentration of glucose and glutamic acid, whereas *M. ramannianus* was not (14).

The degree of parasitism of *M. pusilla* by *P. virginiana* was directly correlated with the amount of soluble nitrogen in the host mycelium (Table 1). Furthermore, it was possible to prevent the growth of this parasite in liquid media for 15 days or longer by maintaining a high concentration of glucose in the culture medium (54). In the same medium with no additional glucose the parasite appeared on the fifth day and became heavy by the ninth day. There is strong evidence that the nitrogen source and concentration of nitrogen are important factors in determining the susceptibility of hosts to the haustorial mycoparasites.

Axenic Growth of Biotrophic Mycoparasites

Since the primary basis of parasitism is nutritional, it becomes necessary to determine the nutritional requirements of the biotrophic mycoparasites before the host-parasite relationships can be understood completely. Each parasite must be cultured axenically away from its host in order to determine which nutrients it cannot synthesize and that must be furnished by the culture medium. Nutrient deficiencies appear to be common but are not evident in the parasitic relationship with the host, which furnishes a full complement of nutrients. Much of the preliminary research in this area has of necessity been an exercise in the addition or substitution of nutrients to a basal medium that does not support sustained axenic growth of these mycoparasites.

We have had marked success in the past two decades in culturing several, but not all, of the biotrophic mycoparasites formerly considered as "obligate" because they made no axenic growth on common laboratory media. Since the contact and the haustorial mycoparasites have had widely different lines of evolutionary development, it is not surprising that their nutritional requirements are different.

Contact Mycoparasites

All of the five species of biotrophic contact mycoparasites discussed above with their hosts are now known to have in common a requirement for a nutrient present in hot water extracts of the host mycelium (10, 23, 31, 50, 64). A crude extract containing this nutrient was obtained by mincing the mycelium of a host (or other selected fungus) in distilled water, boiling, and filtering. It was partially purified by absorbing on activated charcoal, eluting with pyridine, and evaporating to reduce volume. Some efforts have been made to purify this compound further by means of paper chromatography and Sephadex, but its identity remains unknown (33). It is an organic compound of low molecular weight and is active for these mycoparasites in very small quantities. This growth factor may be in the nature of a vitamin but does not satisfy the requirement for any of the known B vitamins in deficient fungi. Because of the need for frequent reference to this nutrient, it has been tentatively called "mycotrophein" (64).

The presence of mycotrophein was demonstrated in all host fungi and many ascomycetes and imperfects which were nonhosts (50, 64). Apparently it is present in the phycomycetes in low concentrations and in only some of the basidiomycetes. The requirement of the contact mycoparasites for mycotrophein was first tested on a glucose-yeast extract agar medium. Spores of most of the parasites germinated readily but extensive mycelium was produced only when the mycotrophein solution was added.

Calcarisporium parasiticum, in liquid synthetic media containing mycotrophein, utilized common hexose sugars but little or no growth occurred on sorbose, xylose, or disaccharides. Several single amino acids were utilized as nitrogen sources, with glutamic acid being the best (33).

Gonatobotrys simplex utilized only the common hexose sugars, but failed to grow when single amino acids were the nitrogen source. Only yeast extract provided a good nitrogen source for growth and sporulation (64).

Gonatorrhodiella highlei differed from the two previous contact parasites in making no growth on a yeast extract medium, even with added mycotrophein (31). The requirement for a second nutrient was demonstrated when a water extract of beech bark was added, resulting in good axenic growth.

Gonatobotryum fuscum did not grow on any of the media that supported growth of other contact mycoparasites. It made substantial axenic growth only when the medium contained unusually high concentrations of thiamine (optimum 4–8 mg/1) and biotin (optimum 0.2–0.4 mg/1) in addition to high mycotrophein (23). Only the common hexose sugars were utilized, but a number of single amino acids were utilized as nitrogen sources. The rate of growth of *G. fuscum* on a favorable medium was greatly increased when 1.2% dimethyl sulfoxide was added, suggesting that a problem of permeability may be involved.

Stephanoma phaeospora failed to grow on a malt extract medium unless a water extract of certain fungi was added (50). It seems likely that the fungus extract contains the same nutrient (mycotrophein) that is required by other biotrophic contact mycoparasites, but this cannot be determined with certainty until the nutrient is identified and tested further.

Haustorial Mycoparasites

Nutritional requirements for axenic growth of the haustorial mycoparasites are more complex than for the contact mycoparasites. None of the species studied were affected by the addition of the mycotrophein solution to the medium. Of the several species successfully cultivated axenically, only *Dispira parvispora* grew well and sporulated on a glucose-yeast extract medium, but no detailed study of this species has been made. Some species have an apparent inability to utilize glucose, or other common sugars as a carbon source.

Syncephalis spp. grew axenically on a complex medium containing beef liver (28). *Dispira cornuta, Dispira simplex, Dimargaris verticillata,* and *Tieghemiomyces parasiticus* made only a trace of growth on a yeast extract or casein hydrolysate medium with glucose, but all grew well when glycerol replaced the glucose as the carbon source (9). Detailed studies in liquid culture have been made only with *D. cornuta* and *T. parasiticus.* Failure of these parasites to make substantial growth on a glucose medium could mean either that the glucose did not enter the mycelium or that the fungus lacked the necessary enzymes to catabolize glucose. One is tempted to use the latter explanation, except for certain experiments performed with *T. parasiticus* (16). Labeled $^{14}CO_2$ was produced from media containing ^{14}C-glucose or ^{14}C-glycerol as single carbon sources, but at a much higher rate from the latter source (16). This indicates a slow or limited entrance of glucose into the mycelium under these conditions. On the other hand, an increase in glucose concentration did not increase the rate of growth of the parasite. The addition of the surfactant Tween 80 to a glucose-casein hydrolysate medium resulted in much greater growth than did Tween 80 without glucose (180 mg vs. 35 mg dry wt per culture in 6 wks) (16). While some of the Tween 80 may have been used as a carbon source, the greater effect was on the rate of entrance of glucose. This action may be similar to that of glycols and Tweens reported as facilitating the entry of nutrients into mycelium of *Claviceps paspali* (48).

In further studies, all of the enzymes for the catabolism of glucose via the Embden-Meyerhof-Parnas and the hexose-monophosphate pathways were demonstrated in cell-free extracts of mycelium of *Tieghemiomyces parasiticus* grown on a glycerol medium (16). Glycerol may play a dual role of serving as an excellent carbon source and of lowering the surface tension of the medium.

The carbon metabolism of *Dispira cornuta* appears to be similar to that of *T. parasiticus* but no study of its enzymes has been made (4).

There seems little doubt that glucose can enter the mycelium of the haustorial parasites under some conditions, but it also appears that the vegetative hyphae serve as a barrier to certain nutrients in media that are readily absorbed by haustoria within the host cells. It is also possible that the presence of the host may influence the permeability of the parasite membranes to certain nutrients.

Growth of several of the haustorial mycoparasites was favored by an unusually high concentration of casein hydrolysate as the nitrogen source (20–40 g/liter) (4, 9, 16). This may be due to a high nitrogen requirement or to the increased concentration of a required amino acid present in low concentrations in casein hydrolysate. Utilization of nitrogen sources by *Dispira cornuta* and by *Tieghemiomyces parasiticus* has been studied and was found to be markedly different. *T. parasiticus* made good axenic growth on a glycerol medium with casein hydrolysate or a mixture of 18 amino acids, but no growth on any single amino acid as the nitrogen source. The conclusion reached from a special study of amino acid utilization was that L-cysteine, L-valine, and L-leucine are extremely important for axenic growth of this fungus (16). The addition of certain other amino acids to these three did not increase the rate of growth greatly. *Dispira cornuta* utilized L-alanine, L-aspartic, and L-glutamic acid in a glycerol medium (4). Certain other amino acids and ammonium sulfate were utilized less efficiently and nitrate nitrogen was not utilized. In general, increased concentrations of single amino acids or ammonium sulfate, to several times the concentrations used for most saprophytic fungi, resulted in increased growth rate. The reason for this is not at all clear and the problem should receive further careful study.

The haustorial mycoparasites frequently show deficiencies for thiamine and biotin (4, 9, 16). High concentrations of thiamine (1 mg or more/liter) have been reported as favorable for *Dispira cornuta* and for *T. parasiticus* on agar media, but not for the latter in liquid media.

The success that has been enjoyed in the axenic culture of several of the haustorial mycoparasites has brought with it the promise that other parasites now considered as "obligate" may likewise be cultured axenically, provided that the appropriate nutritional and physical conditions are met by the medium. However, the fact that great differences exist among species or genera has been brought out by research with *Piptocephalis* spp. After many attempts using a great number of nutritional conditions, we have never succeeded in culturing *Piptocephalis*, except on a living susceptible host. Spores of *Piptocephalis* germinate readily on various media in the absence of a host fungus, but only a limited amount of mycelium is formed, which often sporulates before it ceases to grow (14). This problem remains as a real challenge to the interested researcher.

Discussion and Summary

The success of the necrotrophic mycoparasites in destroying host cells or reducing populations of competing fungi depends on a number of external factors, of which nutrition is of high importance. Yet these parasites have no specific nutritional requirement that is satisfied only by the host. We may conclude that this type of mycoparasitism is exceedingly common among filamentous fungi in nature. There is much to be learned from studies of the production of toxins and enzymes by these parasites and the mode of host destruction.

The mycelium of the susceptible host is known to furnish all nutrients required for growth and development of the biotrophic contact mycoparasites, but the mere presence of a required nutrient does not necessarily result in susceptibility. The parasite must be able, by means of special cells or branches, to absorb the essential nutrients from the living host cell, presumably by altering the permeability of the cell membrane. All five known mycoparasites of this group have one essential nutritional deficiency in common, i.e. a water-soluble, heat-stable, vitamin-like growth factor (mycotrophein) present in all host and some nonhost fungi. Its purification and identification are among the major unsolved problems related to nutrition of the biotrophic contact mycoparasites.

The rate and degree of development of the biotrophic haustorial mycoparasites on their hosts are greatly favored by a substrate high in nitrogen and relatively low in carbon. There is evidence that a high concentration of soluble nitrogen in the host mycelium is directly related to concentration of nitrogen in the medium and to host susceptibility.

It is of interest that the electron microscope has revealed that the haustorium of a mycoparasite (*Piptocephalis virginiana*) is strikingly similar in structure to that of the haustorial parasites of higher plants. One must conclude that its formation and structure are the response to a nutritionally compatible environment which leads to free passage of nutrients into the parasite.

The failure of *Tieghemiomyces parasiticus* to make good axenic growth on a glucose medium, even though all enzymes involved in glucose catabolism were shown to be present, suggests strongly that the rate of penetration of nutrients into vegetative mycelium may be a limiting factor in growth of the haustorial mycoparasites.

The presence and concentration of specific amino acids appear to be of prime importance to axenic growth of some of the haustorial mycoparasites. A similarity with the requirements of the rust, *Melampsora lini*, for a S-containing amino acid was indicated by recent work (25). Other similarities in nutrition of these two groups of parasites may be expected.

Literature Cited

1. Adams, J. F. 1920. *Darluca* on *Peridermium peckii. Mycologia* 12: 309–15
2. Armentrout, V. N., Wilson, C. L. 1969. Haustorium-host interaction during mycoparasitism of *Mycotypha microspora* by *Piptocephalis virginiana. Phytopathology* 59: 897–905
3. Ayers, T. T. 1935. Parasitism of *Dispira cornuta. Mycologia* 27: 235–61
4. Barker, S. M., Barnett, H. L., 1973. Nitrogen and vitamin requirements for axenic growth of the haustorial mycoparasite, *Dispira cornuta. Mycologia.* In press
5. Barnett, H. L. 1958. A new *Calcarisporium* parasitic on other fungi. *Mycologia* 50:497–500
6. Barnett, H. L. 1963. The nature of mycoparasitism by fungi. *Ann. Rev. Microbiol.* 17:1–14
7. Barnett, H. L. 1964. Mycoparasitism. *Mycologia* 56:1–19
8. Barnett, H. L. 1968. The effects of light, pyridoxine and biotin on the development of the mycoparasite, *Gonatobotryum fuscum. Mycologia* 60:244–51
9. Barnett, H. L. 1970. Nutritional requirements for axenic growth of some haustorial mycoparasites. *Mycologia* 62:750–61
10. Barnett, H. L., Lilly, V. G. 1958. Parasitism of *Calcarisporium parasiticum* on species of *Physalospora* and related fungi. *West Va. Univ. Agr. Exp. Sta. Bull.* 420T. 37 pp.
11. Barnett, H. L., Lilly, V. G. 1962. A destructive mycoparasite, *Gliocladium roseum. Mycologia* 54:72–79
12. Benjamin, R. K. 1959. The merosporangiferous Mucorales. *Aliso* 4: 321–433
13. Benjamin, R. K. 1961. Addenda to "The merosporangiferous Mucorales." *Aliso* 5:11–19
14. Berry, C. R. 1959. Factors affecting parasitism of *Piptocephalis virginiana* on other Mucorales. *Mycologia* 51:824–32
15. Berry, C. R., Barnett, H. L. 1957. Mode of parasitism and host range of *Piptocephalis virginiana. Mycologia* 49:374–86
16. Binder, F. L. 1971. *Carbohydrate metabolism and nitrogen nutrition of Tieghemiomyces parasiticus.*

Ph.D. Dissertation, West Virginia University, Morgantown. 89 pp.
17. Bishop, R. H. 1964. *Effects of nutrition on the mycoparasite, Gonatobotryum fuscum.* Ph.D. Dissertation, West Virginia University, Morgantown. 134 pp.
18. Boosalis, M. G. 1964. Hyperparasitism. *Ann. Rev. Phytopathol.* 2: 263–76
19. Brunk, M. A., Barnett, H. L. 1966. Mycoparasitism of *Dispira simplex* and *D. parvispora. Mycologia* 58: 518–23
20. Bushnell, W. R. 1968. In vitro development of an Australian isolate of *Puccinia graminis* f. sp. *tritici. Phytopathology* 58:526–27
21. Butler, E. E. 1957. *Rhizoctonia solani* as a parasite of fungi. *Mycologia* 49:354–73
22. Butler, E. E., McCain, A. H. 1968. A new species of *Stephanoma. Mycologia* 60:955–59
23. Calderone, R. A., Barnett, H. L. 1972. Axenic growth and nutrition of *Gonatobotryum fuscum. Mycologia* 64:153–60
24. Calpouzos, L., Theis, T., Batille, C. M. 1957. Culture of the rust parasite, *Darluca filum. Phytopathology* 47:108–09
25. Coffey, M. D., Shaw, M. 1972. Nutritional studies with axenic cultures of the flax rust. *Melampsora lini. Physiol. Plant Pathol.* 2:37–46
26. De Vay, J. E. 1956. Mutual relationships in fungi. *Ann. Rev. Microbiol.* 10:115–40
27. Dobbs, C. G., English, M. P. 1954. *Piptocephalis xenophila* sp. nov. parasitic on non-mucorine hosts. *Trans. Brit. Mycol. Soc.* 37:375–89
28. Ellis, J. J. 1966. On growing *Syncephalis* in pure culture. *Mycologia* 58:465–69
29. Emmons, C. W. 1930. *Cicinnobolus cesati,* a study in host-parasite relationships. *Bull. Torrey Bot. Club* 57:421–41
30. England, W. H. 1969. Relation of age of two host fungi to development of the mycoparasite, *Piptocephalis virginiana. Mycologia* 61: 586–92
31. Gain, R. E., Barnett, H. L. 1970. Parasitism and axenic growth of *Gonatorrhodiella highlei. Mycologia* 62:1122–29

32. Gilman, J. C., Tiffany, L. H. 1952. Fungicolous fungi of Iowa. *Proc. Iowa Acad. Sci.* 59:99–110

33. Goldstrohm, D. D. 1966. *An investigation of the growth factor and other nutritional requirements of Calcarisporium parasiticum.* Ph.D. Dissertation, West Virginia Univ., Morgantown. 132 pp.

34. Griffith, N. T., Barnett, H. L. 1967. Mycoparasitism by basidiomycetes in culture. *Mycologia* 59:149–54

35. Hansford, C. G. 1946. The foliicolous ascomycetes, their parasites and associated fungi. *Commonwealth Mycol. Inst., Kew, England,* Paper No. 15, 240 pp.

36. Hashioka, Y., Fukita, T. 1969. Ultrastructural observations on mycoparasitism of *Trichoderma, Gliocladium* and *Acremonium* to phytopathogenic fungi. *Rept. Tottori Mycol. Inst. (Japan)* 7:8–18

37. Haskins, R. H. 1963. Morphology, nutrition and host range of a species of *Pythium. Can. J. Microbiol.* 9:451–57

38. Heim, R. 1967. Cle pour la determination des espices banales de champignons fongicoles. *Rev. de Mycol.* 31:393–99

39. Karling, J. S. 1942. Parasitism among the chytrids. *Am. J. Bot.* 29:24–35

40. Karling, J. S. 1960. Parasitism among chytrids. II. *Chytridiomyces verrucosus* sp. nov. and *Phlyctochytrium synchytrii. Bull. Torrey Bot. Club* 87:326–36

41. Kenneth, R., Isaac, P. K. 1963. *Cephalosporium* species parasitic on *Helminthosporium. Can. J. Plant Sci.* 44:182–87

42. Komatsu, M., Hashioka, Y. 1964. *Trichoderma viride,* as an antagonist of wood-inhabiting Hymenomycetes. V. Lethal effects of the different *Trichoderma* forms on *Lentinus edodes* inside log-woods. *Rept. Tottori Mycol. Inst. (Japan)* 4:11–18

43. Kurtzman, C. P. 1967. *Parasitism, spore germination, and axenic growth of Dispira cornuta.* Ph.D. Dissertation, West Virginia Univ., Morgantown. 127 pp.

44. Kurtzman, C. P. 1968. Parasitism and axenic growth of *Dispira cornuta. Mycologia* 60:915–23

45. Linnemann, G. 1968. *Ampelomyces quisqualis* Ces., ein Parasit auf Mucorineen. *Archiv. Microbiol.* 60:59–75

46. Madelin, M. F. 1968. Fungi parasitic on other fungi and lichens, p. 253–59. In *The Fungi.* Ed Ainsworth, G. C., Vol. 3, Academic: New York

47. Manocha, M. S., Lee, K. Y. 1971. Host-parasite relations in mycoparasite. I. Fine structure of host, parasite, and their interface. *Can. J. Bot.* 49:1677–81

48. Mizrahi, A., Miller, G. 1969. Role of glycols and Tweens in production of ergot alkaloids by *Claviceps paspali. J. Bacteriol.* 97:1155–59

49. Nicolas, G., Villanueva, J. R. 1965. Physiological studies on the rust hyperparasite *Darluca filum. Mycologia* 57:782–88

50. Rakvidhyasastra, V., Butler, E. E. 1972. Mycoparasitism by *Stephanoma phaeospora. Mycologia.* In press

51. Rambo, G. W., Bean, G. A. 1970. Survival and growth of the mycoparasite *Darluca filum. Phytopathology* 60:1436–40

52. Scott, K. J., MacLean, D. J. 1969. Culturing of rust fungi. *Ann. Rev. Phytopathol.* 7:123–46

53. Shigo, A. L. 1960. Parasitism of *Gonatobotryum fuscum* on species of *Ceratocystis. Mycologia* 52:584–98

54. Shigo, A. L., Anderson, C. D., Barnett, H. L. 1961. Effects of concentration of host nutrients on parasitism of *Piptocephalis xenophila* and *P. virginiana. Phytopathology* 51:616–20

55. Siebs, E. 1971. Über die Kultivierbarkeit des Kronenrosts (*Puccinia coronata* Cda.) von Futtergrosern in vitro. *Phytopathol. Z.* 72:97–114

56. Slifkin, M. K. 1961. Parasitism of *Olpidiopsis incrassata* on members of the Saprolegniaceae. I. Host range and effects of light, temperature and stage of host infectivity. *Mycologia* 53:183–93

57. Thrower, L. B. 1966. Terminology for plant parasites. *Phytopathol. Z.* 56:258–59

58. Tubaki, K. 1955. Studies on Japanese hyphomycetes. II. Fungicolous group. *Nagaoa* 5:11–40

59. Vincent, M. 1953. A *Chalaropsis* on beech. *Nature* 172:963–64

60. Warren, J. R. 1948. An undescribed species of *Papulospora* parasitic on

Rhizoctonia solani. Mycologia 40: 391–401

61. Watson, P. 1955. *Calcarisporium arbuscula* living as an endophyte in apparently healthy sporophores of *Russula* and *Lactarius. Trans. Brit. Mycol. Soc.* 38:409–14

62. Weindling, R. 1932. *Trichoderma lignorum* as a parasite of other soil fungi. *Phytopathology* 22:837–45

63. Weindling, R. 1959. Role of parasitism in microbial antagonism. In *Recent Advances in Botany. Int. Bot. Congr. IX.* Montreal. 623–26

64. Whaley, J. W., Barnett, H. L. 1963. Parasitism and nutrition of *Gona-*

tobotrys simplex. Mycologia 55: 199–210

65. Williams, P. G., Scott, K. J., Kuhl, J. L., MacLean, D. J. 1967. Sporulation and pathogenicity of *Puccinia graminis* f. sp. *tritici* grown on artificial medium. *Phytopathology* 57:326–27

66. Wong, A. L., Willetts, H. J. 1970. Observations on growth of selected Australian races of wheat stem rust in axenic culture. *Trans. Brit. Mycol. Soc.* 55:231–28

67. Yarwood, C. E. 1932. *Ampelomyces quisqualis* on clover mildew. *Phytopathology* 22:31

GENETIC VARIABILITY OF CROPS ❖ 3575

P. R. Day

Genetics Department, The Connecticut Agricultural Experiment
Station, New Haven, Connecticut

It is fashionable just now for biologists with social concerns to make dire predictions about the consequences of technology. The current concern over reduced genetic variability of crop plants and its consequence of genetic vulnerability (exemplified by the 1970 southern leaf blight epidemic on maize in the U.S.) is not new. The dangers of monoculture were recognized early in the development of modern agriculture. Large areas of uniform crop plants were uniformly susceptible to pests, diseases, and unfavorable weather conditions. The sciences of plant pathology and entomology were spawned out of the need to devise ways to protect vulnerable crops. Plant breeders introduced genetic resistance to restore the balance but only succeeded in making the crops still more uniform in this and other respects as they successfully pursued the search for higher yields and greater quality. Why add my voice to the babel of the prophets of doom? Because I believe that plant breeders and agronomists, caught up in the success of modern methods, should be aware of their long-term genetic consequences to avoid laying foundations for intractable problems in the years ahead.

SCOPE

In this review, I shall limit my remarks mainly to the major crops. Mangelsdorf (66) estimated that man has used at least 3000 species of plants for food during his history; that at least 150 of these have entered world commerce; but that the world's people are today fed by about 15 species. The same forces that caused man to concentrate on less than 0.5% of the food plants he has explored can be expected to have guided and hence narrowed his choice of the variability encountered in each of the 15 major species. These 15 species include the cereals: rice, wheat, maize, sorghum and barley; the sugar plants: sugar cane and sugar beet; the root crops: potato, sweet potato, and cassava; the legumes: common bean, soy bean, and peanut; and two tree crops: coconut and banana. While I do not cover all these crops I do include some others.

I shall briefly survey their variability, discuss the genetic and economic rea-

293

sons why variability is lost, the hazards of genetic uniformity, and the reme-
dies proposed. I have tried to be brief and provocative.

GENETIC VARIABILITY

Earlier Views

Nearly 50 years ago Jones (55) noted that extreme uniformity in size of
growth and time of flowering in corn hybrids could be disadvantageous if all
plants were adversely affected by unfavorable weather at a critical time.
Some 33 years later he (56) pointed out that the greater genotypic variability
of double cross hybrids, compared with single cross hybrids, was advantageous and not the serious objection it was first thought to be. Hartley (45)
saw that the uniformity associated with clonal propagation favors the buildup
of destructive parasite strains and also reduces adaptation to varied local conditions. He advocated planting mixtures of clones rather than single clones.
Some years later Stevens (81, 82) concluded that uniform clonally reproduced and self-pollinated crops are predisposed to attack by plant diseases.
The first extensive and detailed review of variability in crop plants was by
Simmonds (79). His major conclusions are worth restating here. Plant breeding was envisaged as "the current phase of crop evolution." The breeder's
objective, in evolutionary terms, is to produce populations that are better
adapted; that have a greater fitness in a given environment. However, adaptation and adaptability (the capacity for genetic change in adaptation) are antagonistic. As adaptation in one environment is maximized, genetic information for fitness in other environments tends to be lost. Successful plant breeding, while improving adaptation, reduces variability and hence long-term
adaptability.

A consideration of breeding systems and methods of propagation showed
that the techniques of advanced agriculture have reduced variability. Pure
lines have replaced land races, hybrid maize has replaced open-pollinated varieties and clonal propagation has taken the place of seedlings.

Simmonds called the variability immediately available to the breeder
among widely grown cultivars the genetic base of the crop. He noted that as
the genetic base narrows it limits further progress in breeding. He pointed
also to the neglect of "population adaptation." This results from productive
interaction between individuals of heterogeneous populations.

The practical problem faced by breeders is how to combine the benefits of
close adaptation with long-term conservation of variability. Four measures
for conserving variability were outlined; maintenance of germ plasm collections, development of mass reservoirs of variability, use of deliberately heterogeneous populations, and the compromise offered by conservative breeding
methods. The usefulness of each of these measures varies with crops and circumstances.

In the 10 years since Simmonds' review, the trends to a narrower genetic
base have become still more pronounced. Much of the evidence for this was

documented in a recent report (70) on major crops in the United States. Other recent discussions of the same topic are by Adams et al (1) and Smith (80).

The dwarf wheat and rice of the green revolution have been responsible for much more erosion of variability as they have replaced locally adapted cultivars in Asia (36). A new crisis has arisen: the preservation of variability through genetic conservation of germ plasm before it is lost irrevocably.

Estimating Variability

Genetic variability of crops can be partitioned into variability between crop cultivars and variability within a crop cultivar (the genetic differences within the population of plants that make up the cultivar). The extent of variation between cultivars is a function of the number that make up a significant part of the crop acreage and of the genetic differences between them.

Information on the acreages occupied by individual crop cultivars is scattered. For some, such as wheat, barley, potatoes, and sugar cane, annual statistics compiled by local Departments of Agriculture and regulatory agencies are available. Sometimes the information is in another form but can be converted. For example, acreages of certified seed tubers of potato (85) or tonnages of crushed cane (14) are listed by cultivars. For other crops, the data is available only from breeders, seedsmen, and processors.

Acreage data can be transformed by counting the number of cultivars that individually make up more than a given percentage of the crop, presenting this figure with their aggregate acreage as a percentage of the total. Table 1, adapted from (70), shows data for 14 crops in the United States for 1969,

Table 1 Acreage and farm value of major U.S. crops showing extent to which small numbers of varieties dominate. From (70).

Crop	Acres×10⁶	Value×10⁶	Total Varieties	Major Varieties	Acreage %
Beans, dry	1.4	143	25	4	93
Cotton	11.2	1,200	50	4	61
Corn	66.3	5,200	197	5	66
Millet	2.0	?	?	3	100
Peanut	1.4	312	15	2	78
Peas	.4	80	50	2	96
Potato	1.4	616	82	4	72
Rice	1.8	449	14	7	90.2
Sorghum	16.8	795	?	?	?
Soybean	42.4	2,500	62	6	56
Sugar beet	1.4	367	16	5	73
Sweet potato	.13	63	48	1	69
Wheat	44.3	1,800	269	2	25

Table 2 Acreage and farm value of major Australian crops showing extent to which small numbers of varieties dominate.

Crop	Acres×10³	Value×10⁶	Total Varieties	Major Varieties	Acreage %
Barley[a]	2,491		11	4	88
Oats	3,800	54	39	4	62
Peanuts	95	8	2	2	99
Sorghum[b]	896	35	20+	3	69
Sugar cane	640	162	54	4	51
Wheat	16,000	414	49	6	70

[a] South Australia and Victoria
[b] Queensland

using 5% as the cut-off figure. For 8 of the 13 crops shown, 1 to 5 varieties account for 69% or more of the total acreages. The national figures for wheat are somewhat misleading because within the major wheat states a single variety may account for 38% to as much as 99% of the state's acreage (70). Localized uniformity is also true of other widely grown crops such as maize and soy bean where varietal responses to season or day length are important. In spite of the crudeness of these statistics it is clear that a few cultivars predominate in the acreage of most major U.S. crops.

Table 2 shows data for 6 crops in Australia for the 1970–71 season and the same trends are evident. But we can ask if this is new. Was it true 30 or more years ago? Records of tonnages by variety for Queensland sugar cane (14) were first published in 1939, while records for wheat acreages by variety for the principal Australian wheat producing states (20) first appeared in 1946. The data for every fourth year is given in Table 3; the figures for intervening years merely confirm the trends shown in the table.

No systematic reduction in the number of cultivars with breeding progress is shown, although in both crops, pronounced changes of varieties occurred as a result of plant breeding. Over the period shown, the Queensland sugar cane acreage almost doubled, and local preferences and a need for diversification to suit the three major sugar growing districts may have prevented wider adoption of fewer cultivars.

The low numbers of cultivars in advanced agricultures stems from the successes of plant breeding beginning in the middle of the last century (34) and the development of seed and propagation industries to satisfy farmers' requirements. In primitive, undeveloped agricultures a variety of locally adapted and preferred cultivars are propagated by the farmers themselves. The large scale replacement of these forms by modern cultivars has brought about a dramatic reduction in genetic variability and is a prime cause of the crisis in germ plasm conservation.

Table 3 Numbers of cultivars that each make up 5% or more of the crop acreage and their aggregate acreages as percentages of the total. Years represent sowing date for wheat and harvesting date for sugar cane (based on tonnage of crushed cane). From (20) and (14).

| Year | Sugar cane | | Wheat | | | |
| | Queensland | | N.S.W. | | W.A. | |
	No.	%	No.	%	No.	%
1970	8	71.7	7	76.8	5	88.5
1966	8	80.7	6	71.8	5	77.0
1962	6	76.4	6	51.6	6	86.9
1958	6	78.6	5	62.5	6	84.8
1954	5	86.4	3	61.7	7	84.9
1950	4	66.8	4	61.3	4	76.1
1946	7	74.4	2	56.5	3	73.5
1942	6	79.0				
1939	6	76.5				

How Different are Modern Cultivars?

The degree of genetic homology between crop cultivars could be established by comparing the extent to which their DNAs share identical nucleotide sequences. The DNA-hybridization technique (6) has been successfully applied to establish homologies between different cereals (7), between hexaploid, tetraploid, and diploid wheats (8), and different species of *Vicia* (17). To my knowledge it has not yet been used to investigate affinities between varieties although Chooi (17) has reported that closely related varieties of the same species of *Vicia* may have significantly different DNA contents per cell. Other methods compare polymorphic characters. Gel-electrophoresis reveals similarities and differences in polymorphic enzymes (so-called isozymes) from crude extracts of various tissues (40). Comparison of the stained gels, or zymograms, quickly reveals similarities and differences.

In studies with barley the most useful markers are the isozymes of α-amylases and esterases in extracts of endosperm and plumule tissues. The alleles at three of the esterase loci are codominant so that heterozygosity of individual plants can be readily detected from their zymograms without progeny testing (4, 57). Polymorphism between or within cultivars is revealed as patterns of homozygosity and heterozygosity for the loci concerned. It has been assumed that there was no conscious selection for these markers during varietal breeding (31). If true this should allow some estimate of the variability in that part of the genome not subject to conscious selection. Unfortunately the biochemical relationship between these markers and those aspects of plant phenotype that a breeder selects is still unknown. Comparisons among

Table 4   The 6 most important Australian spring wheats in 1970–71 showing ancestral varieties common to 2 or more. Acreages from (20) and pedigrees from (64).

Variety	% of acreage	Common ancestors						
		Gabo	Ghurka	Currawa	Federation	Improved Fife	Gular	Dundee
Gamenya	22.4	X						
Heron	15.9	X	X	X	X	X		
Insignia	10.0		X	X	X	X		
Falcon	8.9				X	X	X	X
Timgalen	6.7	X					X	X
Olympic	6.2		X	X	X			

barley cultivars of north western Europe (72) and Canada (32) clearly reflect the limitations imposed by narrow parentage. For example 9 different esterase patterns were distinguished among 55 Canadian cultivars (32). The commonest, pattern 1, was found in 18 cultivars and, not unexpectedly, was also the commonest pattern among the parental cultivars. Gel-electrophoresis of slow-moving wheat endosperm proteins has also been used by Shepherd (77) to compare genetic homologies. The analysis can be made on a single grain without destruction of the embryo, which can then be grown on for further testing. Closely related wheats have very similar electrophoretic patterns (92). For example endosperm proteins of 6 major Australian spring wheat cultivars (see Table 4) show only minor differences involving one or two out of some 25 different bands (78).

The most general method for establishing varietal affinities is pedigree analysis but it suffers from the difficulty of summarizing and quantifying pedigree records.

Pedigrees of crop cultivars are frequently unpublished and available only from breeder's records. For some crops such as commercial maize and sorghum hybrids, detailed pedigrees and parents of hybrids are trade secrets. Seed producers are reluctant to divulge information that could be used to prepare general statistics showing the use of privately developed inbreds, although U.S. producers have released figures showing their use of maize inbreds developed by the U.S.D.A. and State Agricultural Experiment Stations (70).

Cultivars with common ancestors can be expected to have more of their genomes in common than unrelated cultivars, but we cannot say precisely how much more. Crop cultivars carrying deliberately introduced genes will of course be uniform for those genes. Uniformity for other linked genes will depend on their proximity to the selected genes and on the number of generations of breeding and selection to produce the cultivars.

Examination of crop variety pedigrees shows that the few varieties that dominate a crop's acreage have common ancestors and hence must have genetic information in common. Table 4 shows an example for Australian spring wheats grown in 1970–71. The precise extent to which these varieties are similar is difficult to gauge, even from a pedigree chart. Some varieties are very similar. Heron, Insignia 49, and Insignia differ principally in major genes for rust resistance introduced by back-crossing. I have already noted that their endosperm proteins are very similar. However, the spring wheat variety Gamenya alone occupied 1.4 million hectares in Australia in 1970–71. In the U.S.S.R. the spring wheat Saratovskaja is currently grown on 30 million hectares or 75% of the spring wheat acreage (54). Similarly large areas are planted to successful winter wheats. Two related cultivars, Besostaya 1 and Mironovskaya 808, account for 85% of the total area sown in the U.S.S.R. (61). These figures illustrate the extent to which we accept very large acreages occupied by crop plants that appear to be genetically uniform.

How Much Variation Is There Within Modern Cultivars?

The individual plants of varieties of vegetatively propagated crops are expected to be genetically identical barring spontaneous mutation or the accident of mechanical mixture. The variability among individual plants raised from seed is governed not only by these two factors but by the amount of outcrossing permitted by the breeding system, the effectiveness of the isolation procedures used in growing seed crops, and the selection pressures governing the relative fitness of homo- and heterozygotes. These problems have been considered elsewhere and few of their details will be dealt with here. They have technological solutions that for the most part provide the degree of phenotypic uniformity the modern farmer has come to expect of high yielding varieties.

Allard et al (3) have shown that natural populations of inbreeding species are able to store large amounts of genetic variability for continuously varying characters. This storage is in the form of linkage block heterozygotes that persist in a stable equilibrium because of their strong selective value (18). The persistence of heterozygotes has been demonstrated very clearly in barley cultivars by analysis of isozyme polymorphisms. Allard et al (4) examined a minimum of 36 seedlings of each of 30 varieties maintained in a U.S.D.A. collection of barley varieties. Three tightly linked esterase loci A, B, and C, with 4, 3, and 3 codominant alleles respectively were scored. Only 6 of the varieties were monomorphic for all 3 loci, 7 were polymorphic for 1 locus, 8 were polymorphic for 2 loci and 9 were polymorphic for all 3 loci. Many plants of the polymorphic varieties were heterozygotes. The same authors in a preliminary examination of a much wider range of cultivated barleys found a further 11 band positions in addition to the 4 known for the A zone, a further 14 for the B zone but no additional band positions for the C zone. Thus while a so-called pure line may contain substantial genetic variability with respect to esterase isozymes, this still represents only a fraction of the known variability. Although in some ways it is reassuring that an inbred crop like cultivated barley can maintain heterozygosity this can be no guarantee that the variation that is preserved in the varieties that dominate the acreage would enable the crop to survive some future catastrophe.

In maize, isozymic polymorphism has been demonstrated for 10 different enzymes including peroxidase and acid and alkaline phosphatases as well as amylase and esterase (63). Little or no systematic mapping of these enzymes among different inbreds has been attempted comparable to the work with barley varieties. Other techniques to promote uniformity do not depend on genetic manipulation. For example, pelleting seed creates an artificial size uniformity that makes drilling easier (23, 51). Pelleting can also be used to modify the microenvironment at the time of germination. Shade-grown tobacco illustrates another kind of environmental modification designed to maximize the production of uniformly thin broad leaves for cigar wrappers. Attempts to breed open air shade types have so far been unsuccessful (83).

WHY VARIABILITY IS LOST

Economic Reasons

Modern agriculture depends on high grade uniform produce. Consumers will pay premium prices for high quality and farmers prefer to grow the one variety that will assure them of the highest cash returns for the lowest cost per unit of production. Farming is competitive and no farmer can afford not to grow the best and highest yielding cultivars available to him. As everyone wishes to grow the best to achieve the highest returns, and modern testing methods and communications rapidly tell farmers what is best and where to obtain it, most farmers grow the same cultivars. Industry-imposed standards may work to the same end. Marketing boards dictate which varieties shall be grown. Australian barley marketing boards demand a uniform crop so that maltsters will have less trouble in adjusting their methods to their raw material. When enzymes of microbial origin are used in the malting process the pressures towards varietal uniformity may ease.

Wheat exports in Australia are increasingly of high quality grades (20) that command a better price than average quality grain. This will intensify the trend to growing highly selected varieties with desirable qualities for export markets. In a number of other crops the cultivars that are grown are determined by industry. For example Canadian wheat farmers are obliged to grow only registered wheat cultivars (70). In the San Joaquin Valley in California only certain cotton cultivars may be grown (70). Sugar beet processors and potato processing plants commonly supply seed and seed stock to contracting farmers. The varieties supplied are generally the product of breeding programs sponsored or undertaken by the processors themselves.

While these pressures reduce the number of cultivars grown, other pressures increase uniformity within them. Uniformity in germination, subsequent growth, and time of harvest, are overriding requirements for successful culture, harvesting, and marketing of modern vegetable varieties. They are imposed by the high degree of mechanization that has helped to sustain and increase agricultural productivity (70).

Seed certification programs and plant patent laws impose uniformity by requiring that plant phenotypes conform to easily administered standards. These schemes were designed to protect the farmer, the breeder, and the seed industry but could become an obstacle to commercial introduction of mixtures, multilines, or other variable cultivars.

Genetics and Breeding Reasons

Donald (27) pointed out that there are three kinds of plant breeding programs. The first seeks to eliminate defects. Incorporation of disease resistance, earliness, ozone resistance, and freedom from cracks in tomato fruits are a few examples from just one crop. The second is based on selection for yield mainly through inbreeding and selection for combining ability among F_1 hybrid parents, but also through hybridization of promising lines with selection of high yielding segregants. The third kind is breeding for model plants or ideotypes. This last method has only recently been put into practice. It proposes a theoretically efficient model of plant form and function which is then made a breeding objective (27, 28). Although ideotypes were originally thought of in morphological terms, maximal physiological efficiency is a final objective (60, 93). The successes of dwarf rice and semi-dwarf wheats (5) owe much to breeding towards such objectives even though the procedures that led to them were basically the elimination of such defects as poor response to fertilizer, disease susceptibility, lodging, daylength sensitivity, and undesirable plant architecture.

Following successful breeding, the economic requirements for varietal uniformity are met by clonal reproduction (potato, sweet potato, sugar cane, cassava, and banana); seed reproduction by continued inbreeding (wheat, barley, rice, common bean, peanut, soy bean, and pasture legumes) and production of uniform F_1 hybrids by controlled pollination (maize, sorghum, sugar beet, onion, cucumber, and, more recently, wheat, barley, and sunflower). All three reproductive methods give crop populations that are phenotypically uniform.

In all breeding programs adaptation must be maximized to obtain the highest yields and, where breeding is carried out in one location, adaptability decreases (79). Two recent developments are of particular interest here.

The CIMMYT[1] wheat breeding program showed that varieties with wide adaptation, due largely to their daylength insensitivity, may be selected by testing and screening segregating populations in two environments separated by 10 degrees of latitude and 8,500 feet elevation (9). The success of some of these varieties may well have the effect of reducing genetic variability in wheat still further. Let us see how this could happen.

Since 1960 CIMMYT has run a series of International Wheat Yield Nurseries. For example the International Spring Wheat Yield Nursery (ISWYN) was begun in 1964 and grown in each succeeding year. It includes 50 entries from cooperating wheat breeders, which are multiplied at Sonora, Mexico,

[1] Centro Internacional de Mejoramiento de Maiz y Trigo, Mexico City.

and grown in replicated trials at more than 60 different spring wheat regions in the world. The entries represent the principal varietal types of spring wheat grown in many areas of the world. A proportion of each year's entries has been grown in former nurseries to keep a balance between old and new. A performance analysis is rapidly made available to individual co-operators who can compare their own results with the international average. One year's tests at a number of places around the world are as useful as tests at one location for several years and save much time and effort. So quick an assessment of so much promising material on so large a scale is unprecedented. The comparisons (9, 65) show that the daylength insensitive Mexican varieties are clearly outstanding. Can any breeder afford to ignore such material in his own program? The consequence will surely be that most, if not all, spring wheat breeders will include this material in their breeding programs to secure maximal adaptation and high yield. While local diversification in response to specific defects will relieve uniformity to some extent the spring wheats will become even more uniform.

A similar International Winter Wheat Performance Nursery, co-ordinated by wheat breeders at the University of Nebraska, tests 30 varieties at 45 sites in winter wheat regions (54).

The second development concerns the implications of the concept of breeding for ideotype. Donald (28) stated: "The very diversity of form among currently successful cultivars may indeed suggest that each variety is deficient in one or several characteristics. The narrower array of material to be used in the breeding of models is implicit in the concepts behind such a programme, . . .". Where yield is the selection criterion the variation between cultivars that Donald refers to results from the different ways in which the variables affecting yield are integrated. The successful cultivars represent a series of best compromises.

At first sight the reduction of variability involved in breeding for ideotype appears more hazardous in terms of increasing genetic vulnerability than presently used methods. For example, when efforts to increase plant productivity through biochemical and genetical engineering of photosynthesis and photorespiration begin to pay off, great dependence on a very narrow genetic base could result. In practice, while plant breeding remains as much an art as a science, this hazard is remote. As long as there are different breeders working in different environments there will be different concepts of the theoretically best model and hence different approaches to it. As before, the chief danger is the widespread adoption and cultivation of one phenotype and this danger exists irrespective of the method by which a breeder arrives at his goal.

The realization that further genetic improvement of existing cultivars cannot be made without recourse to new germ plasm is a common crisis in plant breeding. Defect-elimination, although valuable, is a conservative and somewhat wasteful method of gaining improvement. The transfer of a single gene from a wild relative totally neglects the opportunity for radical reappraisal and genetic re-structuring to remove blocks to further progress by enlarging the genetic base of the crop.

HAZARDS OF GENETIC UNIFORMITY

The southern corn leaf blight epidemic was the most recent and in terms of food destroyed, most destructive, example of the dangers of genetic uniformity. In 1970 over 70% of the North American maize crop carried T cytoplasm because a single type of cytoplasmic male sterility (*Tcms*) was used in the production of hybrid seed. Race T of *Helminthosporium maydis,* virulent on maize with T cytoplasm, was present and the weather conditions favored the development of the epidemic. The circumstances and course of the epidemic have been well documented (70, 87). The hazard of cytoplasmic uniformity had been assessed by corn breeders. There were reports (67, 90) of what now appears to have been race T in the Philippines in 1961; these did not go unnoticed in the United States. Duvick (29) considered the evidence for differential sensitivity of *Tcms* lines to *H. maydis* and concluded that reduced plant vigor accentuated by the environment in the Philippines was responsible for their failure. Hooker et al (49) also found no differences using strains from the pathogen from Illinois. No other drawback to the widespread use of *Tcms* for making hybrid seed was noted. The economic advantages of avoiding hand emasculation by detasselling were compelling. The consequence in 1970 was a high degree of cytoplasmic uniformity over more than 60 million acres of cornfields in the United States that resulted in losses due to southern corn leaf blight approaching one billion dollars.

Corn with T cytoplasm was also found to be highly susceptible to a second fungus disease, *Phyllosticta maydis,* causing yellow leaf blight (87), but losses due to this disease were insignificant compared with those from *H. maydis.*

Another example of genetic vulnerability in the oat crop of Iowa and surrounding states in 1946 depended on the uniform distribution of the gene *Pc-2* for resistance to crown rust—*Puccinia coronata.* This gene was released in 1942 in an oat variety called Victoria, which at that time was resistant to all known races of crown rust (69). Several other cultivars making use of the same resistance were introduced and, by 1945, made up 97% of the oat acreage. In 1946 a seedling blight began to damage the oat crop so severely that many farmers had to plough their crops in. The disease, called Victoria blight, was caused by a fungus *Helminthosporium victoriae* and was a problem only on oat cultivars carrying the dominant gene *Pc-2* for crown rust resistance. Within several years a switch to oat varieties protected from crown rust by other genes for rust resistance had completely eliminated Victoria blight as a hazard.

These two Helminthosporium diseases indicate an important trend in genetic vulnerability. Both epidemics were due to the introduction of what appeared to be a single character, in a number of different genetic backgrounds, that soon came to be present in the bulk of the crop acreage. For oats, rather

Table 5   Some examples of crop uniformity for single characters.

Crop	Character
Bean	stringless
Maize	opaque-2 (high lysine)
	Tcms—cytoplasmic male sterility
Rice	dwarfing gene
	photoperiod insensitivity
Sorghum	Milo source of cms
Sugarbeet	monogerm seed
	single source of cms
Tomato	determinate habit
Wheat	dwarfing gene
	photoperiod insensitivity
	cms from *Triticum timopheevi*

few backgrounds were used; only the 4, or so, varieties released carrying *Pc-2*. For corn, however, T sterile cytoplasm was introduced into a much larger number of inbreds of both field and sweet corn (87). Although rather few publicly developed inbreds are used in current hybrids, T sterile cytoplasm and its associated fertility restorer system were sufficiently easy to transfer to other lines and did not hinder diversification in other respects. Clearly vulnerability, like resistance, can depend on a single gene or a cytoplasmic determinant. Equally clearly it will be very difficult or even impossible to devise tests to reveal flaws in other single genes that are carried by all cultivars of a particular crop. Such genes are not uncommon. They should be distinguished from the many genes that varieties, species, and genera may be presumed to share, that govern basic aspects of metabolism such as photosynthesis, respiration, and plant structure. The latter genes were subject to natural selection during evolution and any penalties they may have originally imposed are likely to be balanced by compensating genetic heterogeneity. The hazards imposed by *Pc-2* in oats or T sterile cytoplasm in corn could not be compensated until they were recognized, by which time they had caused very considerable damage. Some spontaneous and induced mutants of *Pc-2*, or a tightly linked locus *Hv*, have partial resistance to *H. victoriae* toxin but retain crown rust resistance (25). Hooker et al (49) have shown that resistance to *H. maydis* race T can be expressed in plants carrying T cytoplasm.

Some examples of single genes that are now, or are likely to be, widely used in certain crops are given in Table 5. Determinate habit in tomato had the defect of increasing susceptibility to *Alternaria solani* (50). This was overcome by additional breeding to introduce resistance. In hybrid seed production plots of sorghum, cytoplasmic male sterile lines are very susceptible to the ergot fungus *Sphacelia sorghi* if pollination is poor (39). Under these conditions individual florets remain open for a long time allowing a much greater opportunity for head infection. Male sterile barley and wheat are similarly susceptible to ergot, *Claviceps purpurea* (73). High-lysine corn is very susceptible to kernel and ear rots, necessitating early harvest and rapid drying to minimize loss (86). Semi-dwarf wheats in India are susceptible to *Alternaria triticina* (70) and in Mexico to leaf blotch caused by *Rynchosporium* (70).

Most of these defects are minor and can be corrected either by additional breeding to introduce resistance or by varying agronomic methods. The importance of such characters as monogerm seed to the sugar beet industry, or stringless to fresh bean growers is so great that the risks in introducing these genes, along with their associated closely linked genes, are worth taking. Rice breeders have justified their investment in the dwarfing gene from IR8 and TN1 on the grounds that introduction of other genetic diversity will minimize the risk (70). The important point is that the introduction of such characters must be recognized as introducing a risk. The more important the resulting technological advance, the more concern there should be to develop genetic alternatives. Sugar beet and sorghum breeders are looking for alternative male steriles. These are available in maize although the extent to which they will be used remains to be seen. Breeders interested in hybrid seed production are also exploring other methods of producing hybrids such as chemical male gametocides. Total dependence on single gene traits must be relieved by research to find other pathways to similar end points. If modern rices should be devastated by an insect or fungus epidemic, or some other defect closely associated with the dwarfing gene, we would be much better prepared if a series of different dwarfs (some induced by mutagens) were available for screening. While deliberately introduced single characters represent a slight but real risk, a more common situation is that crop cultivars are uniformly susceptible to hazards that are either not present or not a problem prior to and during varietal development. Examples include the susceptibility of Irish potato clones to late blight *Phytophthora infestans* in the 1840s, the susceptibility of the Gros Michel clone of banana to Panama disease *Fusarium oxysporum cubense* (68) and, even more recently, the susceptibility of South American coffee to rust *Hemileia vastatrix* (76).

As we know to our cost, the same hazard occurs in natural and cultured host populations as shown by the spread of chestnut blight *Endothia parasitica* in North America and Europe and the current destructive effects of *Phytophthora cinnamomi* on a number of Australian plants (71).

Johnson (53) pointed out that breeders, in introducing genes for resistance, guide the evolution of the parasites they wish to control. Of equal significance is reliance on oligogenic race-specific resistance, which has led to an erosion of less dramatic nonrace specific resistance frequently determined by many genes of individually small effect. Virulent races are even more damaging to such cultivars than they are to the "susceptible" older cultivars. This was called the "Vertifolia effect" by Van der Plank (88) after a late-blight resistant potato cultivar of that name. Lupton & Johnson (62) have described similar effects among wheat varieties varying in resistance to yellow rust *Puccinia glumarum.*

Some Anticipated Problems

Early in 1970 coffee (*Coffea arabica*) was found to be infected with the rust *Hemileia vastatrix* in several coffee producing states in Brazil. South American coffee cultivars are uniform and highly susceptible to rust (48, 76). Many predictions (76) had been made that rust would cross the Atlantic Ocean carried either by prevailing winds or aircraft. The search for resistant varieties was begun in anticipation of their need but is now the subject of emergency programs in reaction to what is seen as the inevitable spread of the disease throughout the coffee growing areas of South America (76).

All planted rubber in Africa and Asia is susceptible to leaf blight disease caused by *Microcyclus ulei*. This disease is so far restricted to South America where it was responsible for the destruction of extensive plantations (48). Malaya, the largest producer, has more than four million acres. The genetic uniformity stems from the fact that the Asian industry was begun from one seed sample obtained from South America via the Royal Botanic Gardens in Kew. Although resistant clones have been developed they have, so far, unsatisfactory yields. As the disease attacks only young leaves, one of the most rapid control measures is the grafting of resistant crowns onto high yielding clones (48). Plans for eradication of infected trees by spraying defoliants from aircraft are among the measures proposed to cope with the threat of outbreaks.

REMEDIES

Two basic breeding approaches may be used to combat the dangers of genetic uniformity. The first is breeding to eliminate defects as soon as they appear. The risk of increasingly uniform cultivars is assumed in the belief that alternative germ plasm can be introduced to meet each crisis as it occurs. The great weakness of this method is the time it takes for breeders to respond to an unanticipated threat. Varietal development takes time. Even the incorporation of a new gene for resistance takes hardly less than 5 years in most crops and for some may be longer than 20 years (70). The second approach is the deliberate introduction and retention of variability in crop populations with the object of ensuring that part of the population will survive each crisis. Both methods, and indeed all breeding, depend heavily on germ plasm resources, so I will first briefly review the present situation and future prospects in this key area.

Genetic Conservation

Plant breeders have always turned to the variation present in primitive and outmoded cultivars and to wild relatives collected from centers of genetic diversity (38, 43, 89) to find new genes for their breeding programs. The genetic diversity of wild and cultivated forms present in these centers is the result of adaptation to a range of macro and micro environments. Cultivation and the variety of methods used by primitive farmers have also helped in producing diversification (39, 94). Introduction of exotic germ plasm has always been vital in the United States for the reason that none of the major food crops were developed from native plants (42).

The role of federal plant introduction in the United States was reviewed by Hodge & Erlanson (46) and, more recently, by Burgess (15). The sources, uses, and maintenance of plant germ plasm were dealt with at Symposia in Chicago in 1959 (47), Columbus in 1968 (58, 59, 74, 75), and in Rome in 1967 (38). The last, published as an IBP handbook in 1970, is an especially valuable and authoritative treatment of all aspects of genetic resources in plants. Since the Rome conference, several other important reviews dealing with germ plasm resources have appeared, including those of Creech & Reitz (21), Timothy (84), Frankel (33–37), and the National Academy Committee (70).

Nearly 20 years ago Harlan (42) warned that the raw materials for breeding programs that had been readily obtained from centers of diversity were threatened by extinction and that they should be preserved indefinitely in world collections. Until quite recently these warnings were largely ignored. Now we are much more aware of the invasion of centers of diversity by modern cultivars and the rapid disappearance of primitive forms, first noted by Harlan. In Asia this has been accentuated by the intense cultivation, fertilizer application, plant protection, weed control, and irrigation practices that characterize the green revolution (5, 36). Wild plant habitats are also being destroyed by roads, drainage schemes, power lines, industrial development, and population growth.

Frankel has pointed to these dangers, stressing the tremendous importance and great urgency in attending to genetic conservation (33, 34). His proposals (35) include (a) surveys of where genetic diversity presently exists in the field; (b) a survey of existing germ plasm collections, now being undertaken by FAO; (c) development of an international program of exploration and collection, with priority given to the most endangered areas and species; (d) the establishment of Genetic Resources Centers that will provide facilities for genetic stock conservation in perpetuity, and also encourage the establishment of working collections for evaluation and utlization as well as conservation; (e) development of international documentation systems for data storage and retrieval.

Although some progress has been made, there is little reason for complacency about existing germ plasm reserves and facilities. Frankel's proposal of Genetic Resource Centers meets the criticism that existing storage facilities are only safe deposits and not living collections undergoing study and evaluation (70). The working collections and facilities in overseas Genetic Resource Centers could also be used to study exotic pests that are potential threats to major crops in other places.

Many problems remain. Allard (38) has shown that guidelines for sampling variation in the field that are based on extrapolation from well studied cases are better than arbitrary guidelines. He has also discussed how selective bulking and special handling can minimize both loss of variation (genetic erosion) and expenditure on maintenance and rejuvenation in germ plasm stored as seed. Short-lived clones (potato, sweet potato etc.), and fruit and forest trees also have special problems (38). One major potato collection is largely maintained as populations planted and harvested by machine (70). While this makes material available to the breeder in a more immediately usable form, the cost in terms of genetic erosion is unknown. The crucial international co-operation needed to define problems, plan and establish a co-ordinating center and genetic resource centers, to take care of staffing and training, and to obtain financial support is now being worked out by FAO (30).

The chief danger is that few scientists will see the broadness and urgency of the crisis. Individual breeders who presently may well have more resource material than they can handle tend to have little concern for the problems of posterity. Too many collections disintegrate when breeders retire or change interests.

Two other sources of genetic variability are being explored. Induced mutants may prove to be of increasing usefulness. Brock (11) has suggested that accumulated induced nucleotide base changes that are individually selectively neutral could lead to a novel source of genetic change and may underlie some successes in selecting for quantitatively inherited variation. Of still greater significance is our growing ability to select for altered metabolic pathways in higher plant tissue cultures using methods that are so successful with micro-organisms. A case in point is Widholm's (91) isolation of spontaneous mutants of cultured cells of *Nicotiana tabacum* that accumulate tryptophan because they possess an anthranilate synthetase that has reduced sensitivity to feedback inhibition. Brock et al (12) have proposed similar methods for selecting high-lysine forms of higher plants based on their success in recovering feedback insensitive mutants in *Escherichia coli*.

The other source of variation is parasexual genetic exchange. This was recently demonstrated by Carlson et al (16) who were able to produce, by vegetative cell fusion and appropriate selection techniques, hybrids between *N. glauca* and *N. langsdorfii* that duplicated the known sexual hybrids. The prospect of overcoming the barriers to sexual recombination and recovering interspecific and intergeneric hybrids was recently discussed by Cocking (19).

Synthetic Variation

The concept of growing mixtures of cultivars rather than pure stands has great intellectual appeal. In the first place, mixtures avoid the uniformity that leads to pest epidemics or catastrophic losses due to unusual weather conditions. Mixtures offer a method of deploying resistance genes in multilines to control pathogen spread and development, perhaps preventing the build-up of complex physiologic races (13). A similar idea is to use mixtures of male sterile cytoplasms of different origins in the production of hybrids (41). Mixtures seem likely to have other benefits. They suggest the possibility of manipulating competition in plant stands by selecting components that complement each other in their demands on the environment. Legume and grass mixtures for forage are to some extent an example. They greatly increase the productivity of nitrogen-poor soils (52). Other mixed crop species have also been studied, mostly as potential feed crops for cattle. Donald (26) and Simmonds (79) have reviewed the productivity of mixed stands.

In general, results from yield trials of mixed stands are disappointing. Mixtures are seldom more efficient than expected on the basis of component means. For example a soybean study (10) showed that although better competitors produced more dry matter, and were more efficient in mixtures than in pure stands, the gain in efficiency was outweighed by the loss in efficiency of the poor competitor in the mixture. A recent study of mixtures of sugar beet lines by Curtis & Hornsey (22) showed that the minimum density of unthrifty plants that a population can support without detectable yield loss was less than 8%.

All yield studies of controlled mixed plant stands perhaps suffer from the fact that finished varieties selected for performance in uniform stands are used as the components. That some combinations are better than others suggests that the true potential of mixtures will only be realized following selection for what Harper (44) called "ecological combining ability." This may be hazardous because it might well lead eventually to genetic uniformity.

For many crops, market demands for uniformity place strict limitations on the degree of variation that can be tolerated in the final product. While it is true that for some crops such as wheat or peanuts (2, 79) blends are used in processing the product it does not follow that the same cultivars can be blended in the field. Not only are yields likely to be lowered but the components may have different maturity dates or cultural requirements. Blending in processing is carried out after quality tests to determine optimum proportions. Control of field blends to optimize the end product will be much less precise.

Multiline cultivars whose components differ by rather few genes have been tested and appear to have promise in disease control (13). But they suffer from conservatism. By the time a number of resistance genes have been introduced into different lines of a single cultivar, usually by backcrossing, and are ready for release, that cultivar may be outmoded by other breeding advances. Another snag is that the variability is polarized; it is directed against only one or a few known threats (24). Multilines offer little or no protection against unforeseen hazards.

Acknowledgement

I thank all those who provided me with information and unpublished manuscripts. I also thank the University of Queensland for facilities, and the Guggenheim Memorial Foundation for the opportunity to complete this review on sabbatic leave.

Literature Cited

1. Adams, M. W., Ellingboe, A. H., Rossman, E. C. 1971. Biological uniformity and disease epidemics. *BioScience* 21:1067–70
2. Allard, R. W., Bradshaw, A. D. 1964. Implications of genotype-environmental interactions in applied plant breeding. *Crop Sci.* 4:503–8
3. Allard, R. W., Jain, S. K., Workman, P. L. 1968. The genetics of inbreeding populations. *Advan. Genet.* 14:55–131
4. Allard, R. W., Kahler, A. L., Weir, B. S. 1971. Isozyme polymorphisms in barley populations. *Barley Genetics* II. *Proc. 2nd Int. Barley Genet. Symp.*, Pullman:1–13
5. Athwal, D. S. 1971. Semidwarf rice and wheat in global food needs. *Quart. Rev. Biol.* 46:1–34
6. Bendich, A. J., Bolton, E. T. 1967. Relatedness among plants as measured by the DNA-agar technique. *Plant Physiol.* 42:959–67
7. Bendich, A. J., McCarthy, B. J. 1970. DNA comparisons among barley, oats, rye and wheat. *Genetics* 65:545–65
8. Bendich, A. J., McCarthy, B. J. 1970. DNA comparisons among some biotypes of wheat. *Genetics* 65:567–73
9. Borlaug, N. E. 1968. Wheat breeding and its impact on world food supply. *Proc. 3rd Int. Wheat Genet. Symp.* Canberra, Aust. Acad. Sci.: 1–36
10. Brim, C. A., Schutz, W. M. 1968. Inter-genotypic competition in soy beans. II. Predicted and observed performance of multiline mixtures. *Crop Sci.* 8:735–9
11. Brock, R. D. 1971. The role of induced mutations in plant improvement. *Radiat. Bot.* 11:181–96
12. Brock, R. D., Frederich, E. A., Langridge, J. 1973. The modification of amino acid composition of higher plants by mutation and selection. *I.A.E.A. Panel Ser.* In press
13. Browning, J. A., Frey, K. J. 1969. Multiline cultivars as a means of disease control. *Ann. Rev. Phytopathol.* 7:355–82
14. Bureau of Sugar Experiment Stations, Brisbane, Queensland. *Annual Reports:* 1939–1971
15. Burgess, S. (ed.) 1971. The national program for conservation of crop germ plasm. *U.S. Dep. Agr., Agr. Res. Serv.,* Washington, D.C. 73 pp.
16. Carlson, P. S., Smith, H. H., Dearing, R. D. 1972. Parasexual interspecific plant hybridization. *Proc. Nat. Acad. Sci. USA* 69:2292–94
17. Chooi, W. Y. 1971. Comparison of the DNA of six *Vicia* species by the method of DNA-DNA hybridization. *Genetics* 68:213–30
18. Clegg, M. T., Allard, R. W., Kahler, A. L. 1972. Is the gene the unit of selection? Evidence from two experimental plant populations. *Proc. Nat. Acad. Sci. USA* 69:2474–78
19. Cocking, E. C. 1972. Plant cell protoplasts—isolation and development. *Ann. Rev. Plant Physiol.* 23: 29–50
20. Comm. Bureau Census & Statistics, Canberra. *The Wheat Industry.* Annual Reports:1946–1971
21. Creech, J. L., Reitz, L. P. 1971. Plant germ plasm now and for tomorrow. *Advan. Agron.* 23:1–49
22. Curtis, G. J., Hornsey, K. G. 1972. Competition and yield compensation in relation to breeding sugar beet. *J. Agr. Sci.* 79:115–19
23. Dawson, C. D. R. 1972. Pelleted seed. *Roy. Hort. Soc. J.* 97:87–90
24. Day, P. R. 1971. Crop resistance to pests and pathogens. In *Pest Control Strategies for the Future,* 257–71. Nat. Acad. Sci. Washington
25. Day, P. R. 1973. *Genetics of host-parasite interaction.* San Francisco: Freeman. In press
26. Donald, C. M. 1963. Competition among crop and pasture plants. *Advan. Agron.* 15:1–118
27. Donald, C. M. 1968. The breeding of crop ideotypes. *Euphytica* 17: 385–403
28. Donald, C. M. 1968. The design of a wheat ideotype. *Proc. 3rd Int. Wheat Genet. Symp.* Canberra, Aust. Acad. Sci. 377–87
29. Duvick, D. N. 1965. Cytoplasmic pollen sterility in corn. *Advan. Genet.* 13:1–56
30. F.A.O. 1970. Fourth Session of the F.A.O. Panel of Experts on Plant Exploration and Introduction, F.A.O., Rome
31. Fedak, G., Rajhathy, T. 1971. A study of α-amylase distribution and DDT response in Canadian barley cultivars. *Can. J. Plant Sci.* 51: 353–59
32. Fedak, G., Rajhathy, T. 1972. Esterase isozymes in Canadian barley cultivars. *Can. J. Plant Sci.* 52: 507–16

33. Frankel, O. H. 1970. Genetic conservation of plants useful to man. *Biol. Conserv.* 2:162–69
34. Frankel, O. H. 1970. Variation—the essence of life. *Proc. Linn. Soc. N.S.W.* 95:158–69
35. Frankel, O. H. 1971. *The significance, utilization and conservation of crop genetic resources.* Crop Ecology & Genetic Resources Unit, F.A.O., Rome. 29 pp.
36. Frankel, O. H. 1971. Genetic dangers in the green revolution. *World Agr.* 19:9–13
37. Frankel, O. H. 1972. Genetic conservation—a parable of the scientist's social responsibility. *Search* 3: 193–201
38. Frankel, O. H., Bennett, E. (eds). 1970. *Genetic Resources in Plants —Their Exploration and Conservation.* IBP Handbook No. 11, Oxford, Blackwell. 554 pp.
39. Futrell, M. C., Webster, O. J. 1965. Ergot infection and sterility in grain sorghum. *Plant Dis. Reptr.* 49: 680–83
40. Gottlieb, L. D. 1971. Gel electrophoresis: new approach to the study of evolution. *BioScience* 21: 939–44
41. Grogan, C. O. 1971. Multiplasm, a proposed method for the utilization of cytoplasms in pest control. *Plant Dis. Reptr.* 55:400–01
42. Harlan, J. R. 1956. Distribution and utilization of natural variability in cultivated plants. *Genetics in Plant Breeding, Brookhaven Symp. Biol.* 9:191–206
43. Harlan, J. R. 1971. Agricultural origins: centers and noncenters. *Science* 174:468–74
44. Harper, J. L. 1967. A Darwinian approach to plant ecology. *J. Ecol.* 55:247–70
45. Hartley, C. 1939. The clonal variety for tree planting: asset or liability? *Phytopathology* 29:9 (abstr.)
46. Hodge, W. H., Erlanson, C. O. 1956. Federal Plant introduction—a review. *Econ. Bot.* 10:299–334
47. Hodgson, R. E. (ed.) 1961. *Germ Plasm Resources.* Am. Assoc. Advan. Sci. Publ. No. 66. 381 pp.
48. Holliday, P. 1971. Some tropical plant pathogenic fungi of limited distribution. *Rev. Plant. Pathol.* 50: 337–48
49. Hooker, A. L., Smith, D. R., Lim, S. M., Musson, M. D. 1970 Physiological races of *Helminthosporium maydis* and disease resistance. *Plant Dis. Reptr.* 54:1109–10
50. Horsfall, J. G., Heuberger, J. W. 1942. Causes, effects and control of defoliation on tomatoes. *Conn. Agr. Exp. Sta. Bull.* 456:181–223
51. Hull, R., Jaggard, K. W. 1971. Recent developments in the establishment of sugar-beet stands. *Field Crop Abstr.* 24:381–90
52. Hutton, E. M. 1970. Tropical pastures. *Advan. Agron.* 22:2–73
53. Johnson, T. 1961. Man-guided evolution in plant rusts. *Science* 133: 357–62
54. Johnson, V. A. Personal communication
55. Jones, D. F. 1925. *Genetics in Plant and Animal Improvement.* New York: Wiley. 568 pp.
56. Jones, D. F. 1958. Heterosis and homeostatis in evolution and in applied genetics. *Am. Nat.* 92:321–28
57. Kahler, A. L., Allard, R. W. 1970. Genetics of isozyme variants in barley I. Esterases. *Crop Sci.* 10: 444–48
58. Knowles, P. F. 1969. Centers of plant diversity and conservation of crop germ plasm : safflower. *Econ. Bot.* 23:324–29
59. Konzak, C. F., Dietz, S. M. 1969. Documentation for the conservation, management, and use of plant genetic resources. *Econ. Bot.* 23: 299–308
60. Loomis, R. S., Williams, W. A., Hall, A. E. 1971. Agricultural productivity. *Ann. Rev. Plant Physiol.* 22:431–68
61. Lukyanenko, P. P. 1972. Some results of winter wheat breeding in the U.S.S.R. and the trends of its progress. *Proc. Winter Wheat Conf.*, Ankara, Turkey. In press
62. Lupton, F. G. H., Johnson, R. 1970. Breeding for mature-plant resistance to yellow rust in wheat. *Ann. Appl. Biol.* 66:137–43
63. Macdonald, T., Brewbaker, J. L. 1972. Isoenzyme polymorphism in flowering plants. VIII. Genetic control and dimeric nature of transaminase hybrid maize isoenzymes. *J. Hered.* 63:11–14
64. Macindoe, S. L., Walkden Brown, C. 1968. Wheat breeding and varieties in Australia. *Science Bull. 76 N.S.W. Dept. Agr.* 255 pp.
65. MacKenzie, D. R. 1971. Results of the fifth international spring wheat yield nursery, 1968–1969. *CIMMYT Res. Bulletin* 19. 13 pp. & 64 tables
66. Mangelsdorf, P. C. 1966. Genetic potentials for increasing yields of

312   DAY

food crops and animals. *Proc. Nat. Acad. Sci. USA* 56:370–75
67. Mercado, A. C., Lantican, R. M. 1961. The susceptibility of cytoplasmic male-sterile lines of corn to *Helminthosporium maydis* Nisikado & Miy. *Philipp. Agr.* 45:235–43
68. Meredith, D. S. 1970. Major banana diseases: past and present status. *Rev. Plant. Pathol.* 49:539–54
69. Murphy, H. C. 1965. Protection of oats and other cereal crops during production. In *Food Quality Effects of Production Practices and Processing*, 99–113. ed. G. W. Irving, Jr., S. R. Hoover. Am. Assoc. Adv. Sci. Publ. 77, 298 pp.
70. National Academy of Sciences, 1972. *Genetic Vulnerability of Major Crops*. Washington, D.C. 307 pp.
71. Newhook, F. J., Podger, F. D. 1972. The role of *Phytophthora cinnamomi* in Australian and New Zealand forests. *Ann. Rev. Phytopath.* 10:299–326
72. Nielson, G., Freydenberg, O. 1971. The inheritance of esterase isozymes in barley. *Barley Genetics II, Proc. 2nd Int. Barley Genet. Symp.*, Pullman, 14–22
73. Puranik, S. B., Mathre, D. E. 1971. Biology and control of ergot on male sterile wheat and barley. *Phytopathology* 61:1075–80
74. Reitz, L. P., Craddock, J. C. 1969. Diversity of germ plasm in small grain cereals. *Econ. Bot.* 23:315–23
75. Rowe, P. R. 1969. Nature, distribution, and use of diversity in the tuber-bearing *Solanum* species. *Econ. Bot.* 23:330–38
76. Schieber, E. 1972. Economic impact of coffee rust in Latin America. *Ann. Rev. Phytopathol.* 10:491–510
77. Shepherd, K. W. 1968. Chromosomal control of endosperm proteins in wheat and rye. *Proc. 3rd. Int. Wheat Genet. Symp.* Canberra, Aust. Acad. Sci. 86–96
78. Shepherd, K., Wrigley, C. W. Personal communication
79. Simmonds, N. W. 1962. Variability in crop plants, its use and conservation. *Biol. Rev.* 37:422–65
80. Smith, H. H. 1971. Broadening the base of genetic variability in plants. *J. Hered.* 62:265–76
81. Stevens, N. E. 1942. How plant breeding programs complicate plant

disease problems. *Science* 95:313–16
82. Stevens, N. E. 1948. Disease damage in clonal and self-pollinated crops. *J. Am. Soc. Agron.* 40:841–44
83. Taylor, G. S. Personal communication
84. Timothy, D. H. 1972. Plant germplasm resources and utilization. In *The Careless Technology; Ecology and International Development*, 631–66, ed. M. T. Farvar, J. P. Milton. Nat. Hist. Press
85. Turnquist, O. C. 1970. Production of certified seed potatoes by varieties—1969. *Am. Potato J.* 47:138–42
86. Ullstrup, A. J. 1971. Hyper-susceptibility of high-lysine corn to kernel and ear rots. *Plant Dis. Reptr.* 55:1046
87. Ullstrup, A. J. 1972. The impacts of the southern corn leaf blight epidemics of 1970–1971. *Ann. Rev. Phytopathol.* 10:37–50
88. Van der Plank, S. E. 1963. *Plant Diseases: epidemics and control.* New York, Academic. 349 pp.
89. Vavilov, N. I. 1949. *The Origin, Variation, Immunity and Breeding of Cultivated Plants*. Transl. K. Starr Chester, Waltham, Chronica Botanica. 364 pp.
90. Villareal, R. L., Lantican, R. M. 1964. The effect of "T" cytoplasm on yield and other agronomic characters in corn. *Philipp. Agr.* 48:144–47
91. Widholm, J. M. 1972. Cultured *Nicotiana tabacum* cells with an altered anthranilate synthetase which is less sensitive to feedback inhibition. *Biochim. Biophys. Acta* 261:52–58
92. Wrigley, C. W., Moss, H. J. 1968. Selection for grain quality in wheat breeding. *Proc. 3rd Int. Wheat Genet. Symp.* Canberra, Aust. Acad. Sci. 439–48
93. Yoshida, S. 1972. Physiological aspects of grain yield. *Ann. Rev. Plant Physiol.* 23:437–64
94. Zohary, D. 1969. The progenitors of wheat and barley in relation to domestication and agricultural dispersal in the Old World. Pp. 47–66 in *The Domestication and Exploitation of Plants and Animals* ed. P. J. Ucko, G. W. Dimbleby. London. Duckworth

SIGNIFICANCE OF SPORE RELEASE AND DISPERSAL MECHANISMS IN PLANT DISEASE EPIDEMIOLOGY[1]

❖ 3576

Donald S. Meredith

Department of Plant Pathology, University of Hawaii, Honolulu, Hawaii

Spore release in fungi has been the subject of classical biological research for more than a century, and Ingold (68, 69, 71–73) has provided comprehensive reviews of the diverse mechanisms involved and factors that affect them. The overall significance of spore release is quite clear to plant pathologists; it is the process that puts the spore in motion towards the ultimate infection site. This review brings together information about how studies on spore release and subsequent dispersal can be specifically related to various aspects of disease epidemiology.

The Nature of Spore Release

Spore release is the process during which a spore becomes set free from the parent tissue (conidiophore, acervulus, pycnidium, ascus, basidium, etc.). Other terms commonly used in the literature are "take off," liberation, detachment, discharge, and projection. In a great many terrestial fungi, spore release is violent or active in that the fungus provides the necessary energy. In others, release is passive so far as the fungus is concerned, and the energy of wind, water, and moving animals is responsible.

VIOLENT SPORE RELEASE   Until quite recently it was thought that spore release in most Fungi Imperfecti is passive, by means of wind, rain, or dew. In the Phycomycetes, release of spores due to hygroscopic twisting movements of conidiophores was described long ago for *Phytophthora infestans* and *Peronospora tabacina* (29, 117). In the 1960s a similar mechanism was observed for *Botrytis cinerea* (75) and species of *Cladosporium, Helminthosporium,* and *Alternaria* (93, 95–99). In all cases, as the air near the conidial apparatus dries out from a near-saturated state, conidiophores undergo twisting movements with varying degrees of violence, and the delicately poised conidia are scattered in all directions. Jarvis (75) concluded that hygroscopic

[1] Journal Series Paper No. 1564, College of Tropical Agriculture, Honolulu, Hawaii.

movements in *B. cinerea* usually do no more than dislodge mature conidia from conidiophores, but this of course facilitates their further removal by air currents. In *Helminthosporium turcicum* and *Cladosporium* sp., the success of spores being projected from the conidiophore is governed, in part, by the violence of hygroscopic movements, which in turn is related to the rapidity and magnitude of the decrease in vapor pressure to which the conidial apparatus is subjected (93, 97). Thus the efficiency of certain pathogens in releasing spores by hygroscopic means may vary from one part of the world to another, and from season to season at a given locality.

Since 1952 several other pathogenic Fungi Imperfecti, apparently more specialized, have been found to release spores actively under conditions of decreasing vapor pressure. They include *Nigrospora sphaerica, Cordana musae, Deightoniella torulosa, Corynespora cassiicola, Alternaria porri, Helminthosporium turcicum, Drechslera gigantea,* other species of *Alternaria* and *Helminthosporium, Curvularia* spp., and *Cladosporium* spp. (91, 92, 95–97, 160). In these, except *N. sphaerica*, energy for spore release appears to derive from sudden changes in the form of the spore or sporophore or both. These changes in form are associated with the sudden appearance of a gas phase inside the spore or sporophore or both, similar to the water-rupture mechanism in the well known example of fern annulus cells (68). In *N. sphaerica* a water-squirting mechanism is involved (160). In some species, particularly *Cladosporium* spp. and *Helminthosporium* spp., both hygroscopic twirling movements and water-rupture occur simultaneously. It is readily seen how a twirling motion, occurring at the time when a jolt weakens or breaks the spore attachment, would facilitate spore release.

In contrast to the above-mentioned fungi, spore release in both *Pyricularia grisea* and *P. oryzae* occurs under conditions of increasing vapor pressure (70, 145). There is strong evidence that spores of *P. oryzae* are discharged, but to a distance of only a fraction of a millimeter, by the bursting of the minute stalk cell by which the spore is attached to its conidiophore (70).

The general mechanism of ascospore discharge is the controlled apical bursting of a turgid living cell, the ascus (72, 73). The spores from an ascus are usually liberated in succession through an elastic aperture in the ascus. An important feature of the exploding ascus is its range, which varies from a few millimeters in many small-spored species, to around 40 cm for *Podospora setosa* (72).

The most controversial aspect of violent discharge of basidiospores is the mechanism involved. Violent spore release in Hymenomycetes is poorly understood and several mechanisms have been suggested (72, 74, 114).

PASSIVE SPORE RELEASE   Although the quantitative effects of wind on spore release have not been investigated fully, some preliminary studies indicate certain basic principles that apply to this passive form of release. In wind-tunnel experiments, Stepanov (142) found that the minimum wind speed required to release spores varied from fungus to fungus; for instance, less than

1.6 km per hr for *Botrytis cinerea,* and about 5.6 km per hr for *Cunningha-mella* sp. Spores of *Phytophthora infestans* and *Fusarium culmorum* were not released by winds of 12 km per hr. Stepanov also found that turbulent wind released more spores than streamlined wind. Zoberi (168), using horizontal tube cultures through which air of known humidity was passed, found that in dry-spored fungi, such as *Trichothecium roseum,* dry air currents removed more spores than damp ones, and faster currents more than slow ones. At winds of 16 km per hr and more, the numbers of spores released decreased rapidly with time, due to exhaustion of the source.

Gregory & Lacey (48) studied the liberation of spores from moldy hay while shaking it gently in a small wind-tunnel. Of relevance to the release of spores in nature was the observation that a sample of hay that had already released 50 million spores per g while subjected to a wind speed 0.8 km per hr for 30 min, released a further 55 million spores per g after a further 31 min at about 17.6 km per hr. Gregory & Lacey suggested that part of the mechanism of spore release by wind is that increasing wind speed decreases the thickness of the boundary layer of air at the leaf surface, thus exposing more deeply immersed spores to the action of eddies. Examples of this in the field are seen in spore trapping results on *Alternaria* species. Thus, both Ro-tem (128) working on *A. solani,* and Meredith (99) working on *A. porri* have recorded successive increases in atmospheric content of these fungi with successive increases in wind velocity.

Smith (136) made a quantitative study of uredospore release in *Puccinia graminis,* using a specially constructed wind-tunnel in which temperature, hu-midity, and light could be controlled. Few spores were released at wind speeds below 4.5 km per hr, but above this speed the increase in uredospore liberation with increasing wind speed was almost linear, suggesting a passive release of spores from the sorus. Smith also found that spore production in-creased with increased wind speed, light, temperature, and possibly relative humidity. He concluded that the rate of spore release was controlled by the rate of spore production and the wind speed past the sorus.

Rain splash is well recognized as an agent in the release and dispersal of spores of many fungi but, again, very few quantitative studies have been car-ried out. Fungi known to be dispersed by water are generally considered to have a restricted distribution from a source, but the possibilities and limita-tions of water dispersal mechanisms are at present only partly understood. Most spores that depend on water for their release are typically "slime-spores," for example, species of *Gloeosporium* and *Phyllostictina.* However, water may also bring about release of certain "dry-spored" fungi such as *Bo-trytis* spp. (75). Spores that are formed in a mucilage are held firmly to the plant surface when dry but are readily released when wetted and become sus-pended in a film of water on the host surface. Experiments by Gregory et al (50) showed that, in still air, a drop of water 5 mm in diameter (the size of a large raindrop) falling vertically at less than its terminal velocity onto a thin film of water containing spores in suspension on a horizontal, smooth surface could disperse more than 5000 droplets to a maximum distance radially of about 100 cm, and raise them to a height of 40 cm. Almost half of the larger reflected droplets (mean 150 μ) were found to carry spores to a mean distance of 20 cm after following definite trajectories without becoming truly airborne. The incident drop and the surface film became intimately mixed in the reflected droplets, so that a drop of water containing spores which falls onto a wet surface free from spores, will lead to a similar distribution of spores. The proportion of reflected droplets that carry spores is related to the energy available on impact. Thus a decrease in the thickness of the surface film, or in size of spore carried, or an increase in size or speed of the incident drop results in more spores being removed from the point of impact. Raindrops falling freely in the open are rarely as large as 5 mm in diameter, but drip from leaves can produce large drops from fine rain or mist consisting of drops that are too small themselves to splash efficiently. One feature of splash dispersal is that a spore can be splashed and resplashed, whereas in dispersal by dry air spore deposit is usually permanent (46).

Gregory (46) suggested that the smallest reflected splash droplets must evaporate rapidly leaving their contained solids suspended in air as droplet nuclei, and the few spores so carried must be regarded as truly air-borne. However, extensive spore-trapping studies in Jamaican banana plantations failed to reveal the presence of airborne conidia of *Mycosphaerella musicola* or *Colletotrichum musae,* both of which fungi are released solely by rain splash (93).

Brunskill (9) demonstrated that water drops bounce in a characteristic manner from the surface of certain leaves. Evans (32) investigated the bouncing characteristics of banana leaves and found that it does not occur from the surface of veins but takes place quite freely from interveinal areas. Consequently water droplets tend to concentrate on the veins. Fulton (39) concluded that this enhances the release of conidia of *Mycosphaerella musicola,* the cause of banana leaf spot, because most sporodochia of the fungus are formed on veins. However, on the basis of the experiments by Gregory et al (50) it might be argued that the failure of bouncing to occur from veins hinders the release of conidia. Clearly, further experiments are desirable.

Field studies on splash release and dispersal of spores of *Gloeosporium perennans* and *G. album* have been conducted in England (27, 28). The role of rainwater in the disease cycle on apple trees appears to be a complex one. In addition to releasing spores from their point of origin (branch cankers) and transporting them along branches to new infection sites, water plays a vital part in the removal of antifungal materials used in control measures. Although release of spores by rainwater makes them readily available for splash dispersal, the manner in which this occurs and the effects of weather, method of pruning, and other factors are imperfectly understood (28).

STUDIES ON THE COMPOSITION OF THE AIR-SPORA   Many studies on the composition of the general air-spora in different parts of the world have been made with the aid of the Hirst (53) volumetric spore trap and similar instruments. In England Gregory, Hirst and co-workers [see (45) for bibliography] studied the incidence of fungus spores in the air above agricultural areas, and in Kansas extensive aeromycological surveys were conducted by Pady, Kramer, Kelly and co-workers [see (116, 138) for bibliography]. Other studies were those of Sreeramulu (138) in India, Meredith (93) in Jamaica, and Dransfield (31) in Northern Nigeria. However, in general, the data obtained were not specifically related to phytopathological problems. There is usually a daily succession of spores in the atmosphere (45, 54, 115). The concentration of some species is maximal between about 7 AM and noon (the forenoon group). The afternoon pattern develops from noon to 4 PM, and the nocturnal group is maximal sometime between midnight and 6 AM. The various diurnal periodicities appear to be a result of interacting biological and physical factors. For example, fungi known to release spores only under conditions of rapidly decreasing vapor pressure (e.g. *Deightoniella torulosa*, 91) are usually present in highest concentrations in the early morning when temperature is increasing and relative humidity decreasing, especially in the tropics (93). Fungi whose spores are dependent on water for release (e.g. many Ascomycetes) are usually common at night when dew is maximal, or during the day shortly after rain (54). In many fungi diurnal periodicity appears to be a result of the effect of environmental conditions in triggering spore release mechanisms (54, 93). In others, especially basidiomycetes and *Cladosporium* (54, 56), the situation is more complex, and diurnal periodicity may well be affected by other conditions, such as atmospheric turbulence and the periodic return of airborne spores to the ground by sedimentation.

Some spore surveys have also demonstrated well-defined seasonal changes in the composition of the air-spora (45), but results have rarely been applied to phytopathological problems. In the following sections more applied aspects of spore release and dispersal studies are discussed.

Spore Release and Dispersal Related to Epidemiology

DISEASE SYMPTOMS RELATED TO MODE OF SPORE RELEASE An example of how the method of spore release affects the nature of disease symptoms is provided by Leach's (83, 84) work on banana leaf spot disease, or Sigatoka, caused by *Mycosphaerella musicola*. Air currents did not release conidia of *M. musicola*. However, large numbers were removed from the surface of leaf spots by dew or rain. In the field, conidium-laden water frequently falls from infected leaves of tall banana plants onto the young folded heart leaves of shorter plants below. This water tends to lodge under the folds of the heart leaf, and resultant lesions are distributed in definite lines on the leaves (line-spotting) in the same pattern as the trapped water. In contrast, ascospores of *M. musicola* are violently ejected when mature perithecia embedded in diseased leaves are wetted (83). Infection by freely airborne ascospores is almost the complete antithesis of conidial infection because ascospores are carried upwards in air currents from the lower older spotted leaves onto the younger leaves above them. Ascospore infection occurs mainly on the lower surface of young leaves, notably at the tip (tip-spotting), and just after the leaves have fully opened. The relation of infection sites to spore type and stage of leaf development is sometimes more complex (147) but, in general, passive water release of conidia of *M. musicola* leads to line-spotting whereas active release of airborne ascospores results in tip-spotting.

Passive release of spores by rain water or dew results in characteristic symptom patterns in several other diseases. In Kenya, Bock (3) found that secondary pustules of coffee leaf rust (*Hemileia vastatrix*) appeared on leaves about 3 weeks after a heavy shower. Uredospores were released by water from primary pustules and transported over the surface of the leaf. Secondary pustules developed only along the paths of flow of spore-laden water, in the form of characteristic streaks. Similar streaking caused by water-borne spores occurs in freckle disease of banana fruit caused by *Phyllostictina musarum* (101) and tear-stain melanose of citrus caused by *Diaporthe citri* (36). Presumably if spores of these pathogens were released by wind, or as a result of an active mechanism, they would become freely airborne and more diffuse symptom patterns would result as, for instance, in the case of banana speckle caused by *Deightoniella torulosa* (89, 90), and pitting disease of banana caused by *Pyricularia grisea* (94).

DISEASE OUTBREAK RELATED TO SPORE RELEASE AND DISPERSAL   Gregory (47) commented, "to understand fully the dispersal of a pathogen from a single infected plant source we should need information verging on the unknowable. The location of each infection derived from that source, no matter how far away it occurred, would be needed to complete the picture." Also, we are usually ignorant of the number of spores released (also known as the strength of the source, or Q). Gregory (47) attempted to interpret the fragmentary data and the few field experiments on plant disease dispersal gradients, using Cammack's (18) results for *Puccinia polysora* in Nigeria, and several workers' data for *Phytophthora infestans* in the United States and Europe. Basic principles were as follows (47): (*a*) A dispersal gradient implies a local source of spores. Spores from afar will not show a dispersal gradient, (*b*) A dispersal gradient requires a population of susceptible host plants. A spore dispersal gradient is not necessarily followed by an infection gradient because of lack of suscepts (e.g., plants protected with fungicides, resistant cultivars), loss of viability, and unfavorable conditions for infection, (*c*) The primary infection gradient is steeper than gradients in which secondary spread has occurred. Secondary spread of spores flattens a primary gradient, as does background contamination by spores from afar. (*d*) When disease is recorded as percentage of infected plants, multiple infections will occur and flatten the primary gradient. It is desirable to apply the multiple infection transformation (43) when disease is more than 20%.

From a computer analysis of over 400 microbial gradients described in the literature, Gregory (47) made the following further interpretations. The geometry of the source of spores plays a large part in determining the slope of the infection gradient. With line, strip, and area sources infection falls off less rapidly with increasing distance from the source than with point or near-point sources. Strip and area sources also produce more disease than point sources at all distances. The amount of disease downwind is usually greater than that upwind, but if there are many fluctuations in wind direction gradients may be similar in all directions.

Stepanov (143) and Sreeramulu & Ramalingam (139) conducted dispersal and deposition experiments with different-sized spores, and with a knowledge of the number of spores released (Q). The fraction of Q deposited per unit area at any distance was greater with the largest spores (*Lycopodium*) and least with the smallest spores (*Podaxis* sp. and *Bovista* sp.). However, the slopes of the respective dispersal gradients did not differ significantly. Steep gradients near the source are often interpreted to mean that the source disseminates only over a short distance, but Gregory (47) pointed out that that is not necessarily so and steep gradients can be associated with enhanced distant dispersal.

Van der Plank (151) suggested that a study of spore dispersal gradients may enable the researcher to pinpoint a "horizon" beyond which infection is negligible, but Gregory (47) doubted this and considered that the most important quantity to know is Q, and how it affects dispersal beyond the immediate experimental area.

Roelfs (125) studied gradients of uredospore dispersal in *Puccinia graminis* and *P. recondita* within and around a 72-m-diam source plot of infected wheat. Spores were collected on 5-mm-diam rod impaction traps 15 m above the canopy and stationed on annuli inside and outside the plot. An estimate of the numbers of spores released (Q) was obtained from spore counts at the center of the source plot. On the downwind axis, Gregory's (47) regression equation described downwind movement of spores quite accurately:

$$\log y = a + b \log x$$

or in conventional terminology:

$$\log Qx = \log Qo + bX$$

where Qx is the predicted number of spores impacted at distance X (m) from the source, b is the slope, and Qo the spore concentration at the source. About 10% of spores released reached a point 100 m downwind. Prediction of omnidirectional horizontal gradients, using the same equation, was far less accurate.

Many Ascomycetes discharge their spores only after being wetted by rain, irrigation water, or dew; examples are *Ophiobolus graminis* (49, 131), *Venturia inaequalis* (11), *Guignardia citricarpa* (87), *Podosphaera leucotricha* (10), *Mycosphaerella pinodes* (23), *M. musicola* (84), *M. fijiensis* (86), and *Eutypa armeniacae* (22). This is a striking and consistent feature of the epidemiology of all the major plant diseases caused by Ascomycetes. An exception is *Epichloe typhina* which can continue to discharge spores by obtaining the necessary water from the transpiration stream of the living grass stalk which it parasitizes (72).

As little as 0.2 mm of rain is sufficient to release ascospores of *Venturia inaequalis* from dead leaves on the ground in apple orchards (11, 60). Dew can lead to the release of some spores (107), but very few compared with the number released by rain (11). Although ascospores become freely airborne, studies by Hirst & Stedman (59, 61) and Burchill (11) indicate a subsequent short-range dispersal, and the level of scab infection developing within an orchard is determined almost entirely by the amount of inoculum produced and released within that orchard. The fall-out of ascospores appears to be rapid, and in one experiment no infection was detected beyond about 17 m from the source (11). Spraying apple trees once in autumn at 0.05%, and again just before bud burst at 0.01%, with phenylmercuric chloride, which prevents the formation of *V. inaequalis* perithecia in fallen apple leaves, reduced scab infection in one orchard by 99% in the following spring. In another orchard which was subject to contamination by ascospores from nontreated apple trees in the same orchard, infection was reduced by only 57% (12, 13). It is obviously important to ensure suppression of ascospore production and release on all trees within an orchard.

Carter & Moller (24) studied the phenology of ascospore release in *Mycosphaerella pinodes* on peas in South Australia. Sufficient inoculum to cause a serious epiphytotic often persists in pea debris from the previous crop. The release of this inoculum is normally brought about by irrigation of the crop between February and March, aided in some seasons by appreciable rainfall. The subsequent development of an epiphytotic is dependent on seasonal conditions favoring maturation of new perithecia on senescent plant tissue of the current crop, and upon appropriate conditions for release of ascospores and for infection, during May and June. Elimination of previous crop debris is an obvious means of delaying the epidemic. In a later study Carter (23) found that maximum ascospore release occurs during or shortly after rain and irrigation and daily in the late afternoon of rainless days when the substrate is sufficiently moist. He concluded that the occurrence of a daily cycle of maximum output, prior to the dew period, ensures that most ascospores deposited on foliage are exposed to the maximum period of surface moisture during the night, thus enabling germination to proceed within a short time after deposition.

Gummy stem blight (*Mycosphaerella citrullina*) is one of the most important diseases of watermelon in the southeastern United States (132). Because of the difficulty in controlling it with fungicides, total crop failures may result. Schenck (132) found a correlation between disease increase and increase in number of ascospores of *M. citrullina* released, and concluded that the importance of ascospores in the epidemiology of the disease had previously been underestimated. However, in England Fletcher & Preece (38) concluded that stem rot of cucumbers, also caused by *M. citrullina,* was chiefly spread in glasshouses by knives during pruning and harvesting, especially when the source of contamination was pycnidia and not ascospores. The numbers of ascospores released into the air of infected houses was, on occasion, very high, but this appeared to be relatively unimportant in bringing about spread of the disease. Watering an infected crop did not influence ascospore release, in contrast to Carter's (23) observations on *M. pinodes*.

In *Mycosphaerella musicola*, ascospore release may occur within 10 min after wetting infected banana leaf tissue, and about 85% of the ascospores may be released after 2 hr (84, 144). Price (118), working in Cameroun, considered that alternate wetting and drying of perithecia may result in repeated periods of ascospore discharge. Stover (144) was unable to confirm this and stated that in a given leaf spot lesion, most ascospores mature almost simultaneously and are discharged within a few hours of each other, leaving the perithecia empty. Stover postulated that widespread outbreaks of leaf spot lesions "result from a specific and almost simultaneous rain-induced ascospore discharge from spots of a previous single infection, and that repeated cycles of ascospore discharge, and lesions, occur at climate-controlled intervals throughout the year." Stover's conclusions concerning cycles of ascospore discharge are no doubt correct, but it does not necessarily follow that streaks will appear after each outburst, because conditions may be unfavorable for infection. This was clearly shown in studies on the closely related fungus, *M. fijiensis*, causing black leaf streak disease of bananas in the Pacific region and Southeast Asia (103, 104). Even at times of abundant ascospore release, disease incidence declined during winter months in Hawaii, probably because low temperatures limited the infection process. Another example was observed by Norse (108) who found that epidemic development of *Alternaria longipes* in glasshouse tobacco plants was not consistently related to the atmospheric spore content. As Hirst & Stedman (61) pointed out, spore trapping usually measures a "relative dose" of spores rather than an "effective dose." The "effective dose," which is an estimate of the proportion of ascospores able to succeed, would be a more valuable measure and should be sought.

Ascospores of *Ophiobolus graminis* are forcibly ejected into the air when perithecia are wetted by rain (131). Experiments by Gregory & Stedman (49) indicated that 0.2 mm of rain produced maximum discharge in the field. Until 1965, in spite of repeated experiments, ascospores of *O. graminis* had not been shown to infect wheat plants in nonsterile sand, and the behavior of the take-all fungus as a crop pathogen was generally accounted for without the intervention of ascospores (40). However, Brooks (8) found that ascospores readily infected the proximal part of the seminal roots of wheat seedlings that had been produced by allowing them to germinate *on the surface* of wet soil. This part of the root is not in direct contact with the soil and so is at first virtually free from a competitive root-surface microflora. Brooks concluded that ascospore infection occurred in the field when wheat seeds were germinating, on the surface of sufficiently moist soil at a time of ascospore release and dispersal. Such conditions were fulfilled by rainy weather around harvest time, when some wheat seeds were shed from over-ripe ears, giving rise to ascospore-infected "volunteer" plants in the next crop. The importance of air-borne *O. graminis* ascospores in the epidemiology of take-all disease was later confirmed by Gerlagh (41) in The Netherlands.

Up to 1950 the role of basidiospores of *Fomes annosus,* causing butt-rot or heart-rot of conifers, was unknown and spread of disease from tree to tree was thought to be caused by mycelium in the soil, or through root contacts. Extensive studies by Rishbeth (122, 123) showed that air-borne basidiospores of *F. annosus* alighting on the freshly exposed stump surface of a felled tree can germinate under suitable conditions, and the mycelium grows down through the body of the stump into the lateral roots and into the roots of adjacent healthy trees through root contacts. The removal of a single tree in a young pine plantation often resulted in a large disease gap after a few years, as a result of basidiospore infection of a single stump. Control measures were developed in which chemicals were applied to stumps immediately after felling to prevent stump infection by basidiospores. Later, Rishbeth (124) found that airborne spores of the common saprophyte *Peniophora gigantea* also infected pine stumps and could compete in the stump to the exclusion of *F. annosus.* An effective method of biological control was devised in which a spore suspension of *P. gigantea* was applied to stumps immediately after felling (40, 124).

In some fungi the relative importance of aerial and splash dispersal is not always clear. Using a Hirst (53) spore trap, Kable (77) demonstrated aerial dispersal of conidia of *Sclerotinia fructicola* causing brown rot of stone fruits in Australia. Jenkins (76) confirmed that conidia can become freely airborne but he detected them only when large inoculum sources were present. Release was brought about by air streams of low rather than high humidity, as demonstrated by Zoberi (168) for mold spores. Using water traps, Jenkins (76) detected the release of considerable numbers of conidia by rainwater. He concluded that droplets containing several conidia represent a most efficient infection unit, as under favorable temperature conditions spore germination and fruit infection would follow their dispersal and deposition. Aerial dispersal involves individual spores, and the likelihood of infection from a single spore is less than that from a clump of spores. Jenkins concluded that splash dispersal is more important than dry air dispersal in the epidemiology of brown rot disease. In contrast, Corbin et al (26), working on the related species *S. laxa* in California, concluded that most aspects of the epidemiology of brown rot of apricots can be related solely to dry air dispersal of conidia, and that conidia of this fungus can be included in Hirst's daytime, dry air-spora grouping (54). This was in agreement with earlier detailed studies of Wilson & Baker (164). The principles of control of brown rot in California are similar to those for *Venturia inaequalis* viz. the application of eradicant fungicides to trees during winter to reduce the number of spores produced and released during flowering time (113). Ogawa et al (113) stated that with detailed knowledge of the life cycle of *S. laxa,* especially the manner of aerial dissemination, "one can conclude that reduction or elimination of the primary inoculum will afford almost complete control . . . in the semi-arid environment of California." The freely airborne nature of conidia further indicated that applying blossom sprays may reduce fruit infections at harvest, and that the application of pre-harvest fungicides may reduce losses from crown rot (113).

Canova (20) suggested that conidia of *Cercospora beticola* on sugar beet were usually released by rain or dew but that dispersal was by wind. Carlson (21) considered rain the principal release and dispersal agent and wind of secondary importance. However, his trapping methods were relatively inefficient for detecting airborne conidia, and were predisposed to collecting conidia in water droplets. Laboratory and field experiments in Nebraska indicate that both wind and rain can bring about release (82, 100) but the relative importance of the two mechanisms was not determined. In Iowa an exponential increase in leaf spot of sugar beet was clearly related to an increase in numbers of airborne conidia of *C. beticola* (159). Obviously, disease development is not always dependent on water-borne conidia.

A paradoxical situation in the epidemiology of white-pine blister rust in North America was resolved by Van Arsdel (150) partly as a result of studies on spore release. Spores of the causal fungus, *Cronartium ribicola,* are released from alternate *Ribes* hosts at night during wet weather. Van Arsdel pointed out that most fungus spores released at night, including those of *C. ribicola,* are smaller than those released during the day. Thus, according to Stoke's Law, these smaller spores tend to settle more slowly in the low velocity nighttime breezes. It was suggested that conventional eddy diffusion theories do not apply to spores released at night. Rather, the spores exist as clouds in a structured atmosphere. The various layers are relatively stable and Van Arsdel suggested that spores move in the same direction as the air in these layers, and in a predictable manner. These theories were tested at various locations using colored tracers in the form of grenade smoke, and confirmation of the structured nature of the nighttime atmosphere was obtained. When smoke was released at night under a temperature inversion it flowed upwards out of the *Ribes* area and thence in a warmer air layer into the crowns of pine trees some 0.4 km away. Assuming that *C. ribicola* spores move away from *Ribes* plants in the same way as smoke, then the observed distribution of blister rust on the same pine trees, and its absence from pines close to the *Ribes* plants, is readily understood. In this case, therefore, the time of spore release affects the distribution of disease. Presumably a different disease distribution pattern would exist if spores were released during the day at a time when dispersal is governed by eddy diffusion.

In several diseases attempts have been made to determine the relative importance of ascospore versus conidium release (and infection) in the disease cycle. For *Nectria galligena* in Northern Ireland Swinburne (149) found that ascospores, violently released from perithecia in response to wetting, were most prevalent in spring and early summer, whereas splash-released conidia were most frequent later in the year. Few ascospores were released in late summer or mid-winter. Most infection courts were exposed in spring and early summer at the junctions of the extension growth of consecutive years. Thus apple canker might be more effectively controlled by fungicidal sprays applied in spring and early summer to control ascospore infections (149). This has yet to be tested.

In Jamaica in the 1940s infection of Gros Michel bananas by splash-borne conidia of *Mycosphaerella musicola* was common (84). However, by 1960 infection of Cavendish cultivars appeared to be almost entirely due to air-borne ascospores of the fungus, and conidial infections on such Gros Michel plants that remained were hard to find. Leach (85) suggested that the genetical nature of the pathogen may have changed between 1940 and 1960. In black leaf streak disease (*M. fijiensis*) in the Philippines and elsewhere, current control measures are almost exclusively designed to prevent ascospore release and infection, and the role of splash-dispersed conidia appears to be relatively minor (Meredith, unpublished). In both *M. musicola* and *M. fijiensis* windborne ascopores can rapidly move disease outward from primary foci of infection, whereas conidial infection is localized mostly on lower canopy leaves below or near spotted leaves (102, 146).

Conidia of *Colletotrichum coffeanum,* which causes coffee berry disease (CBD), are released and dispersed, in the absence of vectors, solely by water-splash over comparatively short distances (109). Medium-range distribution, for example from one plantation to a nearby one, or over appreciable distances within a large plantation can be effected by human agency as a result of the many agricultural operations which involve the handling of plants. During picking, highly infective conidia from infected berries are frequently transferred to healthy ones (109).

SPORE RELEASE AND DISPERSAL STUDIES RELATED TO DISEASE PREDICTION AND CONTROL   Losses from and the cost of controlling certain plant diseases have been greatly reduced by forecasts of probable seasonal incidence in Europe, the United States, Japan, and other temperate countries (105). These systems either give sufficient time for carrying out control measures or indicate that such controls are not necessary. Most existing prediction systems are based purely on a knowledge of the effects of meteorological conditions on disease incidence and severity (4, 153, 154). However, assuming that spores reach susceptible host tissues and conditions are favorable for infection, it can be expected that the subsequent epidemic is proportional to Q (45). Hence, another opportunity for forecasting is presented. A knowledge of Q and conditions under which it is maximal involve studies on the mode and site of formation of inoculum, and on conditions bringing about its release. Van der Plank (151, 152) has stressed the importance of reducing Q as much as possible to delay the onset of an epidemic. Q is also important in relation to dispersal of released spores, and the spread of disease. For example, in attempting to derive formulae to characterize the dilution of a spore cloud as it travels horizontally and is diffused, Gregory (45, 47) suggested using parameters that correspond to environmental factors, and that take Q into account, if known. Q also features in diffusion formulae given by other workers (135, 143). Although Q is usually unknown, except in experimental tests, the various formulae indicate that as it increases so the concentration of the spore cloud at a given distance increases and also the number of spores deposited on the host at that point increases.

Schrödter (135) pointed out that gigantic numbers of spores need to be released for effective dispersal, because of the very steep slope of the spore dispersal gradient near the source. Many fungi no doubt satisfy this requirement. For instance, *Ganoderma applanatum* can release 3×10^{10} spores per day for 6 months (69).

The quantity of primary inoculum available for release has been used as a criterion for forecasting; it is a particularly useful one because of its early appearance. *Venturia inaequalis* overwinters in the perfect stage in fallen leaves and its maturity is dependent on winter and early spring weather (163). Microscopic examination of perithecia ripening in the laboratory, and considerations of winter weather have formed the bases for prediction of the first discharges of ascospores in spring (153). According to Hirst (55) interpretation of spore release data and leaf wetness and temperature data in England during one period enabled both the date and the approximate severity of scab infection to be estimated. For example, in 1955 ascospores of *V. inaequalis* were liberated on 26 days, but the liberation was accompanied by weather suitable for infection on only 2 days. Of these 27 and 28 April appeared outstanding, because many spores coincided with ample opportunity for infection. Growers were advised to apply curative sprays at once. The warning was effective for growers who complied, but those who delayed spraying until after 27 April observed infections that increased in incidence as the delay increased. Hirst (55) cautioned, however, that more information about the interaction of the fungus and environmental conditions is necessary for a successful prediction system that will apply to all weather conditions.

In Australia (67) and England (12, 13) spraying apple leaves in autumn with 5% urea or 0.005% phenylmercuric chloride (PMC) suppressed the production of perithecia of *Venturia inaequalis* the following spring. Of special interest was the subsequent discovery that a second application of 2% urea or 0.005–0.02% PMC in spring prevented release of ascospores from mature perithecia (12). Brook (6, 7), distinguished two phases in ascospore release of *V. inaequalis*. High temperature or high humidity during the conditioning phase favored subsequent release, and the release phase itself was inhibited by absence of red light. The effect of chemicals on these two phases should be investigated because if the mode of action can be elucidated it may have far-reaching implications for disease control in general.

Of special interest is the system worked out by Kerr & Rodrigo (78) for blister blight of tea (*Exobasidium vexans*) in Ceylon. The number of spores of *E. vexans* deposited on susceptible leaves of tea bushes was directly correlated with the number of spores in the atmosphere, as determined by a volumetric suction trap. The following multiple regression equation was developed to predict accurately disease incidence approximately 3 weeks later.

$$Y = 33 + 0.3145\ x_1 - 0.03725\ x_1 x_2$$

where Y = number of blisters per 100 shoots, x_1 = number of spores per unit volume of air during the infection period, and x_2 = mean daily sunshine (hr) during the infection period. The infection period was 15–25 days before disease incidence was measured. The system was tested between May and December 1966 in tea growing under heavy shade. It was possible to determine both when an epidemic of blister blight was likely to occur, and what the level of infection would be. Kerr & Rodrigo (78) claimed that the accuracy of prediction is much higher than for any other air-borne fungus disease of plants. Because the incidence of blister blight is much higher in shaded tea than in unshaded, it is necessary to modify the prediction equations for different levels of shade. In a further study (79), the measurements and calculations required for accurate prediction of blister blight incidence were considerably simplified. Thus, a forecast of disease incidence 2–3 weeks later can be made by measuring (a) the current level of infection in tea leaf brought to the factory, and (b) the mean daily sunshine for the previous 7 days. A simple calculating device, similar to a circular slide rule, was designed for the use of planters. The method can be applied to mature tea and can save many cycles of spraying each year. Although this system does not require spore-trapping data, the latter were essential in the derivation of the final simplified equation.

The use of daily hygrothermograph and spore trap records were reported by Berger (1) to permit accurate predictions of *Cercospora apii* attacks on Florida celery in the 1968 winter growing season. Spore production and release increased progressively with each successive night of 8 hr or more with relative humidity (RH) near 100% and temperatures in the range of 15–30°C (maximum of 700 spores were trapped per day). When nighttime temperatures were below 15°C, 0–20 spores were trapped per day regardless of RH or previous sporulation record. Two or more successive nights of temperatures above 15°C and humidities near 100% RH were necessary for renewed, appreciable spore production and release following exposure to temperatures below 12°C. During times of scanty spore release commercial growers were able to omit a total of 5–15 fungicide applications without disease buildup.

Conidia of *Helminthosporium turcicum* become freely airborne after release (97, 98), and Berger (2) found that spore counts were valuable in determining the threat of sweet corn blight during all seasons in Florida. A forecasting system was developed and it was possible to reduce the number of fungicide applications without disease buildup.

In South Africa, a knowledge of environmental conditions favoring abundant release of ascospores of *Guignardia citricarpa,* the cause of black spot of citrus, resulted in improved timing and a reduction in number of fungicide applications (87). Improved control of citrus greasy spot (*Mycosphaerella citri*) and black spot of plums (*Dibotryon morbosum*) may also result from studies on the phenology of ascospore release (137, 162).

Ono (115) commented that probably the greatest advance in forecasting rice blast disease outbreaks was the spore-trapping procedure first worked out between 1934 and 1949 in the Nagano Prefecture, Japan (81). Close correlations were obtained between numbers of spores trapped and the severity of disease, and reliable forecasting formulae were developed. Other factors taken into account were meteorological data (temperature, relative humidity, wind, moisture, and precipitation) and host resistance. Suzuki (148) noted that the source of primary inoculum of *P. oryzae* is infected rice straw from the previous year. *P. oryzae* spores were released from this source either passively by wind action or actively when conidiophores were wetted by dew or rain. Using a modified Rotorod Sampler (148) Suzuki studied spore release, dispersal, and deposition in *P. oryzae* in Japan and concluded that by counting the number of trapped spores 10 m above ground it is possible to forecast disease outbreaks within a circle of 1 km radius. Corrections had to be applied to counts obtained during periods of rain and of low wind velocity. Kiyosawa (80) recently analyzed the 1938–1949 spore-trapping data of Kuribayashi & Ichikawa (81) and attempted to develop improved equations for forecasting blast development. By plotting the logarithm of numbers of cumulative spores trapped against time during early stages of infection, drawing a regression line through these points, and then extrapolating the line in a specified manner, it was claimed that a forecast could be made of the level of disease at harvest. However, the method has yet to be tested on new disease outbreaks. The relation between rice blast severity and number of trapped spores is not always a simple one, and more information is needed about the effects of environmental conditions and host resistance on the infection rate [r, of Van der Plank (152)].

The course of rust epidemics in North America has been traced for many years by means of greased microscope slides exposed as simple impactor traps (141). Comparisons were made of dates when spores were first caught, date and amount of large numbers, and the general relation of spore numbers to wind direction. Early detection of spores when winds were southerly from areas of heavy rust infection was considered to imply the deposition of wind-borne spore showers of distant origin [see (127) for bibliography]. However, the possibility of local rust infection being a source of spores was not conclusively eliminated. Romig & Dirks (127) adapted some of Van der Plank's (152) methods of epidemic analysis to cumulative daily spore counts of *P. graminis* and *P. recondita* at 12 sites located in the wheat-producing area between the Mississippi River and the Rocky Mountains. The basic supposition was that if the $\log e \, (x/1-x)$ transformation gave a significantly better fit than did log e $(1/1\text{-}x)$, then spores trapped on slides were generated locally for the most part rather than blown in as a concentrated dose from a distant source. In their study, the local inoculum theory seemed to explain best the deposition obtained. Thus, it was recommended that the timing of fungicidal sprays for control should be based principally upon sampling data arising from the fields to be treated, rather than from other areas of the country, and in fact this has resulted in good control (130). It should be pointed out, however, that this study in no way disproved the importance of long-distance spore transport in cereal rust epidemiology.

The relative inefficiency of glass slides in trapping spores was noted by Gregory (44), and vertical rods were found to be much more efficient. Roelfs et al (126) tested 5-mm-diam cylindrical glass rods having a sticky surface, and their use resulted in earlier detection of rust uredospores than glass slides. Consequently, rods replaced slides in the continuing epidemiological studies of cereal rusts in the United States (15, 16, 30, 125–127). Eversmeyer & Burleigh (34) used a stepwise multiple regression computer program to formulate equations to predict severity of wheat rust caused by *P. recondita*. Measurements were made of disease severity, weekly uredospore numbers and cumulative numbers (obtained with 5-mm-diam rods), average maximum and minimum temperatures, duration (hr) of free moisture as dew or rain per day and days of precipitation, recorded during the 7 days immediately preceding the date of prediction. Initially, it appeared that cumulative uredospore numbers could be substituted for disease severity in the prediction model. Later, more accurate equations were obtained when leaf rust severity was used as an inoculum variable (15, 16). Burleigh et al (15) concluded that the principal difficulty with spore numbers is that impaction traps are exposed to spores from endogenous and exogenous sources, and the vicissitudes of weather might cause deposition of more or fewer spores than expected from a given level of infection near the trap.

In Kenya, coffee leaf rust, caused by *Hemileia vastatrix* is controlled by fungicidal sprays (110). Experiments by Nutman et al (112) indicated that uredospores of *H. vastatrix* were not released by wind currents of up to 19 km per hr. Also, very few spores were detected when highly efficient traps were operated among severely infected bushes, even when bushes were mechanically disturbed (however, see p. 333). On the other hand, large numbers of spores were easily detached by water running over the pustules. Dispersal of spores in water was over relatively short distances; that is, from one part of a given leaf to another part, or from one leaf to another on the same bush. Consequently, disease outbreaks were usually highly localized. On the basis of these findings, Nutman & Roberts (110) formulated the following basic principles of control of coffee leaf rust: (*a*) Inoculum that gives rise to a rust outbreak is derived from pustules at very close range, usually within the same tree; (*b*) Inoculum is not spread during dry weather, but only by rainfall in excess of 7.5 mm; (*c*) Residual inoculum is at a minimum at the end of a dry season. It is at this time, as precisely as possible before the onset of the heavy rains of the season, that sprays have their maximum effect.

Corke (28) studied the role of rainwater in release and dispersal of the bitter-rot organisms on apple (*Gloeosporium album* and *G. perennans*). Conidia of both fungi are produced on branch cankers and are released and transported by water to new infection sites. Water traps were developed to determine when release occurred, how much rain was required to bring about release, and to trace the directions of flow of spore-laden water along branches and down the trunk (14, 26). At the same time, control measures for bitter rot were sought using spore suppressants applied to trees throughout the year. Significant reductions in storage losses due to bitter rot have been achieved by various suppressants. Of particular interest was the observation that water traps gave a good indication of when the effect of an earlier application of suppressant was declining and, therefore, when a further application was desirable. This is a good example of how continuous observations on spore release have been used as a basis for timing the application of chemicals.

In the 1930s, the banana industry in the Caribbean and Central and South America was threatened with extinction by leaf spot disease (Sigatoka) caused by *Mycosphaerella musicola* (102). At that time no satisfactory control method was known. Spraying with Bordeaux mixture held some promise, consequently, massive Bordeaux spray installations were set up over more than 40,000 hectares. Spraying was carried out throughout the year at intervals of 2–5 weeks and at rates of up to 2,400 liters/hectare/cycle. Initially, it was thought that Bordeaux mixture was acting as a protectant against conidial infections of the youngest leaves, and most of the spray was directed at them. Leach (84) confirmed that most infections occurred on the unfurled heart leaf, or shortly after it had unfurled. However, he observed that very little spray reached the heart leaf and, if it did, only a small area was covered because of the continual exposure of tissues as unfurling progressed. To obtain complete coverage the heart leaf would have to be sprayed daily. Evidently Bordeaux mixture was not acting entirely as a protectant, although it was probably assisting control to some extent in this way. Earlier workers assumed that conidia of *M. musicola* are released and dispersed chiefly by wind (140), but Leach (84) showed that this is not the case and both processes are brought about by rain water or dew running over spots on older leaves. On sprayed plants, this water becomes toxic from contact with Bordeaux mixture, and conidia released and dispersed by such water fail to germinate. Thus the fungicide inactivated conidia at the time of release, rather than at the time of deposition or later. Leach (84) therefore recommended that special attention be given to spraying the older, heavily spotted leaves to inactivate mature and developing conidia and, in practice, the revised method improved the level of leaf spot control in the West Indies.

Soon after Leach (83) discovered the perfect state of *M. musicola* he found that ascospore infection was as important as conidial infection (sometimes more important) in bringing about the spread of Sigatoka. Ascospores

are violently released from perithecia immersed in leaf spots and, unlike conidia, do not become contaminated with Bordeaux mixture at the time of release. Further, the unfurled heart leaf is a vulnerable target for ascospores.
Consequently, Leach (84) recommended continued spraying to attain a high
level of control of conidial infection before the onset of the rainy seasons,
before ascospore production and release became maximal. By reducing the
number of spots before rain, the eventual ascospore output was reduced. In
this way, even further improvements were made in Sigatoka control.

Eutypa armeniacae, an Ascomycete abundant on dead wood of Prunus
spp., Pyrus malus, Vitis vinifera, and Tamarix sp., is the perfect state of Cytosporina sp., long known as a pathogen of apricot trees in southern Australia
(22). Ascospore release is maximal in spring, summer, and autumn, and
therefore, the chances of pruning wounds on apricot trees becoming infected
by E. armeniacae ascospores are greatest at that time (106). It was recommended that growers should prune in winter immediately after leaf fall is
complete.

Fungicidal control of coffee berry disease (CBD) in Kenya, caused by
Colletotrichum coffeanum, was the subject of a detailed study between 1956
and 1960 (109). Indications were obtained that good control of the disease
could be achieved by a short program of antisporulant sprays applied early in
the season (February to April). The efficacy of the program was believed to
be due to a reduction in water-released inoculum derived from the bark, and
not to protection of the developing crop. However, since 1961 the program
has given erratic and generally unsatisfactory results. Reasons were not
known, but further investigation (47, 51, 52) showed that although inoculum
released from bark might initiate an epidemic, subsequent disease progress
was more dependent upon spores released from diseased berries, which accounted for most of the inoculum released and dispersed during the greater
part of the season. Therefore experiments were conducted to test the effects
of protectant sprays applied throughout the long rains (February to July).
These gave the best control and increased yields threefold (51, 52). Thus in
the case of CBD, information about the site of major spore production and
release (berries and not bark) has resulted in improved control measures.
Furthermore, Gibbs (42) showed that the proportion of the pathogenic
strain of C. coffeanum in the bark is minute compared to that of diseased
berries, which produce and release virtually all pathogenic conidia.

Ogawa et al (113) reported that almost complete control of hop downy
mildew (Pseudoperonspora humuli) was obtained in California by removal
of hop plants as soon as diseased shoots appeared, thereby markedly reducing
the amount of initial inoculum released.

SPORE RELEASE AND DISPERSAL CONSIDERED IN MODEL SYSTEMS   Waggoner &
Horsfall (155) constructed a simulator, EPIDEM, for predicting the course
of epidemics of Alternaria solani causing early blight of tomato and potato.
Later, EPIMAY was constructed as a simulator of Southern corn leaf blight,

caused by *Helminthosporium maydis* (156). In both simulators, detailed consideration was given to the components of the system that would influence disease spread, and the many different ways in which weather and host factors affect different stages in the life cycle of the two fungi were examined. Spore release by both rain-washing and wind action was taken into account, but it was clearly evident that more complete quantification of release was required for increased accuracy of disease prediction. Simulators may play an important part in future studies on spore release and dispersal, by indicating what is the most important information required.

Zadoks (166, 167) presented curves for simulated epidemics using parameters found in potato late blight (*Phytophthora infestans*) and the wheat rusts (*Puccinia graminis, P. recondita,* and *P. striiformis*). A marked effect in accelerating epidemic development was evident as a result of the influx of "effective" inoculum from an outside source. This again indicates the importance of measuring an "effective dose" of spores (p. 321) for accurate disease prediction.

SPORE RELEASE AND DISPERSAL RELATED TO IRRIGATION PRACTICES   Rotem & Palti (129) pointed out that in many countries, diseases that were previously absent from crops grown without irrigation have become common since irrigation, especially in sprinkler form, has been applied. This can be attributed to several factors such as improved microclimatic conditions for infection, increased sporulation, and the triggering of spore release and dispersal mechanisms. Mention is made elsewhere (p. 320) of the effect of irrigation in causing ascospore release in *Mycosphaerella pinodes* in pea debris, and thereby initiating serious epiphytoties (24). In Greece, sprinkling was found to induce zoospore discharge from oonpores of *Plasmopara viticola*, and the number of cases in which infections were recorded corresponded to the number of vineyards irrigated (165).

The effect of sprinkling on spore release and dispersal is particularly pronounced in the case of bitter rot of apples (*Gloeosporium fructigenum*) in Israel (129). The sticky spores are not released by winds of up to 20 km per hr, but are readily released from acervuli by water from overhead sprinklers. Irrigated apple orchards are affected by the disease, therefore, whereas those irrigated by under-tree sprinklers or surface irrigation are not.

Spore release and dispersal by irrigation water is also considered to aid the development of *Corynebacterium michiganense* in tomatoes (25), *Fusarium moniliforme* in bananas (157), *Pseudomonas phaseolicola* in beans (158), *Colletotrichum phomoides* in tomatoes (121), *Phytophthora infestans* in potatoes (58), and *Phytophthora* spp. in citrus (17). The role of sprinkling in simultaneously releasing and dispersing spores, and providing favorable conditions for infection, during a short 3½ hr period of water application on a hot dry day was proved experimentally in Israel for *P. infestans* in potatoes (129).

The release of spores from upper leaves and their dispersal to lower ones by sprinkler water is considered by Rotem & Palti (129) to be partly responsible for the greater infection of lower leaves in some crops. They suggest that removal of spores from the shoot into the soil for infection of underground parts is of lesser importance. However, this type of release and dispersal by sprinkling favors potato tuber infection by *P. infestans,* and may aid infection by *Alternaria solani* (37).

In Honduras overhead irrigation of banana plantations sometimes results in up to 5% increase in Sigatoka disease caused by *Mycosphaerella musicola* (102). Irrigation water results in the release of both conidia and ascospores, which would not normally be released in the absence of rain or appreciable dew. In Jamaica the same disease was reported to be more severe in plantations irrigated by under-tree sprinklers than in surface-irrigated plantations (102).

These examples indicate that the method of irrigation used for a particular crop may have to be changed if major disease problems develop as a result of irrigation. For example, if overhead sprinkling results in abundant spore release from the wetted aerial parts of a crop, and subsequent disease build-up, it may be desirable to change to surface irrigation. Alternatively, it may be possible to improve the timing of sprinkler irrigation cycles so that although spores are released they fail to cause infections because microclimatic conditions at the host surface do not remain favorable for a sufficient time interval after spore deposition.

SPLASH DISPERSAL OF FUNGUS SPORES AND FUNGICIDES   Faulwetter (35) suggested that secondary splash droplets, produced as a result of the impact of a rain drop on a surface bearing a film of moisture, consisted almost solely of water from the surface film. However, Gregory et al (50) showed that the incident drop and the surface film become intimately mixed at impact. Consequently, if the leaf carried a relatively loosely held fungicide deposit then spores would be inactivated during splash dispersal, as in the case of *Mycosphaerella musicola* on bananas (84). Hislop (64) observed the same phenomenon, both in the field and in laboratory experiments, using *Botrytis fabae* and *Venturia inaequalis* as test fungi. The reduction in infectivity of spores in secondary droplets was related to the level and tenacity of the fungicide deposit on the target leaf. When spores were splashed directly onto heavy but tenacious deposits, their viability in secondary droplets was unaffected. It seems likely, therefore, that increasing the tenacity of fungicide formulations beyond certain levels will not necessarily be biologically significant because of their reduced redistribution, as suggested earlier by Evans et al (33).

LONG-DISTANCE SPORE DISPERSAL   Coffee leaf rust (*Hemileia vastatrix*) was first recorded in the New World (Brazil) in 1970 (161), and has subsequently continued to spread from the initial focus (133, 134). The spread of *H. vastatrix* to Brazil has once again raised the somewhat controversial subject of long-distance spore dispersal. The best known and best supported example of distant air dispersal of spores is the annual northward migration of *Puccinia graminis* uredospores from Mexico and the southern United States, to cause wheat stem rust in the wheat belt (141). In this case proof was obtained by trapping migrant spores. In other cases, strong circumstantial evidence has been presented for distant, overland transport of spores in detailed maps illustrating the spread of pathogens from the point of their first introduction. Examples are the spread of tobacco blue mold (*Peronospora tabacina*) in Europe (120), and maize rust (*Puccinia polysora*) in Africa (18, 19).

It was assumed until very recently, probably by analogy with the cereal rusts, that uredospores of *H. vastatrix* are released and dispersed by wind. Rayner (119) suggested that the fungus was blown to Ceylon from Africa by the southwest monsoon, and cautioned that spread and intensification of rust in West Africa in the 1950s was a threat to the New World because uredospores might be carried there by the northeast trade winds. However, in Kenya, Nutman et al (110, 112) concluded that release and short-range dispersal of *H. vastatrix* uredospores by rain (see p. 328) is inconsistent with the concept of intercontinental spread by wind. Short-range dispersal and strict quarantine measures were thought to explain the long freedom of the New World from coffee rust. Possibly the fungus was introduced by plant introductions or by insect or human agencies.

Bowden et al (5) suggested that trade winds could have taken spores of *H. vastatrix* across the Atlantic from Angola to Bahia, Brazil, in 5–7 days, but Nutman & Roberts (111) argued that very few viable spores could be expected to survive the journey. Recent spore-trapping studies in Brazil, using a small airplane, have revealed airborne uredospores of *H. vastatrix* at heights of up to 1000 m at a distance of 150 km from a diseased coffee area (134). The numbers of spores trapped increased as distance from the source decreased. In another study uredospores were found at high altitudes over the state of Parana, which was at that time still free from rust (134). There seems no doubt, therefore, that *H. vastatrix* spores can become truly airborne, at least in Brazil, and it is significant that the fungus has spread from north to south in the same direction as prevailing wind currents (134). The reason Nutman & Roberts (110) failed to detect airborne uredospores of *H. vastatrix* in Kenya, and to bring about their release by wind action, is not known. The situation is further complicated by a statement of Wellman (personal communication, 1972) that he has brought about release of large numbers of uredospores by gently blowing across the surface of leaves bearing pustules. A detailed re-examination of spore release and dispersal of *H. vastatrix* in Kenya is desirable, preferably using wind tunnels and controlled environmental chambers.

Another fungus of interest in relation to long-distance spore dispersal is *Mycosphaerella musicola*. Stover (145) postulated that the great Caribbean epidemic in the 1930s was due to the relatively sudden arrival of airborne ascospores in the latter part of 1933, from eastern Australia. He suggested that large numbers of ascospores were released in that region and carried upward to heights of 2100 m or more. Thence they were supposed to have moved westward in the general direction of the southeast trade winds, at an average speed of about 27 km per hr, reaching the Caribbean region in about 37 days. As Hirst & Hurst (57) commented, Stover's hypothesis is reasonable but because the distances are so great and because the proposal is supported only bibliographically, it is not wholly convincing. An alternative explanation is that *M. musicola* was introduced to the New World on infected banana leaves.

Holliday (66) has given a useful account of important fungal pathogens of tropical crops that are not co-extensive with their respective hosts. At present, *Microcyclus ulei,* the cause of South American leaf blight of rubber, is restricted to the mainland of tropical America, and Trinidad. *Hevea* spp. are the only known hosts. The freely airborne nature of conidia led Holliday (65) to conclude that if once introduced to the large rubber plantations in Asia, further spread would be rapid.

The Phycomycete, *Trachysphaera fructigena,* is of special interest because of its very restricted distribution in a few tropical west and central African countries (88). It causes mealy pod disease of cacao and coffee, and a serious fruit rot (cigar-end) of bananas. It is highly infective, grows rapidly through tissues, and sporulates abundantly on all hosts. It is difficult to explain why it has not spread widely throughout Africa since the first report from Ghana in 1923. Studies on the mode of spore release and dispersal may help to elucidate its confined distribution.

Airplane flights over the North Sea confirmed that spores are often transported far and in great numbers (62, 63). Migrations occurred in many directions, with, against, and across the direction of the prevailing westerly winds and extending more than 1500 km. Concluding a review on long-distance spore transport, Hirst & Hurst (57) stated that "this review . . ., like its predecessors, shows how incomplete are our knowledge and abilities. Our failure to measure viability is one example but scarcely more important than the lack of numbers relating airborne spore concentration to deposition on exposed surfaces."

Conclusions

A knowledge of the mode of spore release and dispersal, and conditions affecting both processes, is just as essential to a full understanding of the epidemiology of a fungus-incited plant disease as information about spore germination, penetration of the host, etc. In this review examples have been given in which studies on release and dispersal have contributed either directly or indirectly to improved disease-control measures. It is vital to know whether spores are released by wind or water and whether they become freely airborne or are dispersed by water. In pleomorphic fungi the two (or more) spore stages may be released and dispersed differently and may differ appreciably in their respective roles in disease outbreaks, as in *Mycosphaerella musicola* for example (84). Control measures effective against one spore stage may be ineffective against another.

Production and release of inoculum occur periodically in many fungi in response to specific environmental influences. Release of this inoculum can be forecast from information on temperature and rainfall, as in *Venturia inaequalis* (153), or other factors such as irrigation dates in the case of *Mycosphaerella pinodes* (24). However, to improve the accuracy of prediction using spore data, it will be necessary to investigate further the size of the spore load needed to start an epidemic under different environmental conditions (105). Also, it may be necessary to develop spore samplers with much lower detection thresholds than those now available. More information is needed on the relation of outbreaks of certain diseases to inoculum carried long distances by air. Much can be learned from aerobiological studies of the rate, direction and distance of transport of different types of inoculum, and the probability of disease establishment and spread after fall-out of inoculum or vectors (105).

In future there will probably be other examples, similar to the situation in tea blister blight (78, 79), whereby studies on spore release and dispersal will help to elucidate disease epidemiology, but eventually, spore sampling will not be necessary for forecasting and control recommendations. Spore sampling and counting can be tedious, and alternate, simpler parameters should be used if possible.

The prevention of spore release, quite apart from the effect of antisporulants, is one possible means of controlling disease that has been little exploited. The effects of urea, phenylmercuric chloride, and other chemicals in suppressing release of *Venturia inaequalis* ascospores (12, 13, 67) should be investigated for other ascomycetes.

Certain basic studies on spore release should continue. Mycologists might look more closely at the ultrastructure of spores, conidiophores, and so on, to elucidate mechanisms further. The process might be investigated in more detail at the biochemical level. Excellent facilities are now available to study release and dispersal under strictly controlled environmental conditions, and this can provide important information for an input into epidemiological models (155, 156). In this respect, it should be borne in mind that spore release is but one facet in the life cycle of a fungus and cannot be dissociated from other vital processes; and both release and dispersal are only two components of a complex interacting system so far as disease outbreak is concerned.

Literature cited

1. Berger, R. D. 1969. Forecasting *Cercospora* blight of celery in Florida. *Phytopathology* 59:1018 (Abstr.)
2. Berger, R. D. 1970. Forecasting *Helminthosporium turcicum* attacks in Florida sweetcorn. *Phytopathology* 60:1284 (Abstr.)
3. Bock, K. R. 1962. Dispersal of uredospores of *Hemileia vastatrix* under field conditions. *Trans. Brit. Mycol. Soc.* 45:63–74
4. Bourke, P. M. A. 1970. Use of weather information in the prediction of plant disease epiphytotics. *Ann. Rev. Phytopathol.* 8:345–70
5. Bowden, J., Gregory, P. H., Johnson, C. G. 1971. Possible wind transport of coffee leaf rust across the Atlantic Ocean. *Nature, London* 229:500–01
6. Brook, P. J. 1969. Stimulation of ascospore release in *Venturia inaequalis* by far red light. *Nature, London* 222:390–92
7. Brook, P. J. 1969. Effect of light, temperature, and moisture on release of ascospores by *Venturia inaequalis* (Cke.) Wint. *N.Z.J. Agr. Res.* 12:214–27
8. Brooks, D. H. 1965. Root infection by ascospores of *Ophiobolus graminis* as a factor in epidemiology of the take-all disease. *Trans. Brit. Mycol. Soc.* 48:237–48
9. Brunskill, R. T. 1956. Physical factors affecting the retention of spray droplets on leaf surfaces. *Proc. 3rd Weed Control Conf., Assoc. Brit. Insecticide Manuf.* 593–603
10. Burchill, R. T. 1964. Seasonal fluctuations in spore concentrations of *Podosphaera leucotricha* (Ell. and Ev.) Salm. in relation to the incidence of leaf infections. *Ann. Appl. Biol.* 55:409–15
11. Burchill, R. T. 1966. Air-dispersal of fungal spores with particular reference to apple scab (*Venturia inaequalis* (Cooke) Winter). In *The Fungus Spore*, ed. M. F. Madelin, 135–140. London: Butterworths, 338 pp.
12. Burchill, R. T. 1968. Field and laboratory studies of the effect of urea on ascospore production of *Venturia inaequalis* (Cke.) Wint. *Ann. Appl. Biol.* 62:297–307
13. Burchill, R. T. 1972. Comparison of fungicides for suppressing ascospore production by *Venturia inaequalis* (Cke.) Wint. *Plant Pathol.* 21:19–22
14. Burchill, R. T., Edney, K. L. 1963. The control of *Gloeosporium album* rot of stored apples by orchard sprays which reduce sporulation of wood infections. *Ann. Appl. Biol.* 51:379–87
15. Burleigh, J. R., Eversmeyer, M. G., Roelfs, A. P. 1972. Development of linear equations for predicting wheat leaf rust. *Phytopathology* 62:947–53
16. Burleigh, J. R., Romig, R. W., Roelfs, A. P. 1969. Characterization of wheat rust epidemics by numbers of uredia and numbers of urediospores. *Phytopathology* 59:1229–37
17. Butler, E. J., Jones, S. G. 1949. *Plant Pathology.* London: Macmillan. 979 pp.
18. Cammack, R. H. 1958. Factors affecting infection gradients from a point source of *Puccinia polysora* in a plot of *Zea mays.* *Ann. Appl. Biol.* 46:186–97
19. Cammack, R. H. 1959. Studies on *Puccinia polysora* Underw. II. A consideration of the method of introduction of *P. polysora* into Africa. *Trans. Brit. Mycol. Soc.* 42:27–32
20. Canova, A. 1959. Richerche su la biologia e l'epidemiologia della *Cercospora beticola* Sacc. Parte III. *Ann. Della Sperimentazione Agr.* N.S. 13:477–97
21. Carlson, L. W. 1967. Relation of weather factors to dispersal of conidia of *Cercospora beticola* Sacc. *J. Am. Soc. Sugar Beet Technol.* 14:310–23
22. Carter, M. V. 1957. *Eutypa armeniacae* Hansf. & Carter, an airborne vascular pathogen of *Prunus armeniacae* L. in southern Australia. *Aust. J. Bot.* 5:21–35
23. Carter, M. V. 1963. *Mycosphaerella pinodes.* II. The phenology of ascospore release. *Aust. J. Biol. Sci.* 16:800–17
24. Carter, M. V., Moller, W. J. 1961. Factors affecting the survival and dissemination of *Mycosphaerella*

pinodes (Berk. & Blox.) Vestergr. in south Australian irrigated pea fields. *Aust. J. Agr. Res.* 12:878–88

25. Cass Smith, W. P., Goos, O. M. 1946. Bacterial canker of tomatoes. *J. Dep. Agr. W. Aust.* 23:147–56

26. Corbin, J. B., Ogawa, J. M., Schultz, H. B. 1968. Fluctuations in numbers of *Monilinia laxa* conidia in an apricot orchard during the 1966 season. *Phytopathology* 58:1387–94

27. Corke, A. T. K. 1958. A trap for water-borne spores. *Plant Pathol.* 7:56

28. Corke, A. T. K. 1966. The role of rainwater in the movement of *Gloeosporium* spores on apple trees. In *The Fungus Spore*, ed. M. F. Madelin, 143–49. London: Butterworths. 338 pp.

29. Debary, A. 1887. *Comparative Morphology and Biology of the Fungi, Mycetozoa and Bacteria.* Oxford Univ. Press, 525 pp.

30. Dirks, V. A., Romig, R. W. 1970. Linear models applied to variation in numbers of cereal rust urediospores. *Phytopathology* 60:246–51

31. Dransfield, M. 1966. The fungal air-spora at Samaru, Northern Nigeria. *Trans. Brit. Mycol. Soc.* 49:121–32

32. Evans, E. 1961. Relationship between the physical characteristics of a banana leaf and the distribution of leaf spot lesions (*Mycosphaerella musicola* Leach). *Trans. Brit. Mycol. Soc.* 44:299

33. Evans, E., Cox, J. R., Taylor, J. H. H., Runham, R. L. 1966. Some observations on size and biological activity of spray deposits produced by various formulations of copper oxychloride. *Ann. Appl. Biol.* 58:131–44

34. Eversmeyer, M. G., Burleigh, J. R. 1970. A method of predicting epidemic development of wheat leaf rust. *Phytopathology* 60:805–11

35. Faulwetter, R. C. 1917. Wind blown rain, a factor in disease dissemination. *J. Agr. Res.* 10:639–48

36. Fawcett, H. S. 1936. *Citrus Diseases and their Control.* New York: McGraw-Hill, 656 pp.

37. Fedderson, H. D. 1962. Target spot of potatoes. *Leaflet Aust. Dep. Agr.* 3678, 10 pp.

38. Fletcher, J. T., Preece, T. F. 1966. *Mycosphaerella* stem rot of cucumbers in the Lea Valley. *Ann. Appl. Biol.* 58:423–30

39. Fulton, R. H. 1962. Some factors governing conidial retention of *Cercospora musae* on banana leaves. *Phytopathology* 52:286–87

40. Garrett, S. D. 1970. *Pathogenic Root-infecting Fungi.* Cambridge Univ. Press. 294 pp.

41. Gerlagh, M. 1968. Introduction of *Ophiobolus graminis* into new polders and its decline. *Meded. Lab. Phytopath. Wageningen,* No. 241

42. Gibbs, J. N. 1969. Inoculum sources for coffee berry disease. *Ann. Appl. Biol.* 64:515–22

43. Gregory, P. H. 1948. The multiple-infection transformation. *Ann. Appl. Biol.* 35:412–17

44. Gregory, P. H. 1951. Deposition of air-borne *Lycopodium* spores on cylinders. *Ann Appl. Biol.* 38:357–76

45. Gregory, P. H. 1961. *The Microbiology of the Atmosphere.* London: Leonard Hill. 251 pp.

46. Gregory, P. H. 1966. Dispersal. The Fungi II. In *The Fungal Organism*, eds. G. C. Ainsworth, A. S. Sussman, 709–32. New York: Academic Press, 805 pp.

47. Gregory, P. H. 1968. Interpreting plant disease dispersal gradients. *Ann. Rev. Phytopathol.* 6:189–212

48. Gregory, P. H., Lacey, M. E. 1963. Liberation of spores from mouldy hay. *Trans. Brit. Mycol. Soc.* 46:73–80

49. Gregory, P. H., Stedman, O. J. 1958. Spore dispersal in *Ophiobolus graminis* and other fungi of cereal foot rots. *Trans. Brit. Mycol. Soc.* 41:449–56

50. Gregory, P. H., Guthrie, E. J., Bunce, M. E. 1959. Experiments on splash dispersal of fungus spores. *J. Gen. Microbiol.* 20:328–54

51. Griffiths, E., Gibbs, J. N. 1969. Early-season sprays for the control of coffee berry disease. *Ann. Appl. Biol.* 64:523–32

52. Griffiths, E., Gibbs, J. N., Waller, J. M. 1971. Control of coffee berry disease. *Ann. Appl. Biol.* 67:45–74

53. Hirst, J. M. 1952. An automatic volumetric spore trap. *Ann. Appl. Biol.* 39:257–65

54. Hirst, J. M. 1953. Changes in atmospheric spore content: diurnal periodicity and the effects of weather. *Trans. Brit. Mycol. Soc.* 36:375–92

55. Hirst, J. M. 1958. New methods for studying plant disease epidemics. *Outlook on Agr.* 2:16–26

56. Hirst, J. M. 1959. Spore liberation and dispersal. In *Plant Pathology. Problems and Progress 1908-1958,* eds. C. S. Holton, G. W. Fischer, R. W. Fulton, H. Hart, S. E. A. McCallan, 529–38. Madison: Univ. Wisconsin Press. 588 pp.

57. Hirst, J. M., Hurst, G. W. 1967. Long-distance spore transport. In *Airborne Microbes. Seventeenth Symp. Soc. Gen. Microbiol.,* eds. P. H. Gregory, J. L. Monteith, 307-44. Cambridge Univ. Press. 385 pp.

58. Hirst, J. M., Stedman, O. J. 1960. The epidemiology of *Phytophthora infestans.* II. The source of inoculum. *Ann. Appl. Biol.* 48: 489–517

59. Hirst, J. M., Stedman, O. J. 1961. The epidemiology of apple scab (*Venturia inaequalis* (Cke.) Wint.). I. Frequency of airborne spores in orchards. *Ann. Appl. Biol.* 49:290–305

60. Hirst, J. M., Stedman, O. J. 1962. The epidemiology of apple scab (*Venturia inaequalis* (Cke.) Wint.). II. Observations on the liberation of ascospores. *Ann. Appl. Biol.* 50:525–50

61. Hirst, J. M., Stedman, O. J. 1962. The epidemiology of apple scab (*Venturia inaequalis* (Cke.) Wint.). III. The supply of ascospores. *Ann. Appl. Biol.* 50:551–67

62. Hirst, J. M., Stedman, O. J., Hogg, W. H. 1967. Long distance spore transport: methods of measurement, vertical spore profiles and the detection of immigrant spores. *J. Gen. Microbiol.* 48: 329–55

63. Hirst, J. M., Stedman, O. J., Hurst, G. W. 1967. Long-distance spore transport: vertical sections of spore clouds over the sea. *J. Gen. Microbiol.* 48:357–77

64. Hislop, E. C. 1969. Splash dispersal of fungus spores and fungicides in the laboratory and greenhouse. *Ann. Appl. Biol.* 63:71–80

65. Holliday, P. 1969. Dispersal of conidia of *Dothidella ulei* from *Hevea brasiliensis. Ann. Appl. Biol.* 63:435–47

66. Holliday, P. 1971. Some tropical plant pathogenic fungi of limited distribution. *Rev. Plant Pathol.* 50:337–48

67. Hutton, K. E. 1954. Eradication of *Venturia inaequalis* (Cooke) Wint. *Nature,* London 174:1017

68. Ingold, C. T. 1939. *Spore Discharge in Land Plants.* Oxford Univ. Press, 178 pp.

69. Ingold, C. T. 1953. *Dispersal in Fungi.* Oxford Univ. Press, 197 pp.

70. Ingold, C. T. 1964. Possible spore discharge mechanism in *Pyricularia. Trans. Brit. Mycol. Soc.* 47: 573–75

71. Ingold, C. T. 1965. *Spore Liberation.* Oxford Univ. Press, 210 pp.

72. Ingold, C. T. 1966. Aspects of spore liberation: violent discharge. In *The Fungus Spore,* ed. M. F. Madelin, 113–32. London: Butterworths, 338 pp.

73. Ingold, C. T. 1971. *Fungal Spores. Their Liberation and Dispersal.* Oxford Univ. Press, 302 pp.

74. Ingold, C. T., Dann, V. 1968. Spore discharge in fungi under very high surrounding air-pressure and the bubble theory of ballistospore release. *Mycologia* 60:285–89

75. Jarvis, W. R. 1962. The dispersal of spores of *Botrytis cinerea* Fr. in a raspberry plantation. *Trans. Brit. Mycol. Soc.* 45:549–59

76. Jenkins, P. T. 1965. The dispersal of conidia of *Sclerotinia fructicola* (Wint.) Rehm. *Aust. J. Agr. Res.* 16:627–33

77. Kable, P. 1965. Air dispersal of conidia of *Monilinia fructicola* in peach orchards. *Aust. J. Exp. Agr. Anim. Husb.* 5:166–71

78. Kerr, A., Rodrigo, W. R. F. 1967. Epidemiology of tea blister blight (*Exobasidium vexans*). III. Spore deposition and disease prediction. *Trans. Brit. Mycol. Soc.* 50:49–55

79. Kerr, A., Rodrigo, W. R. F. 1967. Epidemiology of tea blister blight (*Exobasidium vexans*). IV. Disease forecasting. *Trans. Brit. My-*

col. Soc. 50:609–14

80. Kiyosawa, S. 1972. Mathematical studies on the curve of disease increase. A technique for forecasting epidemic development. *Ann. Phytopathol. Soc. Japan* 38:30–40

81. Kuribayashi, K., Ichikawa, H. 1952. Studies on forecasting of the rice blast disease. *Nagano Agr. Exp. Sta. Rep.* 13:1–279

82. Lawrence, J. S., Meredith, D. S. 1970. Wind dispersal of conidia of *Cercospora beticola*. *Phytopathology* 60:1076–78

83. Leach, R. 1941. Banana leaf spot, *Mycosphaerella musicola*, the perfect stage of *Cercospora musae* Zimm. *Trop. Agr. (Trinidad)* 18: 91–95

84. Leach, R. 1946. *Banana leaf spot (Mycosphaerella musicola) on the Gros Michel variety in Jamaica.* Kingston, Jamaica: Government Printer. 118 pp.

85. Leach, R. 1962. Banana leaf spot. *Outlook in Agr.* 3:203–08

86. Leach, R. 1964. Report on investigations into the cause and control of the new banana disease in Fiji, black leaf streak. *Council Paper, Fiji* 38:1–20

87. McOnie, D. C. 1964. Orchard development and discharge of ascospores of *Guignardia citricarpa* and the onset of infection in relation to the control of citrus black spot. *Phytopathology* 54:1448–53

88. Meredith, D. S. 1960. Some observations on *Trachysphaera fructigena* Tabor & Bunting, with special reference to Jamaican bananas. *Trans. Brit. Mycol. Soc.* 43:100–04

89. Meredith, D. S. 1961. Fruit spot ('speckle') of Jamaican bananas caused by *Deightoniella torulosa* (Syd.) Ellis. I. Symptoms of disease and studies on pathogenicity. *Trans. Brit. Mycol. Soc.* 44:95–104

90. Meredith, D. S. 1961. Fruit spot ('speckle') of Jamaican bananas caused by *Deightoniella torulosa* (Syd.) Ellis. II. Factors affecting spore germination and infection. *Trans. Brit. Mycol. Soc.* 44:265–84

91. Meredith, D. S. 1961. Spore discharge in *Deightoniella torulosa* (Syd.) Ellis. *Ann. Bot. London* 25:271–78

92. Meredith, D. S. 1962. Spore discharge in *Cordana musae* (Zimm.) Höhnel and *Zygosporium oscheoides* Mont. *Ann. Bot. London* 26: 233–41

93. Meredith, D. S. 1962. Some components of the air-spora in Jamaican banana plantations. *Ann. Appl. Biol.* 50:577–94

94. Meredith, D. S. 1963. *Pyricularia grisea* (Cooke) Sacc. causing pitting disease of bananas in Central America. I. Preliminary studies on pathogenicity. *Ann. Appl. Biol.* 52:453–63

95. Meredith, D. S. 1963. Violent spore release in some Fungi Imperfecti. *Ann. Bot. London* 27: 39–47

96. Meredith, D. S. 1963. Further observations on the zonate eyespot fungus, *Dreschslera gigantea*, in Jamaica. *Trans. Brit. Mycol. Soc.* 40:201–07

97. Meredith, D. S. 1965. Violent spore release in *Helminthosporium turcicum*. *Phytopathology* 55:1099–1102

98. Meredith, D. S. 1966. Airborne conidia of *Helminthosporium turcicum* in Nebraska. *Phytopathology* 56:949–52

99. Meredith, D. S. 1966. Spore dispersal in *Alternaria porri* (Ellis) Neerg. on onions in Nebraska. *Ann. Appl. Biol.* 57.67–73

100. Meredith, D. S. 1967. Conidium release and dispersal in *Cercospora beticola*. *Phytopathology* 57: 889–93

101. Meredith, D. S. 1968. Freckle disease of banana in Hawaii caused by *Phyllostictina musarum* (Cke.) Petr. *Ann. Appl. Biol.* 62:329–40

102. Meredith, D. S. 1970. Banana leaf spot disease (Sigatoka) caused by *Mycosphaerella musicola* Leach. *Commonwealth Mycol. Inst., Kew, England, Phytopathol. Paper* 11, 147 pp.

103. Meredith, D. S. 1970. Major banana diseases: past and present status. *Rev. Plant Pathol.* 49:539–54

104. Meredith, D. S., Lawrence, J. S., Firman, I. D. 1973. Black leaf streak disease of bananas (*Mycosphaerella fijiensis*): ascospore release and dispersal. *Trans. Brit. Mycol. Soc.* (In press)

105. Miller, P. R. 1969. Effect of environment in plant diseases. *Phytoprotection* 50:81–94

106. Moller, W. J., Carter, M. V. 1965. Production and dispersal of ascospores in *Eutypa armeniacae*. *Aust. J. Biol. Sci.* 18:67–80

107. Moore, M. H. 1958. The release of ascospores of apple scab by dew. *Plant Pathol.* 7:4–5

108. Norse D. 1971. Lesion and epidemic development of *Alternaria longipes* (Ell. & Ev.) Mason on tobacco. *Ann. Appl. Biol.* 69:105–23

109. Nutman, F. J. 1970. Coffee berry disease. *Pest Articles and News Summaries* (PANS) 16:277–86

110. Nutman, F. J., Roberts, F. M. 1970. Coffee leaf rust. *Pest Articles and News Summaries* (PANS) 16:606–24

111. Nutman, F. J., Roberts, F. M. 1971. Spread of coffee leaf rust. *Pest Articles and News Summaries* (PANS) 17:385–86

112. Nutman, F. J., Roberts, F. M., Bock, K. R. 1960. Method of uredospore dispersal of the coffee leaf rust fungus, *Hemileia vastatrix*. *Trans. Brit. Mycol. Soc.* 43:509–15

113. Ogawa, J. M., Hall, D. H., Koepsell, P. A. 1967. Spread of pathogens within crops as affected by life cycle and environment. In *Airborne Microbes. Seventeenth Symp. Soc. Gen. Microbiol.*, eds. P. H. Gregory, J. L. Monteith, 247–67. Cambridge Univ. Press. 385 pp.

114. Olive, L. S. 1964. Spore discharge mechanism in Basidiomycetes. *Science* 140:542–43

115. Ono, K. 1965. Principles, methods, and organization of blast disease forecasting. In *The Rice Blast Disease*, 173–94. Baltimore: John Hopkins Press. 507 pp.

116. Pady, S. M., Kramer, C. L., Wiley, B. J. 1962. Kansas aeromycology. XII. Materials, methods, and general results of diurnal studies 1959–1960. *Mycologia* 54:168–80

117. Pinckard, J. A. 1942. The mechanism of spore dispersal in *Peronospora tabacina* and certain other downy mildew fungi. *Phytopathology* 32:505–11

118. Price, D. 1960. Climate and control of banana leaf spot. *Span* 3:122–24

119. Rayner, R. W. 1960. Rust disease of coffee. II. Spread of the disease. *World Crops* 12:222–24

120. Rayner, R. W., Hopkins, J. C. F. 1962. Blue mould of tobacco. A review of current information. *Commonwealth Mycol. Inst., Kew, England, Misc. Publ.* 16, 16 pp.

121. Raniere, L. C., Crossan, D. F. 1959. The influence of overhead irrigation and microclimate on *Colletotrichum phomoides*. *Phytopathology* 49:72–74

122. Rishbeth, J. 1955. Root diseases in plantations, with special reference to tropical crops. *Ann. Appl. Biol.* 42:220–28

123. Rishbeth, J. 1959. Dispersal of *Fomes annosus* Fr. and *Peniophora gigantea* (Fr.) Massee. *Trans. Brit. Mycol. Soc.* 42:243–60

124. Rishbeth, J. 1963. Stump protection against *Fomes annosus*. III. Inoculation with *Peniophora gigantea*. *Ann. Appl. Biol.* 52:63–77

125. Roelfs, A. P. 1972. Gradients in horizontal dispersal of cereal rust uredospores. *Phytopathology* 62:70–76

126. Roelfs, A. P., Dirks, V. A., Romig, R. W. 1968. A comparison of rod and slide samplers used in cereal rust epidemiology. *Phytopathology* 58:1150–54

127. Romig, R. W., Dirks, V. A. 1966. Evaluation of generalized curves for number of cereal rust uredospores trapped on slides. *Phytopathology* 56:1376–80

128. Rotem, J. 1964. The effect of weather on dispersal of *Alternaria* spores in a semi-arid region of Israel. *Phytopathology* 54:628–39

129. Rotem, J., Palti, J. 1969. Irrigation and plant disease. *Ann. Rev. Phytopathol.* 7:267–88

130. Rowell, J. B. 1964. Factors affecting field performance of nickel salt plus dithiocarbamate fungicide mixtures for the control of wheat rusts. *Phytopathology* 54:999–1008

131. Samuel, Y., Garrett, S. D. 1933. Ascospore discharge in *Ophiobolus graminis* and its probable relation to the development of white-

heads in wheat. *Phytopathology* 23:721–28

132. Schenck, N. C. 1968. Epidemiology of gummy stem blight (*Mycosphaerella citrullina*) on watermelon: ascospore incidence and disease development. *Phytopathology* 58:1420–22

133. Schieber, E. 1970. Viaje al Brasil y el Africa para estudiar y observar el problema de la herrumbre del cafe. *Rept. Org. Int. Reg. San. Agr.* 109 pp.

134. Schieber, E. 1972. Economic impact of coffee rust in Latin America. *Ann. Rev. Phytopathol.* 10: 491–510

135. Schrödter, H. 1960. Dispersal by air and water—the flight and landing. In *Plant Pathology, Vol. III. The diseased population, epidemics and control,* eds. J. G. Horsfall, A. E. Dimond, 169–227. New York: Academic Press. 675 pp.

136. Smith, R. S. 1966. The liberation of cereal stem rust uredospores under various environmental conditions in a wind tunnel. *Trans. Brit. Mycol. Soc.* 49:33–41

137. Smith, D. H., Lewis, F. H., Wainwright, S. H. 1970. Epidemiology of the black knot disease of plums. *Phytopathology* 60:1441–44

138. Sreeramulu, T. 1962. Spore content of the atmosphere. *Proc. Summer School of Botany,* Darjeeling, 1960, 452–62

139. Sreeramulu, T., Ramalingam, A. 1961. Experiments on the dispersion of *Lycopodium* and *Podaxis* spores in the air. *Ann. Appl. Biol.* 49:659–70

140. Stahel, G. 1937. Notes on *Cercospora* leaf spot of bananas (*Cercospora musae*). *Trop. Agr. (Trinidad)* 14:257–64

141. Stakman, E. C., Harrar, J. G. 1957. *Principles of Plant Pathology.* New York: Ronald Press. 581 pp.

142. Stepanov, K. M. 1935. Dissemination of infectious diseases of plants by air currents (translated title). *Bull. Plant. Prot. (U.S.S.R.)* Ser. II Phytopathology 8:1–68

143. Stepanov, K. M. 1962. *Gribnye epifitotii.* Moscow: Selisk. Lit. Zhurn. Plak. 471 pp.

144. Stover, R. H. 1964. Leaf spot of bananas caused by *Mycosphaerella musicola:* factors influencing production of fructifications and ascospores. *Phytopathology* 54: 1320–26

145. Stover, R. H. 1967. Intercontinental spread of banana leaf spot (*Mycosphaerella musicola* Leach). *Trop. Agr. (Trinidad)* 39:327–38

146. Stover, R. H. 1970. Leaf spot of bananas caused by *Mycosphaerella musicola:* role of conidia in epidemiology. *Phytopathology* 60: 856–60

147. Stover, R. H., Fulton, R. H. 1966. Leaf spot of bananas caused by *Mycosphaerella musicola:* The relation of infection sites to leaf development and spore type. *Trop. Agr. (Trinidad)* 43:117–29

148. Suzuki, H. 1969. Studies on the behavior of rice blast fungus spore and application to outbreak forecast of rice blast disease. *Hokuriku Agr. Exp. Sta. Bull.* 10:1–118

149. Swinburne, T. R. 1971. The seasonal release of spores of *Nectria galligena* from apple canker in Northern Ireland. *Ann. Appl. Biol.* 69:97–104

150. Van Arsdel, E. P. 1967. The nocturnal diffusion and transport of spores. *Phytopathology* 57:1221–29

151. Van der Plank, J. E. 1960. Analysis of epidemics. In *Plant Pathology, Vol. III. The diseased population, epidemics and control,* eds. J. G. Horsfall, A. E. Dimond, 229–89. New York: Academic Press. 675 pp.

152. Van der Plank, J. E. 1963. *Plant Diseases: Epidemics and Control.* New York: Academic Press. 349 pp.

153. Waggoner, P. E. 1960. Forecasting epidemics. In *Plant Pathology, Vol. III. The diseased population, epidemics and control,* eds. J. G. Horsfall, A. E. Dimond, 291–313. New York: Academic Press. 675 pp.

154. Waggoner, P. E. 1968. Weather and the rise and fall of fungi. In *Biometeorol., Proc. Ann. Biol. Coll. 28th, 1967,* ed. W. P. Lowry, 45–66. Corvallis, Ore.: Oregon State Univ. Press.

155. Waggoner, P. E., Horsfall, J. G. 1969. Epidem, a simulator of plant disease written for a computer.

Bull. Connecticut Agr. Exp. Sta.,
New Haven. No. 698:1–80
156. Waggoner, P. E., Horsfall, J. G.,
Lukens, R. J. 1972. Epimay, a simulator of southern corn leaf
blight. *Bull. Connecticut Agr. Exp.
Sta.,* New Haven. No. 729:1–84
157. Waite, B. H. 1956. *Fusarium* stalk
rot of bananas in Central America. *Plant Dis. Reptr.* 40:309–11
158. Walker, J. C., Patel, P. N. 1964.
Splash dispersal and wind as factors in epidemiology of halo blight
of bean. *Phytopathology* 54:140–
41
159. Wallin, J. R., Loonan, D. V. 1972.
The increase of *Cercospora* leaf
spot in sugar beets and periodicity
of spore release. *Phytopathology*
62:570–72
160. Webster, J. 1952. Spore projection
in the hyphomycete *Nigrospora
sphaerica. New Phytol.* 51:229–
35
161. Wellman, F. L. 1970. Rust of
coffee in Brazil. *Plant Dis. Reptr.*
54:355
162. Whiteside, J. O. 1970. Etiology

and epidemiology of citrus greasy
spot. *Phytopathology* 60:1409–16
163. Wilson, E. E. 1928. Studies of the
ascigerous state of *Venturia inaequalis* (Cke.) Wint. in relation to
certain factors of the environment.
Phytopathology 18:375–418
164. Wilson, E. E., Baker, G. A. 1946.
Some aspects of the aerial dissemination of spores, with special reference to conidia of *Sclerotinia
laxa. J. Agr. Sci.* 72:301–
27
165. Zachos, D. G. 1969. Recherches
sur la biologie et l'epidemiologie
du mildiou de la vigne en Grece.
Ann. Inst. Phytopath. Benaki, N.S.
2:193–335
166. Zadoks, J. C. 1971. Systems analysis and the dynamics of epidemics. *Phytopathology* 61:600–10
167. Zadoks, J. C. 1972. Methodology
of epidemiological research. *Ann.
Rev. Phytopathol.* 10:253–76
168. Zoberi, M. H. 1961. Take-off of
mould spores in relation to wind
speed and humidity. *Ann. Bot.
London* 25:53–64

EFFECTS OF ENVIRONMENTAL FACTORS ON PLANT DISEASE

❖ 3577

John Colhoun

Department of Cryptogamic Botany, University of Manchester, Manchester, England

Nearly 50 years ago, Jones (51) appealed for increased attention to the relation of environment to the inception and development of plant disease. He pointed out that with many plant diseases, comparing field with field, and season with season, the loss following identical initial infection may vary widely, perhaps even from nothing to total loss. It is appropriate that we should now ask ourselves how far we have succeeded in responding to this appeal, and it would also perhaps be appropriate for us, as plant pathologists, to ask ourselves how far we subscribe to the idea that our prime function is to enable growers to produce healthy, or at least healthier, crops. Have we in the past accepted too readily that many problems of disease control can be solved by the use of chemicals, and does it now require a growing interest in the preservation of the human environment for us to be provoked into attempting to solve our problems by nonchemical methods?

HISTORICAL

The realization that plant diseases are greatly influenced by weather is of long standing. Indeed, Theophrastus (370–286 BC) considered that cereal crops growing in elevated lands exposed to the wind are not so liable to rust as are those that lie low. At that time, it was generally accepted that the microorganisms observed to be associated with the diseased plants arose spontaneously from the plants, or possibly from the environment. Naturally, the relationship of environmental factors to the causation of disease could not be determined with any reliability until it was proved that microorganisms caused disease. Once this was established, the effects of the pathogens could be separated from those of the environment. However, the necessary discoveries to enable this phase of activity to be set into action were long delayed, and so the theory of spontaneous generation dominated thought, among those interested in plant diseases, for more than 2000 years.

During the Renaissance, as European agriculture became more intensive, plant diseases increased in importance, but it was still accepted that they were

343

inevitably associated with unfavorable soil and climate. In the eighteenth century, Tillet (93) referred to opinions attributing the cause of wheat bunt to evil fogs, winds, excessive soil moisture, soil texture, or improper plant nutrients. Thought along these lines continued into the nineteenth century (Ré, 81; Unger, 95; Meyen, 63). When the potato crop was destroyed by blight in Ireland in 1844 and 1845, cool weather was blamed for having upset the normal growth of the plant so that internal breakdown occurred.

In spite of the general acceptance of the theory of spontaneous generation up to the middle of the nineteenth century, some remarkable advances in knowledge were being made. Although not producing proof, de Tournefort (34) expressed the view that fungi are autonomous organisms, which do not arise by spontaneous generation, and can cause moldiness of plants in humid greenhouses during winter. It remained for Prevost (76) to demonstrate that bunt of wheat is caused by a fungus, and to interpret the role of environment in the etiology of this disease. However, like some pioneers in other fields of study, Prevost was ahead of his time, and his classic paper was largely disregarded at the time of its presentation. The work of Berkeley (4, 5) on potato blight and vine mildew, and the proof by Speerschneider (89) and de Bary (31) that potato blight was caused by the fungus *Phytophthora infestans* (Mont.) de Bary, rendered inevitable the acceptance of the concept that fungi can incite plant diseases.

During the second half of the nineteenth century, plant pathologists devoted attention almost exclusively to the naming and description of the causal organisms of plant diseases. Not much prominence was given to the study of the effects of environment, but Soraurer (88) appreciated clearly the significance of environmental factors in relation to plant diseases and emphasized the part they played in predisposing plants to disease. Soraurer recognized that disease could result from an unfavorable environment as well as from attacks of parasitic microorganisms. Soraurer expressed concern that the emphasis placed on the parasites and their life histories resulted in neglect of the interaction of host and parasite in respect to the development of disease caused by microorganisms.

Hartig had a keen appreciation of the importance of environmental factors when investigating pathological phenomena (41). He stated that a plant contracts disease only when subjected to definite conditions, so that a predisposition or tendency to disease results. Among factors that may predispose plants to disease, Hartig recognized the condition of vegetation in regard to the season of the year, and the amount of water in plants as determined by the weather. He also stated that a plant may be predisposed to one disease because it is suffering from another.

Marshall Ward (99), the celebrated supporter of the concept of predisposition, expressed the opinion that an attempt to understand any disease is hopeless unless we consider the variations in the host plant and the parasite induced by changes in the environment. He appreciated that when the physical environment is unfavorable to the host, it may be favorable to the parasite, and in such instances the resulting disease may assume a more or less pronounced epidemic character. Ward, of course, excluded inherited variations from his concept of predisposition, and concerned himself mainly with the effects of such factors of the environment as temperature, light intensity, and humidity. He believed that the occurrence of unfavorable external conditions for the plant brought about the production of parenchymatous tissue with thinner and softer cell walls, while the protoplasm was also less resistant to invasion by a parasite.

During much of the twentieth century, one of the major fields of study in plant pathology has been that of the effects of environment on plant disease. In publications relating to this work little use is made of the term "predisposition." Gäumann (37) discussed changes in disease proneness due to the environment, and distinguished between the influence of external factors on the establishment of the parasitic relationship and its subsequent course. He therefore was concerned with the effects of environmental factors that act before, during, or after the bringing together of host and pathogen, and that affect the disease proneness of the host. In a recent review, Yarwood (105) has extended the concept of predisposition to include changes in disease proneness that result from external causes, and lead to greater or lesser susceptibility. Previous workers have mainly considered predisposition as including changes towards greater susceptibility, but Yarwood's extension of the definition is both welcome and logical.

VARIABILITY IN DISEASE DEVELOPMENT

Some pathogens are capable of causing disease throughout the world, in areas where plants susceptible to them are grown, for they can tolerate a wide range of environmental conditions. Other pathogens may be less tolerant of environmental variations and are likely to be restricted in their geographical distribution.

Although *Urocystis cepulae* Frost, the cause of onion smut, is continually being distributed throughout the U.S.A. on infected onion sets, it is absent in the extensive southern onion growing regions, but is an established disease in the cooler, northern onion growing areas of America and Europe. This has been explained in terms of temperature effects on the pathogen (Walker & Wellman, 98). Infection occurs readily at 10–12°C, which is nearly as low as will permit the germination and growth of onions. Infection occurs equally well up to 25°, but above that there is a rapid reduction, and at 29° or above, plants are not infected. The minimum temperature for germination of teleutospores and growth of *U. cepulae* is very close to that of the onion plant. The optimum temperatures for germination of teleutospores and hyphal fragments, and for hyphal growth, are between 13° and 22°. Above 25° there is a reduction in germination and hyphal growth. Slightly below 29°, the seedlings grow more rapidly than at lower temperatures, so that they reach the nonsusceptible stage earlier, which is when the cotyledon has attained full size.

Another example of restriction of geographical distribution of a disease by

direct action of the environment on the pathogen is provided by chrysanthemum rust caused by *Puccinia chrysanthemi* Roze. Temperatures above 27°
may deter the establishment of this disease in three ways: (*a*) by reducing
the germination of initially viable uredospores at the infection court, (*b*) by
eradicating established infections, and (*c*) by killing spores in existing sori
(Campbell & Dimock, 11). Low temperature effects may also be restrictive,
for spores do not germinate nor does infection occur at 6°. Although the
pathogen has undoubtedly been introduced repeatedly into inland areas of
the U.S.A. where the summers are hot, because of these temperature effects it
has failed to become established and cause serious damage there, except occasionally during winter months in glasshouses.

Direct effects of environmental factors on the pathogen only are certainly
inadequate to explain the results obtained by Dickson, Eckerson & Link (35)
in respect to attacks of *Gibberella zeae* (Schw.) Petch [*G. saubinetii*
(Mont.) Sacc.] on wheat and maize. These attacks are most severe in wheat
at high temperatures and in maize at low temperatures. In culture the fungus
grows well over a wide temperature range with the optimum between 24°
and 28°, according to the medium used. Wheat grows best at low temperatures and maize seedlings develop best at fairly high temperatures. It therefore seems reasonable to conclude that the disease appears with greatest severity on each host at temperatures unfavorable for seed germination and
growth of the seedlings. At 10°–12° the disease is severe in wheat seedlings
under a low light intensity, although they are healthy if the light intensity is
high. It is difficult to imagine that a fungus such as *G. zeae,* which largely
attacks the seedlings below soil level, can be substantially influenced directly
by light intensity. However, examination of the cell walls of seedlings showed
that their chemical nature was related to the type and proportion of the
building materials available under different temperature conditions. The relatively high content of available carbohydrates in wheat seedlings at low soil
temperatures, and in maize seedlings at high soil temperatures, results in the
development of thickened cellulose walls, which can be expected to offer resistance to penetration by the fungus, and no doubt also induce alterations
that influence growth of the pathogen after penetration. The work of Gäumann (37) in respect to *Fusarium* disease of cereal seedlings supports these
conclusions. He found that as the temperature rises, the cell walls of the seedlings become increasingly soft and more readily soluble by fungal enzymes,
and that the whole seedling expends itself by growth in length, thereby becoming less resistant.

Not only may environment directly influence either host or pathogen, it
may also affect disease incidence and development, because of the interaction
of effects on the host and pathogen. The results obtained by Christensen (12)
on head smut of sorghum, caused by *Sphacelotheca reiliana* (Kühn) Clint.,
can be explained on the basis of such interactions. Moist soil is the most favorable for budding of the resting spores, and for their germination, but it
also favors faster development and greater vitality of the seedlings. The inci-

dence of disease is much greater in dry soils than in moist soils, presumably because in moist soils the seedlings have greater powers of escaping or resisting attack. Moreover, in dry soils the temperature range over which infection occurs is wider than in moist soils.

The influence of environment on the type of attack by a particular pathogen is demonstrated by results obtained with *Peronospora tabacina* Adam, the cause of tobacco blue mould. This disease seriously hinders tobacco production in Australia, where it has occurred primarily as a field disease since the 1890s. In 1921, the disease was first recorded in the U.S.A., where it occurs rarely on tobacco in the field, but is primarily a seed-bed disease. This difference in the type of occurrence can be explained in terms of environmental factors, particularly temperature. According to Clayton & Gaines (14), germination of spores and infection are inhibited by temperatures above 30°, but sporulation is most abundant at a temperature of 16° by night and 25° by day. Little sporulation, if any, occurs in full sunlight in summer, unless the temperature during part of the day falls below 20°. In Australia, the temperatures are lower than in the North American tobacco growing areas, and permit the fungus to continue to spread in the field throughout the season. It was predicted in 1950 that if this disease appeared in Europe, it would become a serious field disease because of the environmental conditions. This indeed occurred, for the disease was first reported in Europe in 1958, and by 1963 it had spread as a field disease over most of the tobacco growing areas there. The constant and heavy sporulation that occurs on diseased plants in Europe has enabled the fungus to spread for a greater distance during a single season than in the United States (Miller, 66).

Results obtained by Clayton & Gaines (14) for the effect of temperature on blue mould of tobacco indicate the dangers of neglecting the interaction of environmental factors. Although in summer little sporulation of *P. tabacina* occurs at 20°, if the plants are grown with reduced light as a result of heavy shading, or if they are grown during winter conditions in glasshouses, then sporulation occurs freely at temperatures as high as 27°. Up to now, much of the research undertaken on environmental effects on disease has been concentrated on studying the effects of single environmental factors, and considerable progress has been made in the field. It is, however, obvious that the controlled conditions under which most of the experiments have been made are far removed from what happens in this field. In the crop we are concerned with a number of factors that are simultaneously undergoing changes. This is quite different from what happens during most laboratory or glasshouse studies, where all factors are held constant except for the one that is altered experimentally. There is, of course, no reason why attempts should not be made in glasshouse experiments to study the effects of interactions of the different factors. The results that can be obtained from such work are very illuminating in the analysis of the effects of environment on disease. It is to be hoped that much more work in future will be undertaken to study such interactions. There is no doubt that such an approach to the problem makes much greater demands on manpower and glasshouse facilities as compared with the position when the effects of single factors only are considered.

Although it is well appreciated that studies of the effects of single environmental factors are of limited value, some of the results that have been obtained are considered in the following pages. Results from such experiments can be regarded as providing a basis for the design of experiments to elucidate the effects of interaction of environmental factors.

The wide variety of factors that combine to produce environmental effects on disease development and expression is obvious when it is borne in mind that such effects, in respect to some diseases at least, demand consideration of some or all of the following aspects of study: (a) survival of inoculum from one season to another, or from one crop to another, (b) provision of suitable conditions for infection, (c) sporulation of the pathogen on the diseased plants and transference of inoculum to healthy plants, (d) defensive mechanisms in host plants, (e) effects of environment on fungus and host, (f) influence of the growth of the host on the microclimate in the crop, (g) effects of interactions between host and pathogen, and (h) interactions between microorganisms in the soil, or on aerial parts of plants, in respect to infection by any specific pathogen. In this paper it is not proposed to deal with all these aspects, but particular attention will be paid to the effects of physical factors and their interactions. The effects of temperature will be looked at in depth.

Environmental factors may bring about disease without the intervention of a pathogen. This aspect of the effects of environment on disease is not considered here.

EFFECTS OF INDIVIDUAL FACTORS

Temperature

Temperature is undoubtedly one of the most important factors influencing the occurrence and development of many diseases. Some diseases are most severe at low temperatures, others at high temperatures, although in culture the causal organisms grow over a wide temperature range. This is because the effects of temperature on disease, like those of some other environmental factors, may be attributed to effects on the pathogen, or on the host, or on interaction between host and pathogen. It should also be borne in mind, when attempting to relate the results of laboratory studies to field conditions, that plant temperatures in the field may be higher than that of the surrounding air.

To simplify the discussion a number of individual topics relating to temperature effects will be considered.

EFFECTS OF PRE-INOCULATION TEMPERATURE   Exposure of plants to high or low temperatures before inoculation may increase or decrease their susceptibility to disease, but sometimes has no effect. Heat-induced susceptibility has been demonstrated by Yarwood (104) in respect to *Uromyces appendiculatus* (Pers.) Unger (= *U. phaseoli*), *Erysiphe polygoni* D.C., *Colletotrichum lindemuthianum* (Sacc. & Magn.) Bri. & Cav., tobacco mosaic virus, spotted wilt virus, peach yellow bud mosaic virus, and apple mosaic virus when leaves of *Phaseolus vulgaris* L. were immersed in hot water for a few seconds. Kassanis (52) has produced convincing evidence that maintaining plants for 6 hr at 36° increased the number of local lesions per leaf in respect to tobacco necrosis virus in *P. vulgaris,* tobacco mosaic virus in *Nicotiana glutinosa* L., tobacco bushy stunt virus in *N. glutinosa,* and tomato spotted wilt virus in *N. tabacum* L. The time taken for the maximal response depended on the age and general physiologic state of the plants, older plants requiring longer treatment.

Apples kept for 17 days at 30°, then inoculated with *Botrytis allii* Munn and stored at laboratory temperatures were susceptible, whereas those kept at laboratory temperatures throughout were not attacked. This increased susceptibility could not be related to changes in acidity, concentration of sugars, or total extractable nitrogen resulting from the heat treatment (Vasudeva, 96). It may perhaps be associated with changes in the phenolic constituents resulting from suppression of an enzyme system (Epton, unpublished data). Pretreatment of potato tubers for 10–21 days at 35° also rendered them more susceptible to attack by *Erwinia aroideae* (Towns.) Bergey et al (= *Bacterium aroideae*), *Erwinia carotovora* (Jones) Bergey et al (= *Bacterium carotovorum*), and *Xanthomonas carotae* (Kendrick) Dows. (= *Phytomonas carotae*) (Gregg, 39). It was suggested that the heat treatment either changed the composition of the middle lamella so that it was more readily hydrolyzed, or lessened the effect of some factor that slowed down enzyme action. The suggestion of change in the composition of the middle lamella was not supported by microscopic examination.

Cold pre-treatment of seeds of bean, pea, maize, or cucumber, in the early stages of germination prior to their being exposed to infection by *Rhizoctonia solani* Kühn, increased their susceptibility (Schulz & Bateman, 86). This cold treatment apparently reduced the vigor of the plants, and so, presumably, they were more prone to being killed before emerging above soil level.

Factors that may be involved in reduced susceptibility of potato tubers to *Fusarium solani* var. *coeruleum* (Sacc.) Booth [= *F. caeruleum* (Lib.) Sacc.] induced by low temperature storage before inoculation (Boyd, 7) may include changes in sugar content and the formation of suberized parenchyma.

EFFECTS OF TEMPERATURE ON KEEPING QUALITIES DURING STORAGE   Cold storage of various plant products has now become commercial practice. The use of temperatures below the optimum for a fungus or bacterium capable of attacking fruits or vegetables will reduce loss in most cases. For example, rotting of sweet potatoes due to *Rhizopus stolonifera* (Fr.) Lind can be prevented by storage for short periods at 2° (Daines, 29). On the other hand, refrigerated storage of peaches for two weeks, under certain conditions, was associated with more decay than in those stored at ambient temperatures (Daines, 30). Prolonged storage at low temperatures may cause increased disease, as in potato gangrene caused by *Phoma exigua* Desm., which was more severe in tubers that had been moved after being kept cold (Malcolmson & Gray, 61).

The formation of a retarding barrier of suberized cells, and eventually of periderm, at low temperatures may prevent the advance of *Bacterium carotovorum* in potato tubers, while at higher temperatures the bacteria spread readily in the tubers (Rudd-Jones & Dowson, 83).

EFFECTS OF TEMPERATURE ON THE SURVIVAL OF PATHOGENS   The survival of pathogens may be influenced by both high and low temperatures. Mention has already been made of the effect of temperatures above 27° on eradication of established infections by *Puccinia chrysanthemi* and on killing of its spores in uredial sori. Low survival of *Podosphaera leucotricha* (Ell. & Everh.) Salm. may result from death of infected apple buds after a cold winter, but freezing of the fungus itself may well play a part (Covey, 27). Similarly, wheat leaves infected by *Puccinia striiformis* West. may be predisposed to low temperature death, or to invasion by secondary fungi, and so a reduction in disease in winter wheat may result (Lloyd, 58).

EFFECTS OF TEMPERATURE ON SPORE GERMINATION   The influence of temperatures lower than the minimum, or higher than the optimum, for spore germination of *Urocystis cepulae* and *Puccinia chrysanthemi* on diseases caused by these species, has been mentioned. However, all spores, even of one type, produced by a fungus may not react similarly to temperature. Some spore collections of *Peronospora tabacina* germinate best at 2–10°, and others at 18–26°, so that a single optimum temperature for spore germination of this species does not exist (Clayton & Gaines, 14). Moreover, spores of a species may germinate in different ways at different temperatures. Sporangia of *Phytopthora infestans* produce germ tubes directly at 25°, but on chilling for 60 min at 10° they produce zoospores. The ability of germ tubes resulting from direct germination of sporangia to infect potato leaves is doubted.

Occasionally high temperatures encourage spore germination. Exposure of ascospores of *Rhizina undulata* Fr. to 37° for 3 days brought about rapid and abundant germination while unheated spores showed erratic and poor germination on agar (Jalaluddin, 49). Stimulation of ascospore germination in a zone beneath, and at the margins of, fires in forests may lead to colonization of such sites by this fungus and to attacks of susceptible hosts near the sites (Jalaluddin, 50).

EFFECTS OF TEMPERATURE ON THE INCUBATION PERIOD   Unfavorable temperatures may lengthen the period between infection and the production of a new crop of spores (the incubation period) and so influence the number of spore generations occurring during a season. This effect on the supply of inoculum may determine whether or not an epidemic can develop. Such an effect is important in relation to rusts, downy mildews, and powdery mildews, which build up each season from small foci. Two examples serve to illustrate the significance of the length of the incubation period. The incubation period of *Pseudoperonospora humuli* (Miyabe & Takah.) G. W. Wilson can vary from 3 days at 21–25° to 23 days at 5°, or 11 days at 29° (Petrlik & Stys, 75), while the time required for the production of uredospores of *Puccinia graminis* Pers. var. *tritici* on a susceptible wheat cultivar can range from 85 days at 0°, to 5 days at 24° (Stakman & Harrar, 91).

EFFECTS OF TEMPERATURE ON SYMPTOMS   In the study of systemic plant virus diseases the effect of heat masking, i.e. the reduction in severity of the disease or the disappearance of symptoms as temperature is altered, is well known. For example, barley yellow dwarf is masked at 32°, but produces marked symptoms at 16°, whereas symptoms of aster yellows on barley are severe at 32° but absent at 16° (Gill & Westdal, 38). Samuel (84) was the first to observe that at 35° tobacco mosaic virus in *Nicotiana glutinosa* produced chlorotic instead of the necrotic local lesions that developed at 21° and 28°, while systemic infection occurred only at the highest temperature. This change from localized to systemic infection is not, however, a general phenomenon applicable to other hosts or other infections, and indeed systemic invasion of plants by certain viruses may apparently be prevented by high temperatures (Kassanis, 52).

Temperature may also influence symptoms of some fungal diseases. Low temperatures result in much pre-emergence or post-emergence death of cereal seedlings attacked by *Fusarium nivale* (Fr.) Ces., while at higher temperatures little pre-emergence death occurs, although the hypocotyls of many seedlings bear lesions (Colhoun, 21; Millar & Colhoun, 64, 65). On the other hand, *F. culmorum* (W. G. Sm.) Sacc. and *F. graminearum* Schwabe cause much pre-emergence or post-emergence death of cereal seedlings at 22°, whereas at 12° less pre-emergence or post-emergence death occurs, but many plants develop lesions on the hypocotyls. In these cereal diseases temperature effects must, however, be considered in relation to soil moisture.

Temperature can affect the rate of wilting in *Fusarium* wilt of tomatoes and peas (Clayton, 13; Schroeder & Walker, 85). Sudden wilting of tomatoes occurs at 25–31°, usually without leaf yellowing, whereas at higher or lower temperatures wilting is much slower and is accompanied, or often preceded, by yellowing of the leaves.

INFLUENCE OF TEMPERATURE ON CULTIVAR REACTION   A cultivar may be susceptible to a pathogen at one temperature and resistant at another, but another cultivar of the same species may show quite different reactions. Thus *Lupinus augustifolius* L. with the recessive gene pair *an an* was immune to attacks of *Glomerella cingulata* (Stonem.) Spauld. & Schrenk at 18° and below, but was killed at 27°, whereas plants carrying the dominant pair *Au Au* gene were immune at 27° and below, but were killed at 32° (Wells & Forbes, 102). Some varieties of tobacco normally susceptible to *Thielavia basicola* (Berk. & Br.) Ferraris become resistant on exposure to high temperatures. This has been attributed to the formation of a phellogen layer at the higher temperature, arresting the spread of the fungus (Conant, 23).

EFFECTS OF TEMPERATURE ON RESPONSE TO DISEASE OF DIFFERENT HOST SPE-CIES   Variations exist in the optimum temperature range for infection of different host species, some of which may be very susceptible at temperatures above the maximum for infection of other species. An example of this is provided by *Agrobacterium tumefaciens* (E. F. Smith & Town.) Conn, which does not cause galls on herbaceous hosts held at 31° for 4 days after inoculation. Nevertheless, galls develop on cherry trees held at 37° for 4–6 days after inoculation (Deep & Hussin, 32).

EFFECTS OF STAGE OF PLANT DEVELOPMENT ON TEMPERATURE RESPONSE   Experiments with wheat seedlings show that on cereal seedlings with one leaf *Cercosporella herpotrichoides* Fron becomes established most readily at 15°, but its spread through the host tissues and incidence of stroma in the second leaf sheath is greater at 20° (Defosse, 33). When comparing the reaction of plant organs at different stages of development, it is usually important to ensure that the organs are attached to the plant, otherwise interesting differences in reaction may be missed. This has been clearly shown for potato leaves (Warren, King & Colhoun, 100).

TEMPERATURE EFFECTS IN RELATION TO STAGE OF DEVELOPMENT OF THE FUNGUS   No disease due to *Tilletia caries* (DC.) Tul. occurs at 25° if the soil is inoculated when wheat seeds are sown, but disease does occur if the soil is inoculated 7 days before sowing (Purdy & Kendrick, 78). These results indicate that if spores are allowed to germinate at fairly high temperatures before seed is sown, the number of plants that escape infection is much reduced.

EFFECTS OF TEMPERATURE THROUGH REDUCTION OF PLANT VIGOR   Root diseases are often more severe when soil factors have adverse effects on plant growth. Patrick & Toussoun (73) have stressed that certain root rots may be due to organisms that are not usually regarded as significant pathogens but can become more aggressive under conditions that weaken the plant. They suggest that changes in the aggressiveness of organisms may be triggered by the formation of toxic products during the decomposition of plant residues in the soil. Results that fit in with this idea were obtained by Rawlinson & Colhoun (80) who found that when oat seeds not contaminated with any known pathogen of this crop were treated with an organo-mercury fungicide, seedling vigor was increased only when the seedlings were grown in natural or simulated winter conditions with periods of frost. Under these conditions, mesocotyls were protected by the fungicide from contamination or invasion by soil-borne fungi, whereas untreated seeds gave rise to less vigorous seedlings with discolored mesocotyls, bearing lesions from which fungi, usually regarded as saprophytes, could be isolated.

Recovery of wheat plants from winter attacks by *Typhula idahoensis* Remsberg is greatest from plants with several tillers, and least from small plants with few leaves (Bruehl & Cunfer, 9).

TEMPERATURE EFFECTS ON SPORE PRODUCTION AND DISCHARGE   The effects of temperature on disease are not confined to influencing the length of the incubation period and spore germination. Indeed at unsatisfactory temperatures no sporulation may occur even when the atmosphere is saturated, as has been shown for *Peronospora tabacina*. The viability of spores of this pathogen and their longevity are influenced by the temperature at which they are formed (Clayton & Gaines, 14).

Temperature can play a part in spore release although it is usually less important than moisture. Brook (8), for example, has shown that when perithecia of *Venturia inaequalis* (Cke.) Wint. have been in a warm or humid atmosphere before being wetted, more ascospores are released than when they have been in a cold or dry atmosphere before wetting.

EFFECTS OF TEMPERATURE ON MUTATION OF FUNGI   Mutation of fungi is influenced by a variety of factors including temperature, nutrients, chemicals, ultra-violet light, radiations, and some bacterial products. In *Ustilago maydis* (DC.) Corda [= *U. zeae* (Bechm.) Unger] the largest number of mutants appear at fairly high temperatures (Stakman, Christensen, Eide & Peturson, 90).

EFFECTS OF TEMPERATURE ON EFFECTIVENESS OF FUNGICIDES   The effects of soil environment on the efficacy of seed disinfectants have not received much attention. Examples of the effects of temperature, and other factors such as soil moisture, on the effectiveness of a range of chemicals applied to wheat seed have been provided by Purdy (77) for *Urocystis agropyri* (Preuss) Schröt, and by Holmes & Colhoun (46) for *Septoria nodorum* Berk.

Moisture

In experimental work it is more difficult to control moisture than temperature, and this, together with differences in terminology employed by different workers, particularly those dealing with soil moisture, has led to some confusion. It is not proposed here to review the different methods of expression of soil moisture, as these have already been discussed at length elsewhere (Griffin, 40; Couch, Purdy & Henderson, 26). As different problems are involved in the study of soil moisture and atmospheric humidity, these are now considered separately although, of course, one should not be studied in isolation from the other.

SOIL MOISTURE   Some diseases are most serious in wet soils and others in dry soils. Obviously it can be expected that diseases such as club root of crucifers, where a motile stage in the life cycle of the causal organism is involved, are favored by wet soils. Such an explanation cannot be offered when it has been found that wet soils are associated with severe attacks caused by the following fungi: *Helminthosporium solani* Dur. & Mont. on potatoes, *Phoma chrysanthemicola* Hollos on chrysanthemum, and *Cercosporella herpotrichoides* on wheat. Diseases that are most severe in dry soils include those caused by *Sclerotium cepivorum* Berk. on onions, *Streptomyces scabies* (Thaxt.) Waksm. & Henrici on potato, *Fusarium* diseases of cereals due to *F. culmorum*, *F. avenaceum* (Fr.) Sacc., *F. graminearum* and *F. nivale*, and *Urocystis agropyri* on wheat.

Pea foot rot due to *Fusarium solani* (Mart.) Appel & Wr. f. *pisi* (Jones) Snyder & Hansen is most severe in soils of intermediate moisture. This has been attributed to poor germination of chlamydospores in dry soil and poor germling survival following good spore germination in wet soil (Cook & Flentje, 24). Cook & Papendick (25) regard the frequent occurrence of wheat foot rot, due to *F. culmorum*, in dry soils as a result of the germ tubes being much less liable to lysis by bacteria in dry than in wet soils. A further factor contributing to the more frequent occurrence of this disease in dry soil, is that wheat seedlings take longer to emerge above soil level in dry soil, so that they are exposed for longer periods in their most susceptible stage (Malalasekera & Colhoun, 60).

Increasing the soil water available to plants before inoculation increases the number of local lesions produced per leaf, and per unit area of leaf, by some viruses in *Nicotiana* spp. (Tinsley, 94). Leaves from plants receiving abundant water showed morphological differences as compared with those from plants kept dry, so they were more likely to suffer injuries and be readily penetrated by virus particles.

ATMOSPHERIC HUMIDITY   Nearly all infections of above-ground parts of plants are affected by moisture. Spore germination usually requires a moisture film on the plant surface, but powdery mildews are an exception in that although their conidia germinate well at a high air humidity (Manners & Hossain, 62) they become nonviable if left for even very short periods in water. When a requirement for leaf wetness exists its duration may be vital. A film of water or water droplets must persist on potato leaves for at least 3–8 hr, depending on the temperature, if substantial infection by *Phytophthora infestans* is to occur (Crosier, 28; Lapwood, 54). On the other hand, *Venturia inaequalis* requires a much longer period of wetness, varying from 26 hr at 5° to 17 hr at 26.5° (Sys & Soenen, 92).

Not only is the sporulation of many pathogens associated with high atmospheric humidity, but the dissemination of spores of many species is achieved by splashing by raindrops, or they are carried in water droplets. When ascospores are forcibly discharged the perithecia often must be wetted before spores are released, but if the water film is too deep, the tips of the asci do not readily break its surface, and the ascospores are released in the water (Brook, 8). Light and temperature often interact with moisture in determining when ascospores are released.

During the development of apple fruits, high atmospheric humidity may modify lenticels, and so influence their resistance to invasion by *Penicillium expansum* Link em. Thom and *Botrytis cinerea* Fr. during fruit storage (Colhoun, 20).

A crop modifies its environment as it grows, and so temperature and humidity may be different within a crop to that prevailing above plant level. Indeed, the ecoclimate at any time depends on the recent and present weather above a crop (Hirst & Stedman, 44). This is of particular importance in relation to disease forecasting.

Light

The effects of light and the duration of periods of light and darkness on fungal sporulation in culture have been studied (Leach, 55–57), so that their importance is now appreciated. Light may also influence spore germination, penetration, and infection type, as well as the release and viability of spores.

Low light intensity reduces the level of club root attack in cabbages except where the soil is heavily contaminated by *Plasmodiophora brassicae* (Colhoun, 19). This may be due to an effect on root diffusates, which in turn may influence spore germination and infection, although attempts to test this hypothesis have not yet led to a definite conclusion (Woods & Colhoun, unpublished data). Day length may also influence disease, for under short, as compared with long days, stelar infection of wheat by *Ophiobolus graminis* (Sacc.) Sacc. is much more severe (Wilkinson, 103).

Uredospores of various rust fungi have quite different light requirements for germination. Those of *Uromyces appendiculatus* (= *U. phaseoli*) germinate equally well in light and darkness (Snow, 87), those of *Puccinia graminis* germinate quicker in darkness (Burrage, 10), but exposure to light intensities of 19 fc or higher for 12 hr prevents spore germination of *Hemileia vastatrix* Berk. & Br. although maximum germination can occur at 2.5 fc or less (Hocking, 45).

Light intensity influences penetration of the host by fungi. High light intensity (8000 lux) is optimum for stomatal penetration by *Septoria tritici* Rob. & Desm. in wheat at 23° (Benedict, 3), but penetration of barberry leaves through the cuticle by *Puccinia graminis* is favored by low light intensity, although after penetration has occurred intense light favors the severity of infection (Lambert, 53). Darkness depresses stomatal penetration of wheat leaves by *P. graminis*. It is suggested that the mechanism regulating penetration by this species may involve such factors as the percentage of CO_2 in the intercellular spaces, which is higher in darkness than in light (Yirgou, 106).

Light may determine plant reaction to infection. When day length exceeds 12 hr, barley plants appear to be extremely resistant to *P. striiformis*, whereas they are completely susceptible at shorter day lengths (Bever, 6) although reaction to light can be modified by temperature. With *Leptosphaerulina brissiana* (Poll) Graham & Luttrell, symptoms on alfalfa were of the resistant type at a light intensity of 450 fc or less and an 8 hr day, but plants became more susceptible as the intensity and day length increased (Pandey & Wilcoxson, 72).

Although light encourages sporulation of most pathogens on the plant, sometimes it has an inhibitory effect. Thus, maximum sporulation of *Bremia lactucae* Regel occurs after exposure to at least 6 hr darkness 7 days after inoculation, but continuous light at this time prevents sporulation (Raffray & Sequeira, 79). At 22.5°, provided an initial dark period is given, the increase in the number of uredospore pustules formed by *Puccinia graminis* is proportional to the duration of the subsequent light period (Burrage, 10). These experiments were made with artificial light.

Low light intensity, or short day length before inoculation, can predispose tomato plants to *Fusarium* wilt (Foster & Walker, 36). With viruses, pre-inoculation dark treatment of plants to increase susceptibility has become common practice, since Bawden & Roberts (1, 2) first showed that such treatment for 24 hr was effective. However, post-inoculation dark treatment had little effect, and it was suggested that the increased susceptibility resulting from pre-inoculation dark treatment was roughly proportional to the reduction in photosynthesis (Bawden & Roberts, 2). Further work by Helms & McIntyre (42, 43) showed that exposure of bean plants to light for 2–123 min after a dark period, prior to inoculation, increased susceptibility to tobacco mosaic virus. They suggested that a substance involved in the establishment of infection is inactivated during darkness and is synthesized or becomes available in the light. During pre-inoculation treatments the wavelengths of the light may be important (Coast & Chant, 15).

Soil Reaction

The occurrence of some diseases is greatly influenced by soil pH. An outstanding example is club root of crucifers, which is very severe in acid soils and can sometimes be controlled by liming. Failure to control the disease by liming has been explained by Colhoun (17) as the result of an interaction of factors, which is discussed later. The spores of the causal organism, *Plasmodiophora brassicae,* are either unable to germinate readily in alkaline soils or else the myxamoebae do not show a high rate of survival under such conditions, and so the effect is achieved by action on the pathogen.

Streptomyces scabies does not readily infect potatoes when the soil pH is below about 5.2; this has been attributed to inhibition of the growth of the fungus by high acidity. Some isolates, however, are able to grow in culture at a pH value as low as 4.8 (Waksman, 97), and this may explain the numerous references to infection occurring at pH 5.0 or below.

With reference to the lowest pH value at which infection by a soil-borne pathogen can occur, it must be borne in mind that soil in the immediate vicinity of a root may differ in pH from the remainder of the soil mass. Care must therefore be taken in associating failure to infect with the pH value determined for samples of the soil mass.

Some soil-borne pathogens may be inhibited by the activity of saprophytic organisms in soil. It is therefore obvious that when soil reaction influences disease it may not be attributable only to direct effects on the pathogen. Moreover, soil reaction may influence the availability of soil nutrients and so affect plant growth and vigor, which in turn, through effecting a change in the microclimate within a crop, indirectly affect infection and disease development.

Soil Nutrients

The major elements in plant nutrition, nitrogen, phosphorus, and potassium, may each sometimes increase susceptibility or sometimes decrease it. Indeed, increased nitrogen may increase susceptibility of the flesh of one variety of apple to invasion by *Cytosporina ludibunda* Sacc. but decrease susceptibility in another variety (Horne, 47). This difference in reaction to nitrogen can be explained as the effect of different acidity levels in the apples (Colhoun, 16). It is at present difficult to detect any clear pattern in the effect of nutrients on disease, and it is really necessary to study each combination of host and pathogen, giving consideration to interactions with other factors. The ratio of one element to another may also be important.

The effects of plant nutrients on disease may be attributed to (a) effects on plant vigor that can influence the microclimate in a crop and so affect infection and sporulation of the pathogen, (b) effects on cell walls and tissues as well as on the biochemical make-up of the host, (c) the rate of growth of the host, which may enable seedlings to escape infection in their most susceptible stage, and (d) effects on the pathogen through alterations in the soil environment. Alterations in plant nutrients may induce susceptibility of plant roots to a pathogen, but greatly increase the ability of a plant to produce new roots. Such an instance occurs with root rot of chrysanthemums caused by *Phoma chrysanthemicola* (Menzies & Colhoun, unpublished data). Commercial control of this disease can be achieved with high levels of nitrogen and phosphorus (Peerally & Colhoun, 74), which enable the plant to regenerate roots rapidly, so that root efficiency is not seriously impaired, and good growth of the above ground plant parts occurs even in heavily infested soil.

Minor elements may influence disease reaction although their effects have not received the same attention as those of the major elements. Reference is made here only to the effects of nickel and cobalt in increasing resistance of potatoes to *Phytophthora infestans, Alternaria solani* (Ell. & Mart.) Sor. and viruses X, Y, and S (Isaeva, 48).

Interaction of Factors

The study of the effects of individual factors can be regarded as the first step in understanding environment in relation to crop diseases. In the field, a physical factor rarely remains constant for any substantial length of time, so the plant pathologist faces a situation where a number of factors are undergoing change simultaneously. If an analysis of environmental factors is to be meaningful in relation to field conditions, it is necessary to design experiments that involve various levels of each of a number of factors. When work of this nature for a disease is undertaken, the aim should be to produce predetermined levels of the disease under controlled conditions. When, for example, it is possible to produce no diseased plants in the presence of the pathogen, or only diseased plants, or any intermediate level of attack, considerable progress will have been made in effecting an analysis of the effects of environment. The information available should then explain why high, or low, or intermediate levels of disease occur in a crop.

An investigation of the epidemiology of club root of Brassicae demonstrates how the study of the interaction of environmental factors can be undertaken. Before this study was commenced it was accepted that in acid soils favorable conditions for infection by *Plasmodiophora brassicae* and development of the disease were a soil moisture content of 70–80% of the maximum water holding capacity (Monteith, 69; Naoumova, 71) and temperature within the range 18–25° (Monteith, 69; Wellman, 101). Disagreements, however, occurred over the effects of the number of spores in soil on infection, and Macfarlane (59) reported that the relationship between severely clubbed plants and spore load is less marked at high spore loads when an early supply of nutrients at a high level is available. In different soils the same spore load sometimes gave rise to very different proportions of infected plants. Confusion also existed with regard to the efficacy of liming as a control measure and to the maximum pH level at which infection occurred. Many workers were unable to produce diseased plants in alkaline soils under controlled conditions, but as a result of pot experiments, Colhoun (17–19) showed how predetermined levels of attack could be produced in either acid or alkaline soils.

Analysis of the effect of environmental factors on club root disease showed (Colhoun, 17) that in an acid soil the disease occurs over a much wider range of conditions than in an alkaline soil. In an alkaline soil it is necessary to have a high temperature (about 23°), a high soil moisture (70% of the maximum water holding capacity (M.W.H.C.) is sufficient), and a high spore load (10^5–10^7 spores per g of soil is usually adequate), if a severe attack is to occur. In an acid soil, severe infection can result at 16° at a spore load of 10^3 spores per g of soil when the soil moisture is at 70% M.W.H.C. Quite severe attacks can occur in dry acid soils at a high spore load and a temperature of 23°, so that the high inoculum level can compensate for unfavorable conditions of soil moisture. In alkaline soils, when fluctuations in temperature are permitted, with other conditions being suitable for infection, the longer the initial period at a mean temperature of about 23°, the more severe the attack, but short periods of only a few days may permit some disease to occur. A day/night fluctuation of 26°–19° ensures that quite a severe attack occurs (Colhoun, 18).

Not only do interactions between soil pH, soil moisture, temperature, and spore load affect the incidence of club root, but even in acid soils light intensity has an important effect (Colhoun, 19). A direct relationship between spore load and intensity of attack exists at low light intensities up to a high level of spore load, whereas at a high light intensity this relationship exists only at fairly low spore loads. At both high and low spore loads in wet, acid soils, with high light intensity, the disease index (combining the number of infected plants and disease severity) is lower when nitrogen is abundant than when it is in low supply. This nutrient effect is usually more marked at the higher dosages of nitrogen. The supply of nitrogen, however, does not influence the number of diseased plants. Phosphates and potassium do not influence either the number of diseased plants or their disease index.

The ability of a high inoculum level in soil to compensate for unfavorable soil moisture, as demonstrated in club root disease, is not unique. This occurs in infection of wheat seedlings by *Fusarium culmorum* (Colhoun, Taylor & Tomlinson, 22) and of chrysanthemum plants by *Phoma chrysanthemicola* (Peerally & Colhoun, 74). This compensatory effect may be quite widespread and will come to light when appropriate studies with other diseases are made. There is also no reason for imagining that it is confined to an interaction between inoculum level and soil moisture.

The importance of the interaction of factors on disease has, of course, been appreciated by a number of workers, but more intensive work would be profitable in respect to many diseases. In studies with leaf stripe and seedling blight of oats caused by *Pyrenophora avenae* Ito & Kuribay, Muskett (70) produced an elegant example of the interaction between soil moisture and temperature. He considers that low temperatures are most favorable for the occurrence of the seedling phase of this disease but in dry soils an increase in temperature does not lead to the decrease in disease incidence that results when soil is moist or wet. Pre-emergence death of seedlings is most marked in cold, wet soils but it can also be severe at high temperatures in dry soils.

In a study of infection of potatoes by *Phytophthora infestans,* Rotem, Cohen & Putter (82) concluded that the minimum, maximum, and optimum levels of temperature, inoculum concentration, and leaf wetness duration each depended on the other two factors, sometimes primarily on one and sometimes on a balance of both. For each factor considered, both the optimum and limiting range of levels were broad when the combinations of the other two factors were favorable, but were narrower for unfavorable combinations. These workers rightly concluded that statements concerning the effects of any one factor on infection are relative and valid only under certain combinations of other factors, which must be specified. It therefore follows that much of the published information concerning the effects of individual factors can be applied to field conditions only with extreme caution.

When the occurrence of disease epidemics is being forecast, it is, of course, necessary to consider the effects of interactions of environmental factors. A considerable amount of attention has been directed to this problem (Miller & O'Brien, 68; Sys & Soenen, 92). It is clear that when environmental factors are cycled and varied, both individually and collectively, we should be able to identify the conditions leading to disease epidemics, whereas this is not possible with continual sampling of natural or random outdoor conditions (Miller, 67). This provides encouragement for studies of environmental factors under controlled conditions.

Literature Cited

1. Bawden, F. C., Roberts, F. M. 1947. The influence of light intensity on the susceptibility of plants to certain viruses. *Ann. Appl. Biol.* 34:286–96
2. Bawden, F. C., Roberts, F. M. 1948. Photosynthesis and predisposition of plants to infection with certain viruses. *Ann. Appl. Biol.* 35:418–28
3. Benedict, W. G. 1971. Differential effect of light intensity on the infection of wheat by *Septoria tritici* Desm. under controlled environmental conditions. *Physiol. Plant Pathol.* 1:55–66
4. Berkeley, M. J. 1846. Observations, botanical and physiological, on the potato murrain. *J. Hort. Soc.* (London) 1:9–34 (Reprinted in *Phytopathol. Classics* No. 8, 1948, 108 pp.)
5. Berkeley, M. J. 1847. *Gard. Chron.* 1847:779
6. Bever, W. M. 1934. Effect of light on the development of the uredial stage of *Puccinia glumarum*. *Phytopathology* 24:507–16
7. Boyd, A. E. W. 1952. Dry-rot disease of the potato. VII. The effect of storage temperature upon subsequent susceptibility of tubers. *Ann. Appl. Biol.* 39:351–57
8. Brook, P. J. 1969. Effects of light, temperature and moisture on release of ascospores by *Venturia inaequalis* (Cke.) Wint. *N. Z. J. Agr. Res.* 12:214–27
9. Bruehl, G. W., Cunfer, B. 1971. Physiologic and environmental factors that affect the severity of snow mould of wheat. *Phytopathology* 61:792–99
10. Burrage, S. W. 1970. Environmental factors influencing the infection of wheat by *Puccinia graminis*. *Ann. Appl. Biol.* 66:429–40
11. Campbell, C. E., Dimock, A. W. 1955. Temperature and the geographical distribution of chrysanthemum rust. *Phytopathology* 45:644–48
12. Christensen, J. J. 1926. The relation of soil temperature and soil moisture to the development of head smut of sorghum. *Phytopathology* 16:353–57
13. Clayton, E. E. 1923. The relation of temperature to the *Fusarium* wilt of the tomato. *Am. J. Bot.* 10:71–88
14. Clayton, E. E., Gaines, J. G. 1945. Temperature in relation to development and control of blue mould (*Peronospora tabacina*) of tobacco. *J. Agr. Res.* 71:171–82
15. Coast, E. M., Chant, S. R. 1970. The effect of light wavelength on the susceptibility of plants to virus infection. *Ann. Appl. Biol.* 65:403–09
16. Colhoun, J. 1948. Nitrogen content in relation to fungal growth in apples. *Ann. Appl. Biol.* 35:638–47
17. Colhoun, J. 1953. A study of the epidemiology of club-root disease of Brassicae. *Ann. Appl. Biol.* 40:262–83
18. Colhoun, J. 1953. Observations on the incidence of club-root disease of Brassicae in limed soils in relation to temperature. *Ann. Appl. Biol.* 40:639–44
19. Colhoun, J. 1961. Spore load, light intensity and plant nutrition as factors influencing the incidence of club-root of Brassicae. *Trans. Brit. Mycol. Soc.* 44:593–600

20. Colhoun, J. 1962. Some factors influencing the resistance of apple fruits to fungal invasion. *Trans. Brit. Mycol. Soc.* 45:429–30

21. Colhoun, J. 1970. Epidemiology of seed-borne *Fusarium* diseases of cereals. *Ann. Acad. Sci. Fenn. Ser. A*, 4, 168:31–36

22. Colhoun, J., Taylor, G. S., Tomlinson, R. 1968. *Fusarium* diseases of cereals. II. Infection of seedlings by *F. culmorum* and *F. avenaceum* in relation to environmental factors. *Trans. Brit. Mycol. Soc.* 51:397–404

23. Conant, G. H. 1927. Histological studies of resistance in tobacco to *Thielavia basicola*. *Am. J. Bot.* 14:457–80

24. Cook, R. J., Flentje, N. T. 1967. Chlamydospore germination and germling survival of *Fusarium solani* f. *pisi* in soil as affected by soil water and pea seed exudation. *Phytopathology* 57:178–82

25. Cook, R. J., Papendick, R. I. 1970. Soil water potential as a factor in the ecology of *Fusarium roseum* f. sp. *cerealis* "*Culmorum.*" *Plant Soil* 32:131–45

26. Couch, H. B., Purdy, L. H., Henderson, D. W. 1967. Application of soil moisture principles to the study of plant disease. *Bull. Virg. Polytech. Inst., Dep. Plant Pathol.* 4, 23 pp.

27. Covey, R. P. 1969. Effect of extreme cold on overwintering of *Podosphaera leucotricha*. *Plant Dis. Reptr.* 53:710–11

28. Crosier, W. 1934. Studies in the biology of *Phytophthora infestans* (Mont.) de Bary. *Cornell Univ. Agr. Exp. Sta. Mem.* 155

29. Daines, R. H. 1970. Effects of temperature and a 2,6-dichloro-4-nitroaniline dip on keeping qualities of 'Yellow Jersey' sweetpotatoes during the post-storage period. *Plant Dis. Reptr.* 54:486–88

30. Daines, R. H. 1970. Effects of fungicide dip treatments and dip temperatures on postharvest decay of peaches. *Plant Dis. Reptr.* 54:764–67

31. de Bary, A. 1861. *Die gegenwärtig herrschende Kartoffelkrankheit, ihre Ursache und ihre Verhütung.* A. Felix, Leipzig, 75 pp.

32. Deep, I. W., Hussin, H. 1965. Influence of temperature on initiation of crown gall in woody hosts. *Plant Dis. Reptr.* 49:734–35

33. Defosse, L., 1967. Étude, en conditions expérimentales, des facteurs qui régissent l'inoculation et l'infection du Fromet per *Cercosporella herpotrichoides* Fron. *Bull. Rech. Agron. Gembloux* N. S. 2: 38–51

34. de Tournefort, J. P. 1705. Observations sur les maladies des plantes. *Mem. Acad. Roy. Sci. Paris,* 1705:332–45

35. Dickson, J. G., Eckerson, S. H., Link, K. P. 1923. The nature of resistance to seedling blight of cereals. *Proc. Nat. Acad. Sci. USA,* 9:434–39

36. Foster, R. E., Walker, J. C. 1947. Predisposition of tomato to *Fusarium* wilt. *J. Agr. Res.* 74:165–85

37. Gäumann, E. 1950. *Principles of Plant Infection.* Transl. W. B. Brierley. London: Crosby Lockwood & Son Ltd. 543 pp.

38. Gill, C. C., Westdal, P. H. 1966. Effect of temperature on symptom expression of barley infected with aster yellows or barley yellow dwarf viruses. *Phytopathology* 56: 369–70

39. Gregg, M. 1952. Studies in the physiology of parasitism. XVII. Enzyme secretion by strains of *Bacterium carotovorum* and other pathogens in relation to parasitic vigour. *Ann. Bot. (London)* N. S. 16:235–50

40. Griffin, D. M. 1963. Soil moisture and the ecology of fungi. *Biol. Rev. Cambridge Phil. Soc.* 38: 141–66

41. Hartig, R. 1882. *Lehrbuch der Baumkrankheiten.* J. Springer, Berlin, 198 pp. (English Transl. 2nd ed. W. Somerville & H. M. Ward, 1894, *Country Life,* London, 331 pp.)

42. Helms, K., McIntyre, G. A. 1967. Light-induced susceptibility of *Phaseolus vulgaris* L. to tobacco mosaic virus infection. I. Effects of light intensity, temperature, and the length of the preinoculation dark period. *Virology* 31:191–96

43. Helms, K., McIntyre, G. A. 1967. Light-induced susceptibility of *Phaseolus vulgaris* L. to tobacco mosaic virus infection. II. Daily variation in susceptibility. *Virology* 32:482–88

44. Hirst, J. M., Stedman, O. J. 1960. The epidemiology of *Phytophthora infestans*. I. Climate, ecoclimate and the phenology of disease outbreak. *Ann. Appl. Biol.* 48:471–88

45. Hocking, D. 1968. Effects of light on germination and infection of coffee rust (*Hemileia vastatrix*). *Trans. Brit. Mycol. Soc.* 51:89–93

46. Holmes, S. J. I., Colhoun, J. 1973. A method for assessing the efficiency of seed disinfectants for the control of seed-borne *Septoria nodorum*. *Ann. Appl. Biol.* In press

47. Horne, A. S. 1939. The resistance of the apple to fungal invasion. *Rep. Food Invest. Bd. (London)* 1938. p. 173. London. H.M.S.O.

48. Isaeva, G. Y. 1969. Influence of nickel and cobalt on the resistance of potato to diseases. *Nauk. Pra. Zhytomyr. Sil's'kohosp. Inst.* 16: 110–12 (Abstr. *Rev. Plant Pathol.* 49:507)

49. Jalaluddin, M. 1967. Studies on *Rhizina undulata*. I. Mycelial growth and ascospore germination. *Trans. Brit. Mycol. Soc.* 50: 449–59

50. Jalaluddin, M. 1967. Studies on *Rhizina undulata*. II. Observations and experiments in East Anglian plantations. *Trans. Brit. Mycol. Soc.* 50:461–72

51. Jones, L. R. 1924. The relation of environment to disease in plants. *Am. J. Bot.* 11:601–09

52. Kassanis, B. 1952. Some effects of high temperature on the susceptibility of plants to infection with virus. *Ann. Appl. Biol.* 39:358–69

53. Lambert, E. B. 1929. The relation of weather to the development of stem rust in the Mississippi valley. *Phytopathology* 19:1–71

54. Lapwood, D. H. 1968. Observations on the infection of potato leaves by *Phytophthora infestans*. *Trans. Brit. Mycol. Soc.* 51:233–40

55. Leach, C. M. 1962. Sporulation of diverse species of fungi under near-ultraviolet radiation. *Can. J. Bot.* 40:151–61

56. Leach, C. M. 1967. Interaction of near ultraviolet light and temperature on sporulation of the fungi *Alternaria, Cercosporella, Fusarium, Helminthosporium* and *Stemphylium. Can. J. Bot.* 45:

1999–2016

57. Leach, C. M. 1971. A practical guide to the effects of visible and ultraviolet light on fungi. In *Methods in Microbiology*, Vol. 4, ed. C. Booth, 609–64. London & New York: Academic

58. Lloyd, E. H. 1969. The influence of low temperature on the behaviour of wheat leaves infected with *Puccinia striiformis* West. *Diss. Abstr.* 29B:3156–57

59. Macfarlane, I. 1952. Factors affecting the survival of *Plasmodiophora brassicae* Wor. in soil and its assessment by a host test. *Ann. Appl. Biol.* 39:239–56

60. Malalasekera, R. A. P., Colhoun, J. 1968. *Fusarium* diseases of cereals. III. Water relations and infection of wheat seedlings by *Fusarium culmorum*. *Trans. Brit. Mycol. Soc.* 51:711–20

61. Malcolmson, J. F., Gray, E. G. 1968. The incidence of gangrene of potatoes caused by *Phoma exigua* in relation to handling and storage. *Ann. Appl. Biol.* 62.89–101

62. Manners, J. G., Hossain, S. M. M. 1963. Effects of temperature and humidity on conidial germination in *Erysiphe graminis. Trans. Brit. Mycol. Soc.* 46:225–34

63. Meyen, J. F. 1841. *Pflanzen-Pathologie. Lehre von dem Kranken Leben und Bilden der Pflanzen.* Hande und Spenersche Buchhandlung, Berlin. 399 pp.

64. Millar, C. S., Colhoun, J. 1969. *Fusarium* diseases of cereals. IV. Observations on *Fusarium nivale* on wheat. *Trans. Brit. Mycol. Soc.* 52:57–66

65. Millar, C. S., Colhoun, J. 1969. *Fusarium* diseases of cereals. VI. Epidemiology of *Fusarium nivale* on wheat. *Trans. Brit. Mycol. Soc.* 52:195–204

66. Miller, P. R. 1967. Plant disease epidemics: their analysis and forecasting. *F. A. Organ. N. Symp. Crop Losses*, 9–38

67. Miller, P. R. 1969. Effect of environment on plant diseases. *Phytoprotection* 50:81–94

68. Miller, P. R., O'Brien, M. 1957. Prediction of plant disease epidemics. *Ann. Rev. Microbiol.* 11: 77–110

69. Monteith, J. 1924. Relation of soil

temperature and soil moisture to infection by Plasmodiophora brassicae. *J. Agr. Res.* 28:549–61

70. Muskett, A. E. 1937. A study of the epidemiology and control of *Helminthosporium* disease of oats. *Ann. Bot. (London)* N. S. 1:763–83

71. Naoumova, N. A. 1933. Contribution to the knowledge of the influence of soil factors on the development of club root in the Cruciferae. *Bull. Plant Prot.* (*Leningrad*) *Ser. II, Phytopath.* 3:32–52

72. Pandey, M. C., Wilcoxson, R. D. 1970. The effect of light and physiologic races on *Leptosphaerulina* leaf spot of alfalfa and selection for resistance. *Phytopathology* 60:1456–62

73. Patrick, Z. A., Toussoun, T. A. 1965. Plant residues and organic amendments in relation to biological control. In *Ecology of Soilborne Plant Pathogens*, ed. K. F. Baker, W. C. Snyder, 440–59. Berkeley: Univ. of Calif. 571 pp.

74. Peerally, M. A., Colhoun, J. 1969. The epidemiology of root rot of chrysanthemums caused by *Phoma* sp. *Trans. Brit. Mycol. Soc.* 52:115–23

75. Petrlik, Z., Stys, Z. 1966. Der Einfluss der Temperatur auf das Ausschwärmen der Zoosporen, Infektion und Inkubationszeit der Hopfenperonospora (*Peronoplasmopura humuli* Miy. et Tak.). *Ceská Mykol.* 20:105–10. (Abstr. *Rev. Appl. Mycol.* 45:451)

76. Prevost, B. 1807. Mémoire sur la cause immédiate de la carie ou charbon des blés, et de plusieurs autres maladies des plantes, et sur les préservatifs de la carie. Paris, Bernard, 80 pp. (English Transl. G. W. Keitt in *Phytopathol. Classics* No. 6, 1939)

77. Purdy, L. H. 1966. Soil moisture and soil temperature, their influence on infection by the wheat flag smut fungus and control of the disease by three seed-treatment fungicides. *Phytopathology* 56:98–101

78. Purdy, L. H., Kendrick, E. L. 1963. Influence of environmental factors on the development of wheat bunt in the Pacific Northwest. IV. Effect of soil temperature and soil moisture on infection by soil-borne spores. *Phytopathology* 53:416–18

79. Raffray, J. B., Sequeira, L. 1971. Dark induction of sporulation in *Bremia lactucae*. *Can. J. Bot.* 49:237–39

80. Rawlinson, C. J., Colhoun, J. 1970. Chemical treatment of cereal seed in relation to plant vigour and control of soil fungi. *Ann. Appl. Biol.* 65:459–72

81. Ré, F. 1807. *Saggio teorica practico sulle mallatie delle piante*. Venice. 2nd ed. Milan, 1817. (English Transl. of 2nd ed. M. J. Berkeley in *Gard. Chron.* 1849–50)

82. Rotem, J., Cohen, Y., Putter, J. 1971. Relativity of limiting and optimum inoculum loads, wetting durations and temperatures for infection by *Phytophthora infestans*. *Phytopathology* 61:275–78

83. Rudd-Jones, D., Dowson, W. J. 1950. On the bacteria responsible for soft rot in stored potatoes and the reaction of the tuber to invasion by *Bacterium carotovorum* (Jones) Lehmann & Naumann. *Ann. Appl. Biol.* 37:563–69

84. Samuel, G. 1931. Some experiments on inoculating methods with plant viruses and on local lesions. *Ann. Appl. Biol.* 18:494–507

85. Schroeder, W. T., Walker, J. C. 1942. Influence of controlled environment and nutrition on the resistance of garden pea to *Fusarium* wilt. *J. Agr. Res.* 65:221–48

86. Schulz, F. A., Bateman, D. F. 1969. Temperature response of seeds during the early stages of germination and its relation to injury by *Rhizoctonia solani*. *Phytopathology* 59:352–55

87. Snow, J. A. 1965. Effects of light on initiation and development of bean rust disease. *Diss. Abstr.* 25:6149

88. Sorauer, P. 1874. *Handbuch der Pflanzenkrankheiten*. Wiegandt, Berlin: Hempel und Parey. 406 pp.

89. Speerschneider, J. 1857. Die Ursache der Erkrankung der Kartoffelknolle durch eine Reihe Experimente bewiesen. *Z. Bot.* 15:122–24

90. Stakman, E. C., Christensen, J. J., Eide, C. J., Peturson, B. 1929. Mu-

tation and hybridization in *Ustilago zeae. Minn. Agr. Exp. Sta. Tech. Bull.* 65:1–108

91. Stakman, E. C., Harrar, J. G. 1957. *Principles of Plant Pathology.* New York: Ronald Press Co. 581 pp.

92. Sys, S., Soenen, A. 1970. Investigations on the infection criteria of scab (*Venturia inaequalis* (Cke.) Wint.) on apples with respect to the table of Mills & Laplante. *Agricultura* (*Louvain*) 18:3–8. (Abstr. *Rev. Plant Pathol.* 50:702

93. Tillet, M. 1755. Dissertation sur la cause qui corrompt et noircit les grains de bled dans les épis; et sur les moyens de préventir ces accidens. Bordeaux: Brun. (English Transl. H. B. Humphrey in *Phytopathol. Classics* No. 5, 1937. 191 pp.)

94. Tinsley, T. W. 1953. The effects of varying the water supply of plants on their susceptibility to infection with viruses. *Ann. Appl. Biol.* 40:750–60

95. Unger, F. 1833. *Die Exantheme der Pflanzen.* Vienna: C. Gerold, 422 pp.

96. Vasudeva, R. S. 1930. Studies in the physiology of parasitism. XI. An analysis of the factors underlying specialization of parasitism, with special reference to the fungi *Botrytis allii*, Munn, and *Monilia fructigena*, Pers. *Ann. Bot.* (*London*) 44:469–93

97. Waksman, S. A. 1922. The influence of soil reaction upon the growth of Actinomycetes causing potato scab. *Soil Sci.* 14:61–79

98. Walker, J. C., Wellman, F. L. 1926. Relation of temperature to spore germination and growth of *Urocystis cepulae. J. Agr. Res.* 32: 133–46

99. Ward, H. M. 1890. On some relations between host and parasite in certain epidemic diseases of plants. *Proc. Roy. Soc. London,* 47:393–443

100. Warren, R. C., King, J. E., Colhoun, J. 1971. Reaction of potato leaves to infection by *Phytophthora infestans* in relation to position on the plant. *Trans. Brit. Mycol. Soc.* 57:501–14

101. Wellman, F. L. 1930. Club root of crucifers. *U. S. Dep. Agr. Tech. Bull.* 181

102. Wells, H. D., Forbes, I. 1967. Effects of temperature on growth of *Glomerella cingulata* in vitro and on its pathogenicity to *Lupinus augustifolius* genotypes *an an* and *An An. Phytopathology* 57: 1309–11

103. Wilkinson, V. 1970. Light and infection of wheat by *Ophiobolus graminis. Trans. Brit. Mycol. Soc.* 54:331–32

104. Yarwood, C. E. 1956. Heat-induced susceptibility of beans to some viruses and fungi. *Phytopathology* 46:523–25

105. Yarwood, C. E. 1959. Predisposition. In *Plant Pathology,* ed. J. G. Horsfall, A. E. Dimond, Vol. 1, 521–62. New York, London: Academic. 674 pp.

106. Yirgou, D. 1965. The role of light in stomatal penetration of wheat seedlings by *Puccinia graminis* f. sp. *tritici* and *P. recondita* f. sp. *tritici. Diss. Abstr.* 26:614–15

INTERACTIONS BETWEEN AIR POLLUTANTS AND PLANT PARASITES

❖ 3578

Allen S. Heagle

Agricultural Research Service, U. S. Department of Agriculture,
Raleigh, North Carolina

Plants often suffer from more than one disease at a time. More than one species of parasite can interact to cause disease. Consequently, pathologists are beginning to study the etiology of multiple-pathogen diseases (61). There is another type of multiple-pathogen interaction that pathologists must consider: the interaction between parasites (biotic pathogens) and air pollutants (abiotic pathogens).

Study of plant diseases caused by air pollution began in the 19th century. The literature now contains numerous reports of pollutant effects ranging from alterations in plant physiology and biochemistry to visible symptoms of chlorosis, necrosis, early senescence, and stunting. The known effects of specific air pollutants or air pollution in general have been reviewed (2, 5, 14, 27, 65, 87, 90).

As man continues to add pollutants to the air they continue to be widely dispersed. As a result, plants, plant parasites and plant parasitism are all being affected. At present, very little is known of the effects of pollutants on parasitic diseases of plants. Instead of solid concepts we have only isolated observations. Air pollutants may affect parasitism in different ways. Parasitism may be increased or decreased through a direct effect of the pollutant on the parasite. Or, the effects may be indirect through pollutant-induced changes in the host plant or through changes in other aspects of the environment.

Little is known; much remains to be discovered. Therefore, this paper is intended more as a stimulus for further research than merely as a chronicle of past discoveries.

SULFUR DIOXIDE

Sources and Effects on Vegetation

Sulfur dioxide is released into the atmosphere during fuel combustion, petroleum refinement, ore smelting, and through the natural sulfur cycle. Sulfur dioxide concentrations (conc) at ground level depend upon the amount and conc of emissions, distance from the source, and meteorological and topographical conditions. In general, conc of SO_2 decrease rapidly with distance from the source and with increased air movement. Sulfur dioxide conc near point sources, such as coal-burning power plants and smelters, with little or no pollution control equipment, may be as high as 1–3 ppm (parts per million of air by volume). In large urban areas SO_2 conc ranging from 0.05–0.40 ppm can be expected about 10% of the time. In most smaller cities and rural areas without major SO_2 sources, the conc normally would be less than 0.05 ppm. Sulfur dioxide is very soluble in water, and rain effectively removes it from the air. As sulfur is an essential plant nutrient, some SO_2 in the atmosphere can be beneficial to plants in areas where sulfur deficiency occurs in the soil.

With all pollutants, the threshold conc required to injure plants is affected by the pollutant dose (duration of exposure × conc of pollutant) and a multitude of interacting biological and meteorological variables. Plant species, even cultivars within species, vary widely in sensitivity to a given pollutant. In general, plants are most sensitive during the daylight hours, under humid conditions with moderate temperature and adequate soil moisture.

For sensitive species, exposure for 8 hr to 0.10–0.50 ppm of SO_2 may cause visible symptoms of injury. For most plants of intermediate sensitivity, exposure to 0.20–2.50 ppm for 8 hr can cause significant injury.

Effects of SO_2 on Parasitism by Fungi

There are several interesting accounts of the effects of industrial emissions on the incidence of fungus diseases in the field (Table 1). Most have resulted from observations where mixtures of pollutants occur but the principal pollutant in most cases was SO_2. Several of the reports resulted from casual observations but others were carefully planned. With field studies, the possibility always exists that something other than SO_2 could have caused some of the reported effects. However, there are many reports that generally agree in supporting the thesis that SO_2 emissions can affect fungus diseases in the field (Table 1).

Sulfur dioxide may act directly upon the fungi or indirectly through some effect upon living or dead plant tissues, soil, or water. In addition to causing visible injury to living plants, SO_2 exposure may result in the accumulation of sulfur, increased acidity, and other alterations in plant physiology and biochemistry. The oxidation and dissolution of SO_2 in water has significantly increased the acidity of precipitation, thereby increasing the acidity of soils, rivers, and lakes in several areas of the world (83). Any one or combination of these alterations could cause the effects on fungus diseases observed in the field.

Fungus diseases differ in reaction to SO_2; rust diseases and wood-destroying fungi are uncommonly sensitive. Wheat stem rust caused by *Puccinia graminis* was reported to be less in Sweden's industralized Kvarntorp area than in nonindustrialized areas (34). Scheffer & Hedgcock (69) found decreased parasitism by species of *Cronartium, Coleosporium, Melampsora, Peridermium, Pucciniastrum,* and *Puccinia* where trees were injured by SO_2. Most other needle- and twig-inhabiting fungi were not affected. With increasing distance from the SO_2 source, the amount of SO_2 injury on trees decreased and the rusts increased. Linzon (43) found that smelter emissions decreased the incidence of *Cronartium ribicola.* Less than 1% of the white pine showed rust symptoms at two sites 19 and 25 miles from a smelter, but 4, 3, and 5% showed symptoms 43, 58, and 71 miles away, respectively.

Jancarik (33) observed the occurrence of sporocarps of various fungi (mostly basidiomycetes) in a severely polluted area and in a relatively clean area of the industrialized Ore mountains of Czechoslovakia. Of the 40 species identified, 12 occurred where trees were slightly injured but were not seen where trees were severely damaged. The reverse was true of 6 species. Fungi seen only in the "clean" area included *Poria* sp., three *Mycena* spp., *Schizophyllum commune* and *Polyporus* (=*Trametes* = *Polystictus* = *Coriolus*) *versicolor.* Less heart-rot occurred in white pines close to a smelter where injury from SO_2 was greatest (43) but another study revealed no effect of smelter emissions on the incidence of heart-rot fungi in lodgepole pine and Douglas-fir (69).

Industrial emissions containing SO_2 also decreased the incidence of foliar diseases caused by various Ascomycetes including *Hypodermella laricis, Lophodermium pinastri, Hypodermella* sp. (69), *Lophodermium juniperi, Rhytisma acerinum* (1), *Hysterium pulicare* (78), and *Venturia inaequalis* (63). In 1935, Köck (36) reported the absence of oak powdery mildew caused by *Microsphaera alni* near a paper mill in Austria. The rose blackspot disease caused by *Diplocarpon rosae* was rarely present in areas where the daily average SO_2 conc was greater than 0.04 ppm during the growing season but was frequently observed in areas with less SO_2 (68). Hibben & Walker (30) observed that lilacs grown in the polluted air of New York City and other urban areas often showed substantially less infection by the powdery mildew fungus, *Microsphaera alni,* than lilacs in rural areas.

The only published evidence that a fungus disease can be inhibited significantly by known conc of SO_2, in doses likely to be found in ambient air, was provided by Saunders (68). He exposed roses inoculated with *Diplocarpon rosae* to low doses of SO_2 for 2 days after inoculation and found that 0.01 ppm SO_2 slightly increased the amount of leaflet area diseased, but that more than 0.04 ppm markedly decreased disease. Whether the effect was due to a decreased number of infections or to decreased invasion was not discussed. Sulfur dioxide (0.30–0.50 ppm–72 hr) decreased germination of *Microsphaera alni* conidia and disease development beyond the appressorium stage but 1 ppm–6 hr did not cause a similar response (C. R. Hibben, personal communication).

The effect of pollutants in decreasing populations of various parasites may at first appear to be a positive effect. Microorganisms however, are an extremely important component in all ecological systems. Thus, the more important consideration is that air pollution can affect all forms of plant life,

Table 1 Effects of sulfur dioxide, ozone, and fluoride on plant diseases[a]

Pollutant and disease affected	Effects	Pollutant Dose	Location	References
Sulfur Dioxide				
Wheat stem rust	Decreased incidence	Ambient	Sweden	34
Tree rusts	Decreased incidence	Ambient	Canada	43, 69
Wood rots	Decreased incidence	Ambient	Czecho-slovakia	33
Needle cast on Juniper	Decreased incidence	Ambient	England	1
Needle cast on Larch and Pine	Decreased incidence	Ambient	Canada	69
Rhytisma on Maple	Decreased incidence	Ambient	England	1
Hysterium on Alder and Birch	Decreased incidence	Ambient	Sweden	78
Apple scab	Decreased incidence	Ambient	Poland	63
Oak powdery mildew	Decreased incidence	Ambient	Austria	36
Lilac powdery mildew	Decreased incidence	Ambient	USA	30
Rose black spot	Decreased incidence	Ambient	England	68
Rose black spot	Smaller lesions	0.04 ppm–48 hr	England	68
Dwarf mistletoe on Larch and Pine	Decreased incidence	Ambient	Canada	69
Armillaria in trees	Increased incidence	Ambient	Canada, Czecho-slovakia, Germany, Poland	69, 33 17, 39
Wood rots	Increased incidence	Ambient	Czecho-slovakia	33
Needle cast on Spruce	Increased incidence	Ambient	Poland	39
Ozone				
Botrytis on gladiolus and chrysanthemum petals	Fewer infections by conidia	Not measured	USA	46, 47
Botrytis on Geranium petals	Decreased pathogenesis	0.35 ppm–4 hr	USA	54

Table 1 (*continued*)

Pollutant and disease affected	Effects	Pollutant Dose	Location	References
Ozone				
Oat crown rust	Smaller pustules	0.10 ppm–6 hr/day–10 days	USA	23
Wheat stem rust	Fewer infections by urediospores	0.06 ppm–6 hr	USA	24
Wheat stem rust	Decreased hyphal growth	0.06 ppm–6 hr/day–3 days	USA	24
Wheat stem rust	Decreased urediospore production	0.06 ppm–6 hr/day–17 days	USA	24
Barley powdery mildew	Fewer infections by conidia	0.10 ppm–24 hrs after inoculation	USA	26
Barley powdery mildew	Fewer infections by conidia	0.25 ppm–8th through 12th hr after inoculation	USA	72
Botrytis on potato leaves	Increased incidence and pathogenesis	Ambient	USA	56
Botrytis on broad bean leaves	More infections and pathogenesis	0.15 ppm–8 hr	USA	44
Botrytis on potato leaves	More infections and pathogenesis	0.15 to 0.25 ppm–6 to 8 hr	USA	56
Botrytis on Geranium leaves	More infections and pathogenesis	0.07 to 0.10 ppm–10 hr/day 15 days	USA	53
Tobacco mosaic virus in tobacco	More infections on Pinto bean	Tobacco exposed to 0.30 ppm–6 hr	USA	7
Fluoride				
Bean rust, bean halo blight, tomato early blight	Fewer infections and pathogenesis	300 ppm in leaf tissue	USA	(D. C. McCune L. H. Weinstein, J. F. Mancini, personal communication)
Tobacco mosaic virus	Fewer infections on Pinto bean	500 ppm in leaf tissue	USA	16
Tobacco mosaic virus	More infections on Pinto bean	100–300 ppm in leaf tissue	USA	16, 88

ª Under ambient conditions pollutants often exist in mixtures. Therefore, some of the effects observed in the field (Ambient) may have been caused by interactions of more than one pollutant.

including micro-organisms, and little is known of the effects that do occur or what the consequences will be.

Sulfur dioxide has rarely been reported to increase the incidence of fungus diseases. Increase of *Armillaria mellea* in trees injured by SO_2 (17, 33, 39, 69) is not surprising because *A. mellea* usually invades weakened trees. In addition, three other wood rotters, *Glocophyllum abietinum* (= *Lenzites abietina*), *Trametes serialis,* and *Trametes heteromorpha* were found where trees were damaged by SO_2 but not in a nearby area where trees were not damaged (33). Unlike the two other *Lophodermium* species mentioned earlier, a high incidence of *Lophodermium piceae* (=*L. abietis*) was found on spruce needles injured by SO_2 (39). An explanation has been suggested for the increased incidence of a needle blight of Japanese red pine caused by the weakly pathogenic *Rhizosphaera kalkhoffii* in an industrial area of Japan (8). *Rhizosphaera kalkhoffii* was unable to produce typical disease symptoms or pycnidia unless needles were injured first. It was not determined whether needle injury from causes other than SO_2 would produce a similar effect (8).

A few partially conflicting reports indicate that more propagules of common soil fungi occur in forest soils near SO_2 sources (58, 80). Sobotka (80) found a more diverse array of soil fungi in polluted areas but Mrkva & Grunda (58) found the opposite and neither reported which species were affected. Mycorrhizal associations in spruce were abnormal in polluted areas (80). The Hartig net was sometimes hypertrophic and the mantle was often thinner, tuft-like, or even absent where trees were injured by SO_2.

Effects of SO_2 on Fungus Growth

Fungus hyphae are quite resistant to SO_2, especially when protected by host tissues. Once an infection is established, SO_2 is not likely to eradicate it although the rate of decay caused by fungi may be decreased temporarily. Sulfur dioxide is used to inhibit postharvest decay of grapes by *B. cinerea, Cladosporium herbarum,* and species of *Alternaria* and *Stemphylium,* but the conc used are more than 200 times those usually found in polluted ambient air. Hyphae of different species vary in sensitivity to large SO_2 doses (59). *Botrytis* sp. (cinerea type) colonies on potato dextrose agar were very resistant. After 11 hr of exposure to 4.0 ppm SO_2, all cultures were still alive although growth was inhibited. *Schirrhia acicola* grew normally and produced viable conidia on agar after exposure to 1.0 ppm SO_2 for 4 hr (21). Growth of *Aspergillus niger, Alternaria brassicicola,* and *Didymellina macrospora* was not affected, but a *Penicillium* spp. was slightly stimulated in nutrient solutions that contained the equivalent of 90 ppm SO_2 (68).

Effects of SO_2 on Spore Germination

Although most field reports indicate that SO_2 decreases the incidence of fungus diseases (Table 1), the mechanism of this effect is unknown. Laboratory evidence tends to support the idea that the effects are mostly indirect, i.e., through predisposing influences on the host or other aspects of the environment. The spores of most fungi tested appear to be very resistant to direct exposure to SO_2. Even massive doses rarely affect germination (12, 28). For example, 20% of the conidia of *Botrytis cinerea* exposed to 36 ppm SO_2 for about ½ hour germinated (12). No effect on germination was observed with spores of 10 saprophytic and parasitic fungi exposed on agar to 10 ppm SO_2 for 1–6 hr (28). The percentage germination of urediospores of *Puccinia striiformis* was less when incubated in ambient air during periods conducive to air pollution than during periods relatively free of pollution (75). Decreased germination was related to increased conc of intermediate size ions in the air. Sulfur dioxide or other common pollutants might have been involved but were not measured (75).

Moist spores apparently are more sensitive to SO_2 than dry spores. Germination of wet conidia of *Alternaria* sp. decreased 60% from exposure to 50 ppm SO_2 for 24 min but 100 ppm SO_2 was required to produce a similar decrease in the dry spores at 98% RH (11). With both *B. cinerea* and *Alternaria* sp., the SO_2 dose required for a given amount of inhibition was inversely related to relative humidity (11, 12).

Spores can be affected by SO_2 dissolved in water (68). Conidia of *Diplocarpon rosae* germinated abnormally and hyphal growth was reduced in aqueous sulfite solutions that contained the equivalent of more than 35 ppm SO_2. Spores of other fungi such as *Aspergillus niger* and *Alternaria brassicicola* were also affected but were more resistant than *D. rosae* (68). But germination of *Schirhia acicola* conidia suspended in water was not affected by exposure to 0.90 ppm SO_2 for 6 hr (21).

No experimental evidence conclusively explains the mode of action of SO_2 on diseases caused by fungi. Any one or a combination of direct or indirect effects of SO_2 could be responsible for the effects observed in the field.

The spores shown to be resistant to gaseous SO_2 are relatively thick-walled and capable of surviving climatic extremes. By contrast, the basidiospores of the rusts and wood-rotting fungi have thinner walls and are comparatively fragile. Basidiospores require especially exacting conditions to survive and cause infection and may be a "weak link" responsible for decreased incidence of some rusts and wood-rotting fungi. The basidiospore "weak link" idea may be true for some fungi, but it does not explain the relative success of *A. mellea* and several other wood-rotting fungi or the decreased incidence of diseases caused by ascomycetes mentioned earlier.

The resistance of some fungus spores and hyphae to direct SO_2 exposure suggests an indirect action, perhaps through effects on flowering plants. Sulfur dioxide can decrease plant vigor or even kill some plant species. The disappearance of certain lichen species in polluted areas is well established and an excellent review of the topic has been prepared by Skye (78). Perhaps the disappearance of some of the heteroecious rusts is simply due to the decreased numbers or disappearance of alternate hosts? The thinning of the forest canopy in areas where plants are severely injured results in greater fluctuations of temperature, moisture, and light in the forest ecosystem. Some needle-infecting rusts require more than one year to become established and produce symptoms. Some ascomycetes successfully invade only 2nd or 3rd year conifer needles. The disappearance of certain of these fungi might be traced to a premature, pollutant-induced death of 2nd or 3rd year needles.

OZONE

Sources and Effects on Vegetation

Ozone (O_3) is the most important plant pathogenic component of photochemical oxidant air pollution. Ozone is formed in nature by electrical discharges in the atmosphere and by the action of ultraviolet radiation. The naturally produced O_3 conc at ground level is generally believed to be less than 0.03 ppm. Increased ambient conc of O_3 result from photochemical reactions on the primary products of fuel combustion, mainly hydrocarbons and nitrogen oxides. The conc of O_3 at a given place and time is dependent upon the amount of primary emissions and numerous meteorological and topographical parameters. Thus, estimation of the conc of ambient O_3 likely to occur at any given place and time can only be approximated. Usually, more O_3 occurs with moderate temperature, high light intensity, and low humidity and wind velocity. An estimate of O_3 conc in major metropolitan areas during episodes of photochemical oxidant pollution is 0.10–0.30 ppm. A maximum hourly O_3 average of 0.74 ppm has been recorded in Riverside, California and 0.35 ppm has been measured in St. Louis, Missouri. Increased conc of O_3 occur even in rural areas because of the presence of primary pollutants there and the movement of O_3 as well as the primary pollutant precursors from urban areas. In most rural areas of the highly populated eastern United States, maximum hourly O_3 conc can reach 0.10–0.15 ppm. In most other rural areas, except where topography prevents the dispersion of hydrocarbons and nitrogen oxides, e.g., the Los Angeles Basin, the conc are somewhat lower. For sensitive plant species, growing under field conditions optimum for injury, an 8-hr exposure to O_3 from 0.04–0.10 ppm O_3 can produce visible foliar injury. For plants of intermediate sensitivity, the required O_3 conc for 8 hr would be about 0.08–0.20 ppm. An excellent discussion of factors affecting plant sensitivity to oxidant pollution is given by Heck (27).

Effects of O_3 on Parasitism by Fungi

Field observations of effects of O_3 on diseases caused by fungi are rare (Table 1), mainly because elevated conc of O_3 are so widespread. There are no point-source emissions of O_3 and it is therefore difficult to find contiguous areas with different amounts of O_3 pollution. The best field evidence linking O_3 with an effect on a fungus disease was provided by Manning et al (56). They observed that *Botrytis cinera* infection of potato was especially severe on leaves with symptoms of O_3 injury and that injured leaf areas served as infection courts. Injured leaves were more readily invaded by *B. cinerea,* and died sooner than, uninjured leaves.

Although only limited field evidence supports the contention that O_3 affects parasitism, evidence from laboratory and greenhouse studies supports it overwhelmingly (Table 1).

There is growing evidence that O_3 can decrease infection, invasion, and sporulation of fungi parasitic on growing plants. Doses of O_3 too low to injure gladiolus flowers, inhibited development of disease in petals inoculated with *Botrytis gladiolorum* whether flowers were exposed immediately, or 24 hr after inoculation (46, 47). Ozone decreased disease of chrysanthemum petals caused by *Botrytis* sp. even though the petals were injured (47). Manning et al (54) studied the effects of low conc of O_3 on the grey mold disease of geranium flowers caused by *B. cinerea*. Flowers were inoculated with conidia and the plants were immediately exposed for 4 hr to 0.20, 0.35, and 0.55 ppm O_3. After 24 hr, the flowers were removed from the plants and incubated in wet plastic bags for 72 hr. At 0.20 ppm, O_3 did not affect disease development. At 0.35 and 0.55 ppm, O_3 did not prevent infection but invasion was restricted. Infection and invasion of poinsettia bracts by *B. cinerea* were not affected and bracts were not injured by O_3 (0.15–0.45 ppm–4 hr) after inoculation (55).

Small doses of O_3 can also inhibit obligate parasitism. Heagle (23) inoculated 10 varieties of oat species with the crown rust fungus, *Puccinia coronata*, and exposed them to O_3 (0.10 ppm–6 hr) on the 10 days after infection. All exposed plants were slightly injured by O_3 and the uredia on them were significantly smaller than on non-ozonated plants. The rust reaction was changed, from susceptible on control plants to resistant on exposed plants of most varieties. Ozone can cause similar effects on bean rust, caused by *Uromyces phaseoli*. Ozone exposure (0.10 ppm–14 hr/day) beginning on the day after incubation slowed pustule growth but there were more primary and secondary pustules on the O_3 treated plants. The effect on spore production was not determined and the greater number of pustules on the exposed plants may have compensated for the smaller pustules (H. M. Resh, V. C. Runnecles, personal communication). Small doses of O_3 (0.06, 0.12, 0.18 ppm–6 hr/day) can inhibit sporulation of the wheat stem rust fungus, *P. graminis;* the greater the O_3 conc, the fewer the spores produced (24). The decreased sporulation appeared to be related to O_3 injury of host mesophyll cells and decreased growth of hyphae (24). The effect of O_3 on bean rust also appears to be through an indirect action resulting from injury of host tissue. Bean rust pustules were smaller in interveinal leaf areas where O_3 injury was most severe than in relatively uninjured areas near leaf veins (H. M. Resh, V. C. Runnecles, personal communication).

The decreased sporulation of stem and crown rust was not accompanied by a decrease in germinability or infectiveness of urediospores; no effects were found whether exposures occurred during urediospore formation (23, 24) or after mature spores were placed on fresh leaves prior to incubation (23).

The process of plant penetration and infection by rust fungi (formation of an infection peg, substomatal vesicle, and infection hyphae) is very complex and the fungus might therefore be expected to be sensitive to disruption by O_3 during these stages. When inoculated wheat plants were exposed during the penetration and infection phase however, no effects on infection or on the specialized infection structures were found, indicating resistance to direct action of O_3 (24). Evidently, O_3 injury, or some some effect of O_3 that occurs 1–2 days after exposure, must be present to decrease infection. For instance, percentage infection was not decreased when plants were exposed to O_3, immediately inoculated, and infected before injury developed (24). Percentage infection was decreased when spore incubation was delayed until O_3 injury of host cells developed (1–2 days).

As with most other stress factors, O_3 can increase phenol conc (31, 57), peroxidase activity (13, 15), and phenylalanine ammonium lyase activity (D. T. Tingey, personal communication) in exposed leaves. Certain fungi can cause similar responses in plants (38). The decreased infection and invasion by rust in leaves exposed to O_3 may be related to increases in phenol conc and/or the activation of oxidative enzymes by O_3 causing accumulation of quinones.

Wheat stem rust and oat crown rust are obligate parasites in the mesophyll cells of host plants. The hyphae of both are well protected within host tissue and both produce relatively thick-walled urediospores. The hyphae of the obligate parasite, *Erysiphe graminis,* the cause of powdery mildew, are exposed on the leaf surface with haustoria in leaf epidermal cells. *E. graminis* conidia have thinner cell walls than urediospores. With these differences between rust and powdery mildew in mind it might be expected that O_3 would have different effects upon them. Several workers have recently found this to be true. For example, doses of O_3 that inhibited growth of rust uredia and rust sporulation (23, 24) stimulated colony growth of *E. graminis* on barley and sporulation was not affected significantly (26). This was true even though barley mesophyll cells were injured. Unexpectedly, O_3 at 0.15 ppm or less did not decrease percentage germination of conidia whether exposures occurred during formation of conidia or during incubation of mature conidia on inoculated leaves (26). Schuette (72) reported similar results with small O_3 doses but found that O_3 at 0.25 ppm or greater during conidia formation decreased germination and caused plasmolysis of some conidia. No effects on germination were observed when large doses of O_3 were applied after spores had matured (72). Conidia of *Microsphaera alni* appear to be more resistant to O_3 than conidia of *E. graminis*. Ozone (1.00 ppm–6 hr or 0.25 ppm–72 hr) had little effect on germination or infectiveness whether exposures occurred while conidia were attached to conidiophores or while on glass or lilac leaf tissue (C. R. Hibben, personal communication).

Exposure of wheat stem and oat crown rust urediospores did not affect subsequent germination or infection (23, 24). With *E. graminis,* low doses of O_3 did not affect percentage germination and appressorium formation but percentage penetration was decreased significantly (26, 72). Penetration of barley leaves was decreased to 84, 84, and 75% of the controls by 6-hr exposures to 0.05, 0.10, or 0.15 ppm O_3, respectively, during formation of conidia (26). Similar decreases occurred when mature conidia were exposed on barley leaves to 0.10, 0.20, or 0.30 ppm O_3 for 24 hr during incubation (26). The penetration phase of mildew parasitism (when penetration pegs are developing) appears to be a critical stage when *E. graminis* is most sensitive to O_3 (72). Germination was not affected by 1.0 ppm O_3 during the first 8 hr of incubation although the formation of appressoria was slowed temporarily. The infection process resumed normally after exposure stopped. But a 4 hr exposure to 0.25 ppm O_3 during the penetration phase (8–12 hrs after incubation began) decreased penetration significantly (72). Laboratory evidence indicates that O_3 can disrupt parasitism of plants by certain fungi in ways that could decrease disease prevalence and severity in the field. No attempts have been made to determine whether such disruptions exist in nature.

Ozone can affect flowering plants in many ways that could alter the processes of plant parasitism by fungi (14, 27, 65). Whether parasitism is stimulated or inhibited by such effects may depend upon many factors. One of the most important seems to be the nature of the parasite. With two rust fungi, O_3 injury of host tissue was accompanied by slower hyphal growth, smaller pustules, and less sporulation (23, 24). With facultative parasites, or facultative saprophytes, different effects would be expected. Strong evidence that O_3 injury can predispose plants to disease caused by facultative parasites is provided by Manning et al (56). Infection and invasion of potato leaves by *B. cinerea* was increased greatly when leaves were injured by O_3 (0.15–0.25 ppm–6–8 hr) before inoculation. In many cases infection originated in lesions caused by O_3. With one geranium cultivar, 90–100% of the leaf area showed symptoms of disease caused by *B. cinerea* when leaves were injured by O_3 (0.07–0.10 ppm–10 hr/day–15 days) before inoculation, whereas only 2–5% of the leaf area was diseased in control plants (53). Detached leaves did not become naturally infected by *B. cinerea* in the greenhouse unless the leaves had been injured previously by O_3 (53). Leaves of broad bean plants inoculated with *B. cinerea* and exposed to O_3 (0.15 ppm–8 hr) before incubation showed significantly more infection and invasion than the controls (44). This was true even when leaves were not visibly injured by O_3. In a parallel experiment, no effects of O_3 were found when the leaves were excised and petioles immersed in water before inoculation and exposure (44).

The results with *B. cinerea* indicate that facultative parasitism will be enhanced when host tissues are injured, but this contention appears to be an oversimplification. For example, the type of host tissue being parasitized apparently is important. The facultative parasite, *B. cinerea,* caused more disease on leaves exposed to O_3 (44, 53, 56), but its pathogenicity on flowers was either decreased (54) or unaffected (55) by O_3 exposure. The wide range in pathogenicity between fungi classified as facultative parasites and facultative saprophytes also should be considered. *Botrytis* probably is not the only fungus that can cause increased disease when host tissues are injured by O_3 but attempts to identify similar effects with other fungi have not been successful. Ozone injury of current-season white pine needles was not sufficient to predispose them to increased colonization by *Lophodermium pinastri*, even though this fungus grows rapidly on senescing or dead needles (10). We have been unable to show any striking relationships between O_3 injury and infection or invasion of corn leaves by *Helminthosporium maydis* or of peanut leaves by *Cercospora arachidicola* (A. S. Heagle & L. W. Key, unpublished data).

Recent studies have shown that O_3 can stimulate mycoflora on leaves and roots either directly or indirectly. Long exposure to small conc of O_3 (0.06 ppm–8 hr/day–28 days) injured the leaves of pinto bean plants, and more fungus propagules occurred on the leaves, but the species composition was not affected (52). Ozone caused similar stimulatory effects on the root surface fungi of pinto bean plants (50). About 20% more colonies of fungi (primarily *Fusarium oxysporum*) were isolated from roots of plants with leaves injured by O_3 than on noninjured plants; these effects occurred before differences in root anatomy were observed (50). If O_3 can increase the population of fungi on leaves or roots it is plausible that disease also might be affected, but proof is lacking. Cabbage seedlings planted in soil infested with *F. oxysporum* f. sp. *conglutinans* and exposed to small doses of O_3 (0.10 ppm–8 hr/day–10 weeks) had slightly but not significantly less disease and the frequency of isolation of the fungus was not changed (49).

Present evidence indicates that O_3 alters parasitism primarily through effects on host plants. Facultative parasites may be affected more than facultative saprophytes such as *H. maydis* and *C. arachidicola*. One of the most important research needs is to determine whether these postulations are correct and whether diseases caused by other fungi are affected significantly by O_3.

Effect of O_3 on Fungus Growth and Sporulation

Fungus colonies on artificial media are rarely killed by exposure to large doses of O_3 although small doses can inhibit colony growth, suppress development of aerial hyphae, and decrease sporulation (29, 32, 37, 40). Reports of differences in species sensitivity to O_3 are common (29, 32, 37, 40). Ozone can increase pigmentation and the number of light-refractive globules in hyphae of some species (37) and can cause morphological irregularities in conidia of *Alternaria solani* (37). Elongation of *A. solani* conidiophores was inhibited and the apical cells of conidiophores swelled abnormally and sometimes collapsed or developed breaks in the pigmented layers of the cell walls when colonies were exposed to O_3 (0.10 ppm–4 hr) (66). But damaged conidiophores resumed sporulation after exposure (66).

Although the cases in which O_3 exerts an inhibitory effect may prove to have ecological as well as epidemiological significance, perhaps of more practical interest to those concerned with disease control are the cases where fungus growth, sporulation, or germination are increased by O_3. On agar media, O_3 caused increased sporulation of *A. oleraceae* (37, 67), *A. solani, Myco-*

sphaerella citrullina (67), and *Colletotrichum orbiculare* (= *C. lagenarium*) (29). A later report showed that sporulation by *A. oleraceae* increased about 10-fold, 25-fold, and 25-fold compared to the controls when colonies were exposed for 4 hr to 0.10, 0.40, and 0.60 ppm O_3, respectively (89). Germination of *A. oleraceae* spores was not affected by O_3 in any of these studies (37, 67, 89). Kuss (40) grew 30 fungi, representative of the Phycomycetes, Ascomycetes, Basidiomycetes, and Fungi Imperfecti on agar in covered petri plates and exposed them to O_3. Sporulation was increased in three *Alternaria* spp., two *Fusarium* spp., two *Glomerella* spp., and a *Helminthosporium* sp. All pycnidium- or perithecium-forming fungi tested produced more fruiting bodies in air containing O_3 than in normal air. Hymenophores of *Lenzites trabea* formed only when ozonated. Germinability was decreased in most species but was increased in *Diaporthe phaseolorum* and *Lenzites trabea* (40). Germination of *B. cinerea* (44) and several other fungi (30) was stimulated when spores were exposed to small doses of O_3. Conidia of *Alternaria solani* attached to conidiophores were induced to germinate by exposure to 1 ppm O_3 for 30 min, but were killed if exposed for 1 hr (66).

The mode of action of O_3 on fungi in vitro is not known and little research has been performed on the topic. Ozone decreased the production of lipids by *Helminthosporium sativum* possibly through inhibited synthesis of fatty acid (62). Ozone decreased the sulfhydryl content of *A. solani* hyphae, possibly through oxidation of sulfhydryl compounds required in lipid synthesis (66).

Large doses of O_3 can also inhibit growth and sporulation of fungi growing on fruit, but most fungi tested are resistant to O_3. Large conc of O_3, in continuous exposures, did not inhibit infection of apples by *Penicillium expansum*. Established colonies were not killed but the enlargement of lesions was retarded, aerial hyphae were suppressed, and sporulation was decreased (71). Although the results were not economically important, continuous exposure to 0.50 ppm O_3 decreased the spread of *Rhizopus stolonifer* and *Monilinia fructicola* in peaches. The threshold O_3 conc for suppression of aerial hyphae was between 0.05–0.10 ppm (81). Invasion of strawberries by *B. cinerea* was not decreased by large doses of O_3 although growth of aerial mycelium was suppressed (81). Fewer numbers of decay, less invasion, and less sporulation occurred in oranges and lemons inoculated with *P. expansum* and *P. digitatum* in open storage boxes exposed to about 1 ppm O_3 for 15 days than in the controls (22).

Effects of O_3 on Spore Germination

Ozone is more effective than SO_2 in decreasing spore germination. Differences in spore sensitivity to O_3 have been found by several investigators. The degree of sensitivity to a given dose of O_3 apparently depends upon fungus species, spore morphology, moisture, and substrate.

Multicelled, pigmented spores and spores with thick cell walls, are usually more resistant to O_3 than single-celled spores or those with hyaline or thin cell walls. For example, percentage germination of the large multicelled conidia of *Macrosporium* sp. was greater than that of the smaller, single-celled spores of *Sclerotinia fructicola* and *Penicillium expansum* after O_3 exposure (0.60 ppm–1–3 hr) on freshly cut apples (79). Hibben & Stotzky (29) exposed detached spores of 14 fungi on agar to O_3 (0.10–1.00 ppm–1–6 hr). The percentage germination of large pigmented spores with relatively thick cell walls (*Chaetomium* sp., *Alternaria* sp., and two *Stemphylium* spp.) was the same as the controls after O_3 exposure (1 ppm–6 hr). But germination of small hyaline spores (*Fusarium* sp. and two *Verticillium* spp.) was inhibited by a smaller O_3 dose (0.25 ppm–6 hr). Spores of other fungi, including *Trichoderma viride, Aspergillus terreus, A. niger, Penicillium egyptiacum, Botrytis allii,* and *Rhizopus stolonifer,* were intermediate in sensitivity. In many cases, ozonated, nongerminating spores failed to swell perceptibly and were stained pink from rose bengal in the medium, indicating that the integrity of cell membranes was impaired (29).

Dry spores are more resistant to O_3 than wet spores. Germination of dry spores on cellophane was not affected in any of the 14 species tested, even by a high O_3 dose (1 ppm–64 hr) (29). It was suggested that spores on agar were presumably surrounded by a thin film of water and, therefore, in the early stages of germination during exposure. The processes of active metabolism that accompany germination may be especially sensitive to O_3 (29).

Fungus spores suspended in liquid media also are quite resistant to large conc of O_3 bubbled through the suspension. Germination of *Sclerotinia fructicola* spores was unaffected by bubbling 1.40 ppm O_3 through a spore suspension in water for 1 hr (91). Similar results were obtained with spores of fungi ordinarily sensitive to O_3 (29, 44). This evidence appears to contradict results showing spores to be more sensitive when wet than when dry. Poor aeration may be one reason for decreased effects on spores in a suspension. Much of the O_3 probably breaks down in solution and never directly contacts the spores. Ozone was less toxic to spores when they were suspended in liquids containing organic materials than in water, possibly because more O_3 reaction sites are present in an organic solution than in water, thereby decreasing the effective O_3 conc (91).

The nature of the growth medium may affect spore sensitivity to O_3 when spores are on a surface during exposure. Ozone decreased percentage germination less when spores were on water agar than when spores were on a yeast extract-rose bengal agar (29). If it can be assumed that the same amounts of O_3 and water were in contact with spores on both media, the substrate may affect spore sensitivity to O_3. If this is true, care must be exercised when using data from in vitro studies to theorize on what might happen when exposures occur in the field.

FLUORIDE

Sources and Effects on Vegetation

Fluoride is widespread in soil, rocks, and minerals. Fluoride is released into the air when these materials are heated or treated with acid. Hydrogen fluoride (HF) is generally considered to be the most important plant pathogenic fluoride. Major fluoride sources are the ore smelting and phosphate fertilizer industries. The factors affecting ground level conc of SO_2 also affect conc of fluoride. Large conc of fluoride occur primarily near emission sources and conc decrease rapidly with increasing distance from the source. In large urban areas, conc up to 3 ppb (parts per billion of air by volume) can occur. Near industrial sources with uncontrolled emissions, conc of about 100–250 ppb have been reported. In areas not exposed to fluoride pollution, the tissues of most plants contain about 2–20 parts per million fluoride. Continuous exposure of small conc of fluoride in air results in the gradual buildup of fluoride in plant tissues. Sensitive plant species can be injured when the fluoride conc in the foliage reaches about 25–50 parts per million. About 100–300 parts per million in plant tissues can injure species with intermediate sensitivity. Some species can accumulate several thousand parts per million with no visible effects.

Effects of Fluoride on Fungi

Industrial emissions may contain small amounts of fluoride, but the relative impact of fluoride from most industrial sources on vegetation is probably minor compared with that from SO_2. Nevertheless, fluoride may contribute to effects on parasitism that have been attributed primarily to SO_2. Numerous point sources of fluoride exist but there are no field reports that specifically link fluoride to effects on parasitism. The only indications come from a few laboratory and greenhouse studies. Colony growth of fungi was inhibited on agar containing sodium fluoride (NaF). More NaF was required to inhibit *Verticillium alboatrum* and *Helminthosporium sativum* than to inhibit *Pythium debaryanum*. Growth of *B. cinerea* and two *Colletotrichum* spp. was stimulated slightly at low conc (86). An isolate of *Verticillium lecanii* grew well unless the NaF conc of the agar exceeded 0.105 M, and then the conidia were larger than those in the controls or on agar containing less NaF (42). Sodium fluoride in nutrient agar decreased sugar utilization and formation of itaconic acid by *Aspergillus terreus* (41). This effect was reversed when pyruvate was added to the medium, leading to the theory that NaF inhibits the normal glycolitic pathway as well as a step in the conversion of pyruvate to itaconic acid.

No published reports describe the effects of HF gas on parasitism of plants by fungi and only one research group (D. C. McCune, L. H. Weinstein & J. F. Mancini, personal communication) is presently considering this problem. Known doses of HF gas, resulting in measured conc of fluoride in leaf tissues, are being used to study the effects on two bean pathogens (*Uromyces phaseoli* and *Erysiphe polygoni*) and one tomato pathogen (*Alternaria solani*). Preliminary results indicate that fluoride in leaf tissue generally decreases foliar infection and invasion by all three pathogens (Table 1). These results were found with a fluoride conc in the leaves of more than 300 ppm.

This is somewhat greater than is usually found in vegetation near emission sources. It is possible that smaller amounts of fluoride in leaf tissue could cause stimulation of disease similar to the stimulation of fungus growth on agar containing NaF (86).

Effects of Pollutants on Bacteria

There is some evidence of pollutant effects on bacteria but most concern human pathogens. Evidently, bacteria are resistant to direct attack by pollutants in conc likely to occur in ambient air. For instance, very large doses of O_3 (19, 20, 32, 73) or SO_2 (59) were required to cause measurable effects on *Escherichia coli* grown on agar media. Large doses of O_3 were needed to decrease significantly the luminescence of *Photobacterium phosphoreum* on agar (74). Species of bacteria pathogenic to humans vary in sensitivity to O_3 (32).

As with fungus spores, bacteria appear to be more sensitive when moist than when dry. Human bacterial pathogens were not greatly affected by O_3 when the relative humidity was less than 50% (18). The same dose killed bacteria when the relative humidity was high or when bacteria were in aerosols.

Only one bacterium capable of invading plant tissue has been shown to be affected by a pollutant. Large doses of O_3 (0.12–0.15 ppm–15 days) over a 3-week period decreased the number of *Rhizobium* nodules on soybean roots (64). Smaller doses of O_3 (0.06 ppm–8 hr/day–20–60 days) decreased the number, size, and weight of *Rhizobium* nodules on pinto bean roots (51). Tingey & Blum (85) found similar results when 3-week-old soybean plants were exposed for 1 hr to 0.75 ppm O_3. They also reported that O_3 decreased the leghemoglobin content in plant roots. If O_3 can so drastically affect the relationships between legume roots and *Rhizobium*, other organisms probably can be affected similarly.

The mechanisms of O_3 action on bacteria are not known, but most likely an indirect effect occurs through the action of O_3 on host plants. With *Rhizobium*, it is unlikely that gaseous O_3 directly affects bacterial cells because O_3 is highly reactive and probably breaks down in the soil. More likely, O_3 affects plant foliage in ways that alter root physiology. Perhaps the effect is due to qualitative or quantitative changes in root exudates or an increased rate of lignification of root cells. When bacterial cells are exposed directly to O_3 the mode of action may be through an attack on cell walls or membranes causing lysis or leaking of cell contents (73).

No effects of SO_2 on plant pathogenic bacteria have been reported. Nevertheless, SO_2 can affect the environment in ways that could be expected to affect bacteria; thus it is probable that effects will eventually be found. Mrkva & Grunda (58) provide one of the few examples where SO_2 emissions are thought to have changed bacterial populations. Fewer numbers, but not types, of bacteria were found in the soil near a smelter where the soil pH, soil potassium, and soil magnesium were decreased. Potential effects on plant pathogenic bacteria were not discussed. Preliminary results of experiments on the effects of simulated acid rainfall (H_2SO_4, pH 3.5) on infectivity and invasion of kidney bean by *Pseudomonas phaseolicola* suggest that the bacterium is resistant but that the host may be affected in ways that predispose it towards greater susceptibility (D. S. Shriner, personal communication).

Current research on the effects of HF on the parasitism of bean by *Pseudomonas phaseolicola* is the only case where the effects of fluoride on a plant pathogenic bacterium have been considered (D. C. McCune, L. H. Weinstein, & J. F. Mancini, personal communication). Preliminary indications are that fluoride may decrease disease caused by *P. phaseolicola*.

Limestone dust deposits on plant leaves can affect bacterial populations. More colonies of unidentified bacteria were isolated from grape and sassafras leaves covered with limestone dust than from nearby leaves relatively free of dust, but the reverse was true for hemlock leaves (48).

Effects of Pollutants on Virus Diseases

Study of pollutant effects on diseases caused by viruses have concerned only tobacco mosaic virus (TMV) (Table 1). Apparently, both fluoride and O_3 either predispose pinto bean leaves to TMV infection or somehow increase the infectiveness of the virus particles. The number of TMV lesions was greater on inoculated bean leaves than on the controls when leaves contained 100–300 ppm fluoride (16, 88) but when leaf tissues contained more than 500 ppm fluoride, the reverse was found (16).

More local lesions developed on pinto bean leaves inoculated with TMV and exposed to O_3 (0.10 ppm–3 hr) than on the controls (7). But the response appears to depend upon the timing of the exposure. The response occurred when exposure began 3 or 24 hr after inoculation but was absent when exposure occurred 0 or 48 hr after inoculation. More TMV lesions developed on pinto bean leaves when tobacco plants were exposed to O_3 (0.30 ppm, 6 hr) before using them to obtain inoculum than when the inoculum came from nonexposed tobacco plants (7). Ozone may have caused increased virus reproduction in tobacco, increased infectiveness of existing virus particles, or both.

The mechanisms of action of fluoride and O_3 on disease caused by TMV are unknown but both pollutants apparently can increase disease under certain conditions. Further research is needed to determine whether or not other pollutants or viruses can interact in causing plant disease.

Effects of Pollutants on Insects

Insects play an important part in the transmission and development of many plant diseases. Thus, any alteration of populations or activities of insects by air pollution may also affect plant diseases or the amount of injury caused by the insects themselves.

Most reports have resulted from field observations of coniferous trees severely affected by SO_2. A common finding is that trees injured and weakened by pollutants are more likely to be attacked by insects that normally require weakened trees for successful reproduction. Thus, the incidence of some species of bark beetles was increased in trees injured by SO_2 (4, 17, 39, 69) or O_3 (82). But other factors must also be considered. For instance Templin (84) reported the decreased incidence of two bark-infesting insects, *Tomicus* (= *Myelophilus*) *piniperda* and *Melanophila* (= *Phaenops*) *cyanea* in trees injured by SO_2 pollution. Cobb et al (9) suggested that even though Ponderosa pines that are severely injured by O_3 may be attacked by large populations of bark beetles, the trees may be too decadent for establishment of successful broods. Less injury from the white pine weevil, *Pissodes strobi*, occurred in trees injured by SO_2 (43). But a high incidence of *Pissodes* sp. weevils has been reported on pine and spruce injured by SO_2 (39). Thus, it appears that many variables interact to determine the effects upon insect populations.

One variable seems to be the habit of insect larvae. The larvae of *Acantholyda nemoralis,* are exposed for long periods and while feeding on contaminated foliage near an SO_2 source the incidence of this pest was decreased (77). On the other hand, larvae of *Exoteleia dodecella* remain concealed within the buds for a long period and its incidence was relatively unaffected by SO_2 (76, 77).

The physical protection of insects by plants may not extend to situations where toxic substances resulting from pollutant injury or the pollutant itself accumulate in the host tissue. In fir damaged by fluoride, bark beetle and weevil populations were decreased, possibly as a result of toxic concentrations of fluoride in plant tissue, while several other species were not harmed (60). On the other hand, accumulation of a pollutant or a substance resulting from pollutant exposure, might directly or indirectly stimulate certain insects. The incidence of a spruce gall louse, Adelges (= *Sacchiphantes*) *abietis* was greater in an area where spruce trees were damaged by fluoride (92).

The relative sensitivity of insect pests and their predators to pollutants must also be considered. For example, insect parasites of larvae of *Rhyacionia buoliana* were fewer in areas where trees were severely affected by SO_2 even though larvae were abundant (84).

The reports presented above relate to forest insects. Virtually nothing is published on the effects of pollutants on insects that attack food crops. There are no reports of pollutant effects on insects relating to an effect upon plant disease caused by micro-organisms. Nevertheless, the fact that pollutants can affect forest plants and insects indicates that such effects can occur, and research on the question is needed.

Effects of Parasitism on Ozone Injury

Just as pollutants can affect disease caused by certain parasites, parasitic disease may affect the development of O_3 injury. This phenomenon occurs with a variety of plants and parasites and usually results in a localized decrease of O_3 injury. The effect is usually similar whether the parasite is a fungus, a bacterium or a virus. It has never been reported with SO_2 or other pollutants.

Yarwood & Middleton (94) found that areas adjacent to rust pustules on bean or hollyhock were not injured by exposure to ozonated olefins (artificial smog) which injured other areas. The effect was first seen 3 days after infection and extended several millimeters beyond the limits of the rust mycelium. The wheat stem rust fungus protected local areas of wheat leaves from O_3 injury (25). This protection occurred before infection—as soon as 2 hr after the start of spore incubation. In the field, the peanut leaf spot fungus, *Cercospora arachidicola,* protected localized areas around fungus lesions from O_3 injury (A. S. Heagle, unpublished observations). Similar protection around parasitic infections has also been reported on broad bean infected with *B. cinerea* (45), on kidney bean infected with *Pseudomonas phaseolicola* (35), on lilac infected with *M. alni* (C. R. Hibben personal communication), and on pinto bean infected with TMV (93). Tobacco plants with systemic TMV infection were not injured by O_3, while noninfected plants were injured severely. However, the effect was seen only in winter (6). Recent preliminary evidence suggests that an opposite and perhaps more important virus-O_3 interaction can occur (R. A. Reinert, G. V. Gooding, unpublished data)— more O_3 injury occurred on burley tobacco, cv. "Kentucky 16" when plants were exposed to O_3 (0.50 ppm–2 hr) one month after systemic infection with tobacco streak virus, tobacco etch virus, or potato virus Y. Tobacco vein banding virus or TMV did not cause this effect.

Although the effect of parasitism in decreasing O_3 injury is certainly not economically important and probably not even ecologically so, this is not true of increased O_3 injury on leaves systemically infected with virus but otherwise apparently normal. The localized protection of leaves from O_3 injury results from parasitism by fungi, bacteria, and viruses; a lack of such protection appears to be the exception. The protective mechanism is not known. The fact that the effect is seen with such a diversity of parasites might be helpful in designing research to study the mechanism whereby O_3 affects flowering plants.

Concluding Remarks and Recommendations

Advances in pollution control technology and the enforcement of new and stricter regulations governing industrial and automotive emissions have significantly contributed to the fight for cleaner air. It is not known whether these efforts will succeed in the face of a rapidly increasing population demanding more energy and mobility. It is not likely that our atmosphere will return to the pristine state enjoyed by man in earlier times. More likely, air pollution will continue at present, increased, or only slightly improved conc for many years to come.

Knowledge of the effects of pollutants on fungus parasitism of plants is limited. Data relating O_3 to parasitism by fungi are just beginning to emerge and little is known of the effects of SO_2 or fluoride. Only two reports exist of the effects of particulate pollution on diseases caused by fungi. Infection by *Cercospora beticola* was greater on beet leaves covered with limestone dust than on nondusted leaves (70). Similar effects were found with a foliar disease of grape caused by *Guignardia bidewellii* and of sassafras caused by *Gloeosporium* sp. (48). Data on pollutant effects on diseases caused by bacteria, nematodes, and viruses are almost nonexistent.

Most present knowledge concerns foliar parasites but there is new evidence that soil organisms also can be affected. Ozone can decrease *Rhizobium* nodulation of bean roots. Ozone and SO_2 can increase the populations of fungus propagules in the rhizosphere. Fewer numbers but a greater variety of nematode species occurred in soils where trees were damaged by SO_2 (3). But these reports are all fragmentary and more research is needed on the effect of pollutants on soil organisms in general and on soil-borne and root diseases.

Plants and plant parasites display a wide range of sensitivity to pollutants. The degree and type of pollutant effect appears to depend upon the pollutant, pollutant dose, host species, and parasite. The relative sensitivity of both plant and parasite to a given pollutant also is important. Preliminary data indicate that obligate parasitism generally is decreased and facultative parasitism generally increased by pollution stress. But only a few pollutant-parasite-host interactions have been studied and additional research is necessary to clarify possible distinctions between the effects on various types of parasitism of various types of plants.

In ambient air, pollutants exist in mixtures. The conc of pollutants fluctuate widely, depending on both rate of emission and meteorological conditions. But most research has considered only single pollutants at given doses. Some pollutant mixtures are known to interact synergistically to increase plant injury. Pollutants might also act synergistically to affect parasitism.

Plant pathologists should realize that air pollution may affect results of both greenhouse and field research. In fact, plants often are more sensitive to pollutants in the greenhouse, especially in those that use evaporative cooling. Consider the situation where plants are screened for resistance to a parasitic disease. If this is done where host plants are injured by pollution, there is ample opportunity for mistaken interpretations. Plant varieties, in addition to possessing gradations in resistance to parasitic disease, may also differ in their resistance to air pollution. So, one must ask, is the observed disease response a matter of resistance to the parasite, to the pollutant, or both? Consider, more specifically, research dealing with rust and other foliage diseases where the size of the disease lesion is the major criterion for identifying plant resistance, the prevalence of various genotypes of the pathogen, and the presence of new genotypes virulent on commercial varieties or breeding lines. Pollutants in relatively small doses can affect lesion size under experimental conditions. The question is, can ambient pollution cause similar effects where such research is being conducted?

One of the most important questions is whether parasitic diseases are affected significantly by air pollution in the field. It is possible that the most important effects are indirect through effects on host plants or through other changes in the environment. Green plants appear to be more sensitive to pollutants than do most parasites. Information relating pollutant dose to injury response is available for some plant species and similar data should be generated for parasitism. Many voids still remain in our knowledge of the effects of pollutants on green plants. In most agricultural areas there are no reliable data on the conc of ambient pollutants. Providing proof of significant effects in the field is a difficult but not impossible job, and certainly is worthy of our attention. For a start we should determine the pollution potential in various agricultural areas and determine what effects are likely on important plant species. Concurrent studies should focus on the types and amount of pollutant injury required to significantly affect parasitism of representative plant species by representative parasites. Once this is accomplished, we will be better able to predict the effects of air pollution on plant parasitism.

Literature Cited

1. "Air Pollution" 1968. The effects of air pollution on nonvascular plants. *Proc. First Eur. Congr. Influence Air Pollution Plants & Animals.* Wagenengen 237–41
2. Barrett, T. W., Benedict, H. M. 1971. Sulfur dioxide. In: *Recognition of Air Pollution Injury to Vegetation: A Pictorial Atlas.* ed. Jacobson, J. S., A. C. Hill. Report No. 1, TR-7 Agricultural Committee, Air Pollution Control Assn.
3. Bassus, W. 1968. Uber Wirkungen von industruexhalaten auf den nematodenbesatz im boden von Kiefernwaldern. *Pedobiologia* 8: 289–95
4. Bösener, Rolf. 1969. Zum Vorkommen ridenbrütender Schadinsecten in rauchgeschädigten Kiefern-und Fichtenbeständen. *Arch. Forstwes. Bd.* 18:1021–26
5. Brandt, C. S., Heck, W. W. 1968. Effects of air pollutants on plants. In *Air Pollution.* ed. A. C. Stern, Vol. 1:401–43, New York: Academic
6. Brennan, E., Leone, I. A. 1969. Suppression of ozone toxicity symptoms in virus-infected tobacco. *Phytopathology* 59:263–64
7. Brennan, E., Leone, I. A. 1970. Interaction of tobacco mosaic virus and ozone in *nicotiana sylvestris. J. APCA* 20:470
8. Chiba, O., Tanaka, K. 1968. The effect of sulfur dioxide on the development of pine needle blight caused by *Rhizosphaera kalkhoffii* Bubok (I). *J. Jap. Forest. Soc.* 50: 135–39
9. Cobb, F. W., Jr., Wood, D. L.,

Stark, R. W., Parmeter, J. R., Jr. 1968. Photochemical oxidant injury and bark beetle (Coleoptera: Scolytidae) infestation of Ponderosa pine. IV. Theory on the relationships between oxidant injury and bark beetle infestation. *Hilgardia* 39: 141–52
10. Costonis, A. C., Sinclair, W. A. 1972. Susceptibility of healthy and ozone-injured needles of *Pinus strobus* to invasion by *Lophodermium pinastri* and *Aureobasidium pullulans. Eur. J. Forest Pathol.* 2:65–73
11. Couey, H. M. 1965. Inhibition of germination of Alternaria spores by sulfur dioxide under various moisture conditions. *Phytopathology* 55:525–27
12. Couey, H. M., Uota, M. 1961. Effect of concentration, exposure time, temperature, and relative humidity on the toxicity of sulfur dioxide to the spores of *Botrytis cinerea. Phytopathology* 51:815–19
13. Curtis, C. R., Howell, R. K. 1971. Increases in peroxidase isoenzyme activity in bean leaves exposed to low doses of ozone. *Phytopathology* 61:1306–07
14. Darley, E. F., Middleton, J. T. 1966. Problems of air pollution in plant pathology. *Ann. Rev. Phytopathol.* 4:103–18
15. Dass, H. C., Weaver, G. M. 1968. Modification of ozone damage to *Phaseolus vulgaris* by antioxidants, thiols and sulfhydryl reagents. *Can. J. Plant Sci.* 48:569–74
16. Dean, G., Treshow, M. 1965. Effects of fluoride on the virulence of tobacco mosaic virus in vitro.

Proc. Utah Acad. Sci., Arts Lett. 42:236–39

17. Donaubauer, E. 1968. Sekundär-schäden in Österreichischen Rauch-schadensgebieten. Schwierigkeiten Der diagnose und Bewertung. *Niedzynarodowej Konf. Wplyw Zanieczyszczen Powietrza na Lasy, 6th,* Katowice, Poland.

18. Elford, W. J., Van Den Ende, J. 1942. An investigation of the merits of ozone as an aerial disinfectant. *J. Hyg.* 42:240–65

19. Fetner, R. H., Ingols, R. S. 1956. A comparison of the bactericidal activity of ozone and chlorine against *Escherichia coli* at 1°. *J. Gen. Microbiol.* 15:381-85

20. Haines, R. B. 1936. The effect of pure ozone on bacteria. *Gt. Brit. Food Invest. Board Rep.* #1935:30-31

21. Ham, D. L. 1971. *The biological interactions of sulfur dioxide and Scirrhia acicola on loblolly pine.* PhD thesis. Duke University, Durham, N. C. 74 pp.

22. Harding, R. P. 1968. Effect of ozone on Pencillium mold decay and sporulation. *Plant Dis. Reptr.* 52:245–47

23. Heagle, A. S. 1970. Effect of low-level ozone fumigations on crown rust of oats. *Phytopathology* 60: 252–54

24. Heagle, A. S., Key, L. W. 1973. Effect of ozone on the wheat stem rust fungus. *Phytopathology* 63:397–400

25. Heagle, A. S., Key, L. W. 1973. Effect of *Puccinia graminis* f. sp. *tritici* on ozone injury in wheat. *Phytopathology* 63: In press

26. Heagle, A. S., Strickland, A. 1972. Reaction of *Erysiphe graminis* f. sp. *hordei* to low levels of ozone. *Phytopathology* 62:1144–48

27. Heck, W. W. 1968. Factors influencing expression of oxidant damage to plants. *Ann. Rev. Phytopathol.* 6:165–88

28. Hibben, C. R. 1966. Sensitivity of fungal spores to sulfur dioxide and ozone. *Phytopathology* 56:880 Abstr.

29. Hibben, C. R., Stotzky, G. 1969. Effects of ozone on the germination of fungus spores. *Can. J. Microbiol.* 15(10):1187–96

30. Hibben, C. R., Walker, J. T. 1966. A leaf roll-necroses complex of li-lacs in an urban environment. *Am. Soc. Hort. Sci.* 89:636–42

31. Howell, R. K. 1970. Influence of air pollution on quantities of caffeic acid isolated from leaves of *Phaseolus vulgaris. Phytopathology* 60: 1626–29

32. Ingram, M., Haines, R. B. 1949. Inhibition of bacterial growth by pure ozone in the presence of nutrients. *J. Hyg.* 47:146–58

33. Jancarik, V. 1961. Výskyt drevokazných hub v kourem poskozovani oblasti Krusných hor. *Lesnictvi* 7:677–92

34. Johansson, O. 1954. Rapport över ett studium av luft och nederbörd omkring Svenska Klifferolje Aktiebologets anlaggningar vid Kvarntorp med speciell hänsyn till tistributionen av svavel och dess inverkan pä växtarna. Lic. dvh. vid K. Lantbrukshögskolan, Uppsala

35. Kerr, E. D., Reinert, R. A. 1968. The response of bean to ozone as related to infection by *Pseudomonas phaseolicola. Phytopathology* 58:1055 Abstr.

36. Köck, G. 1935. Eichenmehltau und Rauchgass-schaden. *Z. Pflanzenkr.* xlv:44–45

37. Kormelink, J. R. 1967. *Effects of ozone on fungi.* MS thesis. Univ. Utah, Salt Lake City. 28 pp.

38. Kosuge, T. 1969. The role of phenolics in host response to infection. *Ann. Rev. Phytopathol.* 7:195–222

39. Kudela, M., Novakova, E. 1962. Lesní skůdci a skody zveri v lesich poskozovaných Kourem. Lesnictvi 6:493-502

40. Kuss, F. R. 1950. *The effect of ozone on fungus sporulation.* MS thesis, Univ. New Hampshire. 27 pp.

41. Lal, M., Bhargava, P. M. 1962. Reversal by pyruvate of fluoride inhibition in *Aspergillus terreus. Biochem. Biophys. Acta,* 68:628–30

42. Leslie, R., Parbery, D. G. 1972. Growth of *Verticillium lecanii* on medium containing sodium fluoride. *Trans. Brit. Mycol. Soc.* 58:351–52

43. Linzon, S. N. 1958. The influence of smelter fumes on the growth of white pine in the Sudbury region. *Can. Dep. Agr. Publ., Ontario Dep.* Lands Forests. 45 pp.

44. Magdycz, W. P. 1972. *The effects of concentration and exposure time on the toxicity of ozone to the*

spores of Botrytis cinerea. MS thesis. Univ. Massachusetts, Waltham. 39 pp.

45. Magdycz, W. P., Manning, W. J. 1973. *Botrytis cinerea* protects broad beans against visible ozone injury. *Phytopathology* 63: In press. Abstr.

46. Magie, R. O. 1960. Controlling gladiolus Botrytis bud rot with ozone gas. *Florida Agr. Exp. Sta. J.* Paper No. 1134:373–75

47. Magie, R. O. 1963. Botrytis disease control on gladiolus, carnations, and chrysanthemums. *Proc. Florida State Hort. Soc.* 76:458–61

48. Manning, W. J. 1971. Effects of limestone dust on leaf condition, foliar disease incidence, and leaf surface microflora of native plants. *Environ. Pollut.* 2:69–76

49. Manning, W. J., Feder, W. A., Papia, P. M., Perkins, I. 1971. Effect of low levels of ozone on growth and susceptibility of cabbage plants to *Fusarium oxysporum* f. sp. *conglutinans. Plant Dis. Reptr.* 55:47–49

50. Manning, W. J., Feder, W. A., Papia, P. M., Perkins, I. 1971. Influence of foliar ozone injury on root development and root surface fungi of pinto bean plants. *Environ. Pollut.* 1:305–12

51. Manning, W. J., Feder, W. A., Papia, P. M. 1972. Influence of long-term low levels of ozone and benomyl on growth and nodulation of pinto bean plants. *Phytopathology* 62:497. Abstr.

52. Manning, W. J., Papia, P. M. 1972. Influence of long-term low levels of ozone on the leaf surface mycoflora of pinto bean plants. *Phytopathology* 62:497. Abstr.

53. Manning, W. J., Feder, W. A., Perkins. I. 1970. Ozone injury increases infection of geranium leaves by *Botrytis cinerea. Phytopathology* 60:669–70

54. Manning, W. J., Feder, W. A., Perkins, I. 1970. Ozone and infection of geranium flowers by *Botrytis cinerea. Phytopathology* 60:1302. Abstr.

55. Manning, W. J., Feder, W. A., Perkins, I. 1972. Effects of Botrytis and ozone on bracts and flowers of poinsettia cultivars. *Plant Dis. Reptr.* 56:814–16

56. Manning, W. J., Feder, W. A., Per-

kins, I., Glickman, M. 1969. Ozone injury and infection of potato leaves by *Botrytis cinerea. Plant Dis. Reptr.* 53:691–93

57. Menser, H. A., Chaplin, J. F. 1969. Air pollution: Effects on the phenol and alkaloid content of cured tobacco leaves. *Tobacco Sci.* 13:169–70

58. Mrkva, R., Grunda, B. 1969. Einfluss von immisionen auf die walkböden und ihre mikroflora im gebiet von Südmahren. *Acta Univ. Agr.,* Brno (Fac. Silv.) 38:247-70

59. McCallan, S. E. A., Weedon, F. R. 1940. Toxicity of ammonia, chlorine, hydrogen cyanide, hydrogen sulfide and sulphur dioxide gases. II. fungi and bacteria. *Contrib. Boyce Thompson Inst.* 11:331–42

60. Pfeffer, A. 1963. Insektenschadlinge an tannen im bereich der gasexhalationen. *Z. Angew. Entomol.* 51: 203–7

61. Powell, N. T. 1971. Interactions between nematodes and fungi in disease complexes. *Ann. Rev. Phytopathol.* 9:253–74

62. Price, H. E. 1968. *The effect of ozone on lipid production in the fungus Helminthosporium sativum.* MS thesis. Univ. Utah, Salt Lake City. 22 pp.

63. Przybylski, Z. 1967. Results of observations of the effect of SO_2, SO_3 and H_2SO_4 on fruit trees, and some harmful insects near the sulfur mine and sulfur processing plant at Machow near Tarnobrzeg. *Postepy Nauk Roln.* 2:111–18

64. Reinert, R. A., Tingey, D. T., Koons, C. E. 1971. The early growth of soybean as influenced by ozone stress. *Agronomy* 63:148. Abstr.

65. Rich, S. 1964. Ozone damage to plants. *Ann. Rev. Phytopathol.* 2: 253–66

66. Rich, S., Tomlinson, H. 1968. Effects of ozone on conidiophores of *Alternaria solani. Phytopathology* 58:444–46

67. Richards, M. C. 1949. Ozone as a stimulant for fungus sporulation. *Phytopathology* 39:20. Abstr.

68. Saunders, P. J. W. 1966. The toxicity of sulfur dioxide to *Diplocarpon rosae* Wolf causing blackspot of roses. *Ann. Appl. Biol.* 58:103–14

69. Scheffer, T. C., Hedgcock, G. G. 1955. Injury to northwestern forest

trees by sulfur dioxide from smelters. *U.S.D.A. Tech. Bull. No. 1117,* 49 pp.

70. Schöenbeck, Helfried. 1960. Beobachtungen zur frage des Einflusses von industriellen immissionen auf die Krankheitsbereitschaft der pflanze. *Ber. Landesanst. Bodennutzungsschutz* 1:89–98

71. Schomer, H. A., McColloch, L. P. 1948. Ozone in relation to storage of apples. *U.S.D.A. Circ. No. 765.* 24 pp.

72. Schuette, L. R. 1971. *Response of the primary infection process of Erysiphe graminis f. sp. hordei to ozone.* PhD thesis Univ. Utah, Salt Lake City. 76 pp.

73. Scott, D. B. M., Lesher, E. C. 1963. Effects of ozone on survival and permeability of *Escherichia coli. J. Bacteriol.* 85:567–76

74. Serat, W. F., Budinger, F. E., Jr., Mueller, P. K. 1966. Toxicity evaluation of air pollutants by use of luminescent bacteria. *Atmos. Environ.* 1:21–32

75. Sharp, E. L. 1967. Atmospheric ions and germination of Uredospores of *Puccinia striiformis. Science* 156:1359–60

76. Sierpinski, Z. 1966. Znaczenie gospodarcze skośnika tuzinka (*Exoteleia dodecella* L.) na terenach uprzemyskowionych. *Sylwan* 110: 23-31

77. Sierpinski, Z. 1967. Einfluss von industriellen lufterunreinigungen auf die populations dynamik einiger primärer kiefernschädlinge. *Proc. 14th Congr. Int. Union Forest. Res. Organ.* (IUFRO Congress) 5:518–31

78. Skye, E. 1968. Lichens and air pollution. *Acta. Phytogeogr. Suec.* 52: 1–123

79. Smock, R. M., Watson, R. D. 1941. Ozone in apple storage. *Refrig. Eng.* 42:97–101

80. Sobotka, A. 1964. Vliv prumyslových exhalátu na pudni Zivenu smrkovjch porostu Krusných hor. *Lesn. Cas.,* Praha 10:987–1002

81. Spalding, D. H. 1966. Appearance and decay of strawberries, peaches, and lettuce treated with ozone. *U.S.D.A. Mkt. Res. Rept. No. 756.* 11 pp.

82. Stark, R. W., Miller, P. R., Cobb, F. W., Jr., Wood, D. L., Parmeter, J. R., Jr. 1968. Photochemical oxidant injury and bark beetle infestation in injured trees. *Hilgardia* 39: 121–26

83. Swedish Royal Ministry for Foreign Affairs and Swedish Royal Ministry of Agriculture. 1971. *Air pollution across national boundaries: the impact on the environment of sulfur in air and precipitation: Sweden's case study for the United Nations Conference on the Human Environment.* Kungle. Boktryckeriet P. A. Norstedt & Söner 710396 Stockholm. 96 pp.

84. Templin, E. 1962. Zur populations dynamik einiger kiefernschadinsekten in rauchgeschaedigten bestäenden. *Wiss. Z. Tech.* Universitaet Dresden 11:631–37.

85. Tingey, D. T., Blum, U. 1973. Effects of ozone on soybean nodules. *J. Environ. Qual.* In press

86. Treshow, M. 1965. Response of some pathogenic fungi to sodium fluoride. *Mycologia* 57:216–21

87. Treshow, M. 1971. Fluorides as air pollutants affecting plants. *Ann. Rev. Phytopathol.* 9:21–44

88. Treshow, M., Dean, G., Harner, F. M. 1967. Stimulation of tobacco mosaic virus-induced lesions on bean by fluoride. *Phytopathology* 57:756–58

89. Treshow, M., Harner, F. M., Price, H. E., Kormelink, J. R. 1969. Effects of ozone on growth, lipid metabolism and sporulation of fungi. *Phytopathology* 59:1223–25

90. Treshow, M., Pack, M. R. 1971. Fluoride. In: *Recognition of Air Pollution Injury to Vegetation: A Pictorial Atlas* ed. J. S. Jacobson, A. C. Hill. Rept. No. 1, TR-7 Agricultural Committee, Air Pollution Control Assn.

91. Watson, R. D. 1942. *Ozone as a fungicide.* PhD thesis. Cornell Univ., Ithaca, N. Y. 74 pp.

92. Wentzel, K. F. 1965. Insekten als Immissionsfalgeschädlinge. *Naturewissenschaft* 52:113

93. Yarwood, C. E. 1959. Virus infection and heating reduce smog damage. *Plant Dis. Reptr.* 43:129–30

94. Yarwood, C. E., Middleton, J. T. 1954. Smog injury and rust infection. *Plant Physiol.* 29:393–95

SYSTEMIC FUNGICIDES: DISEASE CONTROL, TRANSLOCATION, AND MODE OF ACTION[1]

❖ 3579

Donald C. Erwin[2]

Department of Plant Pathology, University of California, Riverside

The field of systemic fungicidal control has enjoyed several breakthroughs within the past 7 or 8 years, which undoubtedly will be landmarks in the history of fungicidal control. The discovery of the oxathiins (211), the pyrimidines (61), and benzimidazoles (51, 201, 216) opened up practical possibilities for the control of plant disease that have increased the intensity of work in the area of control, and subsequently on mode of action and translocation.

Systemic fungicides undoubtedly will cover susceptible foliage and flower parts more efficiently than protectant fungicides because of their ability to translocate through the cuticle and across leaves. For instance Ramsdell & Ogawa (166) reported that, after prebloom spray with benomyl, every susceptible flower part of stone fruit trees contained methyl-2-benzimidazolecarbamate (MBC) and was protected against *Sclerotinia laxa*, the cause of brown rot.

Also, Kaars Sijpesteijn (119) noted that the more specific modern systemic fungicides inhibited biosynthetic processes rather than cell respiration (energy producing processes). This quality should be advantageous in that concentrations of the fungicide too low to affect the relatively slower-growing plant might be active enough to suppress the faster-growing fungus. At least some of the systemics act at the DNA level (40).

The world at this time is particularly concerned about chemical pollution, which might, at first thought, discourage work on systemic fungicides. However, if the total dosage and number of treatments needed for control can be reduced, excess chemical use could be avoided. Also systemics may replace certain dangerous chemicals. For instance Ogawa expects that benomyl may

[1] The literature search was concluded on November 15, 1972. The data reported here do not constitute recommendations for use. Registration requirements for each crop must be obtained from the proper regulatory office in each country.

[2] The writer acknowledges financial support by Cotton Incorporated, Raleigh, North Carolina.

389

replace the undesirable eradicant fungicide, sodium arsenite, which has been necessary to suppress overwintering sporodochia of *Monolinia laxa* (150, 166).

This paper will summarize the most recent research on disease control by systemic fungicides and on factors affecting their translocation, metabolism, and mode of action. The latest information, rather than a historical approach, is given priority. The systemic antibiotics and nonfungitoxic systemics are not covered. The review of Brian (22) and Wain & Carter (213) discussed translocation of several antibiotics. Principles in the 1962 paper of the late A. E. Dimond on objectives in plant chemotherapy are still apropos (52). Because of lack of space, the other excellent reviews prior to 1967 have not been included here.

Current reviewers of the modern fungicides (the oxathiins, benzimidazoles, and pyrimidines) have kept well abreast of the field. Crowdy (46) discussed the translocation of fungicides in plants in 1971 and compared and contrasted the effects of protectant and systemic fungicides at the leaf surface. Erwin (62) reviewed the methods used to detect systemic fungicides by biological techniques, with special emphasis on benzimidazoles for control of *Verticillium* wilt. A supplemental review on progress in the development of systemic chemicals for control of diseases (63) appeared in 1970. The nonfungitoxic systemic compounds that apparently act by increasing resistance represent an area with much potential, but are not yet practical (152). Van der Kerk (121) also encouraged search for nonfungitoxic systemic compounds that interfere with the fungus-plant interaction, as well as for the fungitoxic systemics.

Evans (70) presented the comparative structures of oxathiins, pyrimidines, benzimidazoles, thiophanates, and morpholines, and described their activities in practice. The most complete recent review is that by Kirby (122) who documented disease responses to most well-known systemic fungicides. His treatment of the influence of formulation on effectiveness, interruption of overwintering stages of the fungus, selection of resistant strains to fungicides, and mode of action, make this review a must for the well-informed reader. Woodcock's recent review (220) is a concise and complete treatment of control, chemical structure, mode of action, translocation, and resistance to fungicides. This is an excellent supplement to his earlier review (219). Byrde (31) considered the use of systemics in tropical agriculture. The book edited by Tahori (206) contains many recent papers on systemic fungicides.

BENZIMIDAZOLES

Control

Benomyl, methyl-1-(butylcarbamoyl)-2-benzimidazole carbamate hydrolyzes to methyl-2 benzimidazole carbamate (MBC) (Fig. 1) in water (39) and in foliar tissue from plants to which benomyl had been applied via the roots (156, 187).

The butylamine product of the hydrolysis of benomyl induced the phytoalexin, hydroxyphaseolin, in soybean plants, but MBC did not (172), which may indicate that the butylamine moiety is phytotoxic. Another fungitoxic product was tentatively identified as butylisocyanate (92). Another conversion product of benomyl, 5-triazino-benzimidazole, formed when benomyl was mixed with Bordeaux mixture or with NaOH (pH 12). The new compound was less toxic but had a similar spectrum of activity to benomyl (149).

Benomyl has the widest spectrum of fungitoxic activity of all the newer systemics and is the most effective benzimidazole fungicide. Several important groups of fungus pathogens are insensitive to benomyl, such as the Phycomycetes, some Basidiomycetes, and dark spored members of the Deuteromycetes (19, 57). Control of diseases caused by fungi parallels closely the in vitro sensitivity of these fungi (34, 35, 51). Although benomyl is not considered active against mycoplasma, a report from India indicated that bark placement of benomyl induced a remission of symptoms of the spike disease of sandalwood caused by a mycoplasma, after which the mycoplasma was not detectable by staining (169). More data will be required to substantiate this report.

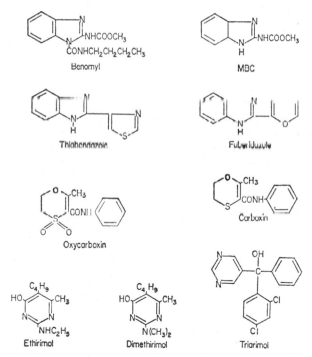

Figure 1 Structures of some systemic fungicides discussed in the text.

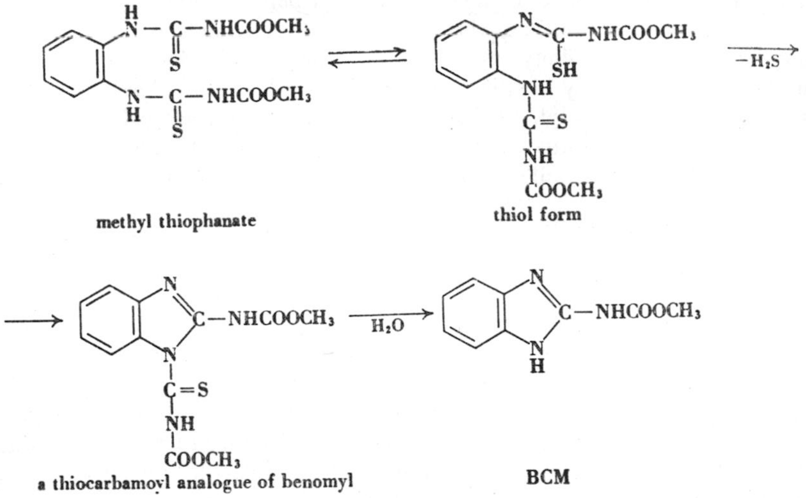

Figure 2 Chemical conversion of thiophanate M to methyl-2-benzimidazole carbamate (BCM). (Reproduced courtesy of Van der Kerk from Figure 17 (121).) See also (184, 212).

Thiabendazole, 2-(4-thiazolyl)benzimidazole, (TBZ) (Fig. 1) has a similar fungitoxic spectrum to that of benomyl, but quantitatively is less effective against diseases that both will control (122). TBZ is taken up and translocated intact, without hydrolysis, to stems and foliage. Weinke et al (216) summarized the diseases best controlled by TBZ.

The fungitoxic activity of thiophanate fungicides is similar to that of benomyl (1). Thiophanate is 1,2-bis(3-ethoxy-carbonyl-2-thioureido)-benzene and thiophanate methyl is 1,2-bis(3-methoxy-carbonyl-2-thioureido)-benzene (thiophanate M.). Another analog is 2-(3-methoxycarbonyl-2-thioureido)a-naline (NF 48). The antifungal spectrum of the thiophanates is similar to benomyl. Benomyl was far the most effective, followed by NF 48, thiophanate methyl, and thiophanate (18). The fungitoxic conversion product of thiophanate M (Fig. 2) is MBC (184, 212) (Fig. 1). Thiophanate converted to the corresponding ethyl-2-benzimidazole carbamate (EBC). Fungus metabolic (212) and plant metabolic activity (139) hastened conversion to MBC.

Fuchs et al (76), studying the conversion of benomyl, thiophanate, thiophanate M, and NF 48 to MBC (or EBC), noted that the rate of transformation (T) decreased: T benomyl > T NF 48 > T thiophanate M > T thiophanate. When concentrations (C) of MBC (or EBC) were measured in plants (cucumber, pea, aster) that had taken up the compounds from a nutrient solution for 2 days, the sequence was: C NF 48 > C benomyl > C thiophanate M > C thiophanate, but the effectiveness (E) in protecting barley plants against *Erisyphe graminis* 20 days later was: E benomyl > E thiophanate M > E NF 48 > E thiophanate. When the compounds were boiled for 1 hour (which converted them to MBC or EBC) effectiveness decreased although the fungitoxic activity in vitro increased. They interpreted this to indicate that benomyl or thiophanate M may be taken up by roots and MBC released gradually. Our unpublished data might also be interpreted in this way. When thiophanate M was applied to roots of cotton plants, thiophanate M was detected in relatively high quantities after 1 day of root uptake, but by 6 days very little thiophanate M and a relatively high quantity of MBC could be detected (29).

There is excellent potential for utilizing benomyl, MBC, or TBZ in selective isolation media. MBC is heat stable (157), as is TBZ; they may be added to media prior to autoclaving. Follin (73), Ponchet et al (160), and Raabe (unpublished) utilized benomyl (15 ppm) as the antifungal component of a differential medium for isolation of Phytophthora. Because the slow-growing Armillaria mellea is less sensitive than the common soil-borne antagonist Trichoderma viride, it was selectively isolated from rotted roots on a benomyl (15 ppm) containing medium (162).

The selectivity of benomyl might change the microfloral balance in soil; however, high dosages (15 g/m²) drenched on soil every 14 days had little effect on the population of the common members of the soil microflora except that the nonsensitive Alternaria, Stemphyllium, Stysanus, Stachybotrys, Mucor, and some species of Penicillium and Aspergillus increased in number. The level of pathogenic activity of Phytophthora did not increase (161).

As benomyl treatment of soil increased the severity of the nonsensitive Alternaria leaf spot of carnation (131) and a turf disease caused by an unidentified Basidiomycete (192), it is possible that selectivity may lead to an increase in severity of some diseases.

Hofer et al (110) reported that benomyl decreased nitrification in soil but respiration values were unchanged. There was little evidence for degradation of benomyl by a large number of microorganisms isolated from soil.

Benomyl, thiophanate M, and thiabendazole have been applied as seed treatments, soil drenches, soil mixtures, and foliar sprays. Systemic activity has been demonstrated for all methods of application. Examples of control by each of the methods will be given in more detail in the section on uptake and translocation. See also other reviews (19, 63, 122).

FOLIAR APPLICATION Powdery mildew has been controlled on cucumbers and other crops by benomyl (51, 154). Control by benomyl of the brown rot disease of stone fruits, which affects petioles, blossoms, and fruit has been remarkable. In addition to blossom blight control, peach fruit sprayed 20 days before picking was protected from a subsequent artificial inoculation by M. fructicola (150, 151, 166, 170). Spraying pastures with benomyl (129) and TBZ (153) controlled Pithomyces chartarum, the cause of facial eczema of sheep. Benomyl controlled pecan scab, caused by Fusicladium effusum, when applied at 4 and 5 week intervals (6). Increased hay and seed yields of alfalfa resulted from control of common leaf spot, caused by *Pseudopeziza medicaginis,* and spring black stem, caused by *Phoma medicaginis,* by spraying weekly with zinc-maneb or benomyl (218). Although control of leaf diseases of alfalfa may not become economical, fungicidal control is a useful tool for determining disease loss and as a comparison with resistant varieties. Foot rot or eyespot of wheat caused by *Cercosporella herpotrichoides* was controlled by spraying with benomyl at 240 g/hectare (72). The reduction in severity of eyespot, and increase in yield, were correlated when benomyl was applied as a seed dressing or as a spray (500 ppm) on barley that had been previously inoculated (48).

Although thiophanate M and NF 48 were effective for control of powdery mildew of barley and cucumber, neither compound effectively controlled apple powdery mildew when applied to roots (142). Benomyl controlled stripe smut caused by *Ustilao striiformis* (96). Robinson & Hodges (175) eradicated *U. striiformis* from infected *Agrostis palustris* grass clones by drenching. Although benomyl almost completely suppressed ascocarp formation by ergot on ryegrass (102), fenazaflor (phenyl-5,6-dichloro-2-trifluoromethyl-1-benzimidazole carboxylate) was even more effective than benomyl in suppressing apothecial formation of *Gloeotinia temulenta* (blind-seed disease on infected ryegrass seeds) (101).

The eradicant action of benomyl against overwintering forms of fungi is a valuable attribute reviewed by Kirby (122). When applied before leaf fall in apple orchards, benomyl completely prevented perithecial formation during the winter (44). Dipping apple leaves infected with *Venturia inaequalis* with benomyl, TBZ, triarimol, or phenylmercuric chloride in England in the fall prevented formation of perithecia. However, dipping leaves in the former three compounds in the spring increased ascospore release. Only phenylmercuric chloride prevented release of ascospores (30).

A cross section of diseases controlled by thiophanate methyl include powdery mildew of apple caused by *Podosphaera leucotricha,* and scab of apple and pears caused by *Ventura inaequalis* and *V. pyrina,* respectively. Late season sprays also controlled the storage-rot disease of apple caused by *Gloeosporium* (42). Diseases controlled in strawberries, black currants, dwarf beans, lettuce, and bulbs and corms include gray mold caused by *Botrytis cinerea,* leaf spot caused by *Pseudopeziza ribis,* powdery mildew caused by *Sphaerotheca mors-uvae,* and tulip fire caused by *Botrytis tulipae* (83).

DIPPING FRUIT OR ROOTS   Control of diseases by immersion of fruit, roots, or bulbs in a suspension of the benzimidazole fungicides has been remarkably more successful than with protectant fungicides. Eckert summarized much of the literature on control of postharvest diseases, including those of citrus and banana (55). TBZ is in commercial use in many countries for control of postharvest citrus fruit decay.

The postharvest banana disease caused by *Gloeosporium musarum* has been controlled with benomyl, TBZ, and thiophanates. The efficacy in most tests is in the following order: benomyl > thiophanate M > TBZ (126). Gutter & Yanko (91) sprayed benomyl, TBZ, and fuberidazole, 2-(2-furyl)-benzimidazole on orange fruit on the tree. After picking, when fruits were washed to remove the residual fungicides, and inoculated with *Penicillium digitatum,* only benomyl protected the fruit from disease. Bioassays indicated that the fungitoxicant penetrated into the flavedo and even into the albedo. *Alternaria, Pleospora, Fusarium,* and *Oospora* were not controlled.

Ceratocystis fimbriata in sweet potato roots was controlled by benomyl and TBZ (47). TBZ was capable of supplanting mercury seed treatment for control of *C. fimbriata, Plenodomus destruens,* and *Monilochaetes infuscans* (133).

Freesia seedling rot caused by *Fusarium oxysporum* and *F. moniliforme,* gladiolus corm rot caused by *Botrytis gladiolorum,* dry rot caused by *Sclerotinia gladioli,* corm rot caused by *Fusarium oxysporum* f. sp. *gladioli,* and tulip bulb rot caused by *Fusarium oxysporum f. sp. tulipae* were controlled with dip treatments with benomyl or TBZ (113).

SEED TREATMENT   Although benomyl has not been particularly effective as a seed treatment to control such damping-off pathogens as *Rhizoctonia solani,* the fungicide has successfully eradicated *Ascochyta pisi* from pea seed (140) and controlled the seed-borne nematode, *Aphelenchoides besseyi,* cause of white tip disease of rice (207). Benomyl controlled *Erisyphe graminis* on barley (71). Take-all disease caused by *Gaeumannomyces graminis* was controlled in the greenhouse by benomyl seed treatment, but not by carboxin, oxycarboxin, ethirimol, or methirimol (85). TBZ is unique in its ability to control the dwarf bunt disease of wheat caused by *Tilletia contraversa* (111).

SOIL APPLICATION   Benomyl has been applied as a drench or by mixing in soil. The advantage of this method is the provision of a continuous supply of MBC to the upper part of the plant. A disadvantage is the large amount required for control in the field. Phymatotrichum root rot (caused by *P. omnivorum*) is lethal only when a lesion on the crown becomes large enough to circumscribe the lower stem. Application of benomyl or TBZ to the crown and surrounding soil, saved cotton plants (108, 127). Diseased trees in Mexico recovered when soil was removed from beneath the affected tree, mixed with benomyl, and replaced (33). Pressure (200 lbs/inch^2) injection of aqueous suspensions of benomyl beneath trees has also induced recovery of infected trees (16).

CONTROL OF VASCULAR WILT DISEASES   Systemic fungicides to control wilt diseases have been studied in Connecticut since the early 1940s. Some of the early work was summarized by Zentmyer et al (223), who reported that 8-hydroxyquinoline sulfate and benzoate, as well as hydroquinone, p-nitrophenol, and benzoic acid, significantly reduced the progress of the Dutch elm disease, and by Dimond et al (53), who summarized the control of Fusarium wilt of tomato. Although there were a few instances of successful application of chemotherapy to treatment or prevention of Dutch elm disease on large trees, these were not common, presumably because chemicals were not as effectively translocated nor as fungitoxic as the benzimidazole systemics. Despite the greater fungitoxicity of the benzimidazoles, control of the vascular wilts has been limited by insufficient root uptake by field-grown plants. This type of disease is one of the most complex and difficult to control (52), primarily because reinfection via the transpiration stream can always recur after the titer of fungitoxicant is reduced.

To control Dutch elm disease, benomyl was injected into the soil under elm trees (520–810 lbs/acre). The fungitoxicant was detected by bioassay in leaves, twigs, and shoots (15), but the disease was only partially controlled. Smalley (191) obtained similar results except that the fungitoxicant was not detected in the upper parts of trees. TBZ was ineffective. These results indicated that control was possible but uptake from soil was extremely inefficient.

Another experiment with soil injection of benomyl (125 g/tree) prior to stem inoculation of elm trees (203) was much more effective than TBZ and captan and by 14 months after treatment foliar symptoms were reduced from 71% to 4%. MBC was detected in foliar tissue as little as 1 day after treatment. When TBZ and benomyl were mechanically mixed with soil around elm trees, TBZ was more effective than benomyl (205). Only one report indicated that spraying trees with benomyl significantly reduced the disease (103).

Himelick (106) injected chemicals in trunks of trees at hydraulic pressures up to 400 lbs psi. Benomyl as a suspension was difficult to inject (100 lbs psi) in elm, oak, and maple trees, but acid hydrolyzed benomyl moved readily (116). Using an apparatus for injecting benomyl (90), the most consistent uptake in oaks was noted for benomyl in 50% aqueous ethanol and benomyl in 21% lactic acid (130). Solubilization of benomyl, MBC, and TBZ in HCl may have a place in tree injection work (27, 28).

Verticillium wilt of cotton was controlled in greenhouse tests by TBZ (68), but TBZ was only about half as effective as benomyl, and exerted little or no control in the field (62, 65, 68). Benomyl as a soil drench was much more effective than TBZ in greenhouse experiments (minimal dosage for control was 10–20 mg/800 g soil in a pot). When leaf tissue of benomyl-treated plants was bioassayed, MBC was detected in the upper parts of the plant (62, 66). Benomyl induced partial control and a significant 20% yield increase in a highly infested field in 2 different years, but only at uneconomical dosages of 20–40 lbs of benomyl per acre when sprayed in the planting beds or in the irrigation furrow at 6–11 inches in depth (65). Benomyl (or MBC) persisted in soil up to 5–6 months (65, 108).

TBZ or benomyl applied as a combination of seed, soil, and late season foliage treatments in the field reduced the incidence of Verticillium wilt and increased the yield, but was not economical (167). Leach et al (125) also controlled wilt in the greenhouse with benomyl, but field trials were unsuccessful (see section on uptake and translocation). Benomyl amended with Tween 20 surfactant, sprayed in the field in soil at a depth of 12 inches, increased the yield of cotton lint and reduced the severity of Verticillium wilt (21).

When benomyl was dissolved in water and HCl (pH 1.7), the water-soluble MBC·HCl salt applied as a stem and foliar spray controlled Verticillium wilt of stem-inoculated cotton plants in the greenhouse, whereas neither non-acidified MBC nor benomyl were effective (27, 28). However, spraying in the field with the MBC·HCl salt or with benomyl (2500 ppm) plus HCl did not control Verticillium wilt. Since MBC was detectable in the plant for only about 10 days after spraying, reinfection probably took place between applications (D. C. Erwin, unpublished).

Benomyl applied as a potato-tuber seed treatment reduced the effect of *Verticillium* from infested tubers (13). However, Cole et al (41) reported TBZ and benomyl to be erratically phytotoxic to potato seed and ineffective in controlling Verticillium wilt. Benomyl as a seed treatment was ineffective in England, but in soil decreased Verticillium symptoms and increased yield (45). Drenching soil with benomyl plus Tween 20 was effective when artificially infested tubers were planted (12).

Benomyl controlled Verticillium wilt of tobacco as a preplant soil treatment, but not as a foliar spray. TBZ in soil was effective in greenhouse tests but not in the field (104). Benomyl and TBZ dusted on strawberry roots before planting, or applied by drenching soil before planting, effectively controlled the disease and increased the yield. The beneficial effects were still evident 2 years after application (117). Benomyl, thiophanate M, and thiophanate reduced the disease severity on tomato. Benomyl was more effective as a drench than when directly mixed in soil (138). Fusarium wilt of tomato (cause *F. oxysporum* f. sp. *lycopersici*) was controlled by a soil drench with benomyl (202). When greenhouse grown tomato plants took up benomyl via the roots, were washed free of benomyl, and then inoculated, little control of disease occurred; however when benomyl was applied after inoculation and not washed off the roots, the incidence of disease was reduced (14). Benomyl as a drench suppressed foliar and internal xylem-browning symptoms of *Fusarium* wilt of tomato in Netherlands (78).

A soil drench with benomyl controlled wilt caused by *F. oxysporum* f. sp. *melonis* and, when applied 3 days before and 3 days after inoculation, also greatly reduced the number of propagules of *Fusarium* in stems of plants (132). Phialophora wilt of carnation was controlled more effectively by benomyl drenched on soil after planting, than when mixed in the soil prior to planting. Thiophanate methyl and thiophanate also reduced the severity of disease (80). Pionnat (157), utilizing a bioautography method for analysis of MBC (155), determined that 0.5 μg of MBC/g of collar tissue was required for control. Drenching infested soil with benomyl (0.37 g/10 liters of water/m²) every 14 days achieved this level of MBC and control. Control of Phialophora and Fusarium wilts of carnations in greenhouse beds by applying benomyl or TBZ has been obtained at economical rates (4).

In summary, control of the vascular wilt diseases requires extremely large dosages, which are not economical except possibly on shallow-rooted high-value crops such as strawberry and carnations. The problems involved will be discussed in the following section.

Uptake and Translocation

The progress in the field of systemic insecticides is worth examining more closely. According to Metcalf (143) the properties of chemicals for systemic action in plants are: ability to penetrate through roots, stems, or leaves; optimum water solubility to permit movement in the transpiration stream; sufficient stability in the plant environment that the compound or its metabolic products may exert the desired degree of residual insecticidal action.

Rombouts (176) used ³⁵S-labeled sulphone compounds as model systemics and showed that the distribution could be changed by shifting the hydrophile-lipophile balance. Hydrophilic compounds were translocated to the margins of leaves more than the lipophilic. Obligate parasites exerted an accumulative sink effect on transpiration fluids. This may explain why powdery mildews are so effectively controlled by systemic fungicides.

Soil type influences uptake. The decreasing amount of MBC detected in leaves of elm seedlings by bioassay after amendment of the planting substrate was: sand > silt loam soil > silt loam soil plus perlite and peat (109, 181). Peat moss added to sand reduced uptake of benomyl in cotton (62). Benomyl was adsorbed to soil, but not to sand in a soil column (78). The greater phytotoxicity of TBZ in sand than in soil was due to greater uptake (222).

After most systemic fungicides are taken up by roots they are translocated upward via the xylem tissue to the leaves. The fungicide that reaches a leaf usually migrates toward the periphery, but usually does not return to the xylem system in the stem to migrate to the leaf above it. The finding that chloroneb moved from a cotyledonary leaf to the opposite leaf may represent an exception (123). Baron (7), assaying MBC in banana leaves treated with benomyl plus oil, also detected MBC by an ethyl acetate-bioassay technique in new leaves, which implied downward movement in the treated leaf. The amount of benzimidazoles translocated downward in most plants is usually small, and in the benzimidazoles is of little physiological or pathological significance. The uptake of systemic insecticidal and fungicidal compounds was related to the degree of transpiration (43, 156). Because MBC from benomyl treatment moves in xylem, very little is deposited in fruit (3, 107, 186).

Meyer et al (144), utilizing the bioautography method of Peterson & Edgington (155), traced movement of both benomyl and MBC upward from treated roots to leaves of bentgrass. This is perhaps the first paper to report detection of benomyl per se in plant tissue. When a second root system was started from a single stolon emanating from the first plant, translocation occurred to leaves and aboveground tissues, but not to the roots of the second plant. When the reciprocal experiment was done by treating roots of the second plant, translocation occurred back to the first plant. They postulated that a reverse flow occurred in the xylem stream to account for this movement. In a similar report, both benomyl and MBC were translocated upward to leaves of strawberry plants and also laterally to all the tissues of a daughter plant grown in a separate nontreated pot of vermiculite (148).

The distribution and metabolic fate of ^{14}C benomyl from roots to foliage was followed in pea plants. ^{14}C MBC, but not ^{14}C benomyl was recovered as the major derivative. Seventy-eight percent of the label was ^{14}C MBC 52 days after treatment, 5% was in water-soluble metabolites, and 14% was bound in plant residue (186). There was no evidence of MBC in seeds of cotton when benomyl was applied to roots or directly on leaves or bolls (3, 107).

Fuchs et al (76) presented the interesting hypothesis that benomyl (an N-1 substituted MBC) and thiophanate M, which probably produces an intermediate N-1 substituted MBC (Fig. 2), entered roots and released MBC gradually to be transported to the upper leaves. NF 48, which was close to MBC, and heated solutions of thiophanate and thiophanate M (transformed almost exclusively to EBC or MBC) moved so rapidly from roots to stems that they did not exert as effective a long-term degree of control of powdery mildew as the parent compound.

Weinke et al (216) indicated that TBZ was taken up by cotton and tomato roots from soil, and in tomato also moved from leaves to roots. However, TDZ and ^{14}C TBZ in cotton moved only toward the apex of leaves (28, 215) TBZ could be detected in xylem and bark of cotton plants that had taken up TBZ from soil but could not be detected consistently in leaves (67). When ^{14}C TBZ uptake was from roots, about 60% of the label that was extracted from plant tissue was authentic TBZ; however, the remaining label consisted mostly of higher molecular weight products, which indicates complexing or binding to plant constituents (215). The translocation of TBZ at least in cotton does not appear to be as free as that of MBC from benomyl.

In soybean ^{14}C TBZ translocated unaltered, and although distribution to all tissues occurred, the highest values were found in roots. Very little accumulated in hypocotyl tissue (89). After seed treatment of soybean, benomyl and chloroneb located in the cotyledons at first, but 4 days later chloroneb moved into the hypocotyl and cotyledons while benomyl moved only into the epicotyl. Carboxin distributed uniformly throughout the seedling, with highest concentrations in the epicotyl (208). The behavior of these chemicals may explain why benomyl has a low degree of effectiveness against seedling pathogens while chloroneb is effective (88). When a hydroxy-ethoxy group was attached to the 1 position of the benzimidazole ring of TBZ (M-75), it was not fungitoxic in vitro, but by breaking down to TBZ in cucumber plants, controlled powdery mildew (59).

Although relatively low dosages of benomyl control vascular diseases in the greenhouse, uneconomically high concentrations must be used in the field. The most precise experimental data explaining the lack of control of Verticillium wilt disease by soil application with benomyl is in a little-noticed paper by Leach et al (125) on the effect of placement and concentration on uptake and control. Naturally-infested field soil in pots was treated with benomyl at 100 ppm, a dosage that previous tests indicated would give protection from wilt for at least 200 days. In other sets of pots in which the same amount of benomyl had been mixed in 50%, 25%, 12%, 6%, and 3% of the soil in pots, the uptake of MBC measured by bioassay 119 days later was highest where 100% of the soil had been treated ranging gradually downward to no uptake where only 3% was treated.

Up to 21.7 lbs/acre were applied in the field by rototilling benomyl in two profile zones of 36-inch^2 and 72-inch^2 dimensions. At the high dosage of 21.7 lb. benomyl/acre, the average inhibition zone (leaf bioassay 26 days after planting) was 360 mm^2 in the 72-inch^2 plots, compared to only 178 mm^2 in the 36-inch^2 plots. After 81 days the inhibition zone was only 18 mm^2 in the 72-inch^2 zone and zero in the 36-inch^2 zone. No control of Verticillium wilt occurred. These data indicated clearly that uptake was dependent on the relative percentage of roots that had an opportunity to take up benomyl (or MBC).

To explain this phenomenon they related their finding to a study in which the pattern of uptake of P^{32} beneath the irrigated cotton row was determined (9). The profile area from which 90% of the P^{35} was extracted by roots rapidly increased from 25-inch^2 on May 1 to 2200-inch^2 by July 18. Thus, theoretically the 36-inch^2 area in which Leach et al (125) incorporated benomyl comprised only 1.5% of the area that the cotton roots explored at the latter date. These data point out the problem in placing a relatively insoluble fungicide like benomyl in the soil in such a way that enough roots have an opportunity to take it up, and indicate that uptake of benomyl (MBC) is too limited to be a practical control for a vascular wilt of deep-rooted crops. Shallow rooted or potted crops are thus much more amenable to systemic fungicidal control.

Foliage application is often the method of choice, and has been most successful for use of all the benzimidazoles. Translaminar movement of benomyl, TBZ, or thiophanate has been detected by treating one side of a sugar-beet leaf with the fungicide, and inoculating the other with *Cercospora beticola* (197). Brown & Hall (25) also report translaminar movement of benomyl and triarimol.

Thiophanate methyl, thiophanate, benomyl, MBC, and TBZ (in that order of efficiency) translocated through cuticle chemically removed from apple leaves (199). Transcuticular movement of acidified MBC and TBZ was four-fold greater than that of the nonacidified compounds. Increased water solubility of HCl salts of MBC and TBZ may be responsible for their increased uptake in plants (27). Benomyl, MBC, thiophanate, and thiophanate methyl moved across the leaf, but captan, captafol, or chlorothalonil did not (199).

When the hydrochloride salts of benomyl and TBZ (water soluble at pH 1.5 and 2.7 respectively) were applied to foliage and stems of cotton plants enough of either MBC or TBZ penetrated 'to control Verticillium wilt induced by stem puncture. Application to stems seemed more important than to leaves. Neither nonacidified TBZ nor benomyl induced control of *Verticillium* (27, 28). The bioassay and chemical extraction data indicated quantitatively that acidification increased the concentration of TBZ or MBC in the xylem and bark tissue.

INCREASE OF UPTAKE WITH ADJUVANTS   How much of the fungicide applied to foliage or soil reaches the place of infection? There is little data, but it is safe to say that uptake is either relatively inefficient or there is considerable room for improvement. According to the review of Foy & Smith (74), surfactants modify the activity of herbicidal sprays. Although increased wetting due to reduction of surface tension favors penetration of chemicals, there are more subtle pesticide-plant interactions beyond the effect of wetting. Since apolar (lipophilic) and polar (hydrophilic) pathways exist in cuticle, certain structural configurations are related to activity. It appears from radioactive studies with herbicides that surfactants enhance penetration only at the point of application and in immediately underlying tissues (74).

Translaminar translocation of benomyl and triarimol was increased by nonionic surfactants (25). Nonionic surfactants, which are less phytotoxic than cationic and anionic types, not only increase the wetting quality of the aqueous spray, but may also have a solubilizing effect: triarimol was soluble in water at 14 ppm, but in 0.1% Tween 20 [polyoxyethylene(20)sorbitan monolaureate] at 23 ppm.

Tween 20 increased the uptake of benomyl applied to soil for control of Verticillium wilt of cotton (171). However, adjuvants RA600, PE510, RS710, AR150, AR200, and Tween 20 increased (in that order) the diffusion of benomyl in agar, suggesting that the adjuvant increased the solubility of benomyl (20). Pitblado & Edgington (158) proved that the solubility of MBC was increased from 10 to 20 ppm with Tween 20, to 840 ppm with RS710, and to 1000 ppm with RA600, part of the effect being due to the acidic nature of the latter two adjuvants. Downward movement of benomyl in soil cores in the field was also increased by these adjuvants. Stipes & Oderwald (204) tested surfactant effects in soil by visualizing the movement of benomyl or TBZ on a soil thin-layer chromatography plate. Movement was increased by the nonionic Tween 80 [polyoxyethylene(20)sorbitan monoleate], but cationic and anionic adjuvants were ineffective.

Nonphytotoxic oils increase the uptake of certain herbicidal chemicals (112). A paraffinic spray oil (10–20%) greatly increased uptake of benomyl in cotton plants (64). However, oil was slightly phytotoxic and reduced yield in the field. Oil increased the efficacy of benomyl as a foliar spray to control Cercospora leaf spot of peanuts (193). The mal secco disease of citrus was controlled by TBZ and carboxin when used with oil on leaves (200).

The manner in which oils increase uptake of herbicidal or other pesticidal materials is not well understood. As only about 10% of the oil applied penetrates to the interior of the leaf, it has been proposed that oil merely softens or solubilizes the cutin layer (178). If fungicides in oils or oil-water emulsions could be applied without undue phytotoxicity by ultra-low-volume spraying (5), perhaps lower dosages and more uniform distribution of fungicides might result in more efficient control. Obviously there is still much to learn.

Toxicology and Mode of Action

Sisler's review in 1969 on the effect of fungicides on protein and nucleic acid synthesis covers earlier work on the relationship of the benzimidazole moiety to toxicity (188). The latest information indicates that MBC (from benomyl) interferes with DNA synthesis and with some post-DNA synthesis aspect of the cell replication process in the sensitive fungi *Ustilago maydis* and *Neurospora crassa* (189).

The similar in vitro fungitoxic spectrum of the various benzimidazole fungicides has already been noted. This, plus the fact that resistant mutants to one of these fungicides (8, 93, 105) are usually resistant to the others, suggests strongly that they share a common mode of action, although there may also be secondary sites of action that condition quantitative differences in sensitivity.

Benomyl, TBZ, and fuberidazole were toxic to *Fusarium oxysporum* f. sp. *melonis* on agar media amended with the fungicides (8). ED_{50} values were in the ratio of 1:4:13 respectively for the 3 compounds. Spontaneous and induced mutants resistant to benomyl and TBZ exhibited cross-tolerance to all 3 fungicides, but the fuberidazole mutant was as sensitive as the wild type to benomyl and TBZ. All 3 compounds appeared to have a common mode of action, but benomyl and TBZ perhaps had an additional mechanism.

When an MBC-resistant isolate of *Neurospora crassa* was tested against TBZ or the herbicidal and fungicidal 4,5 dichloro- and 4,5,6-trichloro-2-trifluoromethyl benzimidazole, the MBC isolate was resistant to TBZ, but not to the latter compounds (189). It was concluded that TBZ and MBC had a common mode of action that was different from that of the 2 trifluoromethyl benzimidazoles. Hammerschlag & Sisler (93) stated that benomyl, TBZ, thiophanate M, and MBC probably shared a common mode of action because the antifungal spectrum of activity was similar for all, and fungal mutants resistant to one were usually resistant to others. There were exceptions, an example of which is the greater toxicity of benomyl over that of MBC to *Saccharomyces pastorianus*.

Depending on their concentration, time of incubation, and type of media, benzimidazole fungicides appear to induce abnormalities in spore germination, germ tube elongation, cellular multiplication, and mycelial growth of sensitive fungi. The ED_{50} values for benomyl ranged between 0.2 ppm for the most sensitive fungi to 125 ppm or more for the most resistant (19). Gottlieb & Kumar (86), studying the effect of TBZ on *Penicillium atrovenetum*, found that 20 μg/ml were required to reduce germination of spores by 80%, but as little as 2 μg/ml prevented elongation or growth of germ tubes. The distorting effect of TBZ on germ tubes was somewhat similar to that illustrated for the effect of MBC on *Botrytis fabae* and on *Neurospora crassa* (40). MBC at 1 μg/ml prevented germ tube development of conidia of *Neurospora crassa*, but permitted a 12-fold increase in dry weight of the control (40). Morphological distortion was evident 4 hours after treatment. MBC was not lethal to conidia of *N. crassa* or sporidia of *Ustilago maydis* incubated in buffer solution, but was lethal to cells of both organisms in a medium supporting growth, which suggests the primary effect could well be on inhibition of mitosis. Inhibition of conidial germination of *Fusarium oxysporum* was logarithmically related to the conc of benomyl (1–50 μg/ml) (49). At a relatively high conc (7 and 10 μg/ml) benomyl prevented an increase in cell number of *Saccharomyces pastorianus* during a 9-hour growing period (93), but there were no morphological changes in the cells after treatment. In a similar experiment with MBC, abnormal growth of the organism occurred in a medium containing 10 μg/ml (approx. ED_{50} value), daughter cells failed to separate, and as a result cells developed in clusters, and some terminal cells became elongated. Benomyl at sublethal concentrations induced swollen and distorted germ tubes in *Botrytis cinerea*. The internal fine structure was an abnormally branched reticulate network in the endoplasmic reticulum. Affected nuclei appeared deeply lobed and convoluted (174).

TBZ, benomyl, and thiophanate M, when applied to barley seeds or to the soil, inhibited appressorium formation but not spore germination or haustorium formation of *Erysiphe graminis* f. sp. *hordei* (180).

The vapor phase of benomyl, ClTBZ, 2-(4-thiazolyl)dichloro-benzimidazole, fentin acetate, and maneb prevented spore germination of various fungi (198) seeded on slides or on sugar beet leaves in sealed containers, while TBZ and thiophanate had no effect on germination, but inhibited elongation of germ tubes. On sprayed (200 μg/ml) leaves the vapor action diminished with time, but after 11 days benomyl still exerted almost complete inhibition of germination. One might speculate that this effect could be due to the volatile compound produced in air over moistened benomyl (tentatively identified as butylisocyanate), which was toxic to *U. maydis* and *S. cerevisiae* (92).

There is some experimental data on the uptake, accumulation, and localization of benzimidazole fungicides in treated fungus cells. TBZ is readily absorbed by spores of *Pencillium atrovenetum* (86). Absorption was directly related to concentration of TBZ and to the time of incubation. TBZ appeared to be present mainly in the cell fluids, with a smaller fraction being in the cell wall, nuclear, mitochondrial, and ribosomal fractions of ground spores.

Maximum uptake of benomyl (10 μg/ml) in conidia of *Fusarium oxysporum* occurred in 60 minutes (49), at which time 46% of the available fungicide had been taken up. ^{14}C-MBC is also taken up rapidly and concentrated by the conidia of *Neurospora crassa* or sporidia of *Ustilago maydis* (40). When spores of either fungus were mixed in a medium containing 2 μg ^{14}C-MBC/ml, the uptake was complete in 30 minutes. The cells contained a higher conc 1.4 μg/g fresh weight) than the medium (0.7 μg/ml ^{14}C-MBC). A portion of the label taken up by the cells was bound or converted to a metabolite that differed from MBC in chromatographic and solubility properties (40).

Clemons & Sisler (40) reported that MBC (1 μg/ml) inhibited DNA synthesis in spores of *Neurospora crassa* in about 3 hours, but not RNA or protein synthesis. DNA, but not RNA or protein synthesis declined in sporidia of *Ustilago maydis* within 60 minutes, after which RNA and protein synthesis also declined. MBC appeared to interfere with DNA synthesis or some closely related process such as nuclear or cell division. Sisler (189) also showed that TBZ and MBC acted by a mechanism other than that of 4,5 dichloro- and 4,5,6-trichloro-2-trifluoromethyl benzimidazole, which not only prevented cell multiplication but also prevented increase in dry weight and stimulated respiration. Benomyl rapidly inhibited glucose or acetate oxidation, RNA and protein synthesis in *Saccharomyces pastorianus* (93); MBC did not inhibit glucose or acetate oxidation, RNA or protein synthesis, but after 1 hour DNA synthesis was curtailed. MBC-treated cells became greatly enlarged and distorted. They suggested that benomyl acted on a "primary phase of metabolism while MBC inhibited cytokinesis or mitosis."

RNA content of *Fusarium oxysporum* conidia was determined after a 7-hour exposure period of 20 μg/ml benomyl (49). RNA content slightly increased during this period, but the net synthesis, measured as the rate of incorporation of ^{14}C-uracil in RNA was significantly less than that of the control. Toxicity of benomyl, TBZ, and fuberidazole against *Fusarium oxysporum* f. sp. *melonis* was reduced in the presence of purines and certain other compounds involved in nucleic acid synthesis (8) suggesting that these fungicides act as antimetabolites. Benzimidazole (1,3 benzodiazole) has been considered to be an antimetabolite of purines, but this role was later discounted by other investigators (40).

Oxygen consumption by *Penicillium atrovenetum* with exogenous glucose was inhibited by TBZ at 20 μg/ml, while inhibition of endogenous respiration required 100 μg/ml TBZ (2). Interference with aerobic respiration could be attributed to the action of TBZ on fungus mitochondria. When the effect of TBZ on various respiratory enzymes present in mitochondria isolated from

Penicillium atrovenetum or beef heart was studied, Allen & Gottlieb (2) concluded that the action of TBZ was primarily on the electron transport system, more specifically between succinate and coenzyme Q, and that the observed decrease in protein synthesis and transaminating properties were secondary effects. Clemons & Sisler (40) questioned this conclusion because the level of TBZ needed to inhibit whole cell respiration was higher than that needed to inhibit growth, which would imply that the mitochondrial site of action may be less sensitive than other sites of action elsewhere in the cell. TBZ did not induce any alteration of respiration of sclerotia of *Phymatotrichum omnivorum* (127).

In summary, the benzimidazole fungicides are selective. Slower uptake, detoxification, or other metabolic factors may explain the tolerance of certain isolates, mutants, or groups of fungi. On the other hand, sensitivity of certain fungi was attributed to alteration of certain physiologic or metabolic processes, particularly those involved in nucleic acid synthesis and respiration. The action of different benzimidazole fungicides, although similar in some instances, is by no means identical. Physical and metabolic transformation of a fungicide to a more or less toxic product may add to the effects of the benzimidazole fungicides on the ultrastructure. Studies of the localization of these fungicides within the cells, and studies on the sensitivity of isolated enzyme systems, may shed more light on the so-called primary site of action of this interesting group of fungicides.

OXATHIINS

The first report on 5,6-dihydro-2-methyl-1,4 oxathiin-3-carboxanilide (formerly, 2,3-dihydro-5-carboxanilido-6-methyl-1,4 oxathiin) (carboxin) and 5,6-dihydro-2-methyl-1,4-oxathiin 3-carboxanilido 4,4-dioxide (oxycarboxin) (see Fig. 1) by Von Schmeling & Kulka (211) established that carboxin as a seed treatment controlled the internally seed-borne loose smut of barley (cause *Ustilago nuda*), and that both chemicals translocated from roots in sufficient quantity to control rust on primary leaves of bean inoculated with *Uromyces phaseoli;* however, when trifoliate leaves were inoculated, only oxycarboxin effectively controlled rust. These chemicals were probably the first of the modern systemic fungicides to succeed on a practical level. Both are much more water soluble, carboxin at 170 ppm and oxycarboxin at 1000 ppm, than benomyl.

Control

Carboxin became uniquely useful as a seed treatment to control the seed-borne loose smut of cereals (146). *Ustilago tritici* was controlled in wheat seed in England only by a soak in 0.2 % carboxin for 6 hours at 30°C, but *U. nuda* was controlled by seed treatment alone (141).

Copper oxyquinolinate and carboxin as a seed treatment acted synergistically for control of *Septoria nodorum*, *Helminthosporium gramineum*, and *Fusarium nivale*. Common bunt (caused by *Tilletia caries* and *T. foetida*) were controlled by seed treatment with benomyl or carboxin but not by oxycarboxin. Dwarf bunt (cause, *T. contraversa*) was not controlled by seed treatment, but was reduced by spraying seedlings and soil in the fall with benomyl or carboxin (173). Carboxin and thiram were more effective together for control of seed-borne *H. gramineum* on barley (124).

The related thiazole fungicide G-696 (2,4-dimethyl-5-carboxanilidothiazole) was more active against *Ustilago striiformis,* but more phytotoxic to bluegrass than oxycarboxin (98). Oxycarboxin was more effective against *Puccinia coronata* of ryegrass (84).

As a cotton seed treatment carboxin and chloroneb tended to reduce emergence and seedling growth slightly, but consistently increased the stand. The interactions with systemic insecticides and protectant fungicides were discussed by Ranney (168). Carboxin and chloroneb, taken up by sugarbeet roots, protected them from attack by *Sclerotium rolfsii* (147). Mal secco of lemon (caused by *Deuterophoma tracheiphila*) was controlled by soil drenches with oxycarboxin, TBZ, benomyl, and cycloheximide semicarbazone. Carboxin and TBZ applied to leaves or the bases of stems in mineral oil also suppressed the disease. The prevention of spore germination at 2 μg/ml was utilized for a bioassay to detect carboxin in plant tissue (200).

Translocation

Carboxin (^{14}C-label) was absorbed by soybean seed and, in contrast to benomyl and chloroneb, did not localize in the cotyledons, but translocated throughout the seedling with the highest concentration in the epicotyl (208). Carboxin and oxycarboxin (^{14}C-label) moved apoplastically in beans, eventually accumulating with no redistribution in the margins of transpiring leaves. Carboxin is transformed to a nonfungitoxic sulfoxide more rapidly than oxycarboxin. The control of bean rust (*Uromyces phaseoli*), in time-course studies after treatment, correlated with chemical data showing transformation of carboxin to the sulfoxide. Translaminar translocation was shown by bean rust inoculations (194). Foliar treatment of almond, chestnut, and peach seedlings with the growth retardants, succinic acid 2,2-dimethylhydrazine or cycocel, increased the uptake of ^{14}C-labeled oxycarboxin applied to a leaf near the tip of the plant. About 10% of the label was recovered in the internal tissues, distributed among stems and roots, as well as in leaves above the treated leaf (114). Perhaps redistribution of systemics can be induced with growth regulators.

Toxicology and Mode of Action

Carboxin and oxycarboxin are most selectively toxic to members of the Basidiomycetes with few exceptions (60). The isomer F 427 (2,3-dihydro-5-ortho-phenylcarboxanilido-6-methyl-1,4-oxathiin) was more toxic to several

Deuteromycetes than was carboxin (56). This indicated that the fungal spectrum could be changed by altering chemical structure.

Chemical structure was related to the inhibition of mycelial growth in agar by carboxin, oxycarboxin, and several substituted analogs (195, 196). Substitution of CH3-, Cl-, or NO_2^- groups in the analino moiety did not change the fungitoxic spectrum. When the 2-methyl oxathiin ring was replaced by thiazole or benzanalide rings, fungitoxicity was not lost, but compounds formed after hydrolysis of the carboxanilide linkage were nontoxic. They proposed a similar mode of action for o-toluanalide to that of carboxin because both have identical antifungal spectra. They presented evidence that different substitutions affecting the carboxamide linkage can alter the fungitoxic spectrum to confer toxicity to certain of the Ascomycetes, Deuteromycetes, Oomycetes, and Zygomycetes. The remarkable observation that 2-aminobenzanilide was also toxic to *Plasmopora viticola,* cause of the downy mildew of grape (196), suggests that systemic fungicides could be prepared that would be active against other Oomycetous fungi, most of which are insensitive. Because the eradicant activity of the oxathiins against bean rust (*Uromyces phaseoli*) correlated with in vitro fungitoxicity tests, it appeared that these fungicides acted on the fungi rather than on host metabolism.

Different analogs of carboxin, including three of those studied by Snel et al (196) affected differently the diseases of grasses caused by *Ustilago striiformis* (stripe smut) (99). The activity of carboxin in potted soil was poor on rust, fair on flag smut, and poor on stripe smut. Oxycarboxin was excellent on rust, good on stripe smut, and eradicated flag smut, caused by *U. agropyri.* A 2',3'dimethyl (on the phenyl ring in the carboxanilide moiety) analog of carboxin improved activity (over carboxin) against flag smut, but decreased activity against rust (99).

Mathre's series of reports on mode of action (134–137) indicate that the sensitivity of fungi to carboxin is due in part to the concentration of the fungicide that accumulates in the cell. Experiments utilizing ^{14}C carboxin and ^{14}C oxycarboxin showed rapid uptake in sensitive *Rhizoctonia solani* and *Ustilago maydis* isolates, but not in the resistant *Fusarium oxysporum* f. sp. *lycopersici* and *Saccharomyces cerevisiae* isolates. Much of the fungicide retained in cells of sensitive fungi was associated with the ribosomal fraction (134). In studies of whole cells of *R. solani* and *U. maydis,* concentrations of carboxin inhibitory to growth also decreased respiration (135, 163).

Although carboxin inhibited succinate oxidation in mitochondria from several systems, the greatest effect was on mitochondria from the susceptible fungus, *U. maydis.* While oxycarboxin had a lesser effect, carboxin appeared to inhibit mitochondrial respiration at or close to the site of succinate oxidation (137). The leakage of phosphate from carboxin-treated cells of *Rhizoctonia solani* was a secondary response resulting from an accumulation of metabolic substrates as the result of inhibition of the TCA cycle, rather than to an alteration of the permeability of membranes (185). The main site of action of carboxin was between succinate and coenzyme Q (209). White confirmed that the site of action appeared to be on succinic dehydrogenase or on an electron carrier component immediate to the dehyrogenase (217).

Damage to the mitochondrial system and vacuolar membrane in fungi sensitive to carboxin was found by electron microscopy within 120 min (128). As Mathre found (137), succinic dehydrogenase is inhibited in mitochondrial systems, and is considered to be the point of attack. Carboxin is oxidized in barley and wheat plants grown from carboxin-treated seed mainly to the sulfoxide and to small amounts of the sulfone. As plants reach maturity, the oxathiin residues are connected to insoluble anilide complexes (38). Carboxin was rapidly degraded by the microflora of a soil suspension. Ninety per cent of the carboxin in a *Rhizopus japonicus* culture was degraded within three weeks to a metabolite identified as butyranilide (214). Carboxin oxidized completely to the sulfoxide in 2 weeks in soil. In water at pH 2 and 4, the sulfone form was also detected (37).

PYRIMIDINES

Control and Translocation

Dimethirimol, 5-n-butyl-2-dimethyl amino-4-hydroxy-6-methylpyrimidine (Fig. 1) was first reported by Elias et al in 1968 (61) to be specifically systemic in cucurbits and fungitoxic to the powdery mildew fungus, *Sphaerotheca fuliginea.* Later Bebbington et al (10) reported that ethirimol, 5-n-butyl-2-ethylamino-4-hydroxy-6-methylpyrimidine (Fig. 1), effectively controlled *Erysiphe graminis,* cause of powdery mildew of barley.

Brooks (23) reviewed the translocation of ethirimol from seed or granule application in soil, and the subsequent control of mildew. A secondary but not surprising effect of control of mildew by ethirimol was the increased vigor and weight of roots produced by the protected plant (24). Edgington et al (58) found that ethirimol was more effective as a seed dressing or as a granular application to soil than as a foliar spray applied 6 weeks after planting.

Triarimol, α-(2-4-dichlorophenyl)-α-phenyl-5-pyrimidinemethanol (Fig. 1), has systemic and curative effects on the apple scab disease (*Venturia inaequalis*) and has a wider antifungal spectrum than the previous pyrimidines (26, 87). Triarimol, thiophanate M, and benomyl controlled stripe smut (*Ustilago striiformis*) on bluegrass (210). An analog EL279, α-(2-chlorophenyl)-α-cyclohexyl-5-pyrimidine-methanol was much more effective than triarimol for eradication of *Urocystis agropyri,* cause of stripe smut of bluegrass (100). Triarimol drench on bluegrass sod eradicated stripe smut (*U. striiformis*) and flag smut (*U. agropyri*) (97). Another pyrimidine, 4-amino-6-chloro-2-(methylthio)pyrimidine, controlled flax rust when applied as an eradicant spray after inoculation, or as a soil drench (75). Triarimol reduced ozone injury to beans (123).

Metabolism and Mode of Action

Considerably more is known about metabolism than mode of action. Both ethirimol and dimethirimol are taken up by roots and readily metabolized. One of the N-methyl groups is rapidly lost from dimethirimol in cucumber plants. This results in an N-desmethyl derivative that is fungicidally active. After the second N-methyl group is lost, a fungicidally inactive $-NH_2$ compound is formed. Ethirimol converts to the same compound by loss of the ethyl group. Residue levels of dimethirimol in cucumber fruit are less than 0.2 ppm (32, 36). Sampson & Slade et al (177, 190) presented evidence that these fungicides are noncompetitive inhibitors of C-1 metabolism. More work needs to be done on the difficult problem of investigating an obligate parasite biochemically.

Ragsdale & Sisler (164) noted that triarimol only slightly reduced the dry weight of sporidia of *Ustilago maydis* after 3 hours, but terminated the multiplication of sporidia. However, nuclear divisions continued, causing sporidia to enlarge, become multicellular and multinucleate. Analysis of sporidia showed that DNA and RNA synthesis was not affected, but that ergosterol and other digitonin-precipitable sterols were reduced markedly. This indicated that triarimol affects sterol biosynthesis (165).

MISCELLANEOUS COMPOUNDS

Triforine, N,N'-bis-(1-formamido-2,2,2 trichloroethyl)-piperazine controls rust on cereals, powdery mildew of cereals, apples, and cucurbits, and apple scab (179). Ebenebe et al (54) reported that triforine applied to roots or leaves was highly effective against stem rust of wheat in a growth chamber, and that it inhibited germination of uredospores on wheat leaves. Triforine on leaves was effective for 9 days. Fuchs et al (77) reported that triforine was effective against powdery mildew on barley and rust on beans. In addition, they obtained control of cucumber scab (cause *Cladosporium cucumerinum*) and tomato scab (cause *C. fulvum*), by preinfectional spraying of foliage.

Triforine is either moderately fungitoxic or nonfungitoxic to most nonobligate parasites, but its effect on spore germination and germ-tube elongation is much more marked on such organisms as *Aspergillus niger* and *C. cucumerinum*. Thin-layer bioautography, using *C. cucumerinum* to detect the inhibitory spots on the thin-layer plate, revealed that triforine was taken up intact by roots of bean, Chinese cabbage, and tobacco. In 1 day after applying ^{3}H-triforine to roots of pea plants in hydroponic solution, about 54% of the label was located at Rf 0.90–1.00 (fungitoxic spot). The remainder of the label was located in other nonfungitoxic spots at different Rf values. After 8 days, only 14% of the label was fungitoxic, while the remainder was found in two other spots. After triforine was taken up by roots from hydroponic culture, the ^{3}H-labeled fungicide was transported to leaves and converted to a number of nonfungitoxic products, the terminal residue being possibly piperazine (79).

Buchenauer & Erwin (unpublished) used triforine as a spray application (10,000 ppm) on leaves of cotton, and as a root drench on sandy loam soil (75 mg/800 g) in a pot. Although disease onset of Verticillium wilt was delayed and symptoms reduced by triforine, the dosage required was much higher than that required for control with benomyl (about 20 mg/800 g soil).

Kaspers & Grewe (120) reported that amidosulfonylmethylphenylthioether was systemic and effective against *Pseudoperonospora humuli,* controlling secondary infections after spraying leaves or applying the fungicide as a soil drench.

2-Methyl-5,6-dihydro-4-H-pyran-3-carboxylic acid anilide was highly effective against the loose smut diseases caused by *Ustilago nuda* and *U. avenae* (115). Organophosphorus compounds for systemic insect control have been widely used (143). O,O-diisopropyl-S-benzyl phosphorthiolate, as a granule in irrigation water, controlled rice blast (*Piricularia oryzae*) (221). N-tridecyl-2,6-dimethylmorpholine (tridemorph) has eradicative and systemic activity against *Erysiphe graminis.* Placement of tridemorph at the base of barley leaves 1 day after inoculation gave complete control. Treatment 2 and 3 days after inoculation resulted in control only in a small area above the treated area, but not below it. Established infections above the treated area did not enlarge. Movement as in most systemic fungicides was acropetal. Mildew was also controlled by soil application (159). Jung & Bedford (118) later reported successful control of mildew and yield increases on spring barley. In winter barley, good control was obtained but there was no positive yield response.

The compound, 4-n-butyl-1,2,4-triazole, has the remarkable effect of being toxic to leaf rust of wheat caused by *Puccinia recondita,* but not to other species of *Puccinia* or *Uromyces* causing rust diseases. It was more effective in the field as a foliar spray than oxycarboxin, giving 100% control at 0.56 kg/hectare under severe epidemic conditions (145).

RESISTANCE TO FUNGICIDES

As fungicides become more specific, one would expect fungi to become resistant, as biochemical specificity or selectivity probably is due to interruption of only one genetically controlled event in the metabolism of the fungus. Thus, either by single-gene mutation or by selection of resistant individuals in a population, a resistant population could arise quickly. Dekker (50) distinguishes between biochemical selectivity and biological selectivity. The latter implies biochemical selectivity that affects the pathogen more than the host.

Evans (69, 70) suggested that the rise in populations of resistant fungi might be alleviated by: (*a*) using the least specific fungicides available; (*b*) adopting mixtures or mixed schedules; (c) reducing the period of contact with the pathogen; and (d) avoiding sub-lethal dosages wherever possible. Kirby (122) feels that, in practice, these suggestions are not likely to have a great impact on the problem, as most of the new fungicides are so specific that using mixtures may empty the armory of alternate fungicides even faster and, considering the stability of the newer fungicides, suggestions (c) and (d) may be unfeasible in practice.

In defense of Evan's suggestions, an example of reducing the period of contact to a fungicide can be found in the citrus packing industry in which treatment with orthophenylphenate during the storage-ripening period selected a resistant population of *Penicillium digitatum*. When these lemons were shipped, the diphenyl fungicide used to control green mold was ineffective; however, on lemons from sheds where this fungicide had not been used diphenyl was effective (55, 81, 94). With the advent of TBZ for *Penicillium* control on citrus fruit, it appears that history is repeating itself (95). Since resistance of *Penicillium italicum* (blue mold) and *P. digitatum* (green mold) to TBZ poses a serious threat, TBZ can no longer be used in packing houses before the storage-ripening period.

Resistance to benomyl was first found in *Sphaerotheca fuliginea* (182). Bollen (17) also found benomyl resistant strains of *Penicillium brevicompactum* and *P. corymbiferum* from cyclamen bulbs treated previously with benomyl. These resistant strains were also cross-resistant to thiophanate methyl and to TBZ and fuberidazole, although the resistance to the latter two fungicides was quantitatively less than to the former.

Bent et al (11) reported that resistance in *Sphaerotheca fuliginea* to dimethirimol occurred in glasshouse culture of cucumbers in Europe in 1970. Since spores of resistant isolates took up ^{14}C-dimethirimol to as great an extent as the susceptible isolates, it appeared that resistance was involved with metabolism or cellular distribution. Interestingly, resistance of some isolates was increased from a tolerance level of 0.32 to 12.0 ppm by successive transfers to leaf discs containing increasing concentrations of dimethirimol.

Benomyl may have a mutagenic effect on fungi, as sectoring of a diploid *Aspergillus nidulans* culture to a haploid form occurred in benomyl-containing media, indicating that benomyl either interferes with chromosome segregation or induces chromosome breaks that lead directly to genetic instability. Hastie & Georgopoulos (105), using *A. nidulans,* about which there is much precise genetic information, isolated UV-irradiated conidia on a medium containing benomyl (1 ppm). Genetic analysis confirmed that benomyl resistance was due to the nonallelic genes, *ben-1* and *ben-2,* which were located in different linkage groups and recombined freely. Both strains were also resistant to MBC and TBZ, although the latter was quantitatively less toxic than benomyl.

Georgopoulos et al (82) obtained a single-gene mutant (oxr) resistant to carboxin within *Ustilago maydis*. The mutation affects the behavior of the succinic dehydrogenase system of mitochondria. This agrees with the proposed site of action (137).

Resistance of fungi to specific systemic chemicals obviously will be a problem to cope with in the future. This phenomenon should be neither surprising nor discouraging to the plant pathologist, because this is to be expected. The technological challenge will have to be met by research that provides data for determining the range of variation to be expected, and the alternative chemical controls that can be used.

Literature Cited

1. Aelbers, E. 1971. Thiophanate and thiophanate-methyl, two new fungicides with systemic action. *Meded. Rijksfac. Landbouwwetensch. Gent* 36:126–34
2. Allen, P. M., Gottlieb, D. 1970. Mechanism of action of the fungicide thiabendazole, 2-(4'-thiazolyl) benzimidazole. *Appl. Microbiol.* 20:919–26
3. Ashworth, L. J., Jr., Hine, R. B. 1971. Structural integrity of the cotton fruit and infection by microorganisms. *Phytopathology* 61:1245–48
4. Baker, R. 1972. Control of Fusarium and Phialophora wilt diseases of carnation with systemic fungicides. *Colo. Flower Growers Assoc. Bull.* 260:1–2
5. Bals, E. J. 1969. The principles of and new developments in ultra low volume spraying. *Proc. Brit. Insectic. Fungic. Conf., 5th,* 1:189–93
6. Barnes, G. L. 1971. Effectiveness of extended interval applications of benomyl for control of pecan scab. *Plant Dis. Reptr.* 55:711–13
7. Baron, M. 1971. Dosage, migration et distribution d'un fongicide systemique (benomyl) dans les feuilles de bananier. *Fruits* 26:643–50
8. Bartels-Schooley, J., MacNeill, B. H. 1971. A comparison of the modes of action of three benzimidazoles. *Phytopathology* 61:816–19
9. Bassett, D. M., Stockton, J. R., Dickens, W. L. 1970. Root growth of cotton as measured by P³² uptake. *Agronomy J.* 62:200–3
10. Bebbington, R. M., Brooks, D. H., Geoghegan, M. J., Snell, B. K. 1969. Ethirimol: a new systemic fungicide for the control of cereal powdery mildews. *Chem. Ind.* 42:1512
11. Bent, K. J., Cole, A. M., Turner, J. A. W., Woolner, M. 1971. Resistance of cucumber powdery mildew to dimethirimol. *Proc. Brit. Insectic. Fungic. Conf., 6th,* 1:274–83
12. Biehn, W. L. 1970. Control of Verticillium wilt of potato by soil treatment with benomyl. *Plant Dis. Reptr.* 54:171–73
13. Biehn, W. L. 1970. Evaluation of seed treatments for control of seedborne Verticillium wilt of potato. *Plant Dis. Reptr.* 54:254–55
14. Biehn, W. L., Dimond, A. E. 1970. Reduction of tomato Fusarium wilt symptoms by benomyl and correlation with a bioassay of fungitoxicant in benomyl-treated plants. *Plant Dis. Reptr.* 54:12–14
15. Biehn, W. L., Dimond, A. E. 1971. Prophylactic action of benomyl against Dutch elm disease. *Plant Dis. Reptr.* 55:179–82
16. Bloss, H. E., Streets, R. B., Sr. 1972. Phymatotrichum root rot in fruit trees. *Agrichem. Age* 15:10
17. Bollen, G. J. 1971. Resistance to benomyl and some other systemic fungicides in strains of *Penicillium* species. *Meded. Rijksfac. Landbouwwetensch. Gent* 36:1188–92
18. Bollen, G. J. 1972. A comparison of the in vitro antifungal spectra of thiophanates and benomyl. *Neth. J. Plant. Pathol.* 78:55–64
19. Bollen, G. J., Fuchs, A. 1970. On the specificity of the in vitro and in vivo antifungal activity of benomyl. *Neth. J. Plant. Pathol.* 76:299–312
20. Booth, J. A., Rawlins, T. E. 1970. A comparison of various surfactants as adjuvants for the fungicidal action of benomyl on Verticillium. *Plant Dis. Reptr.* 54:741–44
21. Booth, J. A., Rawlins, T. E., Chew, C. F. 1971. Field treatments for control of cotton Verticillium wilt with benomyl-surfactant combinations. *Plant Dis. Reptr.* 55:569–72
22. Brian, P. W. 1966. Uptake and transport of systemic fungicides and bactericides. *Simp. Int. Agrochim., 6th,* 237–62
23. Brooks, D. H. 1970. Powdery mildew of barley and its control. *Outlook Agr.* 6:122–27
24. Brooks, D. H. 1972. Observations on the effects of mildew, *Erysiphe graminis,* on growth of spring and winter barley. *Ann. Appl. Biol.* 70:149–56
25. Brown, I. F., Jr., Hall, H. R. 1970. The role of surfactants in foliar sprays of systemic fungicides. *Fungic. Nematic. Tests Results 1970* 26:1–3
26. Brown, I. F., Jr., Hall, H. R.,

Miller, J. R. 1970. EL-273. a lucrative fungicide for the control of *Venturia inaequalis. Phytopathology* 60:1013–14

27. Buchenauer, H., Erwin, D. C. 1971. Control of Verticillium wilt of cotton by spraying foliage with benomyl and thiabendazole solubilized with hydrochloric acid. *Phytopathology* 61:433–34

28. Buchenauer, H., Erwin, D. C. 1972. Control of Verticillium wilt of cotton by spraying with acidic solutions of benomyl, methyl 2-benzimidazole carbamate, and thiabendazole. *Phytopathol. Z.* 75: 124–39

29. Buchenauer, H., Erwin, D. C., Keen, N. T. 1973. Systemic fungicidal effect of thiophanate methyl on Verticillium wilt of cotton and its transformation to methyl 2-benzimidazole carbamate in cotton plants. *Phytopathology* 63. In press

30. Burchill, R. T. 1972. Comparison of fungicides for suppressing ascospore production by *Venturia inaequalis* (Cke.) Wint. *Plant Pathol.* 21:19–22

31. Byrde, R. J. W. 1970. The new systemic fungicides and their potential uses in the tropics. *Trop. Sci.* 12:105–11

32. Calderbank, A. 1971. Metabolism and mode of action of dimethirimol and ethirimol. *Acta Phytopathol.* 6:355–63

33. Castro, J., Rodriquez, A. E. 1970. Pruebas preliminares para el combate de la pudricion texana del durazno en el Bajio. *Instituto Nacional de Investigaciones Agricolas, Sec. Agr. Gan.,* Circ. Centr. Inv. Agr. Bajio No. 34. 11 pp.

34. Catling, W. S. 1969. Benomyl, a broad spectrum fungicide. *Proc. Brit. Insectic. Fungic. Conf., 5th,* 2:298–309

35. Catling, W. S. 1971. The control of powdery mildew, canker and other diseases of apple with benomyl. *Proc. Brit. Insectic. Fungic. Conf., 6th,* 1:103–9

36. Cavell, B. D., Hemingway, R. J., Teal, G. 1971. Some aspects of the metabolism and translocation of the pyrimidine fungicides. *Proc. Brit. Insectic. Fungic. Conf., 6th,* 2:431–37

37. Chin, W. T., Stone, G. M., Smith,
A. E. 1970. Degradation of carboxin (Vitavax) in water and soil. *J. Agr. Food Chem.* 18:731–32

38. Chin, W. T., Stone, G. M., Smith, A. E. 1970. Metabolism of carboxin (Vitavax) by barley and wheat plants. *J. Agr. Food Chem.* 18:709–12

39. Clemons, G. P., Sisler, H. D. 1969. Formation of a fungitoxic derivative from Benlate. *Phytopathology* 59:705–6

40. Clemons, G. P., Sisler, H. D. 1971. Localization of the site of action of a fungitoxic benomyl derivative. *Pestic. Biochem. Physiol.* 1:32–43

41. Cole, H., Mills, W. R., Massie, L. B. 1972. Influence of chemical seed and soil treatments on Verticillium-induced yield reduction and tuber defects. *Am. Potato J.* 49:79–92

42. Cole, R. J., Gilchrist, A. J., Soper, D. 1971. The control of diseases of apples and pears in the United Kingdom with thiophanate methyl. *Proc. Brit. Insectic. Fungic. Conf., 6th,* 1:118–25

43. Coleby, A. W. P., Reynolds, H. T., Metcalf, R. L. 1972. Effects of environmental factors on uptake, translocation, and degradation of a systemic phosphonate insecticide in cotton plants. *Environ. Entomol.* 1:129–36

44. Connor, S. R., Heuberger, J. W. 1968. Apple scab. V. Effect of late-season applications of fungicides on prevention of perithecial development by *Venturia inaequalis. Plant Dis. Reptr.* 52:654–58

45. Corbett, D. C. M., Hide, G. A. 1971. Chemical control of *Verticillium dahliae* and *Heterodera rostochiensis* on potatoes. *Proc. Brit. Insectic. Fungic. Conf., 6th,* 1:258–63

46. Crowdy, S. H. 1971. The control of leaf pathogens using conventional and systemic fungicides, 395–407. In *Ecology of Leaf Surface Micro-organisms.* T. F. Preece, C. H. Dickinson ed. London: Academic. 640 pp.

47. Daines, R. H. 1971. The control of black rot of sweet potatoes by the use of fungicide dips at various temperatures. *Phytopathology* 61:1145–46

48. Davies, J. M. L., Jones, D. G.

1971. The effect of benomyl systemic fungicide and the growth regulator CCC on eyespot disease of barley. *J. Agr. Sci.,* Camb. 77: 525–29

49. Decallonne, J. R., Meyer, J. A. 1972. Effect of benomyl on spores of *Fusarium oxysporum. Phytochemistry* 11:2155–60

50. Dekker, J. 1971. Problems of selectivity in the field of systemic fungicides. *Acta Phytopathol. Acad. Sci. Hungaricae.* 6:329–37

51. Delp, C. J., Klopping, H. L. 1968. Performance attributes of a new fungicide and mite ovicide candidate. *Plant Dis. Reptr.* 52:95–99

52. Dimond, A. E. 1962. Objectives in plant chemotherapy. *Phytopathology* 52:1115–18

53. Dimond, A. E., Davis, D., Chapman, R. A. Stoddard, E. M. 1952. Plant chemotherapy as evaluated by the Fusarium wilt assay on tomatoes. *Conn. Agr. Exp. Sta. Bull.* 557:1–82

54. Ebenebe, Ch., Fehrmann, H., Grossmann, F. 1971. Effect of a new systemic fungicide, piperazin-1,4 - diyl - bis - [1-(2,2,2,-trichloroethyl)formamide], against wheat leaf rust. *Plant Dis. Reptr.* 55: 691–94

55. Eckert, J. W. 1969. Chemical treatments for control of postharvest disenses. *World Rev. Pest Control* 8:116–37

56. Edgington, L. V., Barron, G. I. 1967. Fungitoxic spectrum of oxathiin compounds. *Phytopathology* 57:1256–57

57. Edgington, L. V., Khew, K. L., Barron, G. L. 1971. Fungitoxic spectrum of benzimidazole compounds. *Phytopathology* 61:42–44

58. Edgington, L. V., Reinbergs, E., Shephard, M. C. 1972. Evaluation of ethirimol and benomyl for control of powdery mildew of barley. *Can. J. Plant Sci.* 52:693–99

59. Edgington, L. V., Schooley, J. 1972. The effect of structural changes on the systemicity of thiabendazole. *Proc. Can. Phytopathiol. Soc.,* 39:29–30

60. Edgington, L. V., Walton, G. S., Miller, P. M. 1966. Fungicide selective for Basidiomycetes. *Science* 153:307–8

61. Elias, R. S., Shephard, M. C., Snell, B. K., Stubbs, J. 1968. 5-*n*-Butyl - 2 - dimethylamino - 4 - hydroxy-6-methylpyrimidine: a systemic fungicide. *Nature* 219:1160

62. Erwin, D. C. 1969. Methods of determination of the systemic and fungitoxic properties of chemicals applied to plants with emphasis on control of Verticillium wilt with thiabendazole and Benlate. *World Rev. Pest Control* 8:6–22

63. Erwin, D. C. 1970. Progress in the development of systemic fungitoxic chemicals for control of plant diseases. *FAO Plant Prot. Bull.* 18:73–81

64. Erwin, D. C., Buchenauer, H., Khan, R. 1971. Control of Verticillium wilt of cotton in the greenhouse by foliar sprays with hydrochloric acid solutions and oil emulsions of benomyl and thiabendazole (1). *Proc. Beltwide Cotton Prod. Res. Conf.,* 31st Cotton Disease Council, 83–84 Abstr.

65. Erwin, D. C., Garber, R. H., Carter, L., DeWolfe, T. A. 1969. Studies on thiabendazole and Benlate as systemic fungicides against Verticillium wilt of cotton in the field. *Proc. Beltwide Cotton Prod. Res. Conf.,* 29th Cotton Disease Council, 29 Abstr.

66. Erwin, D. C., Mee, H., Sims, J. J. 1968. The systemic effect of 1-(butylcarbamoyl)-2-benzimidazole carbamic acid, methyl ester, on Verticillium wilt of cotton. *Phytopathology* 58:528–29

67. Erwin, D. C., Sims, J. J., Borum, D. E., Childers, J. R. 1971. Detection of the systemic fungicide, thiabendazole, in cotton plants and soil by chemical analysis and bioassay. *Phytopathology* 61:964–67

68. Erwin, D. C., Sims, J. J., Partridge, J. 1968. Evidence for the systemic, fungitoxic activity of 2-(4'-thiazolyl)benzimidazole in the control of Verticillium wilt of cotton. *Phytopathology* 58:860–65

69. Evans, E. 1971. Problems and progress in the use of systemic fungicides. *Proc. Brit. Insectic. Fungic. Conf., 6th,* 3:758–63

70. Evans, E. 1971. Systemic fungicides in practice. *Pestic. Sci.* 2: 192–96

71. Evans, E., Rickard, M., Whitehead, R. 1969. Effect of benomyl on some diseases of spring barley.

Proc. Brit. Insectic. Fungic. Conf.,
5th, 3:610–19

72. Fehrmann, H., Schrödter, H. 1972. Ökologische Untersuchungen zur Epidemiologie von *Cercosporella herpotrichoides.* IV. Erarbeitung eines Praxisnahen Verfahrens zur Bekampfung der Halmbruchkrankheit des Weizens mit systemischen Fungiziden. *Phytopathol. Z.* 74:161–74

73. Follin, J. C. 1971. L'utilisation du benomyl pour l'isolement selectif des pythiacees. *Coton Fibres Trop.* 26:467–68

74. Foy, C. L., Smith, L. W. 1969. The role of surfactants in modifying the activity of herbicidal sprays. *Advan. Chem. Ser.* 86:55–69

75. Froiland, G. E., Littlefield, L. J. 1972. Systemic protectant and eradicant chemical control of flax rust. *Plant Dis. Reptr.* 56:737–39

76. Fuchs, A., Berg, G. A. van den, Davidse, L. C. 1972. A comparison of benomyl and thiophanates with respect to some chemical and systemic fungitoxic characteristics. *Pestic. Biochem. Physiol.* 2:191–205

77. Fuchs, A., Doma, S., Voros, J. 1971. Laboratory and greenhouse evaluation of a new systemic fungicide, N,N'-bis-(1-formamido-2,2, 2 - trichloroethyl) - piperazine (CELA W 524). *Neth. J. Plant Pathol.* 77:42–54

78. Fuchs, A., Homans, A. L., de Vries, F. W. 1970. Systemic activity of benomyl against Fusarium wilt of pea and tomato plants. *Phytopathol. Z.* 69:330–43

79. Fuchs, A., Viets-Verweij, M., deVries, F. W. 1972. Metabolic conversion in plants of the systemic fungicide triforine N,N'-bis-(1 - formamido - 2,2,2 - trichloroethyl)-piperazine; CELA W 524. *Phytopathol. Z.* 75:111–23

80. Garibaldi, A. 1970. Impiego di sostanze ad azione endoterapica nella lotta contro la fialoforosi del garofano. *L'Agr. Italiana* 70:226–35

81. Georgopoulos, S. G. 1969. The problem of fungicide resistance. *Bioscience* 19:971–73

82. Georgopoulos, S. G., Alexandri, E., Chrysayi, M. 1972. Genetic evidence for the action of oxathiin and thiazole derivatives on the succinic dehydrogenase system of *Ustilago maydis* mitochondria. *J. Bacteriol.* 110:809–17

83. Gilchrist, A. J., Cole, R. J. 1971. Control of diseases of soft fruit and other crops in the United Kingdom with thiophanate methyl. *Proc. Brit. Insectic. Fungic. Conf.,* 6th, 2:332–41

84. Gondran, J. 1970. Action de l'oxycarboxine contre la rouille couronnee du ray-grass. *Phytiatrie-Phytopharmacie* 19:133–40

85. Gorska-Poczopko, J. 1971. Studies on chemical control of *Ophiobolus graminis* Sacc. I. Testing systemic fungicides against *Ophiobolus graminis.* Sacc. *Acta Phytopathol.* 6:393–98

86. Gottlieb, D., Kumar, K. 1970. The effect of thiabendazole on spore germination. *Phytopathology* 60:1451–55

87. Gramlich, J. V., Schwer, J. F., Brown, I. F., Jr. 1969. Characteristics and field performance of a new, broad-spectrum systemic fungicide. *Proc. Brit. Insectic. Fungic. Conf., 5th,* 2:576–83

88. Gray, L. E., Sinclair, J. B. 1970. Uptake and translocation of systemic fungicides by soybean seedlings. *Phytopathology* 60:1486–88

89. Gray, L. E., Sinclair, J. B. 1971. Systemic uptake of ^{14}C-labeled 2-(4'-thiazolyl) benzimidazole in soybean. *Phytopathology* 61:523–25

90. Gregory, G. F., Jones, T. W., McWain, P. 1971. Injection of benomyl into elm, oak, & maple. *US Dep. Agr. Forest Serv. Res. Paper NE-232.* 7 pp.

91. Gutter, Y., Yanko, U. 1971. The protective effect of four benzimidazoles applied preharvest for the postharvest control of orange decay. *Israel J. Agr. Res.* 21:105–9

92. Hammerschlag, R. S., Sisler, H. D. 1973. Benomyl and methyl-2-benzimidazolecarbamate (MBC): Biochemical, cytological and chemical aspects of toxicity to *Ustilago maydis* and *Saccharomyces cerevesiae. Pest. Biochem. Physiol.* 3. In press

93. Hammerschlag, R. S., Sisler, H. D. 1972. Differential action of benomyl and methyl-2-benzimidazolecarbamate (MBC) in *Saccha-*

romyces pastorianus. Pestic. Biochem. Physiol. 2:123–31

94. Harding, P. R. 1962. Differential sensitivity to sodium orthophenylphenate by biphenyl-sensitive and biphenyl-resistant strains of *Penicillium digitatum. Plant Dis. Reptr.* 46:100–4

95. Harding, P. R. 1972. Differential sensitivity to thiabendazole by strains of *Penicillium italicum* and *P. digitatum. Plant Dis. Reptr.* 56:256–60

96. Hardison, J. R. 1968. Systemic activity of Fungicide 1991, a derivative of benzimidazole, against diverse grass diseases. *Plant Dis. Reptr.* 52:205

97. Hardison, J. R. 1971. Chemotherapeutic eradication of *Ustilago striiformis* and *Urocystis agropyri* in *Poa pratensis* 'Merion' by root uptake of α-(2,4-dichlorophenyl)-α - phenyl - 5 - pyrimidinemethanol (EL-273). *Crop Sci.* 11:345–47

98. Hardison, J. R. 1971. Chemotherapy of smut and rust pathogens in *Poa pratensis* by thiazole compounds. *Phytopathology* 61:1396–99

99. Hardison, J. R. 1971. Relationships of molecular structure of 1,4-oxathiin fungicides to chemotherapeutic activity against rust and smut fungi in grasses. *Phytopathology* 61:731–35

100. Hardison, J. R. 1972. Control of *Ustilago striiformis* and *Urocystis agropyri* in *Poa pratensis* 'Merion' by minute dosages of pyrimidine compounds. *Plant Dis. Reptr.* 56:55–57

101. Hardison, J. R. 1972. Prevention of apothecial formation in *Gloeotinia temulenta* by systemic and protectant fungicides. *Phytopathology* 62:605–9

102. Hardison, J. R. 1972. Prevention of ascocarp formation in *Claviceps purpurea* by fungicides applied over sclerotia at the soil surface. *Phytopathology* 62:609–11

103. Hart, J. H. 1972. Control of Dutch elm disease with foliar applications of benomyl. *Plant Dis. Reptr.* 56:685–88

104. Hartill, W. F. T. 1971. Control of Verticillium wilt of tobacco with benomyl. *Plant Dis. Reptr.* 55:889–93

105. Hastie, A. C., Georgopoulos, S. G. 1971. Mutational resistance to fungitoxic benzimidazole derivatives in *Aspergillus nidulans. J. Gen. Microbiol.* 67:371–73

106. Himelick, E. B. 1972. High pressure injection of chemicals into trees. *Arborist's News* 37:97–103

107. Hine, R. B., Ashworth, L. J., Jr., Paulus, A. O., McMeans, J. L. 1971. Absorption and movement of benomyl into cotton bolls and control of boll rot. *Phytopathology* 61:1134–36

108. Hine, R. B., Johnson, D. L., Wenger, C. J. 1969. The persistency of two benzimidazole fungicides in soil and their fungistatic activity against *Phymatotrichum omnivorum. Phytopathology* 59:798–801

109. Hock, W. K., Schreiber, L. R., Roberts, B. R. 1970. Factors influencing uptake, concentration, and persistence of benomyl in American elm seedlings. *Phytopathology* 60:1619–22

110. Hofer, V. I., Bech, T., Wallnöfer, P. 1971. Der Einfluss des Fungizids Benomyl auf die Bodenmikroflora. *Z. Pflanzrankh. Pflanzensch.* 78:398–405

111. Hoffmann, J. A. 1971. Control of common and dwarf bunt of wheat by seed treatment with thiabendazole. *Phytopathology* 61:1071–74

112. Hull, H. M. 1970. Leaf structure as related to absorption of pesticides and other compounds. *Residue Rev.* 31:1–150

113. Humphreys-Jones, D. R. 1971. Control of some diseases of freesias, gladiolus and tulips with benomyl and thiabendazole. *Proc. Brit. Insectic. Fungic. Conf., 6th,* 2.362–66

114. Intrieri, C., Ryugo, K. 1972. Enhanced penetration and translocation of a fungicide following foliar treatment of seedlings with growth retardants. *Plant Dis. Reptr.* 56:590–92

115. Jank, B., Grossmann, F. 1971. 2-methyl - 5,6 - dihydro - H - pyran - 3-carboxylic acid anilide: a new systemic fungicide against smut diseases. *Pestic. Sci.* 2:43–44

116. Jones, T. W., Gregory, G. F. 1971. An apparatus for pressure injection of solutions into trees. *US*

Dep. Agr. Forest Serv. Res. Paper NE-233, 7 pp.

117. Jordan, V. W. L. 1972. Evaluation of fungicides for the control of Verticillium wilt (V. dahliae) of strawberry. Ann. Appl. Biol. 70: 163–68

118. Jung, K. U., Bedford, J. L. 1971. The evaluation of application time of tridemorph for the control of mildew in barley and oats in the United Kingdom. Proc. Brit. Insectic. Fungic. Conf., 6th, 1:75–81

119. Kaars Sijpesteijn, A. 1970. Biochemical modes of action of agricultural fungicides. World Rev. Pest Control 9:85–93

120. Kaspers, H., Grewe, F. 1970. A new systemic fungicide with special activity against phytopathogenic Phycomycetes. Proc. Int. Congr. Plant Prot., 7th. 203

121. Kerk, G. J. M. van der. 1971. Differentiation between transportable fungicides and systemic infection-inhibiting agents, and their relative perspectives. Acta Phytopathol. 6:311–27

122. Kirby, A. H. M. 1972. Progress towards systemic fungicides. Pestic. Abstr. News Sum. Sect. B. Fungic. Herbic. 18:1–33

123. Kirk, B. T., Sinclair, J. B., Lambremont, E. N. 1969. Translocation of ¹⁴C-labeled chloroneb and DMOC in cotton seedlings. Phytopathology 59:1473–76

124. Kline, D. M., Roane, C. W. 1972. Fungicides for the control of Helminthosporium stripe of barley. Plant Dis. Reptr. 56:183–85

125. Leach, L. D., Huffman, E. M., Reische, W. C. 1969. Some factors influencing protection of cotton from Verticillium wilt in benomyl-treated soil based on bioassay and performance. Proc. Nat. Cotton Council, 29th Cotton Disease Council, 36–40

126. Long, P. G. 1971. Evaluation of benomyl, thiabendazole, benzene thiophanate and methyl thiophanate for control of banana stem end rot disease (Gloeosporium musarum). Aust. J. Exp. Agr. Anim. Husb. 11:559–61

127. Lyda, S. D., Burnett, E. 1970. Influence of benzimidazole fungicides on Phymatotrichum omnivorum and Phymatotrichum root

rot of cotton. Phytopathology 60: 726–28

128. Lyr, H., Ritter, G., Casperson, G. 1972. Wirkungsmechanismus des systemischen Fungicids Carboxin. Z. Allg. Mikrobiol. 12:271–80

129. McKenzie, E. H. C. 1971. Seasonal changes in fungal spore numbers in ryegrass-white clover pasture, and the effects of benomyl on pasture fungi. N. Z. J. Agr. Res. 14:379–92

130. McWain, P., Gregory, G. F. 1971. Solubilization of benomyl for xylem injection in vascular wilt disease control. US Dep. Agr. Forest Serv. Res. Paper NE-234, 6 pp.

131. Manning, W. J., Papia, P. M. 1972. Benomyl soil treatments and natural occurrence of Alternaria leaf spot on carnation. Plant Dis. Reptr. 56:9–11

132. Maraite, H., Meyer, J. A. 1971. Systemic fungitoxic action of benomyl against Fusarium oxysporum f. sp. melonis in vivo. Neth. J. Plant. Pathol. 77:1–5

133. Martin, W. J. 1972. Further evaluation of thiabendazole as a sweet potato "seed" treatment fungicide. Plant Dis. Reptr. 56:219–23

134. Mathre, D. E. 1968. Uptake and binding of oxathiin systemic fungicides by resistant and sensitive fungi. Phytopathology 58:1464–69

135. Mathre, D. E. 1970. Mode of action of oxathiin systemic fungicides. I. Effect of carboxin and oxycarboxin on the general metabolism of several Basidiomycetes. Phytopathology 60:671–76

136. Mathre, D. E. 1971. Mode of action of oxathiin systemic fungicides. Structure-activity relationships. Agr. Food Chem. 19:872–74

137. Mathre, D. E. 1971. Mode of action of oxathiin systemic fungicides. III. Effect on mitochondrial activities. Pestic. Biochem. Physiol. 1:216–24

138. Matta, A., Garibaldi, A. 1970. Attivita di fungicidi sistemici contro le verticillosi del pomodoro e della melanzana. L'Agri. Italiana 70:331–40

139. Matta, A., Gentile, I. A. 1971. Activation of the thiophanate systemic fungicides by plant tissues.

Meded. Rijksfac. Landbouwwetensch. Gent 36:1151–58

140. Maude, R. B., Kyle, A. M. 1970. Seed treatments with benomyl and other fungicides for the control of *Ascochyta pisi* on peas. *Ann. Appl. Biol.* 66:37–41

141. Maude, R. B., Shuring, C. G. 1969. Seed treatments with carboxin for the control of loose smut of wheat and barley. *Proc. Brit. Insectic. Fungic. Conf., 5th,* 2:328–32

142. Mercer, R. T. 1971. Some studies on the systemic activity of the thiophanate fungicides in plants. *Pestic. Sci.* 2:214–18

143. Metcalf, R. L. 1966. Absorption and translocation of systemic insecticides. *Simp. Int. Agrochim., 6th,* 151–71

144. Meyer, W. A., Nicholson, J. F., Sinclair, J. B. 1971. Translocation of benomyl in creeping bentgrass. *Phytopathology* 61:1198–1200

145. Meyer, W. C. von, Greenfield, S. A., Seidel, M. C. 1970. Wheat leaf rust: Control by 4-n-butyl-1,2,4-triazole, a systemic fungicide. *Science* 169:997–98

146. Moseman, J. G., compiler. 1968. Fungicidal control of smut diseases of cereals. *U.S. Dep. Agr., Agr. Res. Serv. CR-42-68,* 35 pp.

147. Mukhopadhyay, A. N., Thakur, R. P. 1971. Control of Sclerotium root rot of sugarbeet with systemic fungicides. *Plant Dis. Reptr.* 55:630–34

148. Nicholson, J. F., Sinclair, J. B., White, J. C., Kirkpatrick, B. L. 1972. Upward and lateral translocation of benomyl in strawberry. *Phytopathology* 62:1183–85

149. Ogawa, J. M., Bose, E., Manji, B. T., White, E. R., Kilgore, W. W. 1971. Biological activity of conversion products of benomyl. *Phytopathology* 61:905

150. Ogawa, J. M., Manji, B. T., Bose, E. 1968. Efficacy of Fungicide 1991 in reducing fruit rot of stone fruits. *Plant Dis. Reptr.* 52:722–26

151. Ogawa, J. M., Manji, B. T., Ravetto, D. J. 1970. Evaluation of preharvest benomyl application on postharvest Monilinia rot of peaches and nectarines. *Phytopathology* 60:1306

152. Oort, A. J. P. 1971. Development of systemic fungicides—a historical background. *Meded. Rijksfac. Landbouwwetensch. Gent* 36:59–60

153. Parle, J. N., DiMenna, M. E. 1972. Fungicides and the control of *Pithomyces chartarum*. II. Field trials. *N. Z. J. Agr. Res.* 15:54–63

154. Paulus, A. O., Shibuya, F., Osgood, J., Bohn, G. W., Hall, B. J., Whittaker, T. W. 1969. Control of powdery mildew of cucurbits with systemic and nonsystemic fungicides. *Plant Dis. Reptr.* 53:813–16

155. Peterson, C. A., Edgington, L. V. 1969. Quantitative estimation of the fungicide benomyl using a bioautograph technique. *J. Agr. Food Chem.* 17:898–99

156. Peterson, C. A., Edgington, L. V. 1970. Transport of the systemic fungicide, benomyl, in bean plants. *Phytopathology* 60:475–78

157. Pionnat, J. C. 1971. Mise en évidence du bénomyl et du *Phialophora cinerescens* van Beyma dans les tiges d'oeillets plantés en sols infectés et traités. *Ann. Phytopathol.* 3:207–14

158. Pitblado, R. E., Edgington, L. V. 1972. Movement of benomyl in field soils as influenced by acid surfactants. *Phytopathology* 62:513–16

159. Pommer, E. H., Otto, S., Kradel, J. 1969. Some results concerning the systemic action of tridemorph. *Proc. Brit. Insectic. Fungic. Conf., 5th,* 2:347–53

160. Ponchet, J., Ricci, P., Andreoli, C., Auge, G. 1972. Méthodes sélectives d'isolement du *Phytophthora nicotianae* f. sp. *parasitica* (Dastur) Waterh. A partir du sol. *Ann. Phytopathol.* 4:97–108

161. Ponchet, J., Tramier, R. 1971. Effects du bénomyl sur la croissance de l'oeillet et la microflore des sols traités. *Ann. Phytopathol.* 3:401–6

162. Raabe, R. D., Hurlimann, J. H. 1971. A selective medium for isolation of *Armillaria mellea*. *Calif. Plant Pathol.* 3:1

163. Ragsdale, N. N., Sisler, H. D. 1970. Metabolic effects related to fungitoxicity of carboxin. *Phytopathology* 60:1422–27

164. Ragsdale, N. N., Sisler, H. D. 1972. Inhibition of ergosterol synthesis in *Ustilago maydis* by the fungicide triarimol. *Biochem.*

Biophys. Res. Commun. 46:2048–53

165. Ragsdale, N. N., Sisler, H. D. 1973. Mode of action of triarimol in *Ustilago maydis*. *Pest. Biochem. Physiol.* 3. In press

166. Ramsdell, D. C., Ogawa, J. M. 1973. Systemic activity of methyl-2-benzimidazolecarbamate (MBC) in almond blossoms following prebloom sprays of benomyl + MBC. *Phytopathology* 63. In press

167. Ranney, C. D. 1971. Studies with benomyl and thiabendazole on control of cotton diseases. *Phytopathology* 61:783–86

168. Ranney, C. D. 1972. Multiple cottonseed treatments: Effects on germination, seedling growth, and survival. *Crop Sci.* 12:346–50

169. Rao, P. S., Srimathi, R. A., Nag, K. C. 1972. Response of spike disease of sandal to benlate treatment. *Curr. Sci.* 41:221–22

170. Ravetto, D. J., Ogawa, J. M. 1972. Penetration of peach fruit by benomyl and 2,6-dichloro-4-nitroaniline fungicides. *Phytopathology* 62:784

171. Rawlins, T. E., Booth, J. A. 1968. Tween 20 as an adjuvant for systemic soil fungicides for Verticillium in cotton. *Plant Dis. Reptr.* 52:944–45

172. Reilly, J. J., Klarman, W. L. 1972. The soybean phytoalexin, hydroxyphaseollin, induced by fungicides. *Phytopathology* 62:1113–15

173. Richard, G., Vallier, J. P. 1969. Treatment of cereal seed by the combination of carboxin with copper oxyquinolate. *Proc. Brit. Insectic. Fungic. Conf., 5th,* 1:45–54

174. Richmond, D. V., Pring, R. J. 1971. The effect of benomyl on the fine structure of *Botrytis fabae*. *J. Gen. Microbiol.* 66:79–94

175. Robinson, P. W., Hodges, C. F. 1972. Effect of benomyl on eradication of *Ustilago striiformis* from *Agrostis palustris* and on plant growth. *Phytopathology* 62:533–35

176. Rombouts, J. E. 1971. Factors affecting the distribution pattern of systemic pesticides in plants. *Meded. Rijksfac. Landbouwwetensch. Gent* 36:63–71

177. Sampson, M. J. 1969. The mode of action of a new group of species specific pyrimidine fungicides. *Proc. Brit. Insectic. Fungic. Conf., 5th,* 2:483–87

178. Saunders, R. K., Lonnecker, W. M. 1967. Physiological aspects of using nonphytotoxic oils with herbicides. *Proc. N. Cent. Weed Control Conf.* 21:62–63

179. Schicke, P., Adlung, K. G., Drandarevski, C. A. 1971. Results of several years of testing the systemic fungicide CELA W 524 against mildew and rust of cereals. *Proc. Brit. Insectic. Fungic. Conf., 6th,* 1:82–90

180. Schlüter, K., Weltzien, H. C. 1971. Ein Beitrag zur Wirkungsweise systemischer Fungizide auf *Erysiphe graminis*. *Meded. Rijksfac. Landbouwwetensch. Gent* 36:1159–64

181. Schreiber, L. R., Hock, W. K., Roberts, B. R. 1971. Influence of planting media and soil sterilization on the uptake of benomyl by American elm seedlings. *Phytopathology* 61:1512–15

182. Schroeder, W. T., Provvidenti, R. 1969. Resistance to benomyl in powdery mildew of cucurbits. *Plant Dis. Reptr.* 53:271–75

183. Seem, R. C., Cole, H., Jr., Lacasse, N. L. 1972. Suppression of ozone injury to *Phaseolus vulgaris* 'Pinto III' with triarimol and its monochlorophenyl cyclohexyl analogue. *Plant Dis. Reptr.* 56:386–90

184. Selling, H. A., Vonk, J. W., Kaars Sijpesteijn, A. 1970. Transformation of the systemic fungicide methyl thiophanate into 2-benzimidazole-carbamic acid methyl ester. *Chem. Ind. 1970,* 1625–26

185. Shively, O. D., Mathre, D. E. 1971. Mode of action of oxathiin systemic fungicides. IV. Effect of carboxin on solute leakage from hyphae of *Rhizoctonia solani*. *Can. J. Microbiol.* 17:1465–70

186. Siegel, M. R., Zabbia, A. J., Jr. 1972. Distribution and metabolic fate of the fungicide benomyl in dwarf pea. *Phytopathology* 62:630–34

187. Sims, J. J., Mee, H., Erwin, D. C. 1969. Methyl 2-benzimidazolecarbamate, a fungitoxic compound isolated from cotton plants treated with methyl 1-(butyl-carbamoyl)-

2-benzimidazolecarbamate (benomyl). *Phytopathology* 59:1775–76

188. Sisler, H. D. 1969. Effect of fungicides on protein and nucleic acid synthesis. *Ann. Rev. Phytopathol.* 7:311–29

189. Sisler, H. D. 1972. Mode of action of benzimidazole fungicides, 323–35. In *Herbicides, Fungicides, Formulation Chemistry*, ed. A. S. Tahori, Proc. Int. IUPAC Cong. Pestic. Chem., 2nd, 1971, 5. New York: Gordon & Breach. 565 pp.

190. Slade, P., Cavell, B. D., Hemingway, R. J., Sampson, M. J. 1972. Metabolism and mode of action of dimethirimol and ethirimol, 295–303. In *Herbicides, Fungicides, Formulation Chemistry*. ed. A. S. Tahori, Proc. Int. IUPAC Congr. Pestic. Chem., 2nd, 1971, 5. New York: Gordon & Breach. 565 pp.

191. Smalley, E. B. 1971. Prevention of Dutch elm disease in large nursery elms by soil treatment with benomyl. *Phytopathology* 61:1351–54

192. Smith, A. M., Stynes, B. A., Moore, K. J. 1970. Benomyl stimulates growth of a Basidiomycete on turf. *Plant. Dis. Reptr.* 54:774–75

193. Smith, D. H., Crosby, F. L. 1972. Effects of foliar applications of a benomyl-oil-water emulsion on the epidemiology of Cercospora leaf spot on peanuts. *Phytopathology* 62:1029–31

194. Snel, M., Edgington, L. V. 1970. Uptake, translocation, and decomposition of systemic oxathiin fungicides in bean. *Phytopathology* 60:1708–16

195. Snel, M., Edgington, L. V. 1971. Fungitoxic spectrum and some structure-activity requisites of oxathiin and thiazole fungicides. *Meded. Rijksfac. Landbouwwetensch. Gent* 36:79–88

196. Snel, M., Von Schmeling, B., Edgington, L. V. 1970. Fungitoxicity and structure-activity relationships of some oxathiin and thiazole derivatives. *Phytopathology* 60:1164–69

197. Solel, Z. 1970. Performance of benzimidazole fungicides in the control of Cercospora leaf spot of sugarbeet. *J. Am. Soc. Sugar Beet Tech.* 16:93–96

198. Solel, Z. 1971. Vapour phase action of some foliar fungicides. *Pestic. Sci.* 2:126–27

199. Solel, Z., Edgington, L. V. 1973. Transcuticular movement of fungicides. *Phytopathology* 63:505–10

200. Solel, Z., Pinkas, J., Loebenstein, G. 1972. Evaluation of systemic fungicides and mineral oil adjuvants for the control of mal secco disease of lemon plants. *Phytopathology* 62:1007–13

201. Staron, T., Allard, C., Darpoux, H., Grabowski, H., Kollmann, A. 1966. Persistance du thiabendazole dans les plantes propriétés systémiques de ses sels et quelques donnees nouvelles sur son mode d'action. *Phytiatrie-Phytopharmacie* 15:129–34

202. Staunton, W. P. 1971. Fusarium wilt of tomato. *Meded. Rijksfac. Landbouwwetensch. Gent* 36:1192–201

203. Stipes, R. J. 1973. Control of Dutch elm disease in artificially-inoculated American elms with soil-injected benomyl, captan and thiabendazole. *Phytopathology* 63. In press

204. Stipes, R. J., Oderwald, D. R. 1970. Soil thin-layer chromatography of fungicides. *Phytopathology* 60:1018

205. Stipes, R. J., Weinke, K. E. 1972. Dutch elm disease: Control with soil-amended fungicides. *Plant Dis. Reptr.* 56:604–8

206. Tahori, A. S., ed. 1972. *Herbicides, Fungicides, Formulation Chemistry*. Proc. Int. IUPAC Congr. Pestic. Chem., 2nd, 1971, 5. New York: Gordon & Breach. 565 pp.

207. Templeton, G. E., Johnston, T. H., Daniel, J. T. 1971. Benomyl controls rice white tip disease. *Phytopathology* 61:1522–23

208. Thapliyal, P. N., Sinclair, J. B. 1971. Translocation of benomyl, carboxin, and chloroneb in soybean seedlings. *Phytopathology* 61:1301–2

209. Ulrich, J. T., Mathre, D. E. 1972. Mode of action of oxathiin systemic fungicides. V. Effect on electron transport system of *Ustilago maydis* and *Saccharomyces cerevisiae*. *J. Bacteriol.* 110:628–32

210. Vargas, J. M., Jr. 1972. Evaluation of four systemic fungicides for control of stripe smut in 'Merion'

Kentucky bluegrass turf. *Plant Dis. Reptr.* 56:334–36

211. Von Schmeling, B., Kulka, M. 1966. Systemic fungicidal activity of 1,4-oxathiin derivatives. *Science* 152:659–60

212. Vonk, J. W., Kaars Sijpesteijn, A. 1971. Methyl benzimidazol-2-yl-carbamate, the fungitoxic principle of thiophanate-methyl. *Pestic. Sci.* 2:160–64

213. Wain, R. L., Carter, G. A. 1967. Uptake, translocation, and transformations by higher plants, 561–611. In *Fungicides, an Advanced Treatise*, ed. D. C. Torgeson, 1. New York: Academic. 697 pp.

214. Wallnöfer, P. 1969. Der mikrobielle Abbau des 1,4-oxathiinderivats, 2,3-dihydro-5-carboxanilido-6-methyl-1,4-oxathiin   (DCMO). *Arch. Mikrobiol.* 64:319–26

215. Wang, M. C., Erwin, D. C., Sims, J. J., Keen, N. T., Borum, D. E. 1971. Translocation of ^{14}C-labeled thiabendazole in cotton plants. *J. Pestic. Biochem. Physiol.* 1:188–95

216. Weinke, K. E., Lauber, J. J., Greenwald, B. W., Preiser, F. A. 1969. Thiabendazole, a new systemic fungicide. *Proc. Brit. Insectic. Fungic. Conf., 5th,* 2:340–46

217. White, G. A. 1971. A potent effect of 1,4-oxathiin systemic fungicides on succinate oxidation by a particulate preparation from *Ustilago maydis. Biochem. Biophys. Res. Commun.* 44:1212–19

218. Wilcoxson, R. D., Bielenberg, O. 1972. Leaf disease control and yield increase in alfalfa with fungicides. *Plant Dis. Reptr.* 56:286–89

219. Woodcock, D. 1968. Agricultural and horticultural fungicides—past, present and future. *Chem. Brit.* 4:294–300

220. Woodcock, D. 1971. Chemotherapy of plant disease. Progress and problems. *Chem. Brit.* 7:415–20, 423

221. Yoshinaga, E. 1969. A systemic fungicide for rice blast control. *Proc. Brit. Insectic. Fungic. Conf., 5th,* 2:593–99

222. Zaronsky, C., Jr., Stipes, R. J. 1969. Some effects on growth and translocation of thiabendazole and methyl   1-(butylcarbamoyl)-2-benzimidazolecarbamate   applied to *Ulmus americana* seedlings. *Phytopathology* 59:1562

223. Zentmyer, G. A., Horsfall, J. G., Wallace, P. P. 1946. Dutch elm disease and its chemotherapy. *Conn. Agric. Exp. Sta. Bull.* 498: 1–70

THE ROLE OF CROPPING SYSTEMS   ❖ 3580
IN NEMATODE POPULATION
MANAGEMENT

C. J. Nusbaum and Howard Ferris

Department of Plant Pathology, North Carolina State University,
Raleigh, North Carolina, and Department of Nematology, University
of California, Riverside, California

Introduction

Crop rotation is one of the oldest and most important approaches to the control of nematodes that feed on the roots of annual crop plants. The value of rotations often was recognized long before their effects upon the dynamics of nematode populations and communities were considered. As specific nematode problems were identified and their economic importance demonstrated, attention usually was given to crop rotations as a means of either preventing or reducing crop losses attributed to these pests. In many cases, crop rotation became the conventional method for nematode control and was readily accepted by growers because they previously had been introduced to such management practices as a means of improving soil fertility and crop productivity.

Early rotation schemes were aimed at the control of a single nematode species on a susceptible main crop. They were developed empirically on the basis of knowledge of nematode life histories and host ranges. Field experiments involved the selection of suitable alternate crops and determination of the number of seasons between main crops needed to obtain the desired degree of control. Evaluation of cropping systems was based mainly upon measurements of root disease severity in, and performance of, the main crop at the end of its growth season. Little, if any, attention was given either to the performance of rotation crops or to effects upon nematode populations. Bessey (5) found certain cropping systems to be effective in controlling root knot in susceptible vegetable crops. He concluded that success depended upon how well the rotation crops met these requirements: (*a*) prevent development and reproduction of the parasite; (*b*) at least pay the expense of working the land, as well as rent, taxes, etc.; (*c*) enrich the land or at least not impoverish it; and (*d*) make such vigorous, dense growth as to choke out

423

root-knot susceptible weed hosts. He also recognized certain difficulties and limitations. Because of the wide host range of root-knot nematodes, the choice of alternate crops was limited and many of the most effective crops available could contribute little to farm income. Although these concepts were formulated over 60 years ago, they still apply to the development and evaluation cropping systems for specific nematode problems and land use practices.

In principle, crop rotations are adaptable to a wide variety of nematode problems and offer great flexibility in application. Potential benefits, however, have not always been fully achieved in farming practice. The endemic, complex, variable, and insidious nature of nematode problems often made it very difficult for the farmer to assess their importance and to take proper remedial measures. Crop rotation is only one of the several approaches to control, and its popularity with growers tends to wane as crop production methods intensify due to advances in agricultural technology. For example, when the use of nematicides becomes economical, rotation may be abandoned (53). The development and use of nematode-resistant cultivars may have a similar effect. Crop rotation, chemical treatment, and host resistance are complementary methods. If one should prove to be adequate, however, the others may seem to be superfluous. Reliance upon single methods and limitation of options may have certain temporary economic advantages but can hardly be expected to provide a satisfactory long-range solution. Consequently, increasing attention is being given to the integration of control methods designed for efficiency, reliability, and stability.

Modern concepts of pest management (59), ecosystem analysis (82), nematode population dynamics (68), and nematode control strategy (83) reflect the current trend toward improved pest control within the context of a productive and wholesome environment. Communities of plant parasitic nematodes, as an integral part of the ecosystem, are responsive to environmental factors including those influenced by man. The roots and/or rhizospheres of suitable host plants comprise the ecological niches in which plant-parasitic nematodes develop and reproduce. Food supply, therefore, is one of the most important determinants of seasonal and annual changes in the population structure, density, and distribution (49). In agroecosystems, the food supply is governed largely by the cropping practices employed by the grower and is subject to periodic manipulation. Hence, consideration of cropping effects is basic to any scheme of nematode-population management regardless of the nematode species or crops involved. Management, as used here, implies the integration of various man-imposed actions with those of the ecosystem (60) in reducing and regulating nematode populations. As additional, precise knowledge is acquired on the dynamics of nematode populations in relation to the epidemiology of root diseases involving these pests, effective use of cropping systems in integrated control programs will be enhanced. The term "cropping system" covers all kinds of crop sequences, including continuous monoculture, whereas, "crop rotation" implies an inflexible cycle or a fixed series of crops.

Nematodes Populations and Communities

Root-feeding nematodes of a given species exist and function as members of a reproducing population. Most soils, especially cultivated soils, harbor a mixture of nematode species. The species comprising each community may represent a broad spectrum of types with respect to their behavior, parasitic habits, and effects upon their hosts. Several descriptive categories are recognized. Some species are sedentary during developmental and reproductive periods, whereas others are migratory. Endoparasites enter the roots, ectoparasites remain outside the roots and feed in epidermal and cortical tissues, others are intermediate. Like other kinds of parasites, nematodes cause injury to their hosts by their feeding activities. Many nematodes, however, alter host metabolism and, acting either alone or in combination with other kinds of root pathogens, cause far greater damage than can be accounted for by the mere withdrawal of food substances (49). Some species multiply continuously during active periods, whereas others reproduce in separate generations. In addition to these diverse characteristics, efficient dispersal, polyphagy, relatively weak interspecific competition, and great persistence largely account for the occurrence of nematodes in complex, dynamic communities (46, 51, 52).

The community of plant-parasitic nematodes is an integral part of a complex of soil-inhabiting microorganisms which forms a vital subsystem within the whole ecosystem. Since the individual populations comprising the community occupy the same microhabitat and often rely upon the same sources of food, they tend to coexist in either the same or overlapping ecological niches. Odum (50) defined a niche as a position or status of an organism within its community and ecosystem resulting from the organism's structural adaptations, physiological responses, and specific behavior. According to Gause's Principle (22) species occupying the same niche are competitively exclusive; one species would be better adapted to exploit the niche and the other eliminated. An endo- and an ecto-parasitic nematode feeding upon the same root system occupy niches that are spatially separated; however, they could influence each other since both are dependent upon the same food source. Oostenbrink (52) noted that the concept of the niche may be broad or limited and that the principle of competitive exclusion does not always hold for nematodes. Even so, there is evidence that interactions between species occur in various ways and to varying degrees (12, 13, 20, 21, 29, 31, 32, 43, 44, 64, 69). In general, the highly specialized, sedentary endoparasites appear to occupy niches that are more narrowly defined than those occupied by the migratory ectoparasites which have relatively broad host ranges and tend to provoke less host response. When mixed populations of root-knot nematodes occupy the same niche, one may be better adapted to exploit it than the others (12, 37). Although the interaction mechanisms and the ecological significance of the associations among nematode species are not well understood, the cropping systems employed largely determine the structure of the communities in which they occur.

Ecosystem Stability

The composition of nematode communities and the dynamics of populations comprising them are related to the degree of stability of the ecosystems in which they function. Stability implies the capacity of the system to return to a mean position following disturbance, the mean itself moving in a direction consistent with the development of the system (72). The greater the stability, the more rapid is the return. An absolutely stable ecosystem would remain constant. Preston (58) considers the stability of pest populations in terms of extinction of a species on the one hand and its rise to plague proportions on the other. Fluctuations between these limits represent a more or less stable system. In natural or undisturbed ecosystems, stability is achieved by genetic, structural, and functional diversity of component communities; equilibrium or homeostasis is maintained effectively because of the variety of ecological niches available to species of organisms at all trophic levels (41, 72). Thus the more numerous the kinds of organisms, the greater their numbers and the shorter their generation time, the more associative, competitive, and antagonistic interactions will occur to check unrestrained multiplication and to maintain balance (86). In such ecosystems, therefore, populations of the various species comprising the nematode community should exhibit stability in terms of density fluctuations and genetic constitution, being buffered and governed by dynamic interactions with other components of the system (45).

Agroecosystems, in contrast to natural ecosystems, lack continuity, diversity, and stability (71, 86). Uniformity is usually the goal in agroecosystems. Most cultivated crops consist of the same or similar genotypes within a single species; thus they lack intraspecific diversity. Plant population density is regulated. All plants are of about the same age, nutrients are supplied, and weeds are controlled. Continuous monoculture, lacking both inter- and intra-specific diversity, is the least complex of agroecosystems and is, therefore, the least stable. The latter practice tends to narrow the community spectrum of plant parasitic nematodes to those species most favored by the host (46, 49, 70). Conversely, the abundance of free-living saprophagous nematodes often is more closely related to the amount of organic matter in the soil than to the type of crop (23). There is considerable evidence that the number of species, biomass, and population densities of plant-parasitic nematodes are greater in cultivated than in noncultivated soils (33, 52, 56, 87). A farm survey in the Netherlands revealed that all cultivated fields were infested with four to six genera of plant-parasitic nematodes (54). Parasitic species tend to flourish in cultivated soils where a vigorously growing food source is available, whereas predaceous and buffering species are more sensitive to the disturbances involved. For example, species of predaceous Mononchidae are abundant in forests, fence rows, and stream banks in Oregon, but they tend to disappear from intensely cultivated areas (30).

Outbreaks of pests and diseases, where the population of the pest or pathogen species fluctuates beyond the bounds of dynamic equilibrium, are a common feature of agroecosystems (27, 71). Moreover, unstable ecosystems are more responsive to environmental changes than stable ones. It appears that the nematode communities in cultivated soils lack the biological checks that govern them in undisturbed ecosystems (34, 45).

Although intensive agriculture invites upsets in ecological balance and increases the threat of ubiquitous soil-borne pathogens, crop production would be impossible without the complexity of soil microbial communities even in these disturbed ecosystems (86). Moreover, agroecosystems are stabilized to varying degrees by human inputs. Skillful labor, powerful machines, financial resources, and an array of management systems are employed to modify certain environmental factors and to reduce the danger of pest and disease outbreaks (23, 41, 42, 72, 73). Thus it is possible to maintain or increase crop productivity while encouraging suitable diversity and stability in the ecosystem. In many modern agricultural regimes, stability is being approached, but the cost of man's inputs and their ecological consequences often are limiting factors. When arable lands are not cropped they revert to natural vegetation and the pest communities become increasingly complex and stable (41), but return to a balanced state may require a period of several years.

The attainment and maintenance of stability of root-knot nematode (*Meloidogyne* spp.) populations by human input is well illustrated by the culture of flue-cured tobacco in North Carolina during the past 20 years. In the early 1950s estimates of annual losses due to root knot in this intensively cultivated, high-value crop exceeded $50 million or approximately 10% of the value of the crop. Crop failures due to root knot were not uncommon. An intensive survey by counties showed that an average of 54% of the growers practiced continuous monoculture, whereas 27% used rotation crops occasionally, and the remaining 19% rotated consistently. The use of nematicides was just beginning; fewer than 1000 acres were fumigated in 1951. Yields averaged about 1200 Kg per hectare (= lbs. per acre). The situation has changed drastically since that time. Tobacco is rotated with other crops on more than 85% of the acreage and over 65% of the crop is grown in fields treated with nematicides. Losses due to root knot have declined to less than $5 million annually (76). Average yields have nearly doubled due to improved disease control and production technology. These gains were achieved, however, by considerable investment in both research and educational programs and by increased production costs. Even so, root knot is still a major problem requiring continued, costly inputs by man just to maintain the gains that have been made. Crop production may become unprofitable when yields are limited by a nematode population that is favored by intensive culture (36). It may become profitable, however, where the crop is valuable enough to justify costly control procedures (15).

Nematode-Host Relationships

In considering the use of cropping systems for nematode population management, the interactions between the various nematode species and host plants involved is of primary importance. In each parasite-host combination, characteristics of the nematode may be defined in terms of its parasitic efficiency, i.e. its ability to obtain food from its host, and its ability to cause damage, i.e. its pathogenicity. Host status refers to the suitability of the plant to serve as a substrate for the parasite and its relative vulnerability to damage. All features of this interaction, being interrelated and influenced by environmental factors, comprise a complex, dynamic system. In a recent review, Seinhorst (68) presented the status of quantitative research on the populational aspects of these systems. He has developed concepts and mathematical models that describe the dynamics of nematode population increase and the relationships between nematode density at the time of planting and plant yield.

The status of a crop with respect to a particular nematode species may be defined (67) in terms of a (the maximum reproduction rate of the nematode on that crop) and E (the equilibrium density). The multiplication rate slows with increasing population density until an equilibrium density is reached at which multiplication just suffices to maintain the population density. Relative vulnerability to damage may be determined experimentally by subjecting test plants to a wide range of initial population densities and recording the threshold of tolerance (T), i.e. the density below which no apparent yield reduction occurs. These concepts, considered together, encompass both the susceptibility and tolerance of the crop to a nematode species or pathotype. They are thus indicative of the nematode population changes and crop damage which might be expected under a particular set of environmental conditions. The characterization of crop cultivars in terms of E and a for each nematode species and under various environmental situations then becomes invaluable in planning cropping systems for nematode population management. Since E and a are not necessarily correlated with T, knowledge of the latter is also important (68).

Basically the principles of crop rotation for population management are: (a) the reduction of the initial or preplant population densities (P_i) of nematodes to levels that allow the subsequent crop to become established and to complete its early growth before being heavily attacked, and (b) to preserve competitive, antagonistic, and predaceous nematodes and other organisms at population densities effective in buffering the pathogenic species. The rate of population increase and damage caused by a nematode species is influenced by P_i, a and E for the particular crop, soil community structure, environmental conditions, and time. The initial population density (P_i) is determined by the final density (P_f) of that species at the end of the preceding crop and the mortality rate between crops. Time alone may account for a considerable population reduction through starvation; however, many species are very persistent even over long periods in fallow soil; also in modern intensive agriculture the period between crops is being reduced progressively (7, 61). A low P_i may be achieved by the use of a preceding crop that either is not a host or has a low a or E, and/or by cultural practices and chemical treatments applied between crops. Thus, turning roots onto the soil surface (48, 75), flooding (15, 23), repeated discing to expose the nematodes to the elements (23) are often effective in decreasing nematode populations.

Also, P_i may be reduced by growing cultivars with vertical resistance to the pathogen (80). Vertical resistance is the type involved in most crop varieties into which nematode resistance has been incorporated (26, 28, 63). In this case the initial population is unable to establish a successful relationship with the host plants and is thus reduced. There is a need for caution with this type of resistance due to the intense selection pressure for a vertical pathotype (62, 80). Reduction of the multiplication rate (a) of the pathogen can be achieved through the use of cultivars with effective levels of horizontal resistance (62, 80). Thus, corn hybrids differ in their suitability for species of root-knot nematodes and *Meloidogyne* populations vary in their ability to reproduce in corn roots (1, 2, 47). Because nematodes respond to selection pressure there is a need for caution when horizontally resistant cultivars are used frequently (2, 65) but the danger is not as great as with vertical resistance (35, 62, 81).

The equilibrium density (E) should be considered in conjunction with a, as the overall host status is the determinant of the value of a particular cultivar in a cropping system. Using the categories of host status of Dropkin & Nelson (19) a tolerant host would have a high E value; even if a is low due to horizontal resistance, a crop may be able to support a large nematode population. In this case density-dependent factors governing population increase would not markedly suppress the rate of increase until high densities have been reached. Alternatively, in a susceptible crop where a is large, if the host is intolerant, i.e. has a low E value, density dependent factors will suppress the multiplication rate at low densities so that the rate of population increase slows down early in the season and P_f is low. This is a significant factor when considered in relation to subsequent crops in the rotation. Thus, an intolerant crop with a low E would result in a lower P_f than a tolerant crop with high E value and so favor the subsequent crop. But the use of an intolerant crop in a rotation system would be economically unsound. The effect of a crop on nematode population densities is also governed by time of exposure. Early planting, and the use of early varieties that can become established before conditions are suitable for nematode activity, are frequently recommended (16, 24, 25, 79, 83). Similarly the removal of roots from the soil immediately after harvest may prevent continued reproduction of the nematode and result in a lower P_f, thus favoring the subsequent crop (48, 75).

Although environmental conditions affect nematode populations directly in many ways, indirect effects of host plants on nematode population are perhaps even more profound. Such factors affect plant vigor and thus influence the values a and E which, in turn, are reflected in P_f and plant damage. This may be explained in terms of stress. The plant as a whole is a complex organism comprised of several kinds of organs each of which performs specific functions. All functions, however, are interrelated and interdependent. Root dysfunction caused by nematode attack often impairs the functioning of other organs, thus creating internal stresses. Such stresses may be either ameliorated by environmental conditions favorable for plant growth (85) or intensified by external environmental stresses. Bessey (5) noted that yields of root-knot susceptible crops were markedly increased following certain rotation crops even though the incidence of root galling was not reduced appreciably.

Man has considerable influence on the environment in agroecosystems. The nutrient status of the soil may be altered by the addition of fertilizer. Large amounts of potassium allow cotton to withstand root-knot stress even though the P_f is increased (55). Similarly Bird (6) showed that the P_i at which the plant growth rate was slowed was greater on plants grown with full nutrients than on nitrogen-deficient plants. But *Pratylenchus scribneri* populations were higher in soybean and cotton plots receiving no fertilization than in plots receiving all major elements (14). The rate of development of *M. incognita* was retarded in plants with nutrient deficiencies (18). Man is able to influence soil temperature by plant spacing and soil shading, as well as by irrigation. The amount of oxygen present under *Eragrostis curvula* may be decreased to cause inhibition of *Meloidogyne javanica* (38). Some crops may rapidly deplete the soil of moisture, directly affecting nematode populations; others form a dense cover and reduce evaporation from the soil surface. Soil moisture may be further regulated by irrigation and land drainage, thus reduced irrigation may prevent hatching of root-knot eggs and so reduce reinvasion of the host (83).

Epidemiological Considerations

Root-feeding nematodes damage their hosts and produce symptoms of disease in various ways and to varying degrees. Through their pathogenic activities they also may change the reaction of the host to secondary pathogens (57) which in turn may alter the disease syndrome. When disease increases in a population of host plants, an epidemic is in progress, and theoretically its course is subject to epidemiological analysis. Van der Plank (80) has quantified the science of plant disease epidemiology. He has described mathematically the dynamics of disease increase and spread where appropriate measurements can be made throughout the course of an epidemic, particularly during the early logarithmic phase of the epidemic curve. The number of lesions increases logarithmically until the available infection sites are reduced to an extent that slows the infection rate or results in significant reinfection of established lesions. In the case of systemic infection the individual diseased plant is considered a lesion and the group of infected plants in an area is a focus of disease. Similar concepts of logarithmic increase apply to foci, again until they begin to expand into each other. The dynamics of these increases depend not only on the initial amount of viable inoculum, but also upon the number of available infection sites at the time of initial exposure and at times of reinfection, i.e. the proportion of noninfected tissue $(1-x)$ where x is the proportion of tissue already infected. The quantity $(1-x)$ will increase or decrease depending upon whether the tissue is being infected faster than it is being produced by the growing crop.

In the case of nematodes, the host-parasite interaction takes place below the soil surface and, with some exceptions, pathological processes do not occur in easily recognizable, measurable lesions or units. Even so, it is often possible to study the dynamic aspects of such interactions by periodic measurement of nematode population densities and plant growth responses. Much progress has been made in the development of methods for determining the relationships between P_i and crop performance (68) and information thus obtained is the basis for the establishment of nematode diagnostic and advisory services (4). Van der Plank (82) emphasized the need for measuring the whole epidemic or population dynamic process from start to finish. Although this presents many difficult problems for the phytonematologist, Van der Plank's system of quantitative epidemiology provides certain concepts that are important in developing strategies for nematode population management. A better understanding of the complex patterns of root development and the effects of nematodes upon root morphology and function is needed. The development of roots and shoots is clearly interdependent but not necessarily parallel (40).

Application of modern epidemiological concepts in nematology may be illustrated by situations in which the nematodes invade and develop in specific infection sites (such as root tips), establish specialized relationships with host tissues, and cause discrete lesions. Increase in numbers of foci from a single source occurs so slowly that the individual plant can be considered as a noninfectious or slightly infectious focus. The lesions are then the infected sites within the focus (root) and a given problem area is a group of such foci. If P_i is low, then $(1-x)$ at the beginning of the season will be large and the plant will grow vigorously. If the plant growth rate is greater than the rate of production of propagules by the nematode, $(1-x)$ will remain large and growth will not be inhibited. As root growth ceases with senescence, so $(1-x)$ will become smaller until all available sites are infected or intraspecific competition decreases to a minimum the rate of increase of the nematode population. By this stage the crop may have reached maturity.

The value $(1-x)$ affects not only the dynamics of crop growth, being the amount of root available to supply the plant with nutrients and water, but also the nematode population dynamics as a measure of the number of sites available for infection. Increase in lesion number ceases to be logarithmic once reinfection of established lesions is occurring. This reduction is probably also influenced by decreased growth rate and reproduction by the competing nematodes, and by increased production of males in some nematode species (17, 39, 77, 78). The number of available infection sites and the initial

population density both will be involved in determining the initial value of
$(1-x)$. Anything that can be done to increase $(1-x)$ will favor the plant. Thus,
the establishment of healthy seedlings with vigorous root systems, the use of
horizontally resistant cultivars in which the rate of infection and nematode development is reduced, use of fertilizer and irrigation, enhancement of nematode competitors and antagonists, or the use of nematicides, will all allow
$(1-x)$ to remain large and the plant to grow vigorously. Conversely, if conditions favor the nematode, $(1-x)$ will decrease, and although this will slow
nematode increase it will also depress plant growth rate and yield.

A simplified equation for the rate of increase of infected areas on the root
would be $dx/dt = rP(1-x)$ where dx/dt is the rate of increase of infected
tissue at any point in time, $t;$ r is the maximum infection rate of the nematodes under ideal conditions; P is the number of nematodes available for
infection at time $t:$ $(1-x)$ is the proportion of root tissue still available for
infection. At the beginning of the season the seedling has no infected tissue,
thus, $x = 0$ and infection will occur at a maximum rate if environmental
conditions are ideal. As x increases, $(1-x)$ will be less than unity and have a
slowing effect on the infection rate dx/dt. If the root is still able to grow
rapidly, the proportion x may remain small so that the effect of $(1-x)$ on
the infection rate is negligible. If the proportion of the root infected increases,
$(1-x)$ will decrease so that the rate of subsequent infection will decrease.
The factor $(1-x)$ can also be considered as having a direct effect on the plant
growth rate and on the nematode reproduction rate. When $x = 0$, plant
growth will proceed at a maximum under prevailing conditions; it will slow
down if $(1-x)$ decreases appreciably. Subsequently, reproduction also will
decrease. Thus there are two dynamically interacting components in this simplified system, the nematode population and the roots, each reciprocally
affecting the other through the amount of root tissue infected. If conditions
are sub-optimal for the nematodes but favorable to the plant, $(1-x)$ will remain large and plant growth rapid. Nematode increase will be limited by
environmental conditions rather than availability of infection sites. If environmental conditions favor the nematode, reproduction and infection by
the nematode will be influenced only by the size of $(1-x)$ as it affects infection site availability and plant growth vigor. Integrated control seeks to maximize $(1-x)$ by providing conditions optimal for plant growth and sub-optimal
for the nematode population.

Knowledge of the sequence and timing of events can be employed to determine the critical periods during the course of an epidemic. Using studies of
root knot of flue-cured tobacco in North Carolina as an example, the following
pattern emerges. In most infested fields, preplant densities of infective larvae
are too low to damage young transplants appreciably and the value $(1-x)$
remains near unity during May and June. During the first two months the
host plants grow rapidly while the increase in numbers of galls declines as the
initial inoculum becomes exhausted. In early July the situation begins to
change as reproduction by first generation females reaches a peak resulting

in about a 1000-fold increase in nematode numbers. Thus, it appears that this is a critical period in the epidemiology of root-knot because it comes at a time when the value $(1-x)$ is reduced due to natural decline in host plant vigor and to a great increase in nematode population density. It is the time when the future course and the final outcome of the epidemic are determined by whether the threshold of host tolerance (66) is exceeded and, if so, to what extent. This emphasizes the importance of P_i, the initial population (inoculum) from which the epidemic starts. Presumably, under conditions favorable for larval penetration and nematode development, initial infections x_0 will be proportional to P_i and subsequent increases in x will be related to the value of x_0. Traditional nematode control strategy, therefore, has been aimed at reducing P_i below the economic threshold. Although considerable progress has been made in applying knowledge of nematode population dynamics to the determination of economic threshold values (52, 68), this remains one of the major challenges in population management. When complex cropping systems are employed, especially those involving several kinds of crops, this problem becomes magnified. The occurrence of plant parasitic nematodes in mixed communities, changes in community composition, and fluctuations in population densities of the individual species must be taken into account.

The magnitude of the fluctuations about the mean is important to agroecosystems. Pest or pathogen populations may oscillate above economic threshold densities in unstable situations. With continuous cropping of potatoes, population oscillations of *Heterodera rostochiensis* were large at first but decreased to an equilibrium at which intermediate root size was balanced by intermediate density and P_f/P_i approached unity (36). Thus, a stable situation was reached with plant growth limited by the nematode population and vice versa. Relief of other stresses on crop growth, e.g. soil moisture or fertility, might again unbalance the system by allowing increased plant growth and hence fluctuation of the nematode population above the equilibrium density for the next season (6, 85). The economic threshold density will vary with the amount of stress the plants are under and with their age, more vigorous plants being able to support more nematodes (6, 85). It is probably unnecessary to keep the population low throughout the season; the financial and environmental costs involved would outweigh the benefits derived. A more efficient practice would be to keep densities low while the crop is young and intolerant by using crop rotation, cultural manipulations, and changes in planting time (10, 83).

Crop Selection

Species of plant-parasitic nematodes generally have overlapping host ranges and differ in their sensitivity to environmental manipulation. Many of the ecological features of agroecosystems can be altered by cultural practices and cropping systems, such alterations having both direct and indirect effects upon nematode populations and communities. Although such practices usually provide considerable flexibility in most agricultural regimes, selection of crops may be limited by land use and economic considerations (11). Thus compromises often are dictated by knowledge of both the advantages and limitations of any particular course of action. The achievement of as much intra- and interspecific diversity as possible either in space or in time, is desirable. This may be accomplished by growing multiline varieties, rotating crops, changing cultivars, or "rotating the rotation" to different alternate crops. Multiple cropping in areas having long growing seasons adds organic matter to the soil, increases the cycling of nutrients, and improves soil structure. All of these practices tend to decrease population densities of nematodes on a crop of a particular genotype.

A cropping system should be selected so that one crop does not produce a population of nematodes larger than the economic threshold density of the succeeding crop in the system. Thus, Brodie et al (9), in studying grass-sod and row crop sequences, determined that corn and cotton should not follow each other in soils infested with the sting nematode, *Belonolaimus longicaudatus*. In a tobacco rotation experiment conducted at Rocky Mount, North Carolina from 1954 to 1966, six crops were used in various combinations and sequences based upon 4-year cycles. In each cycle, two different alternate crops were used with tobacco, such as tobacco-corn-tobacco-peanut. The overall nematode community included 9 plant-parasitic species representing 7 genera. After the first cycle of cropping was completed, patterns of community spectra within each system were fairly well established. Although the community structure within each system changed each year, the spectrum was stable and fluctuations of populations within each community were somewhat predictable. Although the community spectrum in each rotation system remained fairly broad (5 to 6 species of plant parasites), that within continuous monoculture plots of either tobacco, cotton, corn, or peanut had only 2 or 3 species. In all cases the stabilizing effects of the rotations benefited the alternate crops as well as tobacco.

Another factor to be considered in crop selection pertains to the effects of crops upon genetic variability within populations of plant parasitic nematodes. The problem of "genetic vulnerability" is beyond the scope of this review, but brief mention seems appropriate. Cropping systems can exert selection pressures upon nematode populations to such an extent that "resistance-breaking" pathotypes can emerge and reduce the effectiveness of the system. Jones et al (35) advocate the rotation of resistant and susceptible cultivars of potato to avoid an intense selection for resistance-breaking pathotypes of *Heterodera rostochiensis*. Populations of *Meloidogyne incognita* initially were suppressed by corn in a tobacco-corn rotation; later they responded to the selection pressure and reached high densities on both crops, negating the value of the cropping system (2, 65). When resistance-breaking pathotypes emerge as the result of selection pressures favoring them, they still may remain at low levels for a period of years as a minority component of the population, gradually building up to damaging proportions. Thus, although the problem of genetic vulnerability with nematodes may not be as serious as with pathogens that spread far and fast, nevertheless, valuable sources of resistance may inadvertently be lost through the continued use of injudicious cropping systems.

Outlook

Concepts of pest management provide a wide variety of options in devising long-range solutions to pest problems (59). They are aimed at the regulation rather than the elimination of pest populations. Strategies and approaches are based upon an understanding of the life systems of pests and upon predictable ecologic and economic consequences (59). Crop rotation is one of the options in regulating nematode populations, but its greatest potential undoubtedly lies in the role it may play in integrated control programs of the future (83). As crop production becomes more intensive, empirical approaches of the past no longer will suffice. Use of crop rotations will be determined by their relative value in complementing other practices in total management systems that are effective, economical, and ecologically sound. Within this context, an appraisal of the present status and future prospects seems appropriate.

POPULATION DYNAMICS   The characteristics and behavior of the major species of plant-parasitic nematodes are well understood. Seinhorst (68) has indicated, however, that studies of nematode population dynamics have been minimal. This also applies to the epidemiology of root diseases caused by nematodes and allied root-infecting pathogens. Last (40) observed that the reasons for our inadequate knowledge of soil borne pathogens can be found in the root pathologists' prediction for single, end-of-the-season observations; and that, to an extent, such investigators have adopted a fixed attitude to their problems and have not developed with the times. In phytonematology, marked changes in attitudes and approaches are evident (68, 83) and encouraging but they also bring deficiencies and difficulties into sharper focus. As Van Gundy (83) observed, predictions of disease potential are based upon knowledge of pathogen population dynamics and the techniques for establishing economic thresholds. This can be done in the relatively few nematode problem areas where advisory services are well established (4). Accuracy of predictions depends largely upon the reliability of the nematode assay data on which they are based. Wide variations in horizontal (3) and vertical distribution of nematodes in the soil create serious sampling problems (4). Assay procedures are often too insensitive to measure populations prior to planting or the infectivity of the nematodes present. Thus, refinement of sampling and assay techniques will play a major role in the determination of economic thresholds.

ECOLOGICAL CONSEQUENCES   Crop rotations, in contrast to continuous monocultures, would be expected to have desirable effects upon agroecosystems because they tend to increase diversity and stability. Although reduction and regulation of nematode populations may be the primary purpose, rotations also may have many intangible side effects. In general, the greater the diversity of crops, particularly if sod crops are interspersed with row crops, and the longer the interval between main crops, the greater the stability. Thus, the need for other kinds of stabilizing inputs is reduced. Where relatively short rotation cycles (2 or 3 crops) are used, stability can be increased by changing periodically either the alternate crops or the cultivars of the main crop, or both. Without careful selection of a cropping system, soil-improvement benefits may be negated by increases in nematode populations (8). Thus the selection of the proper sequence is as important as the choice of the crops. From the viewpoint of soil conservation and the efficient use of land resources as well as pest control practices, available options should be flexible enough to employ a prescription approach to decision making (74).

ECONOMIC CONSEQUENCES   In many nematode problem areas, crop rotation is the best, if not the only, economic means of control, largely because nematode-resistant cultivars are not available and crop values are too low to justify costly treatments. In areas of intensive crop culture, however, this practice is in jeopardy for economic reasons. Most nematologists undoubtedly agree with Van Gundy (83) that regardless of how successful crop rotation is and can be, it may never survive the technological revolution in agriculture unless grower attitudes can be changed. Where farms are located near major urban areas, some of the best agricultural lands are taken out of production as they are diverted to other uses. Because of shortages of skilled farm labor, increased wages, and increased consumer demands for farm products there is a marked trend toward consolidation and centralization of operations into more productive and efficient units. Plant breeders have developed high-yielding cultivars of major food and fiber crops, but, in doing so, they have reduced genetic and cytoplasmic diversity. If these trends continue, long-range management of populations of nematodes and other soil-borne pathogens will require increasingly large inputs in terms of machines, energy, and chemicals at the risk of compromising still further the stability of agroecosystems.

INTEGRATED CONTROL   Because the cropping system is the dominant feature of agroecosystems, it is the foundation upon which integrated pest management systems rest. In the selection of crops and their sequence, ranging from continuous monoculture of a single cultivar to complex, long-term rotations, various degrees of ecosystem diversity in time and space can be obtained. In a broad sense, cropping systems exert a major influence on the dynamics of the whole complex of pests detrimental to plant health and crop productivity. For example, a well designed crop-rotation experiment may provide material for analytical study not only of nematode populations but also of other soil pathogens, soil insects, and weeds. This emphasizes the interdisciplinary nature of pest management, especially where different kinds of pests interact with each other and where a given management practice affects more than one target organism. Hence, the concept of integrated control need not be limited to a combination of complementary practices directed at a single target but rather can be expanded to include pest complexes of varying dimensions. Much progress is being made in breaking down the barriers that impede cooperation between disciplines and agencies. The emergence of broad concepts of pest management and the interest and involvement of all sectors, not only of research and educational institutions, but also of the agribusiness community and the public at large, justify an optimistic outlook.

The ecosystem approach to pest management presents great challenges and opportunities for the future. A stage is being reached where quantification of information on population dynamics of plant parasitic nematodes and the epidemiology of root diseases is possible. Techniques of systems analysis are now being employed in plant pathology (84, 88) and provide a means for synthesizing the wealth of complex information available on integrated control. Models of biological systems are simplifications of the real world and are constructed to assist in its understanding. Simulations based on these models are usually imperfect in predicting the behavior of the system because of knowledge gaps (72, 84). Simulations call attention to these gaps, however, and to interactions between components which formerly were not perceived. The application of systems analysis and simulation in predicting nematode population behavior will provide a better understanding of their role in crop production. It will also aid in planning and evaluating integrated control programs fitting harmoniously into systems of land management designed to maintain the productive capability of the soil (49, 83).

Literature Cited

1. Aliev, A. A. 1961. Infectivity of *Meloidogyne* sp. to varieties of maize. (In Russian) *Trudi Vseso-yuznogo Instituta Zashchiti Rasteni.* 16:89–92
2. Baldwin, J. G., Barker, K. R. 1971. Host suitability of selected hybrids, varieties and inbreds of corn to populations of Meloidogyne spp. *J. Nematol.* 2:345–50
3. Barker, K. R., Nusbaum, C. J. 1969. Horizontal distribution patterns of four plant-parasitic nematodes in selected fields. *J. Nematol.* 1:4–5 (Abstr.)
4. Barker, K. R., Nusbaum, C. J. 1971. Diagnostic and advisory services. In *Plant Parasitic Nematodes* Vol. I. eds. B. M. Zuckerman, W. F. Mai, R. A. Rohde. 281–301. Academic: New York & London. 345 pp.
5. Bessey, E. A. 1911. Root-knot and its control. *U. S. Dept. Agr., Bur. Plant Indus. Bull.* No. 217:89 pp.
6. Bird, A. F. 1970. The effect of nitrogen deficiency on the growth of *Meloidogyne javanica* at different population levels. *Nematologica* 16:13–21
7. Bradfield, R. 1970. Increasing food production in the tropics by multiple cropping. In *Research for the World Food Crisis,* ed. D. G. Aldrich, Jr. 229–42. *Am. Assoc. Advan. Sci., Washington, D.C.* Publ. 92. 323 pp.
8. Brodie, B. B., Good, J. M., Adams, W. E. 1969. Population dynamics of plant nematodes in cultivated soils: effect of sod-based rotations in Cecil sandy loam. *J. Nematol.* 1:309–12
9. Brodie, B. B., Good, J. M., Marchant, W. H. 1970. Population dynamics of plant nematodes in cultivated soil: effect of sod-based rotations in Tifton sandy loam. *J. Nematol.* 2:135–38
10. Brodie, B. B., Dukes, P. D. 1972.

The relationship between tobacco yield and time of infection with *Meloidogyne javanica. J. Nematol.* 4:80–83

11. Browning, J. A. 1969. Introduction. In *Disease consequences of intensive and extensive culture of field crops.* ed. J. A. Browning. *Iowa Agr. Exp. Sta. Spec. Rep.* 64:5

12. Chapman, R. A. 1965. Infection of single root systems by larvae of two coincident species of root-knot nematodes. *Nematologica* 12:89 (Abstr.)

13. Chapman, R. A., Turner, D. R. 1972. Effect of entrant *Meloidogyne incognita* on reproduction of concomitant *Pratylenchus penetrans* in red clover. *J. Nematol.* 4:221 (Abstr.)

14. Collins, R. J., Rodriguez-Kabana, R. 1971. Relationship of fertilizer treatments and crop sequence to populations of lesion nematodes. *J. Nematol.* 3:306–07 (Abstr.)

15. Curl, E. A. 1963. Control of plant diseases by crop rotation. *Bot. Rev.* 29:413–79

16. Daulton, R. A. C. 1952. Towards better tobacco—No. 5. Plant early: beat eelworm. *Rhodesian Farmer* 22:7

17. Davide, R. G., Triantaphyllou, A. C. 1967. Influence of the environment on development and sex differentiation of root-knot nematodes. I. Effect of infection density, age of host plant and soil temperature. *Nematologica* 13:102–10

18. Davide, R. G., Triantaphyllou, A. C. 1967. Influence of the environment on development and sex differentiation of root-knot nematodes. II. Effect of host nutrition. *Nematologica* 13:111–17

19. Dropkin, V. H., Nelson, P. E. 1960. The histopathology of root-knot nematode infections in soybeans. *Phytopathology* 50:442–47

20. Estores, R. A., Chen, T. A. 1972. Interactions of *Pratylenchus penetrans* and *Meloidogyne incognita* as coinhabitants in tomato. *J. Nematol.* 4:170–74

21. Ferris, V. R., Ferris, J. M., Bernard, R. L. 1966. Relative competitiveness of two species of *Pratylenchus* in soybeans. *Nematologica* 13:143 (Abstr.)

22. Gause, G. F. 1934. *The struggle for existence.* Williams & Wilkins: Baltimore. 163 pp.

23. Good, J. M. 1968. Relation of plant parasitic nematodes to soil management practices. In *Tropical Nematology,* eds. G. C. Smart, V. G. Perry. 113–38. Univ. Florida, Gainesville. 153 pp.

24. Grainger, J. 1962. Potato physiology and varietal efficiency in disease behavior. *Eur. Potato J.* 5:267–79

25. Grainger, J. 1964. Factors affecting the control of eelworm diseases. *Nematologica* 10:5–20

26. Hare, W. W. 1965. The inheritance of resistance of plants to nematodes. *Phytopathology* 55:1162–67

27 Harris, D. R. 1972. The origins of agriculture in the tropics. *Am. Sci.* 60:180–93

28. Huijsman, C. A. 1964. The prospects of controlling potato sickness by growing resistant varieties. *Euphytica* 13:223–28

29. Jatala, P., Jensen, H. J. 1972. Interrelationships of *Meloidogyne hapla* and *Heterodera schachtii on Beta vulgaris. J. Nematol.* 4:226 (Abstr.)

30. Jensen, H. J., Mulvey, R. H. 1968. *Predaceous Nematodes (Mononchidae) of Oregon.* Oregon State Univ., Corvallis, 57 pp.

31. Johnson, A. W. 1970. Pathogenicity and interactions of three nematode species on six Bermudagrasses. *J. Nematol.* 4:36–41

32. Johnson, A. W., Nusbaum, C. J. 1970. Interactions between *Meloidogyne incognita, M. hapla* and *Pratylenchus brachyurus* in tobacco. *J. Nematol.* 2:334–40

33. Johnson, S. R., Ferris, J. M. 1971. Nematode community structure of selected deciduous woodlots. *J. Nematol.* 3:315–16 (Abstr.)

34. Jones, F. G. W. 1956. Soil populations of beet eelworm (*Heterodera schachtii* Schm.) in relation to cropping. II. Microplot and field plot results. *Ann. Appl. Biol.* 44:25–56

35. Jones, F. G. W., Parrott, D. M., Ross, G. J. S. 1967. The population genetics of the potato-cyst nematode, *Heterodera rostochiensis:* mathematical models to simulate the effects of growing eelworm-resistant potatoes bred from *Solanum tuberosum* ssp. *andigena. Ann. Appl. Biol.* 60:151–71

36. Jones, F. G. W., Parrott, D. M. 1969. Populations and fluctuations of *Heterodera rostochiensis* Woll.

when susceptible potato varieties are grown continuously. *Ann. Appl. Biol.* 63:175–81

37. Kinloch, R. A., Allen, M. W. 1972. Interaction of *Meloidogyne hapla* and *M. javanica* infecting tomato. *J. Nematol.* 4:7–16

38. Koen, H., Grobbelaar, N. 1965. The detrimental effect of *Eragrostis curvula* on the *Meloidogyne javanica* population of soils. *Nematologica* 11:573–80

39. Koliopanos, C. N., Triantaphyllou, A. C. 1972. Effect of infection density on sex ratio of *Heterodera glycines*. *Nematologica* 18:131–37

40. Last, F. T. 1971. The role of the host in the epidemiology of some nonfoliar pathogens. *Ann. Rev. Phytopathol.* 9:341–62

41. Loomis, A. S., Williams, W. A., Hall, A. E. 1971. Agricultural productivity. *Ann. Rev. Plant Physiol.* 22:431–68

42. Messenger, P. S. 1970. Bioclimatic inputs to biological control and pest management programs. In *Concepts of Pest Management.* eds. R. L. Rabb, F. E. Guthrie. 84–102. N. C. State Univ. 242 pp.

43. Miller, P. M. 1970. Rate of increase of a low population of *Heterodera tabacum* reduced by *Pratylenchus penetrans* in the soil. *Plant Dis. Reptr.* 54:25–26

44. Miller, P. M., Wihrheim, S. E. 1968. Mutual antagonism between *Heterodera tabacum* and some other plant parasitic nematodes. *Plant Dis. Reptr.* 52:57–58

45. Minderman, A. 1956. Aims and methods in population researches on soil-inhabiting nematodes. *Nematologica* 1:47–50

46. Mukhopadhyaya, M. C., Prasad, S. K. 1969. Nematodes as affected by rotations and their relation with yields of crops. *Indian J. Agr. Sci.* 39:366–85

47. Nelson, R. R. 1957. Resistance in corn to *Meloidogyne incognita*. *Phytopathology* 47:25–26 (Abstr.)

48. Nusbaum, C. J. 1959. Effect of cultural practices following tobacco harvest upon root-knot nematode populations. *Phytopathology* 49:547–48 (Abstr.)

49. Nusbaum, C. J., Barker, K. R. 1971. Population dynamics. In *Plant Parasitic Nematodes*, eds. B. M. Zukerman, W. F. Mai, R. A.

Rohde, I:303–33 Academic: New York & London. 345 pp.

50. Odum, E. P. 1959. *Fundamentals of ecology.* Sanders: Philadelphia. 546 pp.

51. Oostenbrink, M. 1964. Harmonious control of nematode infestation. *Nematologica* 10:49–56

52. Oostenbrink, M. 1966. Major characteristics of the relation between nematodes and plants. *Meded. Landbouwhogesch., Wageningen* 66–4:1–46

53. Oostenbrink, M. 1972. Evaluation and integration of nematode control methods. In *Economic Nematology,* ed. J. W. Webster, 497–514. Academic: New York & London. 563 pp.

54. Oostenbrink, M., S'Jacob, J. J., Kuiper, K. 1956. An interpretation of some crop rotation experiences based on nematode surveys and population studies. *Nematologica* 1:202–15

55. Oteifa, B. A. 1953. Development of the root-knot nematode, *Meloidogyne incognita,* as affected by potassium nutrition of the host. *Phytopathology* 43:171–74

56. Oteifa, B. A., Abdel Halim, M. F. 1957. Cropping effect on population dynamics of soil nematodes. *Bull. Fac. Agr., Cairo Univ.* No. 128. 12 pp.

57. Powell, N. T. 1971. Interactions between nematodes and fungi in disease complexes. *Ann. Rev. Phytopathol.* 9:253–74

58. Preston, F. W. 1969. Diversity and stability in the biological world. *Brookhaven Symp. Biol.* 22:1–2

59. Rabb, R. L. 1970. Introduction to the conference. In *Concepts of Pest Management, Conf. Proc.* eds. R. L. Rabb, F. E. Guthrie, 1–5. N. C. State Univ., Raleigh. 242 pp.

60. Rabb, R. L., Guthrie, F. E. 1970. Preface. In *Concepts of Pest Management. Conf. Proc.* Eds. R. L. Rabb, F. E. Guthrie, iii–iv. N. C. State Univ., Raleigh. 242 pp.

61. Radewald, J. D. 1969. The role of agricultural extension in nematology—past, present and future. In *Nematodes of Tropical Crops.* ed. J. E. Peachey 333–40. *Tech. Commun. Commonw. Bur. Helminth.* 40. 355 pp.

62. Robinson, R. A. 1971. Vertical resistance. *Rev. Plant Pathol.* 50:233–39

63. Rohde, R. A. 1965. The nature of resistance of plants to nematodes. *Phytopathology* 55:1159–62

64. Ross, J. P. 1964. Interaction of *Heterodera glycines* and *Meloidogyne incognita* on soybeans. *Phytopathology* 54:304–07

65. Sasser, J. N., Nusbaum, C. J. 1955. Seasonal fluctuations and host specificity of root-knot nematode populations in two-year tobacco rotation plots. *Phytopathology* 45:540–45

66. Schafer, J. F. 1971. Tolerance to plant disease. *Ann. Rev. Phytopathol.* 9:235–52

67. Seinhorst, J. W. 1967. The relationships between population increase and population density in plant parasitic nematodes. III. Definition of the terms host, host status and resistance. IV. The influence of external conditions on the regulation of population density. *Nematologica* 13:429–42

68. Seinhorst, J. W. 1970. Dynamics of populations of plant parasitic nematodes *Ann. Rev. Phytopathol.* 10:131–56

69. Sikora, R. A., Taylor, D. P., Malek, R. B., Edwards, D. I. 1972. Interaction of *Meloidogyne naasi, Pratylenchus penetrans,* and *Tylenchorynchus agri* on creeping bentgrass. *J. Nematol.* 4:162–65

70. Southards, C. J., Nichols, B. C. 1972. Population dynamics of *Meloidogyne incognita* in crop rotations of tobacco, sod and forage crops. *J. Nematol.* 4:234 (Abstr.)

71. Southwood, T. R. E., Way, M. J. 1970. Ecological background to pest management. In *Concepts of Pest Management.* Eds. R. L. Rabb, F. E. Guthrie, 6–29. N. C. State Univ., Raleigh. 242 pp.

72. Spedding, C. R. W. 1971. Agricultural ecosystems. *Outlook Agr.* 6:242–47

73. Stern, V. M., Smith, R. F., Van den Bosch, R., Hagen, K. S. 1959. The integrated control concept. *Hilgardia* 29:81–101

74. Todd, F. A. 1971. System control: A prescription for flue-cured tobacco diseases. *N. C. Agr. Ext. Circ.* 530. 14 pp.

75. Todd, F. A., Bennett, R. R. 1957. Cropping systems for nematode control and tobacco production. *N. C. Agr. Ext. Cir.* 409. 16 pp.

76. Todd, F. A., Nusbaum, C. J. 1972. Flue-cured tobacco summary report of 1972 data. *N. C. Agr. Ext., Plant Pathol. Info. Note* 183, 155 pp. (mimeo)

77. Triantaphyllou, A. C. 1960. Sex determination in *Meloidogyne incognita* Chitwood 1949 and intersexuality in *M. javanica* (Treub, 1885) Chitwood, 1949. *Ann. Inst. Phytopathol. Benaki* N. S. 3:12–31

78. Trudgill, D. L. 1967. The effect of environment on sex determination in *Heterodera rostochiensis* Woll. *Nematologica* 13:263–73

79. Van den Brande, J., D'Herde, J. 1964. Phenological control of the potato root eelworm (*Heterodera rostochiensis* Woll.) *Nematologica* 10:25–28

80. Van der Plank, J. E. 1963. *Plant Diseases: Epidemics and Control.* Academic: New York. 349 pp.

81. Van der Plank, J. E. 1968. *Disease Resistance in Plants.* Academic: New York & London. 206 pp.

82. Van der Plank, J. E. 1972. Basic principles of ecosystem analysis. In *Pest Control Strategies for the Future,* 109–18. *Nat. Acad. Sci., USA* 376 pp.

83. Van Gundy, S. D. 1972. Nonchemical control of nematodes and root-infecting fungi. In *Pest Control Strategies for the Future,* 317–329. *Nat. Acad. Sci., USA* 376 pp.

84. Waggoner, P. E., Horsfall, J. G. 1969. Epidem. A simulator of plant disease written for a computer. *Conn. Agr. Exp. Sta. Bull.* 698. 80 pp.

85. Wallace, H. R. 1970. Some factors influencing nematode reproduction and the growth of tomatoes infected with *Meloidogyne javanica.* *Nematologica* 16:387–97

86. Wilhelm, S. 1965. Analysis of biological balance in natural soils. In *Ecology of Soil-borne Plant Pathogens.* Eds. K. F. Baker, W. C. Snyder, 509–17. Univ. Calif. Berkeley. 571 pp.

87. Yuen, P. H. 1966. The nematode fauna of the regenerated woodland and grassland of Broadbalk wilderness. *Nematologica* 12:195–214

88. Zadoks, J. C. 1971. Systems analysis and the dynamics of epidemics. *Phytopathology* 61:600–10

ENVIRONMENTAL SEX DIFFERENTIATION OF NEMATODES IN RELATION TO PEST MANAGEMENT[1]

❖ 3581

A. C. Triantaphyllou
Department of Genetics, North Carolina State University,
Raleigh, North Carolina

Introduction

The sex ratio of small nematode populations often reflects the end result of the interaction of several parameters including the genetic mechanism of sex determination, the environmental influence on sex differentiation, the relative rate of development, and the differential survival of the members of the two sexes. In terms of population dynamics and potential crop damage, the sex ratio in a particular nematode population is of fundamental significance. It affects not only the potential rate of population increase from generation to generation but it also determines the extent of damage, which most often is proportional to the female population. The latter is true because females require more food to meet the increased requirements of reproduction. Thus it would be advantageous if the sex ratio of a nematode population could be shifted toward the male direction by any conceivable means. In this connection it would be very helpful to have an understanding of the genetic and environmental factors that influence sex determination, sex differentiation, and the final development and survival of the members of the two sexes.

Environmental influence on sex differentiation is a topic of basic biological significance since it relates to the action of the environment on developmental processes in conjunction with, or independently of the genetic constitution of organisms. This general topic has been thoroughly reviewed by Bacci (3) from a genetic and developmental point of view, but developmental sex deviations in plant-parasitic nematodes were covered very lightly.

Recent increased social concern about environmental and ecological problems and strong realization of the need to adopt more natural approaches in pest management have stimulated interest in evaluating the possibilities of

[1] Paper No. 3943 of the Journal Series of the North Carolina State University Agricultural Experiment Station, Raleigh, N.C. This study was supported in part by a grant of the National Science Foundation (GB–29485).

441

exploiting vulnerable aspects of the biology of parasites. The effect of the environment on sex differentiation of insects has already been evaluated by Bergerard (6). At present, the outlook for exploiting sexual peculiarities of nematodes for the purposes of population management is not very encouraging. Some progress can be made in this area, however, if the biological aspects of sex differentiation of plant-parasitic nematodes and the environmental factors that influence sex differentiation are better understood.

Genetic and Environmental Control of Sex Expression in Animals

In gonochoristic animals that develop from fertilized eggs, sex differentiation during embryogenesis and postembryogenesis proceeds according to the genetic constitution of the zygote and can be influenced only slightly by environmental factors. It is generally believed that the genetic material present in the zygote contains two groups of genes, one promoting differentiation toward the male and the other toward the female sex. Their relative balance of strength is stable under a wide variety of environmental conditions, and determines which one of the two groups will dominate. The sex genes are unevenly distributed over the autosomes and specialized sex chromosomes. Depending on the pattern of gene distribution, cytogeneticists recognize various chromosomal mechanisms of sex determination, such as XX♀–XY♂, XX♀–XO♂, ZW♀–ZZ♂, NN♀–N♂, or certain genetic mechanisms of sex determination, such as X♀–A♂, X♀–Y♂ and A♀–Y♂ (29). The definite pattern of distribution of the sex genes in the karyotype and the special nature of meiosis and fertilization usually insure the appearance of males and females in equal numbers in most cross-fertilizing organisms. Often, however, the 1:1 sex ratio can be modified by such mechanisms as meiotic drive (78) and gametic or zygotic selection, by the action of sex-transforming genes, and by environmental factors which affect sex differentiation during embryogenesis and postembryogenesis. The most important environmental factors that directly or indirectly affect sex differentiation of animals, are sex hormones, parasitism, and temperature.

Sex hormones produced under genetic control in the gonads regulate sex differentiation of mammals, birds, amphibians, and fishes. Artificial changes of the hormonal environment of a developing individual decisively affect its sexual differentiation, particularly in the latter three groups of animals (8). To a large extent, hormones or hormone-like substances also appear to control differentiation in many Crustaceans (12), some Mollusks (28), the echiurid worm *Bonellia viridis* (4), the isopode *Ione thoracica* (55), and probably many other invertebrates.

Parasitism has been known to influence sexual differentiation of various insects including some Hymenoptera parasitized by strepsipterous insects (58), Homoptera parasitized by dryinid insects (39), and Diptera parasitized by mermithid nematodes (9, 33, 54). Various degrees of intersexuality appear as a result of parasitism.

Temperature appears to play a significant role in sexual differentiation of various groups of animals. It modifies the balance between feminizing and masculinizing genes in various insects (6). In *Carausius morosus*, a thelytokous parthenogenetic Phasmid, highly masculinized females, or apparently male individuals appear, when eggs are incubated at higher than optimal temperatures (5). Conversely, elevated temperatures induce the transformation of genetic males into intersexes, or into almost normal females in *Aedes stimulans* and in several other species of mosquitoes (1). Similarly, high temperatures induce feminization in triploid intersexes of *Drosophila melanogaster* (20).

Other environmental factors affecting sex differentiation include: lack of food, and amputation as in the polychaete worm *Ophryotrocha puerilis* (32), high population density as in oysters, concentration of certain metallic ions or organic substances in the surrounding medium as in *Bonellia viridis* and the marine copepode *Tigriopus japonicus* (62), and the quality of food as in the cladoceran *Moina rectirostris* (19).

With the exception of hormones and hormone-like substances, which appear to regulate directly sex differentiation and function, the mode of action of all other factors is very little understood. Furthermore, it is suspected that the genetic constitution of the individual animal plays a significant role in sex differentiation even in the most striking cases of known phenotypic sex determination (3).

Sex Determination in Nematodes

Present knowledge of the chromosomal mechanisms of sex determination in plant-parasitic nematodes is limited to a few cases where the cytology has been studied (65). It appears that in most bisexual (gonochoristic) species of nematodes, males and females occur in equal numbers, and the chromosomal mechanism of sex determination is of the XX♀–XO♂ type. Among plant-parasitic nematodes, however, no sex chromosomes have been recognized definitely and it is possible that an XX♀–XY♂ situation prevails with the sex chromosomes not easily distinguishable from the autosomes.

Besides the bisexual species, a number of parthenogenetic species exist among plant-parasitic nematodes. These species comprise females only (thelytoky), or have, in addition, a very small proportion of males produced by a mechanism which is not understood at present (69). Still other parthenogenetic species comprise substantial, and usually variable, percentages of males in addition to females, and are obligatorily or facultatively parthenogenetic. The mechanism by which variable percentages of males develop in such species is not known, although phenotypic sex determination has been implicated in some of these cases (65). Hermaphroditic species of nematodes usually comprise protandric hermaphrodites, although variable percentages of pure males and females may also appear in many species. Knowledge about the cytogenetic mechanisms by which males develop in hermaphroditic species of nematodes is limited to groups other than plant-parasitic ones and has been reviewed previously (46, 69).

Sex Differentiation in Nematodes

Nematode larvae hatching from eggs are morphologically indistinguishable with regard to sex. Only in *Tylenchulus semipenetrans* are male larvae reported to be different from female larvae at the time of hatching (76). Sex of nematodes usually becomes morphologically distinguishable only after the genital primordium starts growing and differentiates into a distinguishable male or female gonad. This occurs in various developmental stages. In the root-knot nematodes it takes place in half-grown second-stage larvae (68), but in the cyst nematodes and in most other plant-parasitic nematodes, this occurs in third-stage larvae.

Development of sex characteristics during postembryogenesis, once initiated, usually proceeds normally and results in the development of normal male and female individuals. In some species, however, development of both primary and secondary sex characteristics may be influenced by the prevailing environmental conditions. Under some circumstances sex reversal may occur and individuals of the opposite sex, or intersexes of various degrees of intersexuality may result. These developmental peculiarities are analyzed below in an attempt to gain a better understanding of the developmental patterns involved and the environmental conditions that promote them.

Environmentally Controlled Sex Differentiation—Early Evidence

The first evidence that environmental factors can influence the direction of sex differentiation of nematodes was provided by studies with members of the family Mermithidae. Members of this family are parasites of invertebrates, primarily insects. As early as 1927, Cobb, Steiner & Christie in the USA (15) and Caullery & Comas in France (10) had observed that several mermithid species found inside nymphs of grasshoppers and larvae of chironomid insects had peculiar sex ratios. In light infestations, when one or few nematodes were found inside an insect host, all or most of the nematodes were females. In moderate infestations males and females were in approximately equal numbers, and in heavy infestations, when a relatively large number of nematodes were parasitizing the same insect, all or most of them were males. Such observations, together with the results of preliminary tests, convinced the above researchers that "environment is a sex determining factor, a factor which becomes potent not during the early embryology of the animal but after a well-developed, highly differentiated larva has been formed" (15). These early convictions were soon substantiated by Christie (14). He fed nymphs of grasshoppers with a known number of embryonated eggs of *Mermis nigrescens,* and later found and identified the sex of the nematodes that developed in each grasshopper. When 4 or 5 eggs were fed to each insect, 72% of them were recovered later as mature nematodes inside the insects. Among the nematodes recovered, 92% were females and this represents 66% of the total number of eggs fed to the insects. In contrast to this, when 20 or 30 eggs were fed to each grasshopper, 86% of them were recovered as mature nematodes and all of them were males. This demonstrated beyond doubt that more than 50% of the nematode larvae developed as females when the number of parasites per host was low and into males when it was high, suggesting that the internal environment of the host influenced sex differentiation of the developing larvae. More recent studies have associated high percentages of males with crowding and adverse nutritional status of the insect hosts in a number of mermithid nematodes (33, 50, 52).

Concomitant with the observation of environmental influence on sex differentiation, the question of intersexuality in nematodes arose again. Intersexes, i.e., individuals of one sex with secondary sexual characters of the opposite sex, had already been observed in several nematode species, including members of the family Mermithidae. An explanation of the development of intersexes in mermithids was naturally attributed to the same environmental causes that directed the phenotypic expression of sex in developing larvae. Thus, Christie (14) stated that when one female develops inside an insect host together with many males, "it is conceivable that this female is under the influence of some environmental force tending to stimulate maleness and which, although insufficient in her case to bring about the development of a functional male individual, is sufficient to induce the development of secondary male characters." This explanation was in contrast to Steiner's (61) view that nematode intersexes, as in the moth *Lymantria* (27), result from hybridization between populations or races with different potency of sex determining factors. The fact that no intersexes were observed in the parthenogenetic *Mermis nigrescens*, whereas many intersexes were found in the amphimictic *Agamermis decaudata* and *Hydromermis contorta* clearly supported Steiner's view, but Christie believed that further confirmation of these observations was needed.

These studies of sexuality of mermithids apparently influenced the thinking of more recent nematologists who in turn attempted to explain unbalanced sex ratios and intersexuality observed in other nematode species, including many plant-parasitic forms.

Environmental Sex Differentiation in the Family Heteroderidae

Unbalanced sex ratios were observed in several members of the family Heteroderidae long before essential aspects of the biology of these animals were known or understood. Molz in 1920 (45) reported that the proportion of males of *Heterodera schachtii*, the sugar beet nematode, varied considerably, and associated the appearance of larger percentages of males with the unfavorable growth condition of the host plant. He was the first in the area of plant nematology to express the opinion that sex differentiation of larvae is influenced by the environment. This view was not shared by Sengbusch (59) who repeated some of Molz's experiments and concluded that increased percentages of males, whenever they occur, result from a differential death rate of larvae of the opposite sexes.

More information implicating the environment as a cause of variation in sex ratio was presented a few years later by Tyler (75) who, at the same time, demonstrated that females of a root-knot nematode species (*Meloidogyne* sp.), could reproduce in the absence of males, i.e., without insemination. She obtained only 0.7% males from single-larva inoculations of tomato seedlings, 16.4% from plants infected with more than one larva, and 56.5% from plants that were re-infected by second generation larvae. The percentage of males was thus well correlated with the crowding of the larvae in the roots and, to some extent, with the nutritional conditions of the infected roots. The mode of action of the environment was not clear, but Tyler suggested that adverse nutritional conditions during the early stages of development could be important enough to change the physiological balance of certain individuals and thus alter the sex ratio. Linford (43) agreed with Tyler that unfavorable conditions cause an increase in percentage of males. He found relatively abundant males in cowpea leaves, especially in small veins and near the margins, where weak development of females indicated poor nutritional conditions. Still, the question of sex differentiation in the root-knot nematodes was very confusing until lately. Thus, Dropkin (21) concluded in 1959 that sex ratios of *Meloidogyne incognita* and *M. arenaria* are not always correlated with suitability of the host plant, or gall size, and that development of males must follow a more complex pattern and must be influenced by specific conditions present in some hosts. More recent studies have partially clarified the mechanism of sex differentiation in the genera *Meloidogyne, Heterodera,* and *Meloidodera.*

GENUS MELOIDOGYNE   Experimental evidence that sex expression is influenced by environmental factors became available following a detailed study of the anatomical features of postembryogenesis of *Meloidogyne incognita* (68), and a study of the effect of starvation on larval development (63). It was tentatively concluded that under favorable conditions of feeding, larvae of *M. incognita* develop normally and become pear-shaped females with two ovaries. Under less favorable conditions, most of the larvae start differentiating as males and develop into vermiform adult males which have one testis, as do males of all other tylenchid nematodes. When the conditions are favorable in the early period of larval development, but become unfavorable later on, the pattern of larval development changes accordingly. The young, second-stage larvae start developing as females, but shift development toward the male sex, as soon as the conditions become unfavorable. Such larvae develop into vermiform males which, however, have two testes that correspond to the double gonad-primordium of the female larvae from which they develop. Anatomical studies have shown that the change in the direction of sexual differentiation occurs in half-grown second-stage female larvae during or shortly after morphological sex differentiation has been initiated (63). It does not occur in fully-grown second-stage or more advanced larvae in which sex differentiation and gonad development are much advanced. The reversal is complete with regard to most morphological features, so that sex-reversed adults look like normal males. However, sex reversal is usually incomplete with regard to gonad development. Adverse conditions cause a masculinizing effect by suppressing development of one of the two gonads of the female larvae and allowing the other to develop and become a normal testis. Depending on the degree of initial development the suppressed gonad may stop developing and remain short, or it may continue developing at a low or almost normal rate and become a testis of small or almost normal length. Thus males derived from female larvae through sex reversal have two testes, one of which is of normal length while the other may be from rudimentary to almost normal (63).

Although these studies were conducted with *M. incognita,* a similar pattern of sex differentiation is believed to operate in many other *Meloidogyne* species. In *M. javanica* intersexual individuals are known to occur in addition to males (13). Intersexes in this species are males functionally and structurally, but they have some secondary female characteristics, such as rudimentary to well-developed vulva and vagina. They are believed to develop from female larvae following incomplete sex reversal in response to adverse environmental conditions (63).

In *M. arenaria* sex differentiation follows the same pattern and is affected by the same environmental factors as in *M. incognita.* However, in at least two populations, no males were produced under various conditions of propagation, suggesting that these populations were completely thelytokous (64). It appears that sex differentiation of *M. arenaria* populations depends to a large extent on their genetic constitution, and that sex expression of certain genotypes in this species is not influenced, or is less influenced by the environment. Similarly, Webber & Fox (77) found that single larva and single egg-mass isolates of *M. graminis* produced different percentages of males under similar environmental conditions. They concluded that larvae differ in their potential to differentiate as males under adverse environmental conditions and that the differences may have a genetic basis. This appears to be substantiated by later observations that a population of *M. graminis* always had an abundance of males and reproduced by amphimixis, whereas other populations had very few males under any environmental conditions and reproduced by meiotic parthenogenesis (67). Observations with *M. hapla* also indicate that, under similar conditions of propagation, males are more abundant in primarily amphimictic populations than in populations that reproduce mainly by parthenogenesis. Apparently the distribution of the male- and female-determining genetic factors in predominantly amphimictic populations is such that the genotypes of the progeny are definitely either male or female and their expression is less sensitive to environmental influences.

GENUS HETERODERA   In amphimictic species of *Heterodera,* males and females usually occur in equal numbers. Unbalanced sex ratios are often ob-

served, however, and are associated with such environmental factors as infection density, nutritional status, and species or variety of the host plant. Studies to elucidate the mechanism by which environmental factors affect sex ratio have been limited to only a few species of *Heterodera* and the results are
not all in agreement.

Ellenby first raised the question of sex determination in the potato cyst
nematode, *H. rostochiensis,* after he found that more males developed on lateral roots than on primary roots, and that the male to female ratio increased
with increasing infection density (22). He attempted to explain this by
considering the possibility that such changes in sex ratio may have resulted
from a differential death rate of the larvae of the opposite sex, or by direct
environmental influence on sex differentiation of developing larvae. He concluded that "sex determination is under environmental influence." Den
Ouden (49) indirectly supported this view, when he recovered many more
females than males from potato seedlings, each inoculated with a single larva
of *H. rostochiensis.* Similarly, Trudgill (71) showed that high infection densities and adverse nutritional conditions resulting from the removal of the
above-ground parts of potato seedlings, or advanced age of the host plants,
cause an increase in percentage of males. He, and later Ross & Trudgill (56),
offered the following interpretation of the mechanism of sex differentiation in
this nematode. Under light infection densities, all or most of the larvae find a
good infection site, with sufficient space to induce development of a large
group of giant cells. Under such favorable feeding conditions, most of the
larvae become females. On the contrary, under heavy infection densities, the
larvae settle close to each other and only limited space is available—sufficient
for the formation of only small groups of giant cells. Under these conditions
most of the larvae become males.

Studies with resistant varieties of potatoes have provided evidence that sex
expression in *H. rostochiensis* is in some way linked to genetic factors (73).
Thus it has been suggested that only larvae that are double recessive (aa) for
overcoming resistance can induce proper giant cells in *Solanum tuberosum*
spp. *antigena* plants, and thus become females. Heterozygous and homozy-
gous dominant larvae (aA,AA) become males. Similarly, more recent studies
indicated genetic differences among various nematode populations with regard to their ability to develop on three resistant potato hybrids and the susceptible variety "banner," and to establish a characteristic sex ratio on these
plants (72). Furthermore, with the exception of two "nematode-host plant"
combinations, where males were more numerous on resistant potatoes than
on "banner," unbalanced sex ratios in the remaining 25 combinations can be
explained by assuming that resistance suppresses development of female larvae to a greater degree than development of male larvae. The same conclusion is reached by evaluating recent data on the development of four nematode populations on eight potato varieties carrying different genes for resistance (74). In all cases, both male and female development was suppressed

in resistant varieties, and without any exceptions, fewer males developed on resistant than on susceptible plants. Since the same number of larvae usually enter resistant and susceptible potatoes (72), the above results can be explained as in the previous tests and without the need to assume that males develop at the expense of females on resistant plants.

In the writer's opinion, the only evidence that sex differentiation in *H. rostochiensis* is environmentally controlled is provided by data indicating that significantly larger numbers of females than males develop in single-larva inoculations and under conditions of low infection densities (49, 71). The increased percentages of males under adverse feeding conditions do not prove direct influence of the environment on sex differentiation. In most cases they can be attributed to differential rate of development and differential death rate of the larvae of the opposite sex, with male larvae being favored in both cases.

Unbalanced sex ratios observed frequently in *H. schachtii* also have been the subject of several investigations with conflicting results as to the mechanism responsible for their occurrence. Apel & Kämpfe (2) found that the sex ratio of this nematode is influenced by infection density and particularly by the species of the host plant. This study was followed up by Kämpfe & Kerstan (36) who demonstrated the effect of various environmental factors, including nitrogen deficiency, unfavorable water supply, and lack of light, on the sex ratio of this nematode. Their studies were not designed to elucidate the underlying mechanism of environmental action. Nevertheless, they suggested that a difference in the mortality rate of prospective male and female larvae in the roots is not a decisive factor, and that the future sex of the larvae is probably influenced by the nutritional status of the host. More recent reports (35, 37) appear to reverse this opinion. According to Kerstan (37) increased male to female ratios, under adverse conditions, can be attributed to high death rate of female larvae rather than to an increase of the absolute number of males at the expense of females. In single-larva inoculations, where conditions are favorable for nematode development, males and females appear in equal numbers, presumably differentiating according to their predetermined sex. Similarly, Johnson & Viglierchio (35) found no evidence that sex of *H. schachtii* is environmentally controlled. In their tests, sex ratios fluctuated considerably, but it appeared that the percentage of developing males was rather constant, whereas that of females fluctuated with environmental and nutritional conditions. Surprisingly in their results, however, the sex ratio under normal conditions was one male to approximately 2.5 females instead of 1:1 as it would be expected in a cross-fertilizing organism with a strong genetic mechanism of sex determination, not influenced by environment. A similar ratio of one male to approximately 3 females had been reported earlier by Apel & Kämpfe (2) on sugar beet. Unless these reports involve technical errors, which conceivably can occur in evaluating sex ratios in these organisms, they will deserve further consideration before the mechanism of sex determination and sex differentiation of *H. schachtii* is clarified. In the meantime, the work of Kerstan (37) supporting environmental influence on sex ratio but not on sex differentiation of *H. schachtii* appears to present a realistic picture of the situation.

Steele (60) suggested recently that the resistance of plant roots to penetration by larvae may be partially responsible for increased percentages of *H. schachtii* males observed in various cases. According to his observations, males of this nematode can develop from larvae with only their head region embedded into the roots of tomato or sugar beet plants, whereas females can only develop from larvae that penetrate deep into the roots and have their heads inside the vascular parenchyma of the roots. Therefore, any factors that restrict deep penetration of the larvae into the roots, such as development of tough periderm in old roots, will restrict development of female larvae and will favor a higher male to female ratio.

Fluctuating sex ratios in *H. glycines* have been equally difficult to interpret as in other *Heterodera* species. Significantly smaller numbers of males than females were found in *H. glycines* cultures maintained at 24°C, whereas the male to female ratio was approximately 1:1 at 28°C according to Ross (57). The low male to female ratio at 24°C was regarded as the normal one, and the relatively increased sex ratio at 28°C was attributed to the influence of the higher temperature on sexual differentiation of female larvae which then developed into males. Further increases in the male to female ratio at 31°C or higher temperatures were associated with degeneration of larvae in the roots and were attributed to death of many female larvae at these high temperatures. Additional studies by Koliopanos & Triantaphyllou (38) on the effect of infection density on sex ratio supported the idea that increased percentages of males are due to differential death rate of larvae of the two sexes under conditions of food stress created by crowding and some other factors. Under noncrowded conditions and in single-larva inoculations, males and females appeared in approximately equal numbers, which was interpreted as indicating development of the larvae according to their predetermined sex. Similarly, recent unpublished data of the author on the development of *H. glycines* on resistant and susceptible soybeans have demonstrated that plant resistance does not influence sex differentiation of the larvae. High percentages of males observed in populations of *Heterodera avenae* were correlated with increased infection densities (42).

The conflicting results presented above allow no definitive statement about the real cause of unbalanced sex ratios in the genus *Heterodera*. Although early work favored an environmental influence on the direction of sexual differentiation of the larvae, most recent work suggests that an excess of males appears only because more female larvae fail to reach maturity under adverse conditions. This may be due to their greater requirements for food as compared to males.

GENUS MELOIDODERA   The most convincing case of environmental influence on sex differentiation in a plant-parasitic nematode is that of *Meloidodera floridensis*. This nematode is a triploid, thelytokous parthenogenetic organism (66), with only an occasional male appearing in natural populations infecting pine roots. Recent studies showed that at least 96 percent of the larvae that entered pine roots grew and developed to adult females following a period of feeding in the roots. The remaining 4 percent apparently failed to establish feeding relationships with the host and developed into very small males (70). Conversely, when larvae were kept in water under prolonged starvation, more than 50 percent became males, while most of the remaining larvae underwent one or two molts, and also differentiated as males, but died in the 3rd or 4th stage. This pattern of sex expression led to the conclusion that sex in *M. floridensis* is controlled environmentally to a large extent.

Environmental Sex Differentiation in Other Plant-Parasitic Nematodes

Unbalanced sex ratios are common in many other plant-parasitic nematodes. Males are usually absent or rare in parthenogenetic and hermaphroditic species (69). Variable percentages of males are observed also in several bisexual, cross-fertilizing species. Two instances in which the sex ratio has been studied and environmental influence on sex differentiation is suspected, are discussed briefly.

The male to female ratio of *Nacobbus serendipiticus* cultured on excised tomato roots at various temperatures varied from 0.14 to 7.33 according to Prasad & Webster (53). It is not clear, however, whether temperature differentially affected development and viability of the larvae of the two sexes, or the direction of sexual differentiation.

Sex differentiation of *Tylenchulus semipenetrans* appears to be influenced by the environment (30, 43a). Information about the mode of reproduction, sex determination, and sex differentiation of this nematode needs further clarification, however. According to van Gundy (76), uninseminated females produced larval offspring, 26 percent of which were identified morphologically as male and 74 percent as female. This means that uniparental reproduction is possible, and that second-stage larvae hatching from eggs are genetically and morphologically determined with regard to sex. Furthermore, preliminary tests suggested that fertilization or lack of fertilization may influence sex ratios; this further supports the genetic nature of sex determination. On the other hand, Macaron & Ritter (43a) found that adult males can develop also from second-stage female larvae after feeding on the roots of a host plant. This means that some female larvae undergo sex reversal and develop into large, adult males, probably under the influence of unknown environmental factors; temperature, soil moisture, and crowding had no significant effect. Furthermore, the same authors confirmed that isolated females can give progeny, probably by parthenogenesis, and that male second-stage larvae develop into small males without feeding. All this tends to clarify some of the questions about mode of reproduction, sex determination, and sex differentiation of *T. semipenetrans,* but additional information is needed.

Environmental Sex Differentiation in Aphelenchus Avenae

Although *Aphelenchus avenae* is not primarily a plant parasite, it is included in this review because it presents a definite case of environmentally controlled sex differentiation in nematodes. Most natural populations of *A. avenae* consist of females only (thelytokous parthenogenetic), with an occasional male appearing sporadically. In axenic cultures, embryonated eggs and freshly hatched larvae of this species developed into adult females when the cultures were maintained at 28°C or lower temperatures, but they became males at 30°C or higher (31). Variable proportions of males and females developed at the intermediate temperatures of 29° and 29.6°C. Also, by increasing the normal concentration of CO_2, Hansen and her associates obtained males even at lower temperatures, such as 28 and 26°C. This clearly demonstrated that high temperatures and increased CO_2 concentrations have a masculinizing effect on the normally female larvae and reverse them to males. Furthermore, it was found that in order to reverse the sex, the environmental influence should start on the egg before the first molt and continue at least until the middle of the second larval stage. Treatments of shorter duration, or in a more advanced stage of development, had no effect and gave rise to females only; no intersexes were produced, probably because morphological sex characters in this nematode begin development after the period of environmental influence on sex has passed.

Another parthenogenetic population of *A. avenae* studied by Evans & Fisher (23) did not produce males at 30°C. This indicates that genetic variation does exist among natural populations of this species with regard to phenotypic sex expression, but the nature and the extent of this variation is not known. In this connection, it should be added that some obligatorily amphimictic populations of *A. avenae* have also been found, which always have an abundance of males (26). *A. avenae,* therefore, represents a species complex in a state of very active evolution, with many parthenogenetic, polyploid forms branching off and evolving independently from the main amphimictic species. The situation is very similar to that in the genus *Meloidogyne,* where again, phenotypic sex determination occurs.

Environmental Factors Affecting Sex Differentiation in the Genus Meloidogyne

HOST NUTRITION   Low rate of development and increased percentages of males of root-knot nematodes were early associated with adverse conditions of nutrition of the host plant (43, 75). But it is difficult to define what the nutritional condition of a host is, especially for the root-knot nematodes, whose feeding relationships with the plant tissues are very complex. Attempts have been made to modify the nutritional status or the biochemistry of a host by subjecting plants to deficiency treatments involving the major mineral elements. In most cases, mineral deficiencies reduced the rate of development of the nematodes, and at the same time increased the percentage of males, although not very strikingly (7, 17). Similarly, lower concentrations of sucrose in the medium increased the percentage of males of *Meloidogyne incognita* developing on excised cucumber roots (44). It is very likely that the availability of organic compounds, serving as food for normal development of root-knot nematodes, is the most important component of what is referred to as nutritional status of the host. Defoliation or complete removal of the above-ground parts of a host plant, which in time is expected to decrease the amount of available organic compounds in the roots, will similarly reduce the rate of development and increase the male percentage of *Meloidogyne* (63).

INFECTION DENSITY   High infection densities usually cause a reduction in the rate of development of nematodes, because they create conditions of food stress and probably an adverse environment resulting from excessive concentrations of waste or metabolic by-products. Such conditions are liable also to affect the process of sex differentiation of root-knot nematodes. High infection densities significantly reduced the rate of development and increased the male to female ratio of *M. incognita* and *M. javanica* developing on tomato plants under greenhouse conditions (16). Infection density did not influence the sex ratio of *M. incognita* developing on excised cucumber roots, however, probably because organic compounds were available to the roots in abundance directly from the nutrient medium (44).

HOST RESISTANCE   No experiments have been conducted to determine specifically the effect of resistant hosts on sex differentiation of *Meloidogyne* species, but incidental observations indicate that the male to female ratio is rather high on moderately resistant plants even under low infection densities. Resistance of *Cucumis ficifolius* and *C. metuliferus* to *M. incognita* is associated with delayed development of larvae and stimulation toward maleness (24). Work conducted with highly resistant plants is not informative in this respect, because such "nonhost" plants do not support development of either males or females.

PLANT GROWTH INHIBITORS OR REGULATORS   Although various chemicals with or without nematocidal effects have been applied to plants infected with nematodes to test their influence on nematode development, their effect on sex differentiation or sex ratio of the nematodes has been studied with only a few such chemicals. Maleic hydrazide [6-hydroxy-3-(2H)-pyridazinone], a potent plant growth inhibitor, was found to inhibit development of *M. incognita* when applied on tobacco and tomato plants (47, 51). Later, it was found that the percentage of males of *M. incognita* and *M. javanica* increased considerably, when maleic hydrazide was applied on nematode-infected tomato plants (18). This was interpreted as an effect of the modified physiology and biochemistry of the host plant on the direction of sexual differentiation of the larvae. More specifically, maleic hydrazide suppresses giant-cell formation which is essential for the nematode to establish and maintain feeding relationships with the plant. Although the system may be very complex, food stress has been regarded as the single main factor directly affecting rate of development and sex differentiation in this case.

Morphactin [methyl-2-chloro-9-hydroxyfluorene-(9)-carboxylate], an antimitotic agent that affects plant morphogenesis, interfered with syncytia formation and significantly reduced the rate of development of *M. javanica* on tomato seedlings (48). The percentage of males among sexually differentiated nematodes increased from zero in the control to approximately 80 and 90 in plants treated with morphactin 2 days before and 2 days after inoculation, respectively. Although morphactin, at the rates used, had a marked injurious effect on tomato plants, "the possibility of treatments with tolerant levels of growth-regulating substances to combat the root-knot nematodes indirectly, should not be overlooked."

AGE OF THE HOST PLANT   As the general metabolism and biochemistry of a host plant may change with age, it was suspected that sex differentiation of the nematodes may be affected accordingly (63, 75). Experimental evidence, however, suggests that rate of development and sex ratio of *M. incognita* and *M. javanica* are not substantially affected by the age of the tomato plants (16). This may be due to the fact that root-knot nematodes establish infections in growing apical meristems of the roots whose physiology and cytochemistry may not be affected substantially by the age of the plants.

IRRADIATION   Gamma radiation reduced the rate of growth of larvae and increased the percentage of males of *M. incognita* (34). When gamma radiation was applied to young second-stage larvae, males that developed from them were small and had one testis. When irradiation was applied to half-grown larvae in which sexual differentiation had already been initiated, the males were large and most had two testes, indicating that they had developed from female larvae following sex reversal. Radiation may directly affect the physiology of sex differentiation of the larvae, or it may act indirectly by decreasing the ability of the nematode to obtain food and utilize it efficiently, or by affecting the capacity of roots of the irradiated plants to provide a favorable nutritional environment for nematode development. The indirect effect is supported by the fact that both growth rate and sex differentiation were influenced by irradiation.

TEMPERATURE   Sex differentiation of *Meloidogyne incognita* was unaffected within a wide range of temperatures, although a slight increase of the male to female ratio occurred at the low temperature of 15°C (16). Similarly, the percentage of males of *M. hapla* was not affected by temperature, but the male to female ratio of *M. javanica* increased significantly at temperatures above 20°C (37a). Also, a high temperature of 32°C during early development increased the percentage of males in cultures of *M. graminis* (41). More information is needed on the effect of short exposures to temperature extremes, particularly during or shortly before the period of sexual differentiation of the larvae. As mentioned earlier, the effect of temperature on sex differentiation has been demonstrated in the fungus-feeding nematode, *Aphelenchus avenae* (31). Also, exposure of eggs of *Ditylenchus myceliophagus* to a low temperature significantly increased the percentage of females among the developing larvae, but the actual mechanism involved needs to be further elucidated (11).

Unbalanced Sex Ratios in Relation to Nematode Population Management—Discussion

The preceding review indicates that unbalanced sex ratios in plant-parasitic nematodes at the time they reach adulthood result from two major causes:

(*a*) Adverse environmental conditions suppress the development of female larvae to a greater degree than that of male larvae so that the male to female ratio increases, while the absolute numbers of males and females that reach adulthood decrease.

(*b*) Adverse environmental conditions affect the direction of sex differentiation so that males develop from larvae which under more favorable conditions would become females.

This distinction is very important not only for clarification of the biological factors involved, but also because it defines the two categories that need to be considered separately in any effort to utilize unbalanced sex ratios as a means of nematode population management. In the first category, unbalanced sex ratios are incidental and of no special importance. The important aspect is the degree of suppression of female development that is exerted by the various environmental factors. A classical example that falls in this category is the use of resistant host or nonhost plants in place of susceptible hosts in order to reduce the final female population of a nematode. The male population may also be decreased, but this has very little effect on the final result. Clearly, such cases are not concerned with sex ratios per se, and therefore lie outside the scope of this discussion.

In the second category, on the other hand, sex ratios are of ultimate importance. The sex ratio automatically defines the female population in relation to the total population which is unaffected by the environment and remains constant during the entire developmental period of each generation. Theoretically, if sex differentiation could be shifted entirely toward the male direction, the female population would be reduced to zero within one generation, and without any change in the final population which, in this case, would consist of males only.

Although the two categories are thus well defined, the preceding review indicates that it is difficult to classify the various nematodes in one or the other category with certainty. Although undoubtedly biased by my own experience in this matter, I tend to believe that sex differentiation in the family Heteroderidae is influenced by environment only in the genera *Meloidogyne* and *Meloidodera*. The genus *Heterodera* falls outside this category, although admittedly low male percentages reported for single-larva inoculations of *H. rostochiensis* present a serious problem which needs to be resolved by additional experimentation. In defense of this view, the biological differences of these genera should be taken into consideration. The four species of *Heterodera* involved in these studies are diploid, obligatorily amphimictic. As is the case in most such animals, if they are to be successful biologically, they are expected to have a strong genetic mechanism of sex determination that will insure the development of approximately equal numbers of males and females in each generation. An influence of the environment on sex differentiation would imply that all larvae of these nematodes are either identical with regard to their genetic sex determining factors, or that genetic differences among larvae are so small that phenotypic expression of these differences is overcome by environmental influences. Such an assumption is not supported by the bulk of experimental evidence available to date, and such an unconventional interpretation cannot be accepted unless further documentation is provided in the future.

Meloidogyne incognita and *M. javanica,* on the other hand, the two species of root-knot nematodes that have been studied most extensively, as well as *Meloidodera floridensis,* are polyploid and obligatorily parthenogenetic. From a cytogenetic point of view, they are expected to produce one kind of eggs or larvae which should develop to females similar to their mothers (thelytoky). Experimental evidence indicates that this holds true under environmental conditions favorable for nematode development, where males are absent or very rare. The same principle applies also to polyploid parthenogenetic species of the genus *Heterodera* such as *H. trifolii, H. galeopsidis,* and *H. lespedezae,* which indeed are completely thelytokous. The appearance of males in *Meloidogyne* and *Meloidodera,* therefore, has to be regarded as a peculiar and unexpected biological event that needs to be explained. As the same genetic factors are present in all the progeny of a single female (mitotic parthenogenesis), differences with respect to sex expression among the progeny must be of external origin. Obviously, this needs no further proof or justification. Instead, further research, particularly in *Meloidogyne*, needs to be directed toward (*a*) defining the external factors that promote a shift in the direction of sex differentiation from the female to the male sex, and (*b*) elucidating the physiological or biochemical mechanism by which such factors exert such an effect.

Our present knowledge in these two areas is very limited, as indicated by the preceding review of the factors affecting sex differentiation in *Meloidogyne.* For this reason our capacity to exploit this weak aspect of the biology of these nematodes is probably also restricted. All the factors that appear to affect sex differentiation can directly or indirectly be related to a common effect of imposing food stress on the nematode. This is rather discouraging, because whatever affects the quantity or quality of food available for the nematode is likely to affect also the physiology and cytochemistry of the host plant, with the undesirable results of reduced plant growth and productivity. Still, some means by which sex differentiation could be affected without harmfully affecting plant metabolism can be visualized and may not lie beyond reach. Plant growth regulators appear to have limited use because they affect plant metabolism and growth pattern, but proper application of certain growth regulators may prove beneficial in reducing root-knot nematode populations under certain circumstances. Feldman and Hanks (25) applied maleic hydrazide to grapefruit seedlings at rates that induced dormancy for up to 6 months in an effort to control the borrowing nematode through starvation. This was not accomplished but if the nematode under consideration had been a root-knot nematode, the results might have been more favorable.

Studies of the physiology of sex differentiation of nematodes could provide information regarding the biochemical agents that regulate it, and thus suggest possible ways by which its course could be modified. It is possible that such chemical agents, which probably are hormones or hormone-like substances, and very likely their precursors, are regular components of plant tissues and that normal nematode sex differentiation depends on their availability. It is already known that some nematodes are unable to synthesize certain sterols which nevertheless are required for normal development (40), and which, because of their chemical relation to various animal hormones, may be essential also in nematode sex differentiation. The fact that *Meloidodera floridensis* larvae develop to females after feeding on a host plant, but to males if they receive no food, suggests that something obtained with the food may be essential for normal sex differentiation.

Besides plant growth regulators and hormones and their inhibitors, a third group of substances that need to be considered are enzymes and enzyme inhibitors. Such substances could partially neutralize the effect of nematode secretions, or modify in some way the normal host response to nematodes. A slight effect in this direction may be sufficient to change the course of sex differentiation of root-knot nematodes without actually rendering the plant resistant to infection or nematode development. Enzyme inhibitors occurring naturally in nematode-resistant plants may indeed exert such an effect which, however, may have been overlooked in some cases, or classified as resistance to development in others. In this context, the concept of plant resistance to nematodes may need to be expanded to include cases where development of root-knot nematodes may not be significantly hindered, but differentiation toward the male sex may be strong enough to secure reasonable control of the nematode population. Although no data are available, I suspect that resistance of some host plants to nematodes is partially operating on this principle.

In conclusion, from a biological viewpoint and, in a teleological sense, phenotypic sex differentiation is quite beneficial to root-knot nematodes. It gradually reduces the female nematode population as the conditions become adverse and thus prevents the occurrence of overpopulation which could lead to population extinction. However, phenotypic sex differentiation of root-knot nematodes is a vulnerable aspect of their biology, which can be exploited to our benefit, but this can be accomplished only through additional research directed toward this goal.

Literature Cited

1. Anderson, J. F., Horsfall, W. R. 1963. Thermal stress and anomalous development of mosquitoes (Diptera: Culicidae). I. Effect of constant temperature on dimorphism of adults of *Aedes stimulans*. *J. Exp. Zool.* 154:67–107
2. Apel, A., Kämpfe, L., 1957. Beziehungen zwischen Wirt und Parasit im Infektionsverlauf von *Heterodera schachtii* Schmidt in kurzfristigen Topfversuchen. *Nematologica* 2:215–27
3. Bacci, G. 1965. *Sex determination.* Oxford: Pergamon Press. 306 pp.
4. Baltzer, F. 1935. Experiments on sex-development in *Bonellia. Collect. Net.* 10:101–08
5. Bergerard, J. 1961. Intersexualité expérimentale chez *Carausius morosus* Br. *Bull. Biol. Fr. Belg.* 95: 273–300
6. Bergerard, J. 1972. Environmental and physiological control of sex determination and differentiation. *Ann. Rev. Entomol.* 17:57–74
7. Bird, A. F. 1970. The effect of nitrogen deficiency on the growth of *Meloidogyne javanica* at different population levels. *Nematologica* 16:13–21
8. Burns, R. K. 1961. Role of hormones in the differentiation of sex. *Sex and internal secretions,* eds. W. C. Young, G. W. Corner, 1:76–158. Baltimore: Williams & Wilkins, 704 pp.
9. Callot, J., Kremer, M. 1963. Intersexués chez des Culicoides (Diptera: Ceratopogonidae). *Ann. Parasitol. Hum. Comp.* 38:113–20
10. Caullery, M. M., Comas, M. 1928. Le déterminisme du sexe chez un nématode (*Paramermis contorta*), parasite des larves de Chironomes. *C. R. Acad. Sci.* 186:646–48
11. Cayrol, J. C. 1970. Influence de la température durant l'embryogénèse sur le sex-ratio de *Ditylenchus myceliophagus,* J. B. Goodey, 1958. *Proc. IX Int. Nematol. Symp.* (Warsaw, 1967) 163–68
12. Charniaux-Cotton, H. 1960. Sex determination. *The physiology of Crustacea.* ed. T. H. Waterman, 1: 411–47. New York: Academic. 670 pp.
13. Chitwood, B. G. 1949. "Root-knot nematodes"—Part I. A revision of the genus *Meloidogyne* Goeldi,

1887. *Proc. Helminthol. Soc. Wash., D.C.* 16:90–104
14. Christie, J. R. 1929. Some observations on sex in the Mermithidae. *J. Exp. Zool.* 53:59–76
15. Cobb, N. A., Steiner, G., Christie, J. R. 1927. When and how does sex arise? *Official Record, U.S. Dept. Agr.* 6(No. 43):6
16. Davide, R. G., Triantaphyllou, A. C. 1967. Influence of the environment on development and sex differentiation of root-knot nematodes. I. Effect of infection density, age of the host plant and soil temperature. *Nematologica* 13:102–10
17. Davide, R. G., Triantaphyllou, A. C. 1967. Influence of the environment on development and sex differentiation of root-knot nematodes. II. Effect of host nutrition. *Nematologica* 13:111–17
18. Davide, R. G., Triantaphllou, A. C. 1968. Influence of the environment on development and sex differentiation of root-knot nematodes. III. Effect of foliar application of maleic hydrazide. *Nematologica* 14:37–46
19. Dehn, M. von. 1955. Die Geschlechtsbestimmung der Daphniden. Die Bedeutung der Fettstoffe untersucht an *Moina rectirostris* L. *Zool. Jahrb.* 65:334–56
20. Dobzhansky, T. 1930. Genetical and environmental factors influencing the type of intersexes in *Drosophila melanogaster. Am. Nat.* 64: 261–71
21. Dropkin, V. H. 1959. Varietal response of soybeans to *Meloidogyne* —a bioassay system for separating races of root-knot nematodes. *Phytopathology* 49:18–23
22. Ellenby, C. 1954. Environmental determination of the sex ratio of a plant parasitic nematode. *Nature* 174:1016
23. Evans, A. A. F., Fisher, J. M. 1970. Some factors affecting the number and size of nematodes in populations of *Aphelenchus avenae. Nematologica* 16:295–304
24. Fassuliotis, G. 1970. Resistance of *Cucumis* spp. to the root-knot nematode, *Meloidogyne incognita acrita. J. Nematol.* 2:174–78
25. Feldman, A. W., Hanks, R. W. 1963. Attempts at eradication of the burrowing nematode (*Rado-*

pholus similis) from grapefruit seedlings by foliage applications of maleic hydrazide. *Plant Dis. Reptr.* 47:27–29

26. Fisher, J. M. 1972. Observations on the effect of males on reproduction and fecundity of *Aphelenchus avenae*. *Nematologica* 18:179–89

27. Goldschmidt, R. 1934. *Lymantria. Bibliogr. Genet.* 11:1–186

28. Gould, H. N. 1917. Studies on sex in the hermaphrodiate mollusk *Crepidula plana*. II. Influence of environment on sex. *J. Exp. Zool.* 23: 225–50

29. Gowen, J. W. 1961. Genetic and cytologic foundations for sex. *Sex and internal secretions.* eds. W. C. Young, G. W. Corner, 1:3–75. Baltimore: Williams & Wilkins, 704 pp.

30. Gutierrez, R. O. 1947. El nematode de las raicillas de los citrus *Tylenchulus semipenetrans* en la Republica Argentina. *Rev. Invest. Agr. Buenos Aires* 1:119–46

31. Hansen, E., Yarwood, E. A., Buecher, E. J. 1971. Temperature effects on sex differentiation in *Aphelenchus avenae*. *J. Nematol.* 3:311 Abstr.

32. Hartmann, M., Lewinski, G. 1938. Untersuchungen über die Geschlechtsbestimmung und Geschlechtsumwandlung von *Ophryotrocha puerilis*. II. Versuche über die Wirkung von Kalium, Magnesium und Kupfer. *Zool. Jahrb.* (Phys.) 58:551–74

33. Hominick, W. M., Welch, H. E. 1971. Synchronization of life cycles of three mermithids (Nematoda) with their chironomid (Diptera) hosts and some observations on the pathology of the infections. *Can. J. Zool.* 49:975–82

34. Ishibashi, N. 1965. The increase in male adults by gamma-ray irradiation in the root-knot nematode, *Meloidogyne incognita* Chitwood. *Nematologica* 11:361–69

35. Johnson, R. N., Viglierchio, D. R. 1969. Sugar beet nematode (*Heterodera schachtii*) reared on axenic *Beta vulgaris* root explants. II. Selected environmental and nutritional factors affecting development and sex-ratio. *Nematologica* 15: 144–52

36. Kämpfe, L., Kerstan, U. 1964. Die Beeinflussung des Geschlechtsverhältnisses in der Gattung *Hetero-*

dera Schmidt. I. Einfluss des physiologischen Zustandes der Wirtspflanze auf *H. schachtii* Schmidt. *Nematologica* 10:388–98

37. Kerstan, U. 1969. Die Beeinflussung des Geschlechterverhältnisses in der Gattung *Heterodera.* II. Minimallebensraum—selektive Absterberate der Geschlechter—Geschlechterverhältnis (*Heterodera schachtii*). *Nematologica* 15:210–28

37a. Kinloch, R. A., Allen, M. W., 1972. Interaction of *Meloidogyne hapla* and *M. javanica* infecting tomato. *J. Nematol.* 4:7–16

38. Koliopanos, C. N., Triantaphyllou, A. C. 1972. Effect of infection density on sex ratio of *Heterodera glycines*. *Nematologica* 18:131–37

39. Kornhauser, S. I. 1919. The sexual characteristics of the membracid *Thelia bimaculata* (Fabr.) I. External changes induced by *Aphelopus theliae* (Gahan). *J. Morphol.* 32: 531–636

40. Krusberg, L. R. 1971. Chemical composition of Nematodes. *Plant Parasitic Nematodes.* eds. B. M. Zuckerman, W. F. Mai, R. A. Rohde, 2:213–34. New York: Academic 347 pp.

41. Laughlin, C. W., Williams, A. S., Fox, J. A. 1969. The influence of temperature on development and sex differentiation of *Meloidogyne graminis*. *J. Nematol.* 1:212–15

42. Lindhardt, K. 1961. Nogle undersøgelser over infektionsgradens indflydelse på havreålens køn (*Heterodera major* O. Schmidt, 1930). *Tidsskr. Planteavl.* 64:889–96

43. Linford, M. B. 1941. Parasitism of the root-knot nematode in leaves and stems. *Phytopathology* 31: 634–48

43a. Macaron, J., Ritter, M. 1972. Inversion de sexe chez le nématode tylenchide phytoparasite *Tylenchulus semipenetrans*. Cobb 1913—Nematoda Tylenchida. *C. R. Acad. Sci. Paris* 274:2676–82

44. McClure, M. A., Viglierchio, D. R. 1966. The influence of host nutrition and intensity of infection on the sex ratio and development of *Meloidogyne incognita* in sterile agar cultures of excised cucumber roots. *Nematologica* 12:248–58

45. Molz, E. 1920. Versuche zur Ermittlung des Einflusses äusserer Faktoren auf des Geschlechtsver-

hältnis des Rübennematoden (*Heterodera schachtii* A. Schmidt). *Landw. Jahrb.* 54:769–91

46. Nigon, V. 1965. Développment et reproduction des Nématodes. *Traité de Zoologie*, ed. P. Grassé, 4 (2):218–386. Paris: Masson, 731 pp.

47. Nusbaum, C. J. 1958. The response of root-knot-infected tobacco plants to foliar applications of maleic hydrazide. *Phytopathology* 48:344 Abstr.

48. Orion, D., Minz, G. 1971. The influence of morphactin on the root-knot nematode, *Meloidogyne javanica*, and its galls. *Nematologica* 17:107–12

49. Ouden, H. den. 1960. A note on parthenogenesis and sex determination in *Heterodera rostochiensis* Woll. *Nematologica* 5:215–16

50. Parenti, U. 1965. Male and female influence of adult individuals on undifferentiated larvae of the parasitic nematode *Paramermis contorta. Nature* 207:1105–06

51. Peacock, F. C. 1960. Inhibition of root-knot development on tomato by systemic compounds. *Nematologica* 5:219–27

52. Petersen, J. 1972. Factors affecting sex ratios of a mermithid parasite of mosquitoes. *J. Nematol.* 4:83–87

53. Prasad, S. K., Webster, J. M. 1967. Effect of temperature on the rate of development of *Nacobbus serendipiticus* in excised tomato roots. *Nematologica* 13:85–90

54. Rempel, J. G. 1940. Intersexuality in Chironomidae induced by nematode parasitism. *J. Exp. Zool.* 84: 261–89

55. Reverberi, G., Pitotti, M. 1942. Il ciclo biologico e la determinazione fenotipica del sesso di *Ione thoracica* Montagu, Bopiride parassita di *Callianassa laticauda* Otto. *Publ. Staz. Zool. Napoli* 19:111–84

56. Ross, G. J. S., Trudgill, D. L. 1969. The effect of population density on the sex ratio of *Heterodera rostochiensis;* a two dimensional model. *Nematologica* 15:601–07

57. Ross, J. P. 1964. Effect of soil temperature on development of *Heterodera glycines* in soybean roots. *Phytopathology* 54:1228–31

58. Salt, G. 1931. A further study of the effects of stylopisation on wasps. *J. Exp. Zool.* 59:133–66

59. Sengbusch, R. 1927. Beitrag zur biologie des Rübennematoden *Heterodera schachtii. Z. Pflanzenkrankh.* 37:86–102

60. Steele, A. E. 1971. Orientation and development of *Heterodera schachtii* larvae on tomato and sugarbeet roots. *J. Nematol.* 3:424–26

61. Steiner, G. 1923. Intersexes in nematodes. *J. Hered.* 14:147–59

62. Takeda, N. 1950. Experimental studies on the effect of external agencies on the sexuality of a marine copepod. *Physiol. Zool.* 23: 288–301

63. Triantaphyllou, A. C. 1960. Sex determination in *Meloidogyne incognita* Chitwood, 1949 and intersexuality in *M. javanica* (Treub, 1885) Chitwood, 1949. *Ann. Inst. Phytopath. Benaki* N. S. 3:12–31

64. Triantaphyllou, A. C. 1963. Polyploidy and parthenogenesis in the root-knot nematode *Meloidogyne arenaria. J. Morphol.* 113:489–99

65. Triantaphyllou, A. C. 1971. Genetics and cytology. *Plant Parasitic Nematodes*. eds. B. M. Zuckerman, W. F. Mai, R. A. Rohde, 2:1–34. New York: Academic 347 pp.

66. Triantaphyllou, A. C. 1971. Oogenesis and the chromosomes of the cystoid nematode, *Meloidodera floridenis. J. Nematol.* 3:183–88

67. Triantaphyllou, A. C. 1973. Gametogenesis and reproduction of *Meloidogyne graminis* and *M. ottersoni* (Nematoda: Heteroderidae). *J. Nematol.* 5:84–87

68. Triantaphyllou, A. C., Hirschmann, H. 1960. Post-infection development of *Meloidogyne incognita* Chitwood 1949 (Nematoda: Heteroderidae). *Ann. Inst. Phytopath. Benaki,* N. S. 3:1–11

69. Triantaphyllou, A. C., Hirschmann, H. 1964. Reproduction in plant and soil nematodes. *Ann. Rev. Phytopathol.* 2:57–80

70. Triantaphyllou, A. C., Hirschmann, H. 1973. Environmentally controlled sex expression in *Meloidodera floridensis. J. Nematol.* In press

71. Trudgill, D. L. 1967. The effect of environment on sex determination in *Heterodera rostochiensis. Nematologica* 13:263–72

72. Trudgill, D. L., Parrott, D. M. 1969. The behavior of nine populations of the potato cyst nematode *Heterodera rostochiensis* towards

three resistant potato hybrids. *Nematologica* 15:381–88

73. Trudgill, D. L., Webster, J. M., Parrott, D. M. 1967. The effect of resistant solanaceous plants on the sex ratio of *Heterodera rostochiensis* and the use of the sex ratio to assess the frequency and genetic constitution of pathotypes. *Ann. Appl. Biol.* 60:421–28

74 Trudgill, D. L., Parrott, D. M., Stone, A. R. 1970. Morphometrics of males and larvae of ten *Heterodera rostochiensis* populations and the influence of resistant hosts. *Nematologica* 16:410–16

75. Tyler, J. 1933. Reproduction without males in aseptic root cultures of the root-knot nematode. *Hilgardia* 7:373–88

76. van Gundy, S. D. 1958. The life history of the citrus nematode *Tylenchulus semipenetrans* Cobb. *Nematologica* 3:283–94

77. Webber, A. J., Jr., Fox, J. A. 1971. Variation in sex differentiation among single larval and single egg mass isolates of *Meloidogyne graminis. J. Nematol.* 3:332–33 Abstr.

78. Zimmering, S., Sandler, L., Nicoletti, B. 1970. Mechanisms of meiotic drive. *Ann. Rev. Genet.* 4:409–36

TRENDS IN BREEDING FOR DISEASE   ❖ 3582
RESISTANCE IN CROPS

Curtis W. Roane
Department of Plant Pathology and Physiology, Virginia Polytechnic
Institute and State University, Blacksburg, Virginia

INTRODUCTION

As I prepare to write this chapter, I cannot avoid reflecting on the vast amount of effort that has been expended in breeding disease-resistant plants. Even though such breeding has been on a genetically sound basis for less than seven decades, plant husbandmen must have pondered the problem for centuries. Since 1900, tons of paper have been expended to spread the gospel of breeding theory, shiftiness of pathogens, nature of resistance, and of measured accomplishment. At the apex of this paper mountain could be placed a list of cultivars bred for resistance to one or several diseases. Certainly it would make an impressive pinnacle, and few could dispute that disease-resistant cultivars have been a major contribution of plant breeders.

How have we gone about breeding disease-resistant cultivars in the past and how do we go about it now? Generally, the methodology of breeding for disease resistance is the same as that used in breeding for any other trait. For those of us associated with this type of work, it is almost facetious to answer any further; yet, even we may need to look into a sort of historical mirror to see that however we approach a problem of disease resistance, we rely on long established principles (90). Thus, recent trends are merely new applications of these principles. Even so, it has been a long rough road from Bolley (12) and Biffen (9) to mutation breeding, multilinial cultivars and recovery from the peril of "T" cytoplasm.

SOME ERAS OF BREEDING FOR DISEASE RESISTANCE

First of all, to satisfy the purists, we actually breed for resistance to pathogens, but "disease resistance" is so well implanted in literature, I will use disease resistance to mean pathogen resistance.

Although effort was spent on breeding disease-resistant cultivars before 1900, the rediscovery of Mendel's work lighted the way for producing them scientifically, and Miffen (9) was the first to see the light. Bolley (12), Orton

463

(90) and others (see 137) provided additional impetus by showing that a wealth of natural resistance occurred among several crops. It soon became apparent that most of this vast resistance could be manipulated according to Mendel's laws. Thus, the Mendelian era of breeding for disease resistance began. Some sources of resistance remained effective for many years and are utilized even today (137), but some were ephemeral.

The elusive nature of resistance was understood only after Barrus (8) reported that two isolates of *Colletotrichum lindemuthianum* possessed differential pathogenicity on bean cultivars. For the next four decades there followed an avalanche of reports triggered by Stakman and his colleagues (117, 118) on the pathogenic or physiologic specialization of plant pathogens and the concomitant breeding of varieties resistant to the physiologic races. The era of race resistant cultivars flourished but all the problems were not overcome. *Puccinia graminis* and other cereal rust fungi conducted their own breeding programs and placed the breeders of rust-resistant cereals on a treadmill (65, 119). Through much of this era, there was a lack of lucid comprehension of the genetic nature of pathogenicity.

There came next an era that placed the breeding of disease-resistant cultivars on its firmest plateau, the gene-for-gene era. Newton, Johnson & Brown (66, 87, 88) made a thorough and sound analysis of the inheritance of pathogenicity in *Puccinia graminis tritici* and found that the various genes for virulence behaved according to Mendel's laws of inheritance. Flor apparently used this work with *P. graminis* as a springboard for his flax rust studies. In 1942, he concluded from studies with flax rust that "the pathogenic range of each physiologic race of the pathogen is conditioned by pairs of factors that are specific for each different resistant or immune factor possessed by the host variety" (42). This was the birth of the gene-for-gene hypothesis that has served as a recent guideline for interpreting the inheritance of host-pathogen interactions, and which is applicable to several host-pathogen systems (45). Newton, Johnson & Brown provided the key and Flor unlocked the door.

Some systems fail to conform to the gene-for-gene concept because either resistance or virulence or both are quantitatively inherited. Van der Plank (130) introduced the terms "vertical" and "horizontal" resistance which generally, but not necessarily, correspond to the long-used genetic terms qualitative and quantitative inheritance of resistance. Although the terms vertical and horizontal resistance have made a great impact on the breeders and pathologists, the actual types of resistance elaborated by Van der Plank under these two terms are not newly recognized but they were given a most careful analysis and appraisal by him (131). Because there have been several attempts by others to make vertical and horizontal resistance synonymous with existing or new terminology, there is much confusion. Robinson has sought to clarify the terminology of disease resistance (99) and the application of the concept of vertical resistance to breeding (100).

From the foregoing comments, one might conclude that each of these eras has begun and ended; obviously, each began and none has ended. The principles evolved during these eras have dominated the trend of breeding disease-resistant cultivars. There are other aspects of the problem, each of which is in some way related to these eras, that may have affected the general trends, or had some effect on breeding certain crops for disease resistance. For convenience these aspects are discussed under biological and nonbiological factors. Under biological factors I shall consider sources of resistance, occurrence of new diseases or physiologic races, vector resistance, new concepts and techniques. Under nonbiological factors I shall consider the influence of changing political, sociological, and economic atmospheres.

TRENDS ASCRIBED TO BIOLOGICAL FACTORS
Seeking and Utilizing Sources of Resistance

The basic requirements for breeding disease-resistant plants are sources of resistance and methodology for combining resistance with commercially or aesthetically acceptable plant types. These requirements have been met in many ways. Early practices of Bolley, Orton, Essary, Jones, and Norton (see 137) are still pursued; variations of these practices are introduced as new methods in breeding and pathology are invented, new sources of resistance are found, and changes occur in cultural practices, harvesting methods, and consumer preference. For these reasons, breeding disease-resistant cultivars remains a continuing process. To produce new cultivars, there must be a continual search for germ plasm with disease resistance traits that meet the changing needs. The search for needed genes goes on in several ways; the emphasis varies with the crop or need in question. Selection, plant introduction, native wild species, and induced mutations represent the primary sources; each has played a significant role.

SELECTION   Selection is a primary aspect of isolating sources of disease resistance. It is a necessary step for isolating resistance from agricultural populations, wild species, plant introductions, and hybrid or mutagen-treated populations. Initially, selection was practiced by isolating resistant survivors of natural epiphytotics. Later, artificial epiphytotics were created in disease nurseries to facilitate selection of resistant individuals rather than escaped suscepts (55). The pioneer methods were and still are effective for diseases incited by soil-borne pathogens, but innovative scientists have developed parallel techniques for creating artificial epiphytotics of aerially disseminated pathogens, especially rusts (27, 68, 102).

Precision of selection methods is improved by coupling artificial inoculation with controlled environment. Walker (135) described several applications of this premise. Wheeler & Luke (141) were able to undertake a mass screening of oats by utilizing controlled environment and a specific phatotoxin. Their work will be discussed later. One of the obvious pitfalls of selection is to choose only those individuals with conspicuous vertical resistance.

As mentioned previously, such resistance is often overcome by the appearance of previously undetected races of a pathogen with corresponding vertical virulence. Van der Plank (131) and others (24, 104) have pointed out the need for greater use of horizontal resistance, general resistance, or tolerance, but recognition and utilization of these traits is more difficult than it is for vertical resistance. Despite these difficulties, the development of a technique for control of soil temperature led to the separation and utilization of both vertical and horizontal resistance to fusarium wilt of cabbage and tomato (134, 139). The commercial production of environmental control chambers has favored the discovery of small increments of useful resistance where formerly none was detectable. Simply by varying salient features of the environment one may observe genotype-environment interactions of minor magnitude (37).

Regardless of the technique used and however fastidious the equipment, availability of large germ plasm collections are necessary prerequisites to present-day applications of selection. Conversely, we should not overlook the advice of Welsh & Johnson (140) following their discovery of new stem rust resistance genes in the old oat cultivar 'Hajira.' "Through natural crossing, or through mutation, genes for resistance to rust may arise in any variety. The discovery of the resistant plants in Hajira is no isolated occurrence but can be duplicated in other cereals and probably in many other types of plants as well." After 50 years, the premises and methods of Bolley and Orton retained their usefulness. We continue to see evidence of this in published reports each month.

PLANT INTRODUCTION   Frequently, genes needed to control a particular disease or race of a pathogen cannot be found in domestic cultivars. Collections of related materials from other countries, particularly from areas where the pathogen and host species may have co-evolved, sometimes provide rich pools of resistance genes (71). The natural occurrence of resistance genes has been one of the inducements for plant exploration, collecting, and introduction. There is continuing evidence that, as breeding criteria change, researchers re-examine germ plasm collections for undiscovered resistance. Many of the resistance genes now employed in United States agriculture have been found by screening germ plasm collections of foreign origin. Reviews concerning germ plasm collections have been published recently (33, 46, 71).

Creech & Reitz (33) give many examples of plant introductions that provided sources of disease resistance in modern breeding programs. Their discussion also highlights the fact that introductions need not be of the same species that is the subject of improvement, but that related species and genera may also be tapped for resistance. Orton (89) pioneered the use of exotic germ plasm as a source of disease resistance when he transferred the fusarium wilt resistance of the African citron to watermelon. Orton's successful use of exotic germ plasm may have been preceded by Cole who, about 1905, crossed *Triticum aestivum* with newly introduced *T. durum* and *T. monococ-cum* in an attempt to transfer the stem rust resistance of durum and emmer to common wheat (77). Although nothing came of Cole's efforts, McFadden, from a cross made in 1915, produced 'Hope' and 'H-44' wheats via the same routes. These two cultivars were used in many subsequent crosses as sources of resistance to stem rust, leaf rust, loose smut, and bunt. Hayes, Parker & Kurtzweil produced 'Marquillo' from a cross of *T. aestivum* × *T. durum* (see 78). These wide crosses have been paralleled many times with solana-ceous crops until today resistance may even be transferred from related gen-era into the subject species. This greatly broadens the scope of plant explora-tion and augments the need to maintain collections of diverse germ plasm. Interspecific and intergeneric hybridization as a trend will be discussed later.

INDUCED MUTATIONS   It matters little whether needed resistance genes origi-nate as natural or induced mutations. In germ plasm collections, many muta-tions for resistance are yet undiscovered. Lacking adequate natural muta-tions, one may attempt to induce appropriate mutations. The shortcomings of induced mutations are that they may only duplicate action of naturally occur-ring genes; the disease or a pathogenic race must appear before one can set about screening for mutations for resistance. Resistance imparted by an in-duced mutation is apt to exhibit monofactorial behavior and, consequently, its usefulness may be shortlived (96). In the screening processes, one would tend to overlook low levels of horizontal resistance, while favoring high verti-cal resistance because of ease of recognition of the latter (96). Finally, the inducement of the sought-after mutation is literally a hit or miss affair and thus may be unsuccessful. One should not be dissuaded from attempting to induce mutations for resistance because some events may occur that counter-act all of the foregoing forebodings.

As sources of resistance, induced mutations may be obtained by treatment of commercial cultivars with the hope of adding specific disease resistances when they are needed but lacking. If mutagens could be directed to produce the needed mutations this would be a wonderful and quick means of produc-ing disease-resistant cultivars, but because of their haphazard nature, more detrimental than beneficial mutations are induced. If a desired mutation is induced, it may be necessary to sort it out from the detrimental ones by back-crossing or other breeding techniques. Although Muller established mutation genetics in the late 1920s and plant breeders immediately perceived the utili-tarian possibilities of the procedure, little attention was given to inducing dis-ease-resistance mutations until the mid-1950s. In 1942, Freisleben & Lein (47) reported the appearance of mildew-resistant plants among the progeny of plants grown from X-irradiated barley seeds. In 1951, Bandlow (6) re-ported the occurrence of X-irradiation-induced mutations for resistance to common races of *Erysiphe graminis hordei*. These may have been serendipi-tous findings. Konzak (70) summarized the status of mutation breeding for disease resistance through 1958. His paper indicates that an avalanche of studies in mutation breeding for disease resistance must have started in the mid-fifties. I recall that most cereal workers were in some way attempting to induce disease-resistance mutations especially among premier cultivars or advanced breeding lines. Mutagenic agents induced sterility and outcrossing was common among progenies produced from irradiated seeds. Some nurseries of these materials were not sufficiently isolated from nurseries of plants having recognized genes for resistance. As a result, when a disease-resistant plant was isolated from mutagen-treated stocks, the so-called mutant genes either could not be distinguished from previously identified genes, the resistance may have come from outcrossing, or else mutations were no more effective than known genes. The problems related to screening the useful mutants from the enormous populations that were generated, coupled with the tedious chore of proving that the mutant genes were actually new, soon drove the faddists from the field of mutation breeding for disease resistance. To their credit, a hard core of workers remained in the field and continued to develop it into a highly respectable, imaginative branch of plant genetics and breeding. Reports of some recent symposia (60, 61) and the literature cited in them attest to this fact.

Sigurbjörnsson & Micke (112) list 77 cultivars that have been developed as a direct consequence of induced mutations; 16 of these were cited as having improved disease resistance. There were 3 wheat, 2 barley, 4 oat, and 7 bean cultivars. Mutants for resistance included 2 in wheat for leaf rust and 1 for stem rust; 2 in barley for powdery mildew; 2 in oats for crown rust and 2 for stem rust; 3 in bean for bean common mosaic, 2 for anthracnose, and 2 for bacterial blights. One parent bean variety, 'Michelite,' was X-irradiated in 1938 and was known to have resistance to mosaic and anthracnose. According to Down & Andersen (38), the mutant line from 'Michelite' was a non-vining or bush type which allowed better air movement among the plants. Less sclerotinia wilt occurred in cultivars of this type. Other resistance mutations have been induced for barley powdery mildew (67), wheat bunt (16), wheat *Septoria nodorum* (73), peppermint verticillium wilt (83), peanut leaf spot, and wheat stripe rust (70). It is not clear whether these have been incorporated into commercial cultivars.

There is great need to expand the research on inducing mutations for disease resistance. Plant pathologists have utilized mutagens to increase and analyze variability in pathogens (109) but have been much less avid than other plant manipulators in applying the techniques to the host. We should be taking the leadership to induce new genes for resistance because disease control is the life blood of our existence. We have allowed others to shoulder our share of the burden.

INTERSPECIFIC AND INTERGENERIC HYBRIDIZATION   Although cereal breeders and pathologists have usually been successful in finding new rust-resistance genes, no satisfactory resistance is known for many crop diseases among the present germ plasm banks. Even in wheat and barley, scab resistance sources are virtually nonexistent; in soybean there is no known resistance to brown stem rot; and there is very little resistance in any crop against *Sclerotium rolfsii* and *Rhizoctonia* (and all of its synonyms). For many diseases, resistance is so inadequate that topical procedures are exigent. Breeders often have turned to wild relatives of cultivated plants to find new sources of resistance and have isolated stocks possessing the desired resistance; however, they frequently have not been able to transfer the resistance because of incompatibilities between subject species or because of sterility in the hybrid. Sears wrote, "The transfer of characters from one species or genus to another is not only of great potential practical importance, but is of considerable genetic interest as well. In general, the wider the transfer, the more interesting it is genetically. From a practical standpoint, too, the greater distance over which transfers can be made, the greater the possibility of introducing useful characters not present in the host species. It is therefore important to extend the limits of transfer as far as possible. Significant progress has already been made in this direction" (111). Although wheat breeders may have been the first to exploit interspecific hybridization in an effort to control plant diseases (78), breeders of tobacco, tomato, potato, cotton, sugarcane, and raspberry also utilized the technique early in their programs. Evidence for this is presented in several chapters of the 1953 USDA Yearbook of Agriculture (128). The success of these exploitations and rapidity of progress have varied with the crop. It is no trick to make interspecific crosses between *Hordeum vulgare* and *H. distichon*, or *Lycopersicon esculentum* and *L. pimpinellifolium*, because in such crosses both parents behave as compatible diploids. Thus, because the taxonomists have seen fit to make a taxon of each parent, interspecific hybrids are easily obtained. In *Triticum* and certain other genera, interspecific hybridization may be difficult because of differences in chromosome numbers and homologies. Breeders have come to rely on special cytogenetic techniques and embryo culture to facilitate the more difficult crosses. These same techniques have allowed us to broaden the base of crosses such that many intergeneric crosses have been made and disease resistance genes from remotely related species have been transferred into agronomically acceptable cultivars.

Generally, for resistance genes, one is interested in transferring only a small chromosome segment from related species into the subject species. I will not discuss the techniques devised to make such transfers because the methodology is beyond the scope of this review; they are well documented (18, 21, 41). Some major contributions will be cited.

Wheat breeders first sought to transfer the resistance genes of 14-chromosome *Triticum* spp. into 21-chromosome common wheats, *T. aestivum* (55), but later methods were developed for reaching out to 7-chromosome *Triticum* spp. and the related genera *Aegilops, Agropyron, Haynaldia,* and *Secale* for genes providing resistance to leaf and stem rust, and wheat streak mosaic (56, 82). Thus, hybrids within *Triticum* and between *Triticum* and related genera are currently being exploited. It can be expected that numerous useful disease resistance genes may be transferred into common wheat backgrounds.

Tomato breeders have made use of several *Lycopersicon* spp. to bring together resistance to a large number of diseases in a single cultivar (138). The following species contributed genes for resistance to the diseases indicated: *L. pimpinellifolium*—fusarium, verticillium, bacterial and spotted wilt, bacterial canker, gray leaf spot, and leaf mold; *L. hirsutum*—septoria leaf spot, tobacco mosaic, leaf mold; *L. peruvianum*—beet curly top, root knot, spotted wilt, septoria leaf spot; *L. chilense*—anthracnose (120, 138).

Similarly, potato breeders have made liberal use of wild *Solanum* species to create new disease resistant cultivars. The *Phytophthora infestans* problem stimulated intensive searches for blight resistance genes among the various *Solanum* spp. The *R* genes for late blight resistance came primarily from *S. demissum* (59) but Stevenson & Jones (120) list nine species in which immunity probably exists. The latter also list several species possessing resistance to common scab and virus diseases; Howard (59) cites *S. multidissectum* and *S. vernei* as sources of resistance to the potato cyst nematode. Current literature indicates that many of these sources have been utilized in breeding disease resistant *S. tuberosum* but that the ephemeral nature of the immunity genes for late blight resistance has brought about a more careful search for field or horizontal resistance among potato breeding stocks.

Because the immunity genes failed to control late blight, there has been a major shift in sources of resistance to potato diseases and in the philosophy of breeding disease-resistant potatoes. The treadmill of breeding rust resistant cereals and late blight resistant potatoes created the spark for the new philosophy (136). Certainly much fuel has been added by Van der Plank and his "horizontal-vertical resistance" terminology and by the proponents of general resistance and spokesmen for tolerance (24, 104, 126, 130). More seems to have been accomplished in potato breeding within the framework of horizontal (= generalized = field) resistance philosophies than in other crops. Schafer's review furnishes evidence that tolerance also has played an effective, useful role in the control of cereal disease (104).

Tobacco workers furnished an early example of interspecific hybridization by transferring mosaic resistance from *Nicotiana glutinosa* directly to *N. paniculata* (57) and then indirectly to *N. tabacum* from *N. glutinosa* by the synthetic bridge species, *N. digluta* (58). As stated by Clayton (30), the feasibility of interspecific hybridization in *Nicotiana* had already been demonstrated: "By fortunate chance, the genus *Nicotiana* was early selected for fundamental genetic investigations. Prior to about 1930, these studies were conducted with little thought as to possible applications to tobacco breeding. However, from investigations of Setchell, Goodspeed, Clausen, East, Kostoff and their many associates, there came a comprehensive knowledge of the genus, and the species relationships; also, the discovery that tobacco was an allopolyploid derived from a cross between *N. sylvestris* and a member of the tomentosa group." Workers to the present are continuing to extract genes for resistance to tobacco diseases from relatives of *N. tabacum*. They have utilized *N. debneyi* for resistance to blue mold and black root rot, *N. longiflora* for wildfire resistance, *N. plumbaginifolia* for black shank resistance, and of course *N. glutinosa* for tobacco mosaic resistance. These are generally considered distantly related to *N. tabacum*. Valuable resistance generally has not been found in *N. sylvestris* and *N. tomentosa* which are closely related to *N. tabacum* (29). New viruses continue to plague tobacco, and resistance to some other diseases is as yet unsatisfactory; thus, the search for new sources of resistance continues.

In addition to the preceding examples, interspecific hybridization has been used as a means of extracting resistance from wild or alien species of *Allium*, *Avena*, *Beta*, *Capsicum*, *Castanea*, *Fragaria*, *Rubus*, *Saccharum*, *Vigna*, and *Vitis*. The possibilities seem unlimited.

MULTILINE CULTIVARS   It has long been recognized that crop homogeneity is a condition favoring epiphytotics. An example confirming this observation was the widespread use of Victoria blight-susceptible spring oats in our midwestern states during the mid-1940s. By 1945, Victoria derivatives occupied an estimated 98% of the oat acreage of Iowa and 80% of that of the United States. In 1945, *Helminthosporium victoriae* appeared in the spring oat belt and in 1946 losses were so severe that Victoria derivatives were soon replaced with resistant cultivars (113). Another reminder that homogeneity is hazardous stems from the dramatic outbreak of *H. maydis* on maize in 1970. Most of the commercial maize seed in the United States was being produced by breeding a cytoplasmic male-sterility factor into the inbreds to be used as seed parents in hybridization. Although a number of sterility sources were known, exclusive use of the "Texas male sterility" source (*T-ms*) provided a type of homogeneity to four million acres of maize. In 1970, a strain of *H. maydis* virulent on all plants with the *T-ms* source swept northward from Florida, Georgia, and Alabama into the corn-belt and caused extreme reduction in yield and quality of the U. S. corn crop (81). There was a mad scramble to eliminate the predisposing *T-ms* factor and return to "normal cytoplasm" and subsequent production of hybrid seed by detasseling.

The southern leaf blight episode thoroughly shocked our agricultural scientists and policy makers. A committee of the National Academy of Sciences studied the genetic vulnerability of major crops and found that uniformity resting on a single genetic character occurs in many cultivars of sorghum, sugar beet, onion, dwarf wheat, and rice. The committee also found that a few cultivars, suggesting genetic similarity, comprise the bulk of the acreage in many other important crops (85, 133). Even though the hazards of genetic homogeneity and the buffering effect of crop mixtures against disease loss was noted as early as 1894 (19), we have suffered through several crop catastrophies because we ignored established principles.

The rumblings of change were heard following the oat Victoria blight era. Rosen (101) proposed mixed populations of oat crosses to provide a buffering effect against crown rust and Victoria blight. He cited the cultivar 'Traveler' as being a 70:30 mixture of crown rust resistant-blight susceptible: crown rust susceptible-blight resistant plants. In 1952, Jensen (64) proposed the development of multiline oat cultivars by blending several compatible pure lines of different genotypes. The blend would be heterogeneous for disease resistance. In 1953, Borlaug (13) described a backcross program to develop a series of lines of wheat differing in genes for stem rust resistance. Seed of several isogenic wheat lines were developed and mechanically blended. As different races of rust appeared, susceptible lines could be withdrawn from the mixture and reserve lines could be added (14, 15). 'Miramor 63' wheat was the first multiline cultivar to be released for commercial production. It was a mixture of 10 of the best of 600 near isogenic lines giving resistance to stripe and stem rust in Colombia. "Within two years, two lines were attacked by stem rust, were withdrawn and replaced by other resistant lines from the original 600. This illustrates the plasticity inherent in multiline cultivars" (19).

In the United States, Suneson (122) has released the barley cultivar 'Harland' in California and Frey, Browning, & Grindeland (49) have released a pair of multiline oat varieties for Iowa. These latter cultivars were bred to fit the general concepts of crop uniformity; they are subject to yearly revision and are really a series of cultivars numbered with a yearly suffix (50). With 'Harland' barley, Suneson breaks many traditions deeply ingrained in crop certification and the seed trade. "As released, Harland will have a trace of 2-row, black, naked seed. The population will show a two-week range of maturity. The seeds and heads will be predominantly large, but diverse in size, shape, and color. Its composite reaction to diseases and shatter will be impressive" (122). That statement ought to shake up the gods of crop uniformity! It remains to be seen whether 'Harland' will become an acceptable cultivar or remain a novelty; the maturity range jeopardizes its success.

The theory and practice in the use of multiline cultivars as a means of disease control have been thoroughly reviewed by Browning & Frey (19). A few multiline cultivars are in commercial production. There are several ways by which heterogeneity for disease resistance may be utilized but one fact appears certain from the knowledge gained thus far; there is no need to practice selection and reselection to obtain beautifully uniform cultivars merely for the sake of uniformity. Such practices delay the utilization of needed disease resistance. It would be wise now to seek uniformity of ripening, and such characters as milling quality in wheat, malting quality in barley, harvestability and flavor in many vegetable crops. Heterogeneity of disease resistance is becoming more necessary for grains, soybeans and a few other field crops because of their expansive distribution than for fruits and vegetables, which are grown in comparatively isolated pockets.

MULTIPLE DISEASE RESISTANCE   A number of publications whose titles begin with "Multiple disease resistant . . ." have appeared recently. One is led to believe that the concept of multiple disease resistant cultivars is new. Actually, resistance to two or more diseases has been bred into individual cultivars since Orton succeeded in combining resistance to fusarium wilt and root knot in cowpea and cotton (90). 'Hope' and 'H-44' wheats combined resistance to leaf rust, stem rust, loose smut, and covered smut (5). Ausemus (5) cites many other examples of cereal cultivars with multiple disease resistance. Multiple resistance has long been a breeding achievement in tobacco, sugar beet, corn, beans, and many other crops (128). It is probable that the present trend is to work with crops where combined resistance may have been neglected in the past. Some noteworthy accomplishments have been reported recently for cabbage (144), cucumber (7), sugar beet (51) and tomato (34). Additional descriptions of cultivars with multiple disease resistance may be found in the journals *Crop Science* and *Seed World*.

VECTOR RESISTANCE   Resistance to insects has been described for many crops (91). In several of these cases, the insects are vectors of certain plant pathogens; yet there has been relatively little effort to control plant diseases by breeding for resistance to vectors. Painter suggested that, "In addition to resistance to insect attack as an adjunct to other means of insect control, there is an excellent possibility that it may aid in the control of some of the diseases that are carried by insects" (91, p. 5). Few examples furnish evidence that breeding for vector resistance has been fruitful.

There are several complicating factors that would deter one from breeding for general vector resistance as a means of disease control. If pathogen-bearing insects are generated on diseased weed hosts or adjacent crops and they invade a field of a disease-susceptible crop, vector resistance would have to be vector-immunity, or else as with stylet-borne viruses, the vector could probe and transmit a virus before discovering that the host is incompatible. Also, as suggested by Yamaguchi & Welch (145), several lettuce cultivars growing in adjacent plots may be differentially preferred by the vector and, consequently, be differentially infected. On the other hand, a field of a nonpreferred cultivar selected from these plots offers no differential feeding choice and the disease may become just as severe in it as it would if the cultivar were the preferred host of the vector. This observation notwithstanding, vector nonpreference is considered to be the factor of resistance to the curly-top virus in the tomato cultivars 'C5' and 'CVF4' (125). Vector preference as a means of disease control merits further investigation.

Examples of disease control through vector resistance include the 'Lloyd George' raspberry which is resistant to the aphid vector of the raspberry mosaic virus (108). This trait has been bred into several other cultivars (120). Broadbent (17) cites other cases; the cotton cultivar 'Lambert' is susceptible to the leaf-curl virus, but the white fly vector fails to transmit the virus to the cultivar in the field (124); the peanut cultivar 'Mwitunde' has field resistance to rosette virus because aphids acquire the virus from plants of this cultivar with difficulty and therefore do not spread the virus within the field (40). Swenson (123) describes other complications involved in resistance to vector-borne viruses.

The reports cited above indicate that differential response to vector-borne virus is in some way related to host-vector interactions. It is not clear whether some of these traits, once recognized, were utilized further in breeding disease resistant varieties. Thomas & Martin (125) indicated that they will exploit leafhopper nonpreference in breeding additional curly-top resistant tomatoes. Breeding insect resistant plants is a very active field, but in the majority of cases concomitant control of plant disease is an incidental objective.

Impact of New Diseases and Physiologic Races

From time to time, crops are attacked by a newly recognized pathogen or race of a pathogen. The occurrence of these new pathogens comes about in several ways; namely, introduction, mutation, genetic recombination of various sorts, release of a disease prone cultivar, or control of major diseases thereby unmasking formerly minor diseases. There may be other means. Whatever the case, the consequences are similar; the disease is characterized, its potential and probable control measures are evaluated. For most field crops, sources of resistance are sought and breeding programs are established or revised. This is a familiar cycle which was invented by Bolley, Orton (12, 90) and other pioneers. What has changed is the degree of understanding of the nature of disease and the technology of control. Two familiar examples illustrate the trend.

RUST OF WHEAT   Breeding for wheat stem rust resistance in the Great Plains of North America is a well documented tale of success and frustration—not frustration because the yields of wheat failed to increase, but because the master plan to control rust as conceived and implemented did not fulfill the expectations of the planners. The fight against rust began with modest participation and great anticipation for short duration, but developed into a battle of attrition; finally, it was recognized that the antagonist could not be overpowered. Only then could new policies for co-existence be established.

Briefly, cultivars of common wheat with resistance to stem rust proved inadequate during the 1920s and genes were extracted from *T. durum* and *T. dicoccum* (5). Durum and emmer resistances were useful through the 1940s but in 1953, race 15B of *Puccinia graminis tritici* forced their retirement (119). 'Selkirk' became the leading spring wheat and the reason it remained free from stem rust is not known for certain but is apparently due to a complex of genes (74). Sources of resistance from Kenya wheats and 'Khapli' emmer were in use through 1971 to control wheat stem rust (48).

A similar history can be written for the spring oats-crown rust relationship (19). These vicious cycles precipitated changes in strategy and implementation. Multiline cultivars were conceived and introduced in an attempt to break the cycle (see p. 471). General resistance was recognized and its use promoted (see p. 477). Regional deployment of resistance genes has been suggested (20, 69). Whatever the methods, genetic diversity has become the watchword (85).

SOUTHERN LEAF BLIGHT OF CORN   Nothing has opened Pandora's box like the use of the Texas source of cytoplasmically-inherited male sterility (*T-ms*) in corn, which eliminates the necessity to detassel in the production of F_1 hybrid corn seed. Susceptibility to two diseases, southern leaf blight and yellow leaf blight, occurs in plants having *T-ms*. Philippine scientists first demonstrated the susceptibility of *T-ms* corn lines and hybrids to southern corn leaf blight in 1961 and later called attention to the hazard it presented to corn seed production (1, 79, 132). The forewarning made no impact on United States corn scientists. By 1970, when 85% of the U. S. corn crop was produced from seed with *T-ms*, the stage was set for a wholesale spread of *T-ms*-virulent *Helminthosporium maydis* followed by a forecast for another epiphytotic in 1971. The anatomy of the epiphytotic is described elsewhere (85, 97).

The impact of *H. maydis* race T on the national economy and on corn breeding was colossal (127). There was a crash program to return to the production of hybrid corn seed by use of resistant non-*T-ms* cytoplasm and consequent detasseling. Pathologists and breeders alike, foretelling doom, shook the money tree in order to expedite research on *H. maydis* race T. In retrospect, it was soon discovered that the T-race, which attacks primarily *T-ms* corn as compared with the O-race which is generally prevalent in the southern states and which attacks both normal and *T-ms* corn albeit less flamboyantly, had been collected and documented before 1970 in the United States (86, 98).

There are many sidelights to this corn blight episode that bear upon other phases of plant pathology; the following deserve mentioning. We committed a major infraction of the principles of plant pathology by breeding into our major cereal crop a common character that became a common weakness. A genetic triumph became a peril. A second infraction of the principles of plant pathology was to import normal type seed corn from remote areas of the world and risk the introduction of exotic pathogens (127). Perhaps we luckily survived that incident unscathed, but we may not have had time to recognize a new pathogen if it were imported. We committed a third infraction of the laws in that we frantically shipped inbred stocks to warm climes including Hawaii and apparently some *H. maydis*-infected *T-ms* seed also were included. Evidently, *H. maydis* race T was established in Hawaii as a result of this action (3). New Zealand, which claims not to have had *H. maydis,* intercepted seed of inbred lines of popcorn infected with *H. maydis* and thereupon placed an embargo on corn seed imports from the United States (110).

The greatest immediate impact on maize breeding and pathology was the urgent world-wide shift back to normal cytoplasm for the production of hybrid corn seed. This resulted in total cooperation of public institutions and private concerns. The program was so successful that in 1972, no damage was attributed to *H. maydis* race T in the United States. This attests to the ingenuity of U.S. plant breeders and to the flexibility of the U.S. seed trade. Other male sterile factors are being evaluated for commercial application to corn seed production (114) and if one type becomes predominant, we will probably paint ourselves into another corner. As some wag has suggested, the paint had better be fast-drying. Grogan has suggested multiplasm, multiple sources of cytoplasmic male sterility, as a possible means of avoiding this (54).

Of long lasting impact, the blight epiphytotic precipitated a study sponsored by the National Research Council on the genetic vulnerability of major crops to pests (85). Perhaps the following exhortations to the scientists portray the message of the study: "They must exercise constant vigilance in detecting new hazards. Second, they must expand and refine their resources for combating disease by providing new parental material. An excess as well as diversity in breeding stocks is the surest measure. In addition, the scientist must push forward his understanding of the basic principles of parasitism. Lastly, he should strive to elucidate the pathway between genes and the specific attributes of resistance." The loudest and clearest message is—diversify the genes in crop varieties (85).

Many other crops could be cited to show the impact of new pathogens on breeding. Outbreaks of stalk rot, northern leaf blight, and maize dwarf mosaic have resulted in waves of breeding corn for resistance before southern corn leaf blight exploded. No matter which example is used, the general outcome has been similar in each case; much breeding pressure was applied, usually with success, to rectify the problem.

New Concepts

THE GENE-FOR-GENE CONCEPT   When Flor (42) introduced the gene-for-gene concept, he said what should have been obvious to us much sooner; that is, a physiologic race of a pathogen is an assemblage of virulence genes each of which can be easily recognized by culturing the race on the appropriate assemblage of host plants. For too long, the physiologic race was recognized as an entity characterized by reaction on a collection of host plants that by trial and error methods had been demonstrated to be useful differential varieties. We were too engrossed in the number of differential hosts attacked by a pathogen to dissect the differentials genetically. The gene-for-gene concept, although regularly repeated in Flor's publications from 1942 on, was ignored or overlooked. Black and others employed the gene-for-gene concept without reference to Flor's work to explain the relationship between *R* genes of potato and races of *Phytophthora infestans* (10, 11). After Flor (43, 44) reviewed the concept, it gained broader acceptance and became a model for interpreting the genetics of many host-parasite interactions. Flor's gene-for-gene concept was first applied to the *Solanum:Phytophthora infestans* relationship by Person (95).

The gene-for-gene concept was recently re-examined by Flor (45). He stated, "One of the most successful means of controlling plant diseases has been the development of varieties with major or vertical resistance genes. This type of resistance is easily manipulated in a breeding program and is effective until strains of the pathogen to which it does not confer resistance become established. Then, if another gene that conditions resistance to the new strains of the pathogen is available, this resistance gene may be incorporated into the variety by the plant breeder. In doing this, the breeder either consciously or unconsciously is applying the principle of the gene-for-gene hypothesis." Thus, Flor gave a name to a practice that had been followed since the 1920s. Why did prominent breeders and pathologists not recognize the model Flor laid before them? I speculate that not enough of them read flax literature.

THE VERTICAL-HORIZONTAL RESISTANCE CONCEPT   According to Flor, "The gene-for-gene concept is applicable only to host-parasite systems in which resistance is conditioned by major (vertical resistance) genes . . ." (45). Van der Plank introduced the terms "vertical" and "horizontal" resistance in 1963 (130). Immediately, a controversy developed because it could be shown from the literature that the two terms were adequately covered by qualitative and quantitative inheritance, standard terms in genetics. Apparently, many other terms were used synonymously for qualitative and quantitative inheritance of resistance because the authors either did not comprehend these terms or else they did not think the terms adequately described the situation they analyzed. Van der Plank was careful to point out that there may be exceptions to horizontal and vertical resistance as he pictured them (131). Robinson reviewed the synonymy (99) and reasonably accepted or rejected many of the terms. Van der Plank's introduction of the terms "horizontal" and "vertical" stimulated a great deal of interest in breeding for resistance conferred by the specifically acting (vertical) genes in Flor's gene-for-gene concept and for those that confer some resistance to the entire population of a pathogen (horizontal), irrespective of the virulence genes present in that population. Thus, regardless of whether one finds the terminology palatable, Van der Plank focused attention on a difficult problem and set in motion a new philosophy of breeding for disease resistance.

Breeding for vertical resistance is much simpler than breeding for horizontal resistance, a fact that all who write about the subject freely admit. As pointed out previously, lines carrying genes for horizontal resistance are apt to be discarded when they are in competition with lines carrying only vertical genes. Van der Plank described this as the "vertifolia effect" (131). Thus, those with horizontal genes would probably survive selection only because they have other desirable traits. The problems involved in breeding for vertical and horizontal resistance and simultaneous use of the two has been reviewed recently (2). Much thought is being given to this problem. Success in using tolerance, a term which Robinson (99) states is not within the scope of horizontal resistance, but which is lately still included within the scope of horizontal resistance by some authors (2), has been cited as a means of providing measurable control of soil-borne oat mosaic (22), sugar beet yellows (31), maize rust (39), barley yellow dwarf (63), and several other diseases (104). Since the degree of control observed here is nonspecific, why is it not horizontal resistance? In view of the confusion that exists as a result of author preferences, two categories of terms (maybe more) are needed—general terms and specific terms—and there needs to be international agreement on their use. The International Congress of Plant Pathology should begin giving consideration to plant pathological nomenclature and terminology. The synonyms for vertical and horizontal resistance furnish ample evidence for such a need.

Miscellaneous Biological Factors

UTILIZING TOXINS IN BREEDING PROGRAMS   A number of pathogens produce toxins that may be used to facilitate selection in breeding programs. In such cases, the crude or refined extracts induce symptoms at least partially characteristic of the syndrome observed when the host is invaded by the pathogen. Thus, toxin sensitivity and disease susceptibility are highly correlated. These toxins are variously described as "pathotoxins" (142), "primary determinants" (105) and "host-specific toxins" (32).

The association of susceptibility to *Helminthosporium victoriae* and resistance to *Puccinia coronata avenae* in Victoria-derived oats is well known. Wheeler, Luke & Wallace used the toxin from *H. victoriae* to isolate from 100 million seedlings, two blight-resistant mutants with resistance to crown rust but otherwise of parental type (75, 141). Although other toxins suitable for such procedures have been known for some time, they have been utilized only recently. From a study with *Sorghum bicolor* to determine the inheritance of reaction to the toxin produced by *Periconia circinata*, the cause of milo disease, the implications for use of the toxin in the sorghum breeding program became obvious (106). A toxin from *H. maydis* race T has been used in programs to facilitate selection of maize resistant to the southern corn leaf blight fungus (53, 72, 143); one from *H. sacchari* was likewise used to select eyespot resistant sugarcane (23).

Other pathogens known to produce toxins potentially applicable to breeding programs include *Alternaria kikuchiana, A. mali,* and *H. carbonum* (105). The diseases caused by these fungi may be of insufficient importance to warrant the use of toxin methods in breeding. It is noteworthy that among the producers of primary determinants, *Helminthosporium* spp. constitute the majority, but it is probable that several other toxins, as yet of unrecognized potential, may be utilized in breeding programs in lieu of the actual pathogen.

ADAPTATION TO PESTICIDES   The ability of plant pathogens to become tolerant of toxic compounds has been recognized for some time (4, 28, 52, 116).
A new fungicide is not long in use before there is a report of acquired tolerance
to it (92–94, 107, 115). Strains of *Pyrenophora avenae* have developed resistance to organic mercury fungicides in several remotely spaced countries
(35). There is clear-cut evidence that the mechanisms for acquiring and
maintaining tolerance to fungicides are similar to those for acquiring virulence for resistant cultivars (52, 76). If this is true, the proponents of resistance and of chemicals as means of disease control are on equal footing, as
there should be no more fear of a pathogen adapting to a resistant cultivar
than of one becoming tolerant of an antimicrobial substance. Consequently,
the demonstration that pathogens acquire tolerance to prevailing protectant
chemicals is incentive enough to continue or to initiate programs for breeding
disease-resistant cultivars.

TRENDS ASCRIBED TO NON-BIOLOGICAL FACTORS

There are some political, economic, and sociological factors that influence
crop breeding; some of these influence breeding for disease resistance directly, but some affect it only indirectly. They are related to the problems of
people and thus they are of biological origin but they seem to have lost their
identity with biology.

MECHANICAL HARVESTING   Labor shortages, union confrontations, regional
and farm specialization have accelerated the development of machinery for
harvesting horticultural crops (25). One needs only to examine the USDA
Yearbook of Agriculture, *Contours of Change* (129) to get a glimpse of the
mechanization of harvesting. In some cases, however, the crop plants had to
be completely redesigned to make mechanical harvesting feasible. Coupled
with the need for plant re-design is the need to maintain existing disease resistance and to add new resistances. Thus, the pathologist, breeder, and geneti-
cist now work together with a relatively new member of the team, the agricultural engineer. Notable examples of recent progress include development
of tomato (34, 121), cabbage (144), snap bean, and cucumber (85) cultivars suitable to mechanical harvesting and with multiple disease resistance;
many others could be cited (25). Even though each breeding innovation requires selection anew for disease resistance, rapid strides are being made with
these programs.

GOVERNMENTAL POLICIES   Several governments throughout the world have
overtly dedicated themselves to improving the quality of the environment.
For several years, there has been unified recognition that certain types of pesticides are extremely deleterious and that their use in agriculture should be
discontinued. We credit Rachel Carson with attracting our attention to the
problem (26). Following the lead of Sweden in an effort to improve the envi-

ronment, our government has banned the use of alkyl mercury fungicides (103). This action seriously affected the cotton and cereal growers and seed industries, which depended heavily on alkyl mercury products to prevent losses from several seed and soil-borne diseases. Probably this action was justified for "Most chemicals used for agricultural pest control have been selected on the basis of optimum effectiveness and maximum persistence. Relatively little thought has been given to their safety or selectivity with regard to higher animals or humans and to the ultimate quality of the environment. The severe criticism and public disenchantment with the use of pesticides proves that this must change" (80). The real danger lies in governmental action in response to public pressures, especially from outspoken, impassioned, ill-informed environmental do-gooders. Each case requires careful consideration of the alternatives available in the absence of the chemical of choice whose use may be cancelled out by governmental action. In some cases, the alternatives are to breed resistant cultivars or to seek replacement chemicals. If governmental action is sudden, a crop may be vulnerable to one or several destructive diseases in the interim required to provide the replacement chemical or resistant cultivar. In addition, a cultivar with resistance to the biotypes in its region of adaptation may be vulnerable to a biotype in another region. This alien biotype in the absence of the cancelled-out seed treatment chemical, may be easily transported, if it is seed-borne, through routine seed trade and research channels where formerly it was destroyed by the disinfesting chemical. Races of the wheat bunt (*Tilletia* spp.), barley scald (*Rhynchosporium secalis*), and stripe (*Helminthosporium gramineum*) fungi, for example, may be more easily disseminated because of the ban on alkyl mercury seed treatment chemicals. Obviously, there are many other examples. The burden is upon the pathologists and breeders to decide which route to take—that of chemicals or that of resistance. Recent experience suggests that the quickest solution with governmental acceptability will achieve permanency. However, the situation may precipitate a re-examination of germ plasm banks for sources of resistance and a realignment of breeding programs in order to control some diseases formerly held in check by chemicals.

THE PLACE OF PLANT BREEDING IN BIOLOGICAL CONTROL   The public pressure to get pesticides out of the environment will, no doubt, bring about some governmental action that is not supported by facts. Nevertheless, there is a big push for biological control of pests. Current literature on this topic is lopsidedly concerned with insects and weeds (36, 62, 84). It is almost as though breeding for resistance to pathogenic organisms were not biological control. For the individual interested in controlling crop diseases by genetic means, this is the era he has long known; he practices the epitomy of biologic control and he should be a front runner in the environmental protection phase of his profession. Yet, since he has been there all along, he will be taken as a matter-of-fact because his contributions do not reek of newness.

To the public, the introduction of a new, disease-resistant cultivar makes as big a splash as a downy feather settling on a pond. But the time is ripe for a new surge of emphasis on breeding disease-resistant cultivars to meet the challenge of environmental protection by biological control of plant disease. The breeders of disease-resistant cultivars deserve a bigger ripple on the pond.

Acknowledgements

Before writing this review, I asked many plant pathologists and breeders for suggestions concerning the topic. All generously shared their concepts, several proffered specific suggestions, and most furnished reprints, which spared me much toil among the library stacks. Although I was unable to use all the material, each person in some way influenced the molding of the manuscript. To these contributors I remain eternally grateful.

I appreciate the opportunity given me by the editors of *Annual Review of Phytopathology* and I am especially grateful to Drs. J. A. Fox, Martha K. Roane, T. M. Starling, R. J. Stipes, and Sue A. Tolin, all of Virginia Polytechnic Institute and State University, for their critical reviews of the manuscript.

Literature Cited

1. Aala, F. T. 1964. The corn leaf blight disease: A problem in the production of hybrid corn seed involving male sterility. *Philipp. J. Plant Ind.* 29:115–22
2. Abdalla, M. M. F., Hermsen, J. G. Th. 1971. The concept of breeding for uniform and differential resistance and their integration. *Euphytica* 20:351–61
3. Aragaki, M., Bergquist, R. R. 1971. Occurrence of *Helminthosporium maydis* in Hawaii. *Plant Dis. Reptr.* 55:392–93
4. Ashida, J. 1965. Adaptation of fungi to metal toxicants. *Ann. Rev. Phytopathol.* 3:153–74
5. Ausemus, E. R. 1943. Breeding for disease resistance in wheat, oats, barley and flax. *Bot. Rev.* 9:207–60
6. Bandlow, G. 1951. Mutationsversuche an Kulturpflanzen. II. Züchterisch wertvolle Mutanten bei Sommer- and Wintergesten. *Züchter,* 21:357–63
7. Barnes, W. C. 1961. Multiple disease resistant cucumbers. *Proc. Am. Soc. Hort. Sci.* 77:417–23
8. Barrus, M. F. 1911. Variation of varieties of beans in their suscepti-
bility to anthracnose. *Phytopathology* 1:190–95
9. Biffen, R. H. 1905. Mendel's laws of inheritance and wheat breeding. *J. Agr. Sci.* 1:4–48
10. Black, W. 1952. A genetical basis for the classification of strains of *Phytophthora infestans. Proc. Roy. Soc. Edinb.* (Sect. B) 65:36–51
11. Black, W., Mastenbroek, C., Mills, W. R., Peterson, L. C. 1953. A proposal for an international nomenclature of races of *Phytophthora infestans* and of genes controlling immunity in *Solanum demissum* derivatives. *Euphytica* 2:173–78
12. Bolley, H. L. 1901. Flax wilt and flax sick soil. *N. D. Agr. Exp. Sta. Bull.* 50
13. Borlaug, N. E. 1953. New approach to the breeding of wheat varieties resistant to *Puccinia graminis tritici. Phytopathology* 43:467
14. Borlaug, N. E. 1959. The use of multilineal or composite varieties to control airborne epidemic diseases of self-pollinated crop plants. *Proc. First Int. Wheat Ge-*

482 ROANE

net. Symp., Winnipeg, 1958:12–26
15. Borlaug, N. E., Gibler, J. W. 1953. The use of flexible composite wheat varieties to control the constantly changing stem rust pathogen. *Agron. Abstr.* 81
16. Bozzini, A. 1971. First results of bunt resistance analysis in mutants of *durum* wheat. *Mutation Breeding for Disease Resistance* 131–37. Vienna: Int. At. Energy Ag. 249 pp.
17. Broadbent, L. 1969. Disease control through vector control. *Viruses, Vectors and Vegetation.* ed. K. Maramorosch. Interscience: New York. 666 pp.
18. Brookhaven National Laboratory. 1956. Genetics in plant breeding. *Brookhaven Symposia in Biology* 9. 236 pp.
19. Browning, J. A., Frey, K. J. 1969. Multiline cultivars as a means of disease control. *Ann. Rev. Phytopathol.* 7:355–82
20. Browning, J. A., Simons, M. D., Frey, K. J., Murphy, H. C. 1969. Regional deployment for conservation of oat crown-rust resistance genes. *Iowa Agr. Exp. Sta. Spec. Rept.* 64:49–56
21. Burnham, C. R. 1966. Cytogenetics in plant improvement. *Plant Breeding, A symposium held at Iowa State University,* ed. K. J. Frey. Iowa State Univ. Press. Ames:139–87
22. Byrd, B. W., Jr., Graham, D., Byrd, W. P. 1971. Inheritance of tolerance to soil-borne oat mosaic virus in oats. *Crop Sci.* 11:875–77
23. Byther, R. S., Steiner, G. W. 1972. Use of helminthosporoside to select sugarcane seedlings resistant to eye spot disease. *Phytopathology* 62:466–70
24. Caldwell, R. M. 1968. Breeding for general and/or specific plant disease resistance. *Third International Wheat Genetics Symposium,* ed. K. W. Finlay, K. W. Shepherd, 263–72, New York. Plenum: 479 pp.
25. Cargill, V. F., Rossmiller, G. E., eds. 1969. *Fruit and Vegetable Harvest Mechanization Technological Implications.* Rural Manpower Center, Michigan State Univ., E. Lansing. 838 pp.
26. Carson, Rachel. 1962. *Silent*

Spring. Houghton Mifflin Co., Boston. Crest Book Reprint, N. Y. 304 pp.
27. Cherewick, W. J. 1946. A method of establishing rust epidemics in experimental plots. *Sci. Agr.* 26: 548–51
28. Christensen, J. J., Daly, J. M. 1951. Adaptation in fungi. *Ann. Rev. Microbiol.* 5:57–70
29. Clayton, E. E. 1954. Identifying disease resistance suited to interspecific transfer. *J. Hered.* 45: 273–77
30. Clayton, E. E. 1958. The genetics and breeding progress in tobacco during the last 50 years. *Agron. J.* 50:352–56
31. Cleij, G. 1970. Breeding for tolerance to virus yellows in sugar beet. *Inst. Int. Rech. Betteravieres* 4:225–26
32. Comstock, J. C., Scheffer, R. P. 1972. Production and relative host-specificity of a toxin from *Helminthosporium maydis* race T. *Plant Dis. Reptr.* 56:247–51
33. Creech, J. L., Reitz, L. P. 1971. Plant germ plasm now and for tomorrow. *Advan. Agron.* 23:1–49
34. Crill, P. et al. 1971. Florida MH-1, Florida's first machine harvest fresh market tomato. *Fla. Agr. Exp. Sta. Cir.* S-212. 12 pp.
35. Crosier, W. F., Waters, E. C., Crosier, D. C. 1970. Development of tolerance to organic mercurials by *Pyrenophora avenae. Plant Dis. Reptr.* 54:783–85
36. DeBach, P. 1964. *Biological Control of Insect Pests and Weeds.* Reinhold: New York 844 pp.
37. Dimock, A. W. 1967. Controlled environment in relation to plant disease research. *Ann. Rev. Phytopathol.* 5:265–84
38. Down, E. E., Andersen, A. L. 1956. Agronomic use of an X-ray-induced mutant. *Science* 124:223–24
39. Ellis, R. T. 1954. Tolerance to the maize rust *Puccinia polysora* Underw. *Nature* (London) 174:1021
40. Evans, A. C. 1954. Rosette disease of groundnuts. *Nature* (London) 173:1242
41. Finley, K. W., Shepherd, K. W., eds. 1968. *Proc. Third Int. Wheat Genet. Symp.* Plenum: New York. 479 pp.
42. Flor, H. H. 1942. Inheritance of

pathogenicity in *Melampsora lini*. *Phytopathology* 32:653–69

43. Flor, H. H. 1956. The complementary genic systems in flax and flax rust. *Advan. Genet.* 8:29–54

44. Flor, H. H. 1959. Genetic controls of host-parasite interactions in rust diseases. *Plant Pathology Problems and Progress 1908–1958*, ed. C. S. Holton 137–144. Madison: Univ. Wisconsin Press. 588 pp.

45. Flor, H. H. 1971. The current status of the gene-for-gene concept. *Ann. Rev. Phytopathol.* 9:275–96

46. Frankel, O. H., Bennett, E., eds. 1970. *Genetic Resources in Plants —Their Exploration and Conservation.* Davis: Philadelphia. 554 pp.

47. Freisleben, R., Lein, A. 1942. Über die Auffindung einer mehltauresistenten Mutant nach Röntgenbestrahlung einer anfälligen reinen Linie von Sommergerste. *Naturwissenschaften* 30:608

48. Frey, K. J., Browning, J. A. 1971. Breeding crop plants for disease resistance. *Mutation Breeding for Disease Resistance.* Vienna: Int. At. Energy Ag. 45–54

49. Frey, K. J., Browning, J. A., Grindeland, R. L. 1970. New multiline oats. *Iowa Farm Sci.* 24:3–6

50. Frey, K. J., Browning, J. A., Grindeland, R. L. 1971. Implementation of oat multiline cultivar breeding. *Mutation Breeding for Disease Resistance.* Vienna: Int. At. Energy Ag. 159–69

51. Gaskill, J. O., Mumford, D. L., Ruppel, E. G. 1970. Preliminary report on breeding for combined resistance to leaf spot, curly top, and *Rhizoctonia*. *J. Am. Soc. Sugar Beet Technol.* 16:207–13

52. Georgopoulos, S. G., Zaracovitis, C. 1967. Tolerance of fungi to organic fungicides. *Ann. Rev. Phytopathol.* 5:109–30

53. Gracen, V. E., Grogan, C. O. 1972. Reactions of corn (*Zea mays*) seedlings with non-sterile, Texas male-sterile, and restored Texas male-sterile cytoplasms to *Helminthosporium maydis* toxin. *Plant Dis. Reptr.* 56:432–33

54. Grogan, C. O. 1971. Multiplasm, a proposed method for the utilization of cytoplasms in pest control. *Plant Dis. Reptr.* 55:400–01

55. Hayes, H. K., Immer, F. R., Smith, D. C. 1955. *Methods of Plant Breeding.* 2nd ed. McGraw-Hill: New York 551 pp.

56. Heyne, E. G., Smith, G. S. 1967. Wheat breeding. *Wheat and Wheat Improvement.* K. S. Quisenberry, L. P. Reitz, eds. Am. Soc. Agron. Mono. 13:269–306

57. Holmes, F. O. 1936. Interspecific transfer of a gene governing type of response to tobacco-mosaic infection. *Phytopathology* 26:1007–14

58. Holmes, F. O. 1938. Inheritance of resistance to tobacco-mosaic disease in tobacco. *Phytopathology* 28:553–61

59. Howard, H. W. 1970. *Genetics of the Potato: Solanum tuberosum.* New York: Springer-Verlag. 126 pp.

60. International Atomic Energy Agency. 1969. *Induced Mutations in Plants.* Vienna: Int. At. Energy Ag. 748 pp.

61. International Atomic Energy Agency. 1971. *Mutation Breeding for Disease Resistance.* Vienna: Int. At. Energy Ag. 249 pp.

62. Irving, G. W., Jr. 1970. Agricultural pest control and the environment. *Science* 168:1419–24

63. Jedlinski, H. 1972. Tolerance to two strains of barley yellow dwarf virus in oats. *Plant Dis. Reptr.* 56:230–34

64. Jensen, N. F. 1952. Intra-varietal diversification in oat breeding. *Agron. J.* 44:30–34

65. Johnson, T. 1961. Man-guided evolution in plant rusts. *Science* 133:357–62

66. Johnson, T., Newton, M. 1940. Mendelian inheritance of certain pathogenic characters of *Puccinia graminis tritici. Can. J. Res., C,* 18:599–611

67. Jørgensen, J. H. 1971. Comparison of induced mutant genes with spontaneous genes in barley conditioning resistance to powdery mildew. *Mutation Breeding for Disease Resistance* 117–24. Vienna: Int. At. Energy Ag. 249 pp.

68. Király, Z., Klement, Z., Solymosy, F., Vörös, J. 1970. *Methods in Plant Pathology With Special Reference to Breeding for Disease Resistance.* Akadémeae Kiadó: Budapest, 509 pp.

69. Knott, D. R. 1971. Can losses from wheat stem rust be eliminated in North America? *Crop Sci.* 11:97–99

70. Konzak, C. F. 1959. Induced mutations in host plants for the study of host-parasite interactions. *Plant Pathology Problems and Progress, 1908–1958*, ed. C. S. Holton 202–14. Madison: Univ. Wisconsin Press. 588 pp.

71. Leppik, E. E. 1970. Gene centers of plants as sources of disease resistance. *Ann. Rev. Phytopathol.* 8:323–44

72. Lim, S. M., Hooker, A. L., Smith, D. R. 1971. Use of *Helminthosporium maydis* race T pathotoxin to determine disease reaction of germinating corn seed. *Agron. J.* 63:712–13

73. Little, R. 1971. An attempt to induce resistance to *Septoria nodorum* and *Puccinia graminis* in wheat using gamma rays, neutrons and EMS as mutagenic agents. *Mutation Breeding for Disease Resistance* 139–49. Vienna: Int. At. Energy Ag. 249 pp.

74. Loegering, W. Q., Johnston, C. O., Hendrix, J. W. 1967. Wheat rusts. *Wheat and Wheat Improvement.* K. S. Quisenberry, L. P. Reitz, eds. Am. Soc. Agron. Mono. 13:307–35

75. Luke, H. H., Wheeler, H. E., Wallace, A. T. 1960. Victoria-type resistance to crown rust separated from susceptibility to Helminthosporium blight in oats. *Phytopathology* 50:205–09

76. MacKenzie, D. R., Cole, H., Nelson, R. R. 1971. Qualitative inheritance of fungicide tolerance in a natural population of *Cochliobolus carbonum*. *Phytopathology* 61:458–62

77. Martin, J. H., Salmon, S. C. 1953. The rusts of wheat, oats, barley, rye. *Plant Diseases, the Yearbook of Agriculture.* U. S. Dept. Agri. 329–43

78. McFadden, E. S. 1949. New sources of resistance to stem rust and leaf rust in foreign varieties of common wheat. *U. S. Dept. Agr. Cir.* 814. 16 pp.

79. Mercado, A. C., Lantican, R. M. 1961. The susceptibility of cytoplasmic male-sterile lines of corn to *Helminthosporium maydis* Nisikado & Miyake. *Philipp. Agr.* 45:235–43

80. Metcalf, R. L. 1972. Development of selective and biodegradable pesticides. *Pest Control Strategies for the Future.* Nat. Acad. Sci. U. S. A. Washington, D.C.:137–56

81. Moore, W. F. 1970. Origin and spread of southern corn leaf blight in 1970. *Plant Dis. Reptr.* 54:1104–08

82. Morris, R., Sears, E. R. 1967. The cytogenetics of wheat and its relatives. *Wheat and Wheat Improvement.* K. S. Quisenberry, L. P. Reitz, eds. Am. Soc. Agron. Mono. 13:19–87

83. Murray, M. J. 1971. Additional observations on mutation breeding to obtain *Verticillium*-resistant strains of peppermint. *Mutation Breeding for Disease Resistance* 171–99. Vienna: Int. At. Energy Ag. 249 pp.

84. National Academy of Sciences. 1972. *Pest Control Strategies for the Future.* Nat. Acad. Sci. U. S. A. Washington, 376 pp.

85. National Academy of Sciences. 1972. *Genetic Vulnerability of Major Crops.* Nat. Acad. Sci. U. S. A. Washington, 307 pp.

86. Nelson, R. R., Ayers, J. E., Cole, H., Petersen, D. H. 1970. Studies and observations on the past occurrence and geographical distribution of isolates of race T of *Helminthosporium maydis*. *Plant Dis. Reptr.* 54:1123–26

87. Newton, M., Johnson, T. 1932. Specialization and hybridization of wheat stem rust, *Puccinia graminis tritici*, in Canada. *Can. Dept. Agr. Bull.* 160, ns

88. Newton, M., Johnson, T., Brown, A. M. 1930. A study of the inheritance of spore colour and pathogenicity in crosses between physiologic forms of *Puccinia graminis tritici*. *Sci. Agr.* 10:775–98

89. Orton, W. A. 1907. A study of disease resistance in watermelon. *Science* 25:288

90. Orton, W. A. 1909. The development of farm crops resistant to disease. *U. S. Dept. Agr. Yearb.* 1908:453–64

91. Painter, R. H. 1951. *Insect Resistance in Crop Plants.* Univ. Kansas Press. Lawrence, Kans. 520 pp.

92. Parry, K. E., Wood, R. K. S. 1958. The adaptation of fungi to fungicides: Adaptation to copper and mercury salts. *Ann. Appl. Biol.* 46:446–56

93. Ibid. 1959. Adaptation to captan. *Ann. Appl. Biol.* 47:1–9

94. Ibid. 1959. Adaptation to thiram, ziram, ferbam, nabam and zineb. *Ann. Appl. Biol.* 47:10–16

95. Person, C. 1959. Gene-for-gene relationships in host:parasite systems. *Can. J. Bot.* 37:1101–30

96. Person, C., Sidhu, G. 1971. Genetics of host-parasite interrelationships. *Mutation Breeding for Disease Resistance.* 31–38. Vienna: Int. At. Energy. Ag. 249 pp.

97. Plant Disease Reporter. 1970. Southern corn leaf blight. Special issue. *Plant Dis. Reptr.* 54:1099–1136

98. Robert, A. L. 1956. *Helminthosporium maydis* on sweet corn ears in Florida. *Plant Dis. Reptr.* 40:991–95

99. Robinson, R. A. 1969. Disease resistance terminology. *Rev. Appl. Mycol.* 48:593–606

100. Robinson, R. A. 1971. Vertical resistance. *Rev. Plant Pathol.* 50:233–39

101. Rosen, H. R. 1949. Oat parentage and procedures for combining resistance to crown rust, including race 45, and Helminthosporium blight. *Phytopathology* 39:20

102. Rowell, J. B., Hayden, E. B. 1956. Mineral oils as carriers of urediospores of the stem rust fungus for inoculating field-grown wheat. *Phytopathology* 46:267–72

103. Ruckelshaus, W. D. 1972. Notice to manufacturers, formulators, distributors, and registrants of economic poisons. *Environmental Protection Agency Press Release* PR72–5

104. Schafer, J. F. 1971. Tolerance to plant disease. *Ann. Rev. Phytopathol.* 9:235–52

105. Scheffer, R. P., Pringle, R. B. 1967. Pathogen-produced determinants of disease and their effects on host plants. *The Dynamic Role of Molecular Constituents in Plant Parasite Interaction,* ed. C. J. Mirocha, I. Uritani: 217–36. St. Paul: Assoc. Services Inc. 372 pp.

106. Schertz, K. F., Tai, Y. P. 1969. Inheritance of reaction of *Sorghum bicolor* to toxin produced by *Periconia circinata. Crop Sci.* 9:621–24

107. Schroeder, W. T., Provvidenti, R. 1969. Resistance to benomyl in powdery mildew on cucurbits. *Plant Dis. Reptr.* 53:271–75

108. Schwartze, C. D., Huber, G. A. 1939. Further data on breeding mosaic-escaping raspberries. *Phytopathology* 29:647–48

109. Schwinghamer, E. A. 1959. Induced mutations of pathogens for the study of host-parasite interactions. *Plant Pathology Problems and Progress 1908–1958,* ed. C. S. Holton 191–201. Madison: Univ. Wisconsin Press. 588 pp.

110. Scott, D. J. 1971. The importance to New Zealand of seedborne infection of *Helminthosporium maydis. Plant Dis. Reptr.* 55:966–68

111. Sears, E. R. 1956. The transfer of leaf-rust resistance from *Aegilops umbellulata* to wheat. *Genetics in Plant Breeding.* Brookhaven Nat. Lab. Symp. in Biol. 9:1–22

112. Sigurbjörnsson, B., Micke, A. 1969. Progress in mutation breeding. *Induced Mutations in Plants* 673–98. Vienna: Int. At. Energy Ag. 748 pp.

113. Simons, M. D., Murphy, H. C. 1961. Oat diseases. *Oats and Oat Improvement,* ed. F. A. Coffman. Am. Soc. Agron., Madison 330–90

114. Smith, D. R., Hooker, A. L., Lim, S. M., Beckett, J. B. 1971. Disease reaction of thirty sources of cytoplasmic male-sterile corn to *Helminthosporium maydis* race T. *Crop Sci.* 11:772–73

115. Stall, R. E., Thayer, P. L. 1962. Streptomycin resistance of the bacterial spot pathogen and control with streptomycin. *Plant Dis. Reptr.* 46:389–92

116. Stakman, E. C., Harrar, J. G. 1957. *Principles of Plant Pathology* 152–53. New York: Ronald. 581 pp.

117. Stakman, E. C., Levine, M. N. 1922. The determination of biologic forms of *Puccinia graminis* on *Triticum* spp. *Minnesota Agr. Exp. Sta. Tech. Bull.* 8

118. Stakman, E. C., Piemeisel, F. J. 1917. A new strain of *Puccinia graminis. Phytopathology* 7:73

119. Stakman, E. C., Rodenhiser, H. A. 1958. Race 15B of wheat stem rust—what it is and what it means. *Advan. Agron.* 10:143–65

120. Stevenson, F. J., Jones, H. A. 1953. Some sources of resistance in crop plants. *Plant Diseases, the Yearbook of Agriculture,* U. S. Dept. Agr. 192–216

121. Strobel, J. W., et al. 1969. Breeding development for fruit, vegetables. *Fruit and Vegetable Harvest Mechanization. Technical Implications.* B. F. Cargill, G. E. Rossmiller, eds. Rural Manpower Center, Michigan State Univ. E. Lansing. 838 pp.

122. Suneson, C. A. 1968. Harland barley. *Calif. Agr.* 22:9

123. Swenson, K. G. 1969. Plant susceptibility to virus infection by insect transmission. *Viruses, Vectors, and Vegetation.* K. Maramorosch, ed. New York: Interscience 666 pp.

124. Tarr, S. A. J. 1951. *Leaf curl disease of cotton.* Commonwealth Institute of Mycology, Kew

125. Thomas, P. E., Martin, M. W. 1971. Vector preference, a factor of resistance to curly top virus in certain tomato cultivars. *Phytopathology* 61:1257–60

126. Thurston, H. D. 1971. Relationship of general resistance: Late blight of potato. *Phytopathology* 61:620–26

127. Ullstrup, A. J. 1972. The impacts of the southern corn leaf blight epidemics of 1970–1971. *Ann. Rev. Phytopathol.* 10:37–50

128. United States Department of Agriculture. 1953. *Plant diseases, the Yearbook of Agriculture.* 940 pp.

129. United States Department of Agriculture. 1970. *Contours of Change. Yearbook of Agriculture.* 366 pp.

130. Van der Plank, J. E. 1963. *Plant diseases: Epidemics and control.* New York: Academic 349 pp.

131. Van der Plank, J. E. 1968. *Disease Resistance in Plants.* New York: Academic 206 pp.

132. Villareal, R. L., Lantican, R. M. 1965. The cytoplasmic inheritance of susceptibility to Helminthosporium leaf spot in corn. *Philipp. Agr.* 49:294–300

133. Wade, N. 1972. A message from corn blight: The dangers of uniformity. *Science* 177:678–79

134. Walker, J. C. 1930. Inheritance of Fusarium resistance in cabbage. *J. Agr. Res.* 40:721–45

135. Walker, J. C. 1965. Use of environmental factors in screening for disease resistance. *Ann. Rev. Phytopathol.* 3:197–208

136. Walker, J. C. 1965. Disease resistance in the vegetable crops. III. *Bot. Rev.* 31:331–80

137. Walker, J. C. 1969. *Plant Pathology.* New York: McGraw-Hill. 3rd ed. 819 pp.

138. Walter, J. M. 1967. Hereditary resistance to disease in tomato. *Ann. Rev. Phytopathol.* 5:131–62

139. Wellman, F. L. 1939. A technique for studying host resistance and pathogenicity in tomato Fusarium wilt. *Phytopathology* 29:945–56

140. Welsh, J. N., Johnson, T. 1951. The source of resistance and the inheritance of reaction to 12 physiologic races of stem rust, *Puccinia graminis avenae* (Erikss. and Henn.). *Can. J. Bot.* 29:189–205

141. Wheeler, H. E., Luke, H. H. 1955. Mass screening for disease-resistant mutants in oats. *Science* 122:1229–30

142. Wheeler, H., Luke, H. H. 1963. Microbial toxins in plant disease. *Ann. Rev. Microbiol.* 17:223–42

143. Wheeler, H. E., Williams, A. S., Young, L. D. 1971. Helminthosporium maydis T-toxin as an indicator of resistance to southern corn leaf blight. *Plant Dis. Reptr.* 55:667–71

144. Williams, P. H., Walker, J. C., Pound, G. S. 1968. Hybelle and Sanibel, multiple disease-resistant F₁ hybrid cabbages. *Phytopathology* 58:791–96

145. Yamaguchi, M., Welch, J. E. 1955. Varietal susceptibility of celery to aster yellows. *Plant Dis. Reptr.* 39:36

THE DEVELOPMENT AND FUTURE OF EXTENSION PLANT PATHOLOGY IN THE UNITED STATES

❖ 3583

Arden F. Sherf

Department of Plant Pathology, New York State College of
Agriculture, Cornell University, Ithaca, New York

> "The Americans have all a lively faith in the perfectibility of man, they judge that the diffusion of knowledge must necessarily be advantageous, and the consequences of ignorance fatal: they all consider society as a body in a state of improvement, humanity as a changing scene, in which nothing is, or ought to be, permanent; and they admit that what appears to them today to be good, may be superseded by something better tomorrow." DeTocqueville, Democracy in America, Chapter 18.

This keen perception of DeTocqueville regarding Americans is an appropriate preamble for a paper that outlines the origins, development, current status, and future of one of the world's successful ventures in adult education. Cooperative Extension in the United States is a system that presupposes that men and women can improve their lot in life through education in making a living, in homemaking and family life, in understanding relationships with their local, national, and international neighbors, and especially how to work together for the common good. Extension's motto of "Helping others to help themselves" projects far beyond the common understanding of the word "extension" by most research plant pathologists and others who have had limited contact with this organization. Extension plant pathologists have played an important role in past and current extension activities. They can be proud of their service in interpreting and integrating new knowledge of plant diseases into the work and life of rural citizens. The purpose of this paper is to outline the origins and development of informal education in agriculture, the role of the extension plant pathologist in the system today, and to venture predictions regarding his future role in disseminating information about plant diseases and their control. In addition, the considerable changes in orientation of cooperative extension programs, personnel, staffing patterns, and especially types of audiences being reached will be presented.

487

Although this is primarily a review of the Cooperative Extension movement in the United States, many of the ideas regarding the specialist's functions and future roles apply in other countries as well. Extension or advisory services are well developed in Great Britain, The Netherlands, Germany, France, South America, and Israel. Funding, staffing patterns, and major audiences vary substantially from country to country but all have the objective of helping to reduce losses from plant disease in food, fiber, and ornamental crops (10, 15).

After 59 years of success in informal education, Extension, like most other institutions and systems, is presently undergoing rapid changes in operating methods, programming, personnel, audiences, and funding. Since it is people who make Extension succeed or fail, considerable space will be devoted to describing the roles and characteristics of the men and women professionals and lay people who presently make Extension the educational unit that it is. Subject-matter specialists are extremely important in determining whether programs succeed or merely plug along and suffer an untoward and early death. It is in the specialist's role that most professional phytopathologists serve and it is here where the biggest changes are projected in the next 25 years.

Extension is strongly people-oriented and people change rapidly in attitudes and educational needs. As the United States matures, Cooperative Extension likewise is subject to the pull and tug of changing life styles, public philosophies, patterns of affluence and poverty, political pressures, etc. Because people at the local level largely determine the program and policies of Extension, these national patterns of change are soon reflected in demands for new programs and new audiences. This is exactly what is happening throughout the 50 states—Extension is changing more rapidly in some states than in others but all are changing markedly. Maintaining the status quo is the easiest route and some people prefer to proceed with little change. Others insist that changes and improvements are essential to vitality in any organization, and especially one that serves the public. With new faces and ideas appearing each year, extension plant pathologists are having to change in many ways. Some against their wills and better judgment perhaps, but change anyhow. In some ways Extension has had a religious fervor, with its hope to improve the lot of people in all walks of life by helping them to better themselves.

Initially the farmer and his family were the main audience but increasingly, urban and suburban citizens have been incorporated in most state Extension programs and are now active participants; some willingly, others unsuspectingly.

What Is Cooperative Extension?

In simple terms, Cooperative Extension is the educational arm of the Land-Grant Colleges and the Department of Agriculture in the U. S. It is a unique educational system that takes knowledge developed in research centers and interprets and disseminates it to local people and industries who can utilize it. In most states, the Agricultural Experiment Station and College of Agriculture and Life Sciences are the major sources of new knowledge and the place where Extension specialists and administrators have their offices. Extension is financed cooperatively by county, state, and federal governments and in some states also by citizens who enroll in the local organization. During the past six decades, this four-way partnership has brought about a very large and highly successful educational program.

How did this organization arise and develop? Until recent years the United States was an agricultural society with many of its citizens on subsistence farms. This situation changed dramatically during two revolutions that took place more or less simultaneously. One was the Industrial Revolution and the other, less recognized, was the Agricultural Revolution which changed subsistence farming into commercial agriculture. The Industrial Revolution most likely could not have occurred if rapidly applied agricultural technology had not produced sufficient food and fiber to release rural manpower for tasks in the developing industrial cities. The manpower that drained from the farms was replaced by mechanization and this trend is still continuing today. Presently one U. S. farmer produces enough food and fiber for himself and 41 others living in the cities and suburbs, and for seven overseas. Credit for this achievement rightfully is due to a steady flow of useful inputs from researchers in the U. S. Department of Agriculture and the Land-Grant Colleges, an aggressive agri-business complex of seed, fertilizer, pesticide, and machinery manufacturers and suppliers, the profit motive characteristic of the U. S. economy, the work-ethic of farmers, and the constant educational efforts of Cooperative Extension.

The Land Grant College Movement and Legislation

The original settlers of what is now the United States brought little farming knowledge with them from Europe; many were tradesmen and merchants possessing no agricultural skills. History recounts the near famines, the movements to new lands, the battles against insects and diseases, and the catastrophies of floods, hail, and drought. In 1800 farming was about the same as in 1620, with people having little formal education but many prejudices and superstitions. Some even believed in witchcraft. By the early 1800s, many good lands were becoming less productive and not all farmers could move west as some did. Scientific agriculture was just developing in England, Germany, and France but when this new information did slowly arrive in the U.S., it remained mainly with a few wealthy farmers and didn't get out to most of the expanding population. In the 1830s some advances came about such as the steel plow, the mowing machine, the reaper, and crude threshers and fanning mills.

In May 1862, through the efforts of a few agricultural leaders and President Lincoln's support of agrarian reform, the U. S. Department of Agriculture was established (1). Parallel with this decision, urgent needs were recognized for agricultural research at the state level and especially for institutions that could teach practical subjects to the sons and daughters of the industrial and farming classes. Up to this time, a college education in the U. S., as in Europe, was available mainly to a social and economic elite which aspired to the professions of law, medicine, and theology. Education of ordinary citizens was considered unimportant. In 1862 only 1 in every 1900 persons entered college; agriculture, engineering, and industrial arts were considered unsuitable subjects for college education.

Support for changes in these conditions did not come from widespread interest among farmers and the working classes. Most farmers were quite sure they could teach their sons without help! Some leaders such as Benjamin Franklin and Thomas Jefferson advocated education for all classes and lent their support to early efforts to expand college enrollments. According to Curtis (3), major credit for pushing the land-grant philosophy of education and its enactment into law must go to Jonathon Turner, a professor of Latin and Greek at Illinois College in Jacksonville, Illinois, and to Senator Justin Morrill of Illinois. Turner proposed a plan for a special kind of educational institution for the industrial classes. After 5½ years of seemingly endless argument and discussion in several states, a bill originally introduced in the Congress by Senator Morrill in 1856, finally became a federal law on July 2, 1862. This law became known as the Morrill Act; it provided for "the endowment, support and maintenance of at least one college in each state where the leading object shall be, without excluding other scientific and classical studies, and including military tactics, to teach such branches of learning as are related to agriculture and the mechanical arts, in order to promote the liberal and practical education of the industrial classes in the several pursuits and professions of life."

The inducement offered by the Federal government to start this new venture in education was the allocation of 30,000 acres of public land to each state for each representative it had in Congress. This land was to be sold, the income invested, and the interest used for the maintenance of the college. States were quick to accept this financial help—New York did so in 1865. By 1870, 37 states had land-grant colleges with many students from the middle and lower classes. College no longer was only for the elite. By 1970, 1 out of every 4 persons entered college compared to 1 out of 1900 in 1862.

As these new so-called "Land Grant Institutions" developed, it was soon found that insufficient knowledge was available on the practical problems of agriculture and the mechanic arts. In 1887 the Federal government began making appropriations through the Hatch Act for the support of agricultural research in the Colleges of Agriculture and Experiment Stations in each state. Soon other Federal as well as State laws were passed that added to the funds for research. Thus, the Morrill Act of 1862, plus the formation of the U. S. Department of Agriculture in the same year, provided the firm foundation upon which Cooperative Extension was built.

The Extension Movement and Legislation

As the new agricultural colleges got underway they began furnishing speakers at county fairs, farmer's institutes, and demonstrations, and locally arranged meetings with interested farmers or women's groups interested in learning some of the basics of home economics and family and child health care. College staff did these things in addition to their regular teaching and research duties. Their expenses were sometimes paid by sponsoring local groups. In this way interest in informal adult education for farm people developed in the states and makeshift methods were used by the colleges to begin to meet the needs of people who were not interested in a regular four-year course of study.

Each of the states of that era had its own early history of extension development. A few statements concerning early extension activities at my home institution, Cornell University, may provide an example. Cornell is the Land-Grant Institution of New York State. Extension work in agriculture started here almost with the beginning of the University (7); Professors Caldwell, Prentiss, and Law took time from their resident teaching to hold meetings with farmers out in the counties. Professor Roberts conducted the first Farmer's Institute in 1877. This was the beginning of a long series of such Institutes. In the 1890s a series of winter short courses was taught in general farming, butter making, livestock health, etc. These courses helped disseminate new knowledge from the College to practicing farmers. In 1907 an Extension Department was established in the College of Agriculture to carry the teachings of the College directly to the farmer. Earlier, the railroads established a special rate for farmers attending educational meetings, but now the process was to take the college to the farmers. In 1908 the College ran the first "farm special" train over the Erie Railroad through western New York counties. The special train left Ithaca with 20 specialists and in the course of their trip they reached 10,000 people, sometimes holding 10 meetings in one day beginning at 7 a.m. In other states, Seaman A. Knapp initiated a successful series of farm demonstrations in Terrell, Texas to show farmers how to control the cotton boll weevil and Professor P. G. Holden conducted notable seed corn and other agronomic demonstrations in Sioux County, Iowa. These are only examples of the extension programs of teaching that were underway throughout the nation in the early 1900s.

One of the earliest men to catch the vision of what extension work might mean to agriculture was Kenyon Butterfield, President of Rhode Island State College. In 1904 he made a brilliant speech before the National Land Grant College Association dealing with the future development and roles of the state agricultural colleges. Under his leadership, a national extension bill was drafted, approved by the Executive Committee of the Land-Grant College Association, and introduced in the House of Representatives by J. C. McLaughlin of Michigan in 1909. The bill provided a basic appropriation of

$10,000 to each state and added grants-in-aid to the states based on their rural population. These funds were "to be used by these colleges in giving instruction and demonstrations in agriculture, home economics, and similar lines of activity to persons not resident in these colleges with subject matter to be with reference to improvement in rural life." Senator J. P. Dolliver of Iowa presented a similar bill to the Senate on January 5, 1910. Few bills related to agriculture have received such thorough and long discussions as did the Cooperative Extension Act. The bill in its final form was introduced in the Senate by Senator Smith of Georgia and in the House by Congressman Lever. Political pressures at state and national levels finally brought passage on May 8, 1914, authorizing establishment of the Cooperative Extension Service whose function was to "aid in the diffusing among the people of the United States useful and practical information on subjects relating to agriculture and home economics and to encourage the application of the same." It appropriated $4,100,000 payable to states over an 8-year period. All funds with the exception of the initial $10,000 were required to be matched by state and local funds. Allocations were based on the ratio of the state's rural population to the nation's rural population. The Smith-Lever Act was unique in establishing a cooperative educational effort between federal, state, and local governments.

The extension idea grew rapidly, with the colleges building central extension staffs, the counties hiring male county agents and later female home-demonstration agents, and rural citizens requesting and participating in a growing and needed educational effort. Today there are approximately 17,000 professionals and several million nonpaid local leaders involved in programs for men, women, and youth.

Staffing Patterns in Extension

THE 1911–1968 PERIOD   The early county agent was truly a man of many talents; he needed some technical knowledge of agriculture, a warm insight into human nature, good speaking ability, and especially ambition. Most pioneer agents had farm backgrounds and some formal schooling in agriculture. Many farmers tended to be suspicious of "book learning," especially when expounded by a young person just out of college. For this reason, it required much patience and an intuitive knowledge of psychology to get new farming ideas understood and adopted. With time, more and more counties hired one or more agents and by the 1920s many counties had two or more agents working with commercial farmers and youth. At that time female home-demonstration agents also were being employed to work with the female audience on home and family education including nutrition, sewing, personal health, child development, and numerous other projects of importance to homemakers. The day of women's liberation had not yet arrived and women were very necessary to keep the family unit properly fed, clothed, and disciplined. Women's programs have continued to be very important components of

Cooperative Extension at the county, state, and federal levels. Their development of the local-leader concept of grass roots teaching was most useful in rural United States in the early part of this century. Youth programs developed around the 4-H Club concept and also served Cooperative Extension audiences well in channeling the talents and interests of young boys and girls into character- and personal-confidence-building undertakings. These youth programs paid off as these boys and girls carried their training into adulthood. Most counties today employ at least one male and one female agent with responsibilities for such 4-H programs. Although initially working almost entirely with rural youth, in recent years 4-H has moved into the suburbs and urban areas with different programs attractive to these young citizens. The county agent has truly been the cornerstone of Cooperative Extension and his or her abilities and ambition have determined the educational thrust that extension has exerted on the citizens of America—the other components including specialists and administrators have been only secondary.

THE SPECIALISTS   Although the agents working in the county agricultural, home economics, and youth programs had to meet their local audience day to day, right from the beginning there was a group of extension workers called specialists who resided on the campuses of the Land-Grant Colleges. They possessed advanced or special training in the agricultural disciplines of Agronomy, Animal Science, Plant Pathology, Entomology, Farm Economics, Home Economics, etc. These were the backup staff for the county agents developing special programs, diagnosing specific problems, participating in county meetings and demonstrations, and in other ways being of help. In the early years, most of these specialists devoted full time to extension with little time for research; in recent years, however, many specialists have split appointments and are involved in research, classroom teaching, or both, in addition to their extension responsibilities. Many men and women have found useful and highly rewarding careers as specialists and have contributed much to the fine record of extension achievements.

As stated in the "Means of Understanding" each Land Grant Institution must maintain a director of Extension to allocate funds from all sources, help with personnel activities at state and county levels, play a role in program planning, and accept responsibility for reporting to Washington on the extension activities of the state. At the federal level a modest staff of discipline-oriented specialists is maintained. Their role is to keep the college specialist appraised of federal and state programs in his field of competence, to help with exchange of publications, and, upon request, to aid with personnel problems in the states. Literally millions of nonpaid volunteer workers must also be mentioned. They worked on advisory committees, executive committees, Home Economics and 4-H committees, and as local leaders helping with the myriad jobs in Extension. Certainly without this vast group of helpful volunteers, the varied educational programs would not have been possible.

THE 1968–PRESENT PERIOD   Since 1968, basic staffing patterns have been changing rapidly at all levels. This has been brought about by great changes in the types of programs, target audiences, financial sources, national objectives, and other factors. Many counties now participate in agricultural programs on a multi-county basis where one agent specializes in one or possibly two commodities but works in a 2- to 6-county geographic region. This permits the hiring of better-trained, better-salaried staff and gives more job satisfaction to the agent. It also provides the higher quality service and advice demanded by modern agriculturists. Each county still maintains a general agent, usually the senior man who assumes the administrative and personnel responsibilities of the office and who can answer most general questions but refers more specific ones to his multi-county specialists. Home Economics Departments at the county level have shifted from their earlier emphasis on programs for middle class farm and suburban women to low-income "inner-city" women who need help with food, clothing, and budget problems. New federal funds have permitted the employment of new types of agents—the so-called para-professional and program-aide whose role is to work with this low-income audience on a 1:1 or 1:2 basis. These new staff members are not college educated but are proving very effective in communicating with this new audience.

As discussed later in this paper, the role of the specialists at the colleges also are changing; they are assuming more applied research responsibilities and do more writing and informal training of field agents. These changes are occurring because of the increasing numbers of well-trained, multi-county specialists located out in the state. At the state and federal administrative level, few staffing changes have been required to meet changes in the field.

Extension Educational Programs—Past, Present, Future

From the beginning, Extension's most important audience was the farmer, his wife, and family. Early programs were directed almost exclusively toward this group. So many ideas, skills, and methods were needed by farmers that early agents found themselves culling chickens, treating seed, testing soils, and vaccinating pigs rather than educating farmers. Since "to learn by doing is to learn forever" new practices were adopted gradually.

Ferguson (4) divides the development of Extension into three eras. During the *Era of the Skeptic,* book learning was pitted against folklore and tradition. This passed in time and was succeeded by the *Era of Confidence* when people began to have confidence in science, research, and Extension and began to feel that here was something they could use on their farms and in their homes. The college graduate moved into a position of stature and was accepted. Then the third era arrived—the one we are in today. This is the *Era of Dependence* and is a time when the best informed people seek the most help and leadership assumes a new role. People look to professionals for facts, interpretation of them, and for guidance. This era of dependence is demanding new standards of proficiency in Extension personnel.

As the Extension agent became better known and accepted and as his or her organization at the state and national levels became better staffed and proficient, their programs and services could change. The educational and financial level of the farm audience changed also, which in turn brought about important changes in Extension that involved all three divisions: agriculture, home economics, and youth. Other important factors involved in the changing emphasis include a declining number of farm families, changing national priorities, the need to raise the standards of living of poor people in urban as well as rural areas, and, most recently, the important environmental and ecological problems of pollution, housing, population, and crime.

In the late 1950s and throughout the 1960s, many self studies of Extension, both at the national and state levels [Miller (13), Crooks (2), Swan (19), Watts (21)] recognized that agriculture should still remain a major concern not only in production but also in processing and marketing as well. Many different categories of homemakers were recognized—the farmwife, the urban homemaker, the rural nonfarm resident, the young homemaker, etc. The youth program was urged to serve a much wider audience than in earlier years when most programs were for farm youth. In essence, Extension was urged to devise programs to reach a larger segment of the total population than it had in its earlier years.

In an organization as large and democratic as Extension, sweeping changes were not made without protest and difficulty. With its early years of experience and funding so firmly tied to agriculture, some leaders were reluctant to change. A majority of Extension personnel at both the county and state levels were reared on farms, trained in colleges of agriculture, and had only limited expertise in some of the nonagricultural subject areas they were being forcefully asked to enter, namely, sociology, public policy, etc. This pulling and tugging has created a furor in Extension Services in many states, especially those with large urban-suburban populations and rapidly declining numbers of farmers. The Extension Services of many states are altering their programs and personnel to meet the new challenges as follows. The needs of farmers, and some food processors and agri-business people, are being served by multi-county specialists operating in groups of counties where the specific commodities are important, be it fruit, vegetable, dairy, or poultry. The "Jack of all trades" generalist is being replaced by highly specialized personnel who can speak with authority on today's agricultural technology to well-educated farmers. Home Economics programs are being expanded to meet the needs of young homemakers, working mothers, retired persons, and are giving special emphasis to the nutritional problems of low-income families. Youth programs are being adapted to the needs of urban, suburban, and rural communities by providing television programs and field days for schools and special programs dealing with such diverse interests as horseback riding, science, and pollution control.

Many present day extension programs involve all three traditional divisions —Agriculture, Home Economics, and Youth. For example the problems of maintaining attractive home grounds are of interest to adults, youth, farmers and other rural residents, and urban or suburban dwellers alike. Problems of the inner-city, such as crime and drug abuse, have influences in nearby suburban and rural areas. Land no longer in agricultural production or presently owned by rural residents presents problems of a different nature from those of commercial farmers. Conflicts between farmers and nonfarm rural residents often result from lack of understanding on the part of both. Water systems in rural areas often pose problems for many citizens. The complexity of such concerns points to the need for land-use planning for the best interests of all parties. The need for help in understanding the problems arising in rural-area development has brought about useful Extension programs in this growing subject in many counties of the United States.

The future of Cooperative Extension is receiving much attention at present for several reasons. These include: (*a*) Demands by nonrural citizens for educational programs to meet their special needs. (*b*) Close scrutiny of all public expenditures to verify that they meet society's present priorities. (*c*) Newly arising urgent problems of drugs, pollution, overpopulation, poor diet, teenage crime, etc. (*d*) Declining numbers of farmers who have been influential spokesmen for Extension. Some critics are saying that Extension has served the United States well, helping it to agricultural abundance, but that its job is now largely out-moded, or being handled by private commercial agencies or by other governmental units better suited to meet the priority needs of a modern urban-suburban and industrial society. In my opinion, two types of Extension services are needed. The Extension needs of our society cannot be filled by a single agency designed to provide voluntary educational programs in all segments of our society. At present the new "people programs" of Extension are being hindered by traditional allegiance to commercial agriculture and, conversely, the proven and still needed agricultural programs are being weakened in many counties by the pressures of the people programs which are of direct interest to the majority of the population. The establishment of two separate but related Extension services seems very feasible to me. One could be the "Agricultural Extension Service" whose avowed audience would be farmers, ranchers, orchardists, animal feed-lot operators, and all components of the agri-business complex. Most of its agents and specialists would be college graduates with a true love of agricultural research and extension. Technical backstopping would come from the state land grant college system. It would be closely related to and serve as the educational arm of the U. S. Department of Agriculture.

The "Citizen's Extension Service," would serve all audiences outside of commercial agriculture including urban and suburban men, women, and youth, rural and small community residents, and retired people. Instead of farmers, the clientele would be Parent-Teachers Associations, League of Women Voters, labor unions, housing associations, etc. All of the new programs for low-income people, child development, drug education, rural area development, home horticulture, and household management would be the

Figure 1 An early effort to control plant diseases with fungicides recommended by Extension—1925

Figure 2 The farmers meeting—a basic method of early Extension work—1922

Figure 3 Extension pathologists demonstrating at a state fair—1928

Figure 4   Today's extension pathologists devote time to applied research

domain of this new organization. The youth division could well fit here. The ratio of city to rural youths in this program is increasing rapidly and there is no longer a need for close attachment of youth work to agriculture. The professional agents of this Citizen's Extension Service would be men and women holding degrees from Colleges of Arts & Science with major interests and training in sociology, education, languages, economics, engineering, criminology, business, home economics, etc. Finally, and very importantly, the backstopping of this new system could come primarily from state supported colleges and universities with their large broadly-educated faculties having expertise in sociology, educational psychology, law, architecture, engineering, city planning, business, and economics. Expertise in these and other disciplines is needed desperately by present Cooperative Extension personnel to carry out the new programs requested of it by federal and state governments. Federal funding for the Citizen's Extension Service should come from the Department of Health, Education and Welfare, or from Housing and Urban Development in a relationship similar to that of the Department of Ariculture to the proposed Agricultural Extension Service.

In recent months the Land Grant Colleges and the U. S. Department of Agriculture have come under severe criticism because of their ties to commercial agriculture, supposedly at the expense of more deserving marginal and part-time farmers and other segments of society (9). One advantage of the dual Extension Services would be that the proposed Agricultural Extension Service could be more responsive to the needs of commercial agriculture with its great task of providing food and fiber for 209 million U. S. citizens and many hungry people overseas. If these two clear-cut educational organizations were operational, it seems obvious that new sources of financial support could be tapped with willing donors including farm commodity groups, agribusinessmen, bankers, etc. in support of the agricultural service, and labor unions, consumer groups, electric and gas utilities, philanthropic foundations, etc. supporting the Citizen's Extension Service.

The proper structure of the Cooperative Extension Service of the 1980s and beyond will continue to be debated and may be determined by the taxpayers and politicians before long. In the meantime Extension will continue to change and expand programs "helping people to help themselves."

Extension has engaged in periodic self-appraisal studies including the Kepner Report of 1948 (11), the Scope Report of 1958 (13), and various subcommittee reports of the 1960s, and most recently in 1968, "A People and A Spirit" (21). Out of this most recent report will come many of the future objectives, audiences, and programs of Extension. Its recommendations include: (a) the Land Grant Universities should continue to administer extension functions, (b) these Universities should develop mechanisms of support from all its departments having needed competencies, (c) county extension offices should be strengthened, (d) Extension should serve as the educational arm of many government agencies in addition to the U. S. Department of Agriculture, (e) present county-government relationships should be maintained but more city governments should contribute to the financial support of programs for urban audiences, (*f*) Extension should strengthen and expand its effort in commercial agriculture by 25% by 1975 with increased emphasis in marketing and farm business management, (*g*) work with low-income farmers should be increased, (*h*) double by 1975 the efforts with natural resources and the environment, pollution, and land-resource use, (*i*) employ more specialized area agents, (*j*) use more joint research-teaching-extension personnel, (*k*) upgrade staff competence by increasing their training and development, and (*l*) increase the effectiveness of staff by use of new electronic and programmed teaching and communications systems.

The changes advocated above (21) will come about in varying degrees within the states depending upon the political pressures, funding procedures, state priorities, the role of the U. S. Department of Agriculture in federal government, and many other factors. Basing my projections on 27 years of experience in Cooperative Extension work in three states, I offer the following: The trend toward fewer but larger-acreage farm units will continue, with family farms continuing to decline in numbers. The regionalization of well-trained agent-specialists should serve these large farming units well and will be continued. Advisory programs for home grounds and insect and plant disease problems around the home will become more popular. Programs for rural-area and community development will increase and their audiences will expand in numbers and composition. Home-economics programs will continue to attract federal funds to support work with low-income families of urban and rural areas using new communications techniques not required with traditional middle class audiences. Youth programs will thrive as they attract young people of all income and racial groups. Television will be used more extensively in youth programs because young people will continue to enjoy this medium of education. Finally, counties that have an important agricultural component will continue to need an agent knowledgeable in various changing state- and federal-agency programs who can relate them to the particular problems of his constituents. Many agency personnel cannot do this with the skill of a competent experienced county agent.

The Specialist—His Past and Future Role in Extension

It is through the role of specialist that most plant pathologists make their contributions to the Cooperative Extension effort. Before the county-agent system was established, college-based staff called specialists began taking old and new research out to the farms, homes, and small villages without the help of local representatives. After county agents were established, corps of specialists continued to operate as direct disseminators of information but were greatly aided by county extension personnel who helped devise educational programs with definite objectives and continued to provide advice when the specialists had departed. As the College research enterprize became divided into disciplines such as plant pathology, entomology, etc., each discipline soon discovered that they required their own representatives to elicit, simplify, and disseminate the latest research findings—thus a small corps of specialists came into being and increased in numbers in direct proportion to the growth of the state and federal Extension Services. The specialist's role is of such prime importance in the overall system that it warrants special consideration.

DEFINITION   A specialist may be defined as "one who devotes himself or herself to some special branch of an activity, interest, business, knowledge, art, or science, and who possesses technical and professional qualifications not ordinarily a part of the normal equipment of every member of an organization and whose duties require the use of such qualifications" (Warren, 20).

Warren also states, "The place of the specialist in the extension organization is chiefly advisory rather than executive. A specialist performs a staff function and is not usually responsible for administrative matters. He is the advisor, the consultant, the source of information for line personnel. He is the resource person upon whom the line personnel must rely. While his is a staff function, he exercises an authority that is no less real than the line authority—even though it includes no right to command, for his is the authority of ideas."

FUNCTION   Obviously the function of a specialist in 1973 is quite different from that of the pioneer specialist of 1914. As the educational level of the audiences improved, so did the degree of sophistication of the educational process. This in turn changed the functions and duties of the specialist. Although functions vary somewhat with subject specialty, numbers of colleagues, and the college institution (16), the most important month-to-month duties of a specialist include:

1. Keeping field agents supplied with technical information and its application.
2. Supplying background information in his field to aid in planning educational programs.
3. Acting as a resource person for agents working on specific problems.
4. Evaluating and communicating the research needs of people in the counties back to the college research staff and administrators.
5. Maintaining a favorable public-relations image for his College and the Extension Service. Since he may be the only representative visible to the citizen, he must leave good impressions.
6. Maintaining two-way relationships with the commercial and government agencies, keeping them informed about extension recommendations and in turn gathering and disseminating their information to the field audience.
7. Developing and supplying program ideas and audio-visual aids in the specialist's field that can be utilized by agents.
8. Helping the agents plan new educational programs and evaluating strengths and weaknesses of current programs.

Someone has said "A specialist is like a fire department, ready to go where there is need, but when not in the field he is busy keeping himself and his equipment in good condition."

PERSONAL ATTRIBUTES OF AN IDEAL SPECIALIST Although few people achieve perfection in their chosen life's work, there are certain key attributes that make one specialist more successful than others. Abilities or traits that prove to be real assets are:

1. A love of people and the ability to get along with people of all ages, educational levels, and social status (including one's colleagues).
2. A keen student of his discipline and an authority in his field with ability to search out, select, and simplify research findings that will be of value to his audience.
3. A good writer, speaker, and actor who can attract and hold his reader's or listener's attention.
4. An ability to integrate his subject matter with those of other specialists in related fields.
5. A real understanding of the Extension philosophy of off-campus education and adaptability in meeting the changing needs of the citizens.
6. Excellent health and an understanding and tolerant wife or husband.

The paramount ability of a good specialist is his skill in human relationships —the ability to inspire and maintain the respect of all persons. Fifield (5) emphasizes the requirement of high professional ethics and conduct in a person representing Extension in the Land Grant College. Somewhat in contrast to his research colleagues, a specialist must not become so specialized that he loses sight of the broad problems of agriculture, of society, or of the total scope of his educational objectives. A good specialist must develop breadth as well as depth in his speciality. In fact, Warren (20) noted that the best ones are really more generalists than the term "specialist" implies. Even though they have mastered a given body of the subject matter, they must make conscientious efforts to keep abreast of developments in other related fields. Specialists need depth to gain authority and breadth to make optimum application of their subject matter.

The Specialist in Plant Pathology Extension

Most of the preceeding description applies well to the extension plant pathologist. Of necessity he deals in a nonexact field of biology that frequently involves gray areas of knowledge rather than hard facts. His is a science that protects life's basic industry, agriculture, and the production of green plants upon which human lives depend.

Our professional people were among the first to venture out from their campuses into rural communities to explain the nature of plant diseases and their control to poorly educated, and often suspicious and doubting farmers.

Most recipients of such information had little or no formal education in Botany and the plant sciences, including insect and disease phenomena, and were suspicious of "book farming" taught by youngsters from the college campus. Thus the Extension pathologists had to be good psychologists and salesmen as well as biologists. The first organized state Extension project in plant pathology was established at the New York State College of Agriculture, which appointed M. F. Barrus to be Assistant Professor of Extension work on September 20, 1911 (Haskell, 8). At that time New York had no county agents, but Professor Barrus lectured at Farmer's Institutes and Grange Meetings, put on exhibits at fairs, gave talks and demonstrations on agricultural train-

Table 1 Growth in Extension Plant Pathology[a]

	1953	1965	1969	1972
Plant Pathologists employed in Extension (EPP)	48	102	123	150
States with at least 1 EPP	26	47	48	47
States with more than 1 EPP	6	20	23	32
Full time (100% extension)	23	54	66	75
Major time (50–90% extension)	17	20	32	47
Part time (5–50% extension)	8	28	25	28
Education of Extension Plant Pathologists: Ph.D. (%)	—	79	93	95
M.S. (%)	—	21	7	5

[a] Portions of this table are from Sharvelle (17).

ing, and wrote plant-disease leaflets. Wisconsin was next to give official status to extension plant pathology; during the summers of 1912 and 1913, R. E. Vaughan was authorized to visit farmers and give spray demonstrations in connection with his studies of the blight of peas. He was appointed Extension Plant Pathologist on July 1, 1915. With the passage of the Smith-Lever Act, several other states appointed plant disease specialists, and their numbers have grown steadily over the years until 1972, when 47 states had at least one position and many states had 3 or more usually dividing responsibilities on a commodity basis (Table 1).

The steady increase in numbers of well-trained "general practitioner" plant pathologists from 1953 to 1972 attests well to the importance placed upon our discipline by citizens who face disease problems with their plants. County, state, federal, and college administrators had to be convinced of the need before these new positions were funded, and this comes only as a result of expressed needs by many people. Even in the recent months of reduced professional job opportunities in the sciences, including phytopathology, new extension positions have continued to develop and be filled with well-qualified scientists. I believe this enviable situation will continue in the foreseeable future as biological pressures on man's food and fiber plants intensify.

In addition to conducting educational programs involving the survey, identification, and control of plant diseases, Extension pathologists in recent years are helping to solve air- and water-pollution problems, assisting in youth training, and operating diagnostic clinics for plant diseases primarily for urban-suburban citizens.

In a discussion of the origins and development of the Plant Pathology component of Cooperative Extension, several pioneers and leaders must be mentioned in addition to M. F. Barrus and R. E. Vaughan, who held early appointments. R. J. Haskell was the first federal specialist working out of Washington, D. C., helping to promote and organize nationwide educational programs and convincing state administrators of the need for at least one extension plant pathologist on their college staffs. He was a forceful leader and a good writer. Others who pioneered extension pathology work in their respective states include R. C. Rose (Minnesota), R. S. Kirby (Pennsylvania), G. L. Zundel (Washington), C. Chupp (New York) C. E. Scott (California), Donald Porter (Iowa), Charles Gregory (Indiana), O. C. Boyd (Massachusetts), E. A. Stokdyk (Kansas), and E. C. Sherwood (West Virginia).

The Changing Role of the Extension Plant Pathologist

Until very recent years, efforts of extension plant pathologists were directed almost exclusively toward aiding commercial farmers, ranchers, greenhouse operators, and food distributors in keeping their plant products in a healthy state. Benefits from such help have been reflected in abundant food available at moderate prices. Plant pathologists have played a most important role in the success of U. S. agricultural production and distribution. It has been good to have been a part of this winning team. Extension, working with colleagues in the Land-Grant College systems of our states, has served by taking agricultural technology as it developed, farm tested it, simplified the data and made it understandable and available to farmers who then proceeded to use it in their operations. Production records attest to the success of this system. Obviously workers in agronomy, horticulture, entomology, forestry, agricultural engineering, and economics were involved as were seedsmen, agricultural chemical and machinery personnel, and others. The rapidity of movement of new agricultural research through the adaptive stage and out to the farms has been a major achievement of the specialists and county agents of the Extension Service.

When the U. S. was still an agrarian society and many of our people were struggling to supply food consistently and cheaply to city and nonfarm citizens, the farmer was appreciated and his numbers were sufficiently great to assure adequate support for the agricultural research and extension functions. As the farmers succeeded in producing ever larger amounts of food and fiber, many families left the farm to go to the cities where other types of employment were less arduous and usually better paying. Today the farmer's importance to society is often forgotten and his needs neglected, especially as other pressing problems have developed in our increasing urban society. In 1973, farmers comprise only 5% of the U. S. population and in certain states only about 3%. This, combined with the one-man, one-vote decree of our courts has resulted in a rapid decline in the political power enjoyed by the farmer in past decades. These changing developments are exerting a real influence on Cooperative Extension today and are altering the tasks and role of the extension plant pathologist.

In the late 1940s as affluence became more common, many families reversed the earlier trend and moved from urban situations in flats or small homes to the suburbs or even into the open country. In such a setting some knowledge of plant life and disease and insect control became a necessity. As a result, a whole new clientele has arisen that needs advice about plant diseases on a variety of house plants, lawns, flowers, vegetables, and shade trees. In some states the numbers of this clientele constitute a majority of the population. Even in sparsely populated states they rapidly build up to a sizeable audience.

Under these circumstances, some of the old extension techniques of one-to-one house calls, the use of demonstration plots, and personal correspondence are impractical or unworkable. Specialists who have ventured to reach this audience usually have done so with a hesitating "toe in the water" approach only to find that their efforts were more appreciated and accepted than by their familiar conventional commercial agricultural audiences. Good progress has been made in developing effective procedures for extending agricultural information on horticultural and insect- and plant-disease control matters. New techniques including telephone tape message centers, computerized advisory services, diagnostic clinics, new types of radio and television programs, information networks for garden-supply dealers, and employment of part-time para-professional aides are just a few of the new approaches being put to use in reaching this mass audience. Politically, working with these people provides excellent public relations value for Cooperative Extension Services, assuming the programs are well run and successful. With farmer numbers declining rapidly each year and likewise their influence with legislative bodies at the local, state, and national levels, Extension needs a broader audience supportive of its programs.

Sherf (18) points out that in addition to serving as a troubleshooter on horticultural problems, the extension plant pathologist must and will assume a role as an impartial mediator in public controversies that develop involving subjects in which he has expertise. His knowledge and proper assessment of data and arguments on both sides of the issue will be most helpful to his urban-suburban audience in their deliberations on public controversies such as air pollution, pesticides, herbicides, organic gardening, organic foods, etc. He is negligent in his duties if he permits only physicists, historians, ministers, and housewives to write and speak on these topics.

The present day extension plant pathologist is as different from the pioneer of 1911 as the present research pathologist with his enzymes, mycoplasmas, and electron microscopes is from the mycological botanist and amateur mi-

croscopist of those early years. In most states now the farmer's needs from
Extension have changed largely from one of gaining information in a usable
form for his farming operation to one of need for help in adapting and integrating multiple disease-, insect-, and weed-control practices on his particular
farm or for help in complying with legislative edicts involving pesticides, pollution, food grades, food quality, etc.

WORKING AS A TEAM MEMBER   Recommendations for agriculture today can
evolve only as a result of group effort involving plant pathology, entomology,
agronomy, horticulture, agricultural engineering, and agricultural economics.
These teams often serve as impartial judges of the most economic ways to
control plant diseases. No longer does the specialist work alone nor can he
for long consider only diseases when he deals with the most efficient farm
technologists of the 1970s and 1980s. Only knowledge developed by a team
of specialists looking at all aspects of a new farm practice, making compromises when necessary, and with necessary field testing can be utilized by today's farmers. Our audience is no longer the farmer alone—rather it involves
society and other members of the agribusiness partnership, i.e., the banker,
seedsmen, farm-implement and chemical dealer, U. S. Department of Agriculture personnel, and others. Total recommendations come harder now than
formerly since farming is more competitive, more costly, and provides a narrower margin for error. Thus, recommendations must be based only on careful research.

HIS ROLE IN APPLIED RESEARCH   As noted by Frolik (6), public tax funds
historically have been provided to the Agricultural Experiment Stations to
support applied research expected to benefit agriculture and home economics.
Since about 1962 the trend in research in plant pathology, as well as in most
other biological disciplines, has been toward basic studies that in themselves
have little, if any, immediate application to crop protection. Unless these basic findings are picked up and used by applied researchers and translated into
information and practices that can be utilized now, the value of basic research may not be apparent and thus no longer supported financially. The era
of "What does it do for me?" is with us.

Some of the necessary applied research is being done by private industry
and by the few Experiment Station researchers still willing to devote their
efforts to it. However, the extension pathologist in recent years and considerably more so in the future, will assume duties involving mission-oriented research along with his normal extension duties. Table 1 indicates that about
half of the extension plant pathologists in the U. S. hold appointments that
will allow them to devote up to 50% of their time to research. Many of the
others also conduct plot, laboratory, and greenhouse experiments to provide
reliable answers to pressing problems and demonstrate their recommendations. This research involvement has real benefits both to the Extension worker
and to his clientele. The upgrading of the educational status and abilities of

the new multi-county specialists has greatly aided this changing role. He is assuming some of the duties formerly assigned to the college-based specialist, thus freeing the latter for part-time research. This involvement will increase in prevalence so that in the future fewer extension plant pathologists will devote 100% of their time to extension. Other Extension plant pathologists agree with this prediction, including Sharvelle (17), O'Reilly (14), Mac-Daniel (12), Horne,[1] and Worf.[1]

The college Extension pathologist and the multi-county specialist will work very closely as a team not only in applied research but in the entire agri-business production scheme. The pathologist's immediate audience will be these regional personnel who look to him for subject matter resources regarding developing disease problems. In many states with large agricultural production, I predict that multi-county assignment of specialists by commodities will have taken place by the mid 1970s, thereby freeing the college specialist for more effective mass media functions and the "teacher of teachers" role. This role he will assume with vigor and with a talented, interested audience.

LABORATORY DIAGNOSTICIAN   Increasingly, Extension plant pathologists are offering as a free service, the diagnosis of plant diseases and the prescription of appropriate controls. Not only does the farmer who grows plants for a livelihood need this service but the urban-suburban householder with yard, house and garden flowers, fruits, and vegetables also requires help periodically. To facilitate the handling of large numbers of specimens, some agricultural colleges have established diagnostic laboratories for receiving, recording, and diagnosing disease problems for the citizens of their state. Such facilities have many merits including prompt service, accurate recording of disease prevalence, favorable public relations, graduate student training, and a saving of time for the research-staff. The Extension pathologist is the logical person to direct such a laboratory and generally does so. With the rapid trend toward urbanization, nearly all states will establish such facilities, which will require considerable amounts of time of the specialist in future years.

A TRAINER OF PESTICIDE SPECIALISTS   In the 1970s a new role is developing for extension plant pathologists and may become of considerable importance in the near future—one that may eventually mean the partial transfer of public plant pathology to the private sector. With the furor over the use of pesticides for control of pests in both agricultural and urban-suburban situations, several states have passed legislation requiring that certain pesticides be available and sold only on written prescription by licensed professionals. The task of training, testing, and licensing this corps of personnel will fall largely on Extension, and the next step will be for former extension plant pathologists themselves to enter this new service field for a fee—somewhat in the manner of the professional veterinarians. This would take our profession out of the

[1] Personal communication.

realm of "socialized plant medicine" and into the private arena as with medical services in the U. S.

THE PRIVATE PRACTITIONER   Another factor giving impetus to such a move is the need for frequent and time-consuming services by large farm cooperatives who must have precise recommendations on disease control in their farming operations and who are very willing to pay well for such private service. This, too, may be a profitable profession in the near future. So service plant pathologists in the 1980s may well be of three types; namely, private diagnosticians available on a fee basis for urban, suburban, rural citizens; private salaried pathologists working for farm cooperatives, large commodity farm groups, etc.; and traditional extension patholoists paid by federal-state funds working on urban, suburban, tree, fruit, vegetable, flower, ornamental, and grass problems or, in some states where commercial agriculture still justifies their services, they will devote their efforts to controlling diseases of food and fiber crops. This latter type of extension worker will probably be supported partially by farm and agri-business grants.

ROLE IN INTERNATIONAL PLANT PATHOLOGY   Presently the world food crisis is an ever-growing problem with populations increasing faster than the ability to produce food in many of the developing countries of the world. In spite of large expenditures of money and the shipment of food by many countries of the developed world, it has become obvious that these efforts are just not adequate. Rather, the self-help programs sponsored by the United Nations, The Rockefeller Foundation, Food and Agriculture Organization of the United Nations, the Ford Foundation, and others that are developing and teaching new technology and Extension methods are finally yielding results. The so-called "Green Revolution," brought about through the efforts of these organizations, resulted in the Nobel Peace Prize being awarded to our very deserving fellow plant pathologist, Dr. Norman Borlaug.

The talents of the extension plant-disease specialist are sorely needed in these food-deficient countries. There he can take information developed both in the outside world and within the country out to the farmers for immediate use, generally working with native pathologists whom he has trained. Plant pathologists skilled in applied research and especially in extension will find attractive opportunities to apply their talents to the compelling needs of developing countries. Young and especially mid-career pathologists in good health are needed rather than retirees, too-highly specialized research types, or former College administrators. This will require men with a true missionary spirit since monetary and professional recognition most likely will be meager. I predict that in the future a modest but increasing number of young plant pathologists will accept these challenging job opportunities overseas.

ROLE IN SOCIALLY-ORIENTED "PEOPLE" PROGRAMS   Although Cooperative Extension most likely will continue to make considerable changes in programs and audience emphasis because of new societal needs such as those forcefully requested in "Hard-Tomatoes—Hard-Times" (9), and "Failing the People" (22), I can see few ways in which a highly trained plant pathologist can contribute directly to the welfare of low-income urban or rural people. True, many of his efforts will benefit the poor indirectly just as they have in the past, and that is good; but beyond such indirect aid, many of the needs of low-income people must be provided by other state and federal government agencies.

Summary

Cooperative Extension in the United States has had a short 59 year history of service to people, primarily those with responsibilities for producing food and fiber for the rapidly expanding population of this new country. Women's and youth programs until recently have also been oriented to improving the diets, home life, and general well being of rural citizens. By helping to develop, prepare, and extend new agricultural technology originating in the Land-Grant Universities of the states, extension workers have played a most important and generally recognized role. Recent changes in priorities among social needs are eliciting changes in extension programs and audiences; to some of these plant pathology has little relevance. Agriculture-based expertise is still greatly in demand, however, and will be obtained through Extension in the foreseeable future. Certain new roles will develop for plant pathologists in Extension work but the basic challenge will remain the same as in the past—wherever crops are to be grown, plant diseases will be in contention.

Also as long as new agricultural technologies are developed, extension specialists will be required to adapt and transmit it to the ultimate users. Man will always obtain contentment and aesthetic value from his flowers, trees, shrubs, and grass and will place high value on maintaining them in a state of good health. Thus he will always need practical plant pathologists to advise him. The ratio of urban-surburban citizens to farmers in the U. S. may change still further but the Extension tasks in 1990 will remain strikingly similar to those to which extension plant pathologists are devoting their life's work in 1973.

Literature Cited

1. Cooley, H. D. 1962. *U.S.D.A.'s First Century.* U.S.A. Congressional Record. Wed., Feb. 21, 4 pp.
2. Crooks, P. B. 1965. Extension's Responsibility to Commercial Farmers and Ranchers. Report of Project III Comm. to the Extension Comm. on Organization and Policy. *Fed. Ext. Serv. USDA*
3. Curtis, C. M. 1962. *The Land Grant Story.* Cornell Univ., Alumni News 64, 7–10
4. Ferguson, C. M. 1959. The Challenge Ahead for the Cooperative Extension Service. *Presented at New York State Ann. Ext. Conf.,* Ithaca, N.Y.
5. Fifield, W. M. 1960. Professional ethics and conduct of an Extension worker. *Presented at State Conf. Agr. Ext. Serv.* Chattanooga, Tenn.
6. Frolik, E. F. 1967. What does the Agr. Exp. Station expect from Cooperative Extension? *Presented at Land Grand College Meetings joint session.* Columbus, Ohio
7. Gibson, A. W. 1962. One Hundred Years of Land Grant Education. *Presented at 130th Ann. Meet. the N.Y.S. Agr. Soc.* Albany, N.Y.
8. Haskell, R. J. 1953. Extension Work in Plant Pathology. *Plant Dis. Reptr.* 37:570–74
9. Hightower, J. 1972. *Hard Tomatoes—Hard Times.* Agri-business Accountability Proj. 1000 Wisc. Ave. N.W., Wash., D.C.
10. Jones, W. E. 1963. The Role of Advisory Services in a Changing World. 16th Amos Memorial Lecture, Part III. *Res. Rep. Rev., E. Malling Res. Station* 44–49
11. Kepner, P. V. 1948. Joint Committee report on Extension Programs, Policies and Goals. U.S.D.A. & *Assoc. Land Grant Coll. Univ.* Govt. Print. Off. 72 pp.
12. McDaniel, M. C. 1964. The present status & future of Extension Plant Pathologists in Arkansas & the Nation. *Plant Pathol. Sem.,* Fayetteville, Ark.
13. Miller, P. A. 1958. The Cooperative Extension Service Today—A statement of scope and responsibility. *Fed. Ext. Serv. U.S. Dept. Agr.* 16 pp.
14. O'Reilly, H. J. 1966. The increasing role of Extension plant pathology in applied research. *Presented at Western Regional Spray Conf.,* Portland, Ore.
15. Paul, L. C. 1962. A Look at Agricultural Advisory Services in the British Isles, Sweden, Denmark and the Netherlands. *Can. Soc. Rural Extension,* Ottawa, Canada
16. Schenemann, C. N. 1959. *The Functions & Procedures of Subject Matter Specialists in the Missouri Cooperative Extension Service.* Ph.D. Thesis. Univ. of Wisc., Madison
17. Sharvelle, E. G. 1968. The Extension Plant Pathologist: Programs and Activities for 1985. Presented at Dept. Heads Meeting. *Am. Phytopathol. Soc. Ann. Meet.,* Ohio State Univ., Columbus
18. Sherf, A. F. 1972. Making the Extension Recommendation: Responsibilities to Different Clientele Groups. Presented at 54th Ann. Meet. *Am. Phytopathol. Soc.,* Mexico City
19. Swan, J. C. 1966. How New York State Cooperative Extension Can Best Serve the Needs and Interests of Commercial Agricultural Producers in the Years Ahead. *Task Force Rept. Dir. Cooperative Extension.* Cornell Univ., Ithaca, N.Y. 56 pp.
20. Warren, H. M. 1952. *Functions and Responsibilities of Subject Matter Specialists in the Extension Service,* M.S. thesis, Cornell Univ., Itaca, N.Y.
21. Watts, L. H. 1968. *A People and a Spirit.* A report of the Joint U.S. Dept. of Agr.—Natural Assoc. State University & Land Grant Colleges Extension Study Comm., Colorado State Univ., Fort Collins, 95 pp.
22. Watson, L., Gatehouse, M., Dorsey, E. 1972. *Failing the People.* Agr. Policy Accountability Proj. 1000 Wisc. Ave., N.W., Wash., D.C., 20007, 27 pp.

SOME RELATED ARTICLES IN OTHER ANNUAL REVIEWS

REPRINTS

The conspicuous number aligned in the margin with the title of each article in this volume is a key for use in ordering reprints.

Available reprints are priced at the uniform rate of $1 each postpaid. Payment must accompany orders less than $10. A discount of 20% will be given on orders of 20 or more. For orders of 100 or more, any Annual Reviews article will be specially printed and shipped within 6 weeks.

The sale of reprints of articles published in the Reviews has been expanded in the belief that reprints as individual copies, as sets covering stated topics, and in quantity for classroom use will have a special appeal to students and teachers.

AUTHOR INDEX

CHAPTER TITLES VOLUMES 7-11